Knowledge and Time

Hans Primas

Knowledge and Time

Edited and annotated by Harald Atmanspacher

Springer

Author
Hans Primas *(deceased)*
Lab. für Physikalische Chemie
Eidgen. Technische Hochschule Zürich
Zürich, Switzerland

Editor
Harald Atmanspacher
Collegium Helveticum
Eidgen. Technische Hochschule Zürich
and University of Zürich
Zürich, Switzerland

ISBN 978-3-319-47369-7
ISBN 978-3-319-47370-3 (eBook)
DOI 10.1007/978-3-319-47370-3

Library of Congress Control Number: 2017942818

Printed on acid-free paper

This Springer imprint is published by Springer Nature
The registered company is Springer International Publishing AG
The registered company address is: Gewerbestrasse 11, 6330 Cham, Switzerland

The proper understanding of a subject and its misunderstanding do not entirely exclude one another.

Franz Kafka, The Process, 1925

Though it would be wrong to persistently hold on to a preconceived bias during an investigation, such a bias is not only innocuous but even mandatory at the start of the investigation.

Heinrich Hertz, Inaugural Dissertation, 1880

Editor's Preface

When Hans Primas died in October 2014 at age 86, he left behind an extensive manuscript on which he had worked for the last five years of his life. It addressed two main topics: first, *knowledge* and the different ways it can be acquired, and second, the role that different notions of *time* play in knowledge acquisition, creatively and rationally. The fields of human scholarship that he touches upon during this challenging journey in thirteen chapters culminate in a proposal for a new understanding of the relationship between the mental and the physical—one of the oldest unsolved mysteries in both science and philosophy.

The key tools he uses in his wide-ranging explorations are concepts such as complementarity and entanglement, and the logic of partial Boolean algebras—all well known to be at the heart of modern quantum physics. However, Primas applies them not only to physics problems. This positions his work in the tradition of Niels Bohr, Wolfgang Pauli and others, who were always convinced that the mathematical structures underlying quantum theory are significant and viable far beyond the limits of quantum physics.

Primas was interested in time from various viewpoints and for many years, for instance in the role of time in dynamical models of measurement in quantum physics. Yet his first publication explicitly discussing time in the context of the mind-matter problem appeared only in 2003: "Time-Entanglement Between Mind and Matter" (Primas 2003). Several years later he published a refined version of that article as "Complementarity of Mind and Matter" (Primas 2009).

In both articles he argued that a more profound understanding of the many facets of time could be useful for our conceptions of the mental, the physical, and their relationship—even more so since our experience of time and the physical concept of time are radically different, despite having important features in common. The philosophical perspective that he proposed falls into the class of *dual-aspect* approaches, in which the mental and the physical are neither substances on their own (as in Cartesian dualism) nor are they reducible to one another (as in physicalist or idealist frameworks). Rather, they are considered as two aspects, as it were, of one underlying "neutral" domain that itself has no mind-matter distinction.

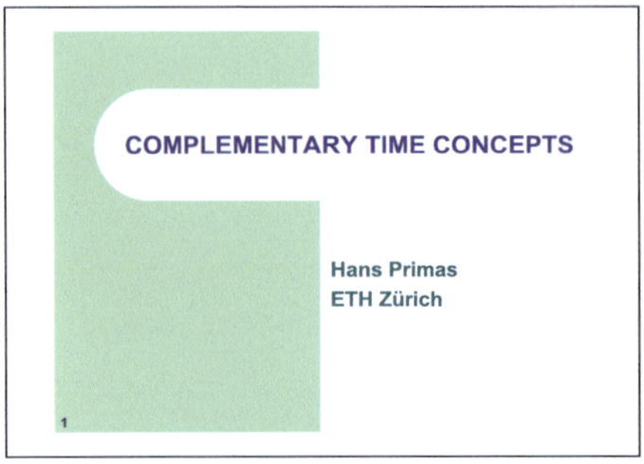

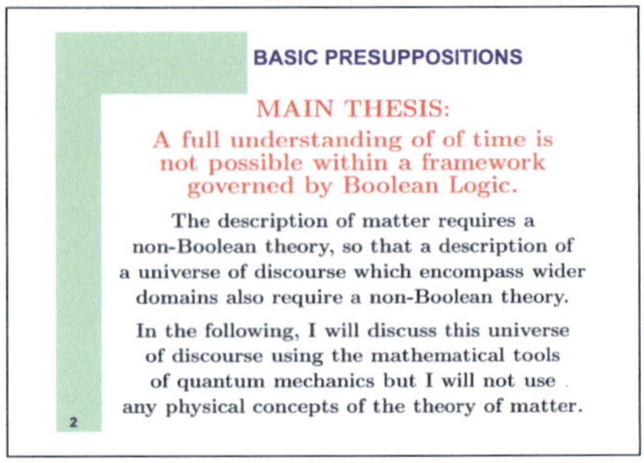

The first two slides of the presentation that Hans Primas gave at a workshop of the Parmenides Foundation in Munich in April 2010

Shortly after his 2009 paper had appeared, Primas began drafting first versions of a manuscript more comprehensive than the two precursor publications. At about the same time he was invited to give a presentation at a workshop about "the forgotten present" of the Parmenides Foundation in Munich in the spring of 2010. This presentation was based on his ideas about various perspectives on time at that juncture. Its slides are available online* and may serve as a quick guide to what the reader can expect from Part II (on time) of the more elaborate and more sophisticated material contained in this final monograph.

With one major interruption due to serious health problems in 2011, Primas worked continuously on the manuscript right up until his death in 2014. His latest adjustment was a reorganization of the material on bottom-up and top-down theories in physics, two sections that now conclude Part I (on knowledge) of the monograph in its present form. They outline his highly original account of how to understand the structure and development of our knowledge about the physical world—with a pronounced focus on quantum theory.

Since I was in constant contact with Hans Primas over the preceding two decades, it is likely that I am more familiar than anyone else with the intentions and the themes of his work and his ways of thinking in these years. So it came as no great surprise to me when his family asked me to examine the manuscript and come to a decision about whether and how it could be brought to the attention of an audience for which it would be of interest. It was evident to me from the outset that there is undoubtedly such an audience. Hence, the actual question was whether the material was prepared well enough to permit an editing process leading to its coherent exposition.

According to first-hand information from his family, Primas, in his last days, had considered the manuscript as essentially complete but largely unedited. What

*The slides of his presentation can be downloaded at www.parmenides-foundation.org/fileadmin/redakteure/events/PW_19/Hans_Primas.pdf. Primas did not submit a manuscript for the proceedings of the workshop, which later appeared as "Re-Thinking Time at the Interface of Physics and Philosophy" (von Müller and Filk 2015), but obviously the feedback he received at the workshop inspired him to work out his ideas to much greater depth.

needed to be done first was to retrieve the relevant files from his computer and locate those which contained the most recent, finished version of each section. It soon turned out that the overall manuscript was in a state ready for editing, but it was also clear early on that this would be a job far less straightforward than editing typical collections of essays or journal issues. With some unavoidable intermissions, it took about 18 months of work from the first steps in the process to the submission of the final manuscript for publication.

The style of the text will not surprise anyone who knows the monograph that Primas had published 35 years earlier on "Chemistry, Quantum Mechanics, and Reductionism" (Primas 1981). The present book is again a composition of pivotal conceptual insights, quotes that illustrate important points, critical remarks about misunderstandings in the literature, the subtle mathematics of algebraic quantum theory, detailed knowledge about engineering and systems theory. It exhibits the personality of a scholar who combines deep foundational interests, formidable mathematical skills, and exceptionally broad scientific and engineering background in a unique way.

Last but not least, the book reflects a respectful appreciation of the groundbreaking work by pioneers such as Araki, Bohr, Connes, Dirac, Einstein, Gödel, Haag, Hertz, Holevo, Kolmogorov, Ludwig, Pauli, Poincaré, Strauss, von Neumann, Weyl, Wiener, and Wigner. This list highlights just the most frequent authors in a bibliography of more than 1000 references. As a special service, readers will find, together with each reference in the bibliography, the particular section(s) in the monograph where it is referred to.

Among the extensive appendices that Primas added for further details and background, only the one about algebras of observables was retained in the published book. It contains many important results whose compact and succinct presentation is hard to find in the standard literature. Other appendices, presenting material about stochastic processes, harmonic analysis, fluctuation-dissipation theorems, etc. have been dropped since there are good reviews or textbooks that can easily be consulted about them.

The text contains several long passages with illustrative engineering examples, for instance detailed discussions of various kinds of filters in Chaps. 3 and 12. Although one might have regarded these as digressions, they were retained because engineering examples are so typical of Primas' way of illustrating his ideas. Moreover, readers will notice his often meticulous mathematical notation, for instance his insistence on specifying the domain of variables (e.g., $t \in \mathbb{R}$) or the expression of functions as $t \to f(t)$ rather than simply $f(t)$. This has become somewhat unusual in standard textbooks, but it is an integral part of Primas' style of presentation, and so it, too, was left untouched in the editing process.

As sketched above, "Knowledge and Time" consists of two parts, the first entitled *Knowledge*, the second *Time*.** Part I begins with introductory remarks about why

**Insiders will realize that a book with almost the same title exists in German language: "Zeit und Wissen" (Time and Knowledge) by the late Carl Friedrich von Weizsäcker (1992). Though its content and style are very different from the present book, the title testifies that Primas and a proper understanding of time cannot be achieved within a Boolean framework of thinking (compare the "main thesis" on the second slide on page viii above) as it underlies virtually all of classical science. Chapters 2 and 3 outline the nature of Boolean versus non-Boolean descriptions. While a Boolean two-valued logic with truth values "true" and "false" is best characterized by the famous "rule of the excluded middle" (or *tertium non datur*), non-Boolean logic violates this rule. The consequence is incompatible descriptions, which are central to the notion of *complementarity*. Originally imported into quantum physics by Bohr, complementary descriptions are formally related to *non-commutative algebras of observables*.

Chapter 4 addresses creativity and rationality or, more precisely, creative insight and its rational reconstruction, two complementary modes of knowledge, as Primas puts it, that are both needed for sound and innovative scientific progress. He relates them to two complementary modes of *non-sequential* and *sequential* processing, which will become the prototypes for non-sequential and sequential time in Part II. Chapter 5 discusses numerous examples of creativity and rationality in mathematical knowledge, for instance the extraordinary insights of the mathematician Ramanujan and the seminal contributions by the logician Kurt Gödel. Part I ends with Primas' account of quantum theory considered in terms of the familiar bottom-up approach (Chap. 6) as contrasted with a less discussed top-down point of view (Chap. 7).

Ultimately, the message of Chaps. 6 and 7 is that both top-down and bottom-up thinking are needed in combination, such as in the proverbial phrase "it's top down all the way from the bottom up". Remarkably, the two directions "up" and "down" may be considered as indicating two different versions of *emergence*—a concept that plays an important role in Primas' work. The traditional bottom-up type of emergence describes how irreducible higher-level properties can be understood to arise from lower-level properties. On the other hand, top-down emergence typically is based on restrictions and resulting symmetry breakings that a higher level imposes on lower levels.

Chapter 8, the first section in Part II, ought to be seen as the conceptual core of "Knowledge and Time". Based on the observation that time as such is neither mental nor material, Primas posits a tripartite structure of the total universe of discourse with a temporal domain, an atemporal mental domain, and an atemporal material domain. Assuming that these domains are non-Boolean, they are holistically correlated, very much in the spirit of quantum *entanglement*. If all three of them are holistically correlated, any strict distinction between time, mind and matter gets lost, and we regain the idea of an undivided universe for which Primas (2003, 2009) used the term *unus mundus* in his earlier publications on the subject.[***] Bilateral correlations between the mental and the material yield mind-matter correlations, and

von Weizsäcker shared a common view on the philosophical foundations of science, which they frequently discussed when they met.

[***]The notion of an *unus mundus* in this sense derives from the dual-aspect framework of thinking that Pauli and Jung developed in the mid-20th century (cf. Atmanspacher 2012 for details). Interestingly, Primas' original manuscript files do not mention the term *unus mundus*, nor does he tell his readers how it is replaced or, more to the point, generalized.

bilateral correlations between each of these two domains and the temporal domain introduce mental (psychological) and material (physical) time.

Subsequently, Primas unpacks his technical repertoire on group theory and algebras of observables by introducing and employing temporal groups and algebras, their representations, and corresponding descriptions of temporal phenomena. Of major relevance for the notions of sequential and non-sequential time are the affine time group and the affine frequency group (Chap. 9), which can be realized by time and frequency operators referring to past and future as well as positive and negative frequencies, respectively. They generate four semigroups, whose representations (for simplicity in Hilbert space) are addressed in Chap. 10, with particular emphasis on the representation of temporal states. In Chap. 11, the complementarity of sequential and non-sequential semigroups in the time and frequency domains is further elaborated. Examples of temporal distributions, expressing states that are extended in time, are indicated.

For the non-expert, these sections will be hard to follow, but they are related to an important conceptual lesson that Primas discovered in his study of algebraic quantum theory in the late 1980s (see Primas 1990a). The distinction between algebras of observables and their (contextual) representation reflects a crucial difference between *ontic and epistemic interpretations* of the properties of a system as they are formalized by observables. While ontic frameworks typically exhibit high degrees of symmetry (and corresponding formal simplicity and transparency), epistemic frameworks require broken symmetries in order to describe the multitude and complexity of properties in the world of empirical facts. Chapters 9–11 adopt this insight with respect to temporal structures.

The concluding chapters of "Knowledge and Time" are devoted to the *relevance* of sequential and non-sequential time, respectively. Sequential time is the time concept mostly used in the sciences, a parameter time measuring the temporal distance between facts with a clock (Chap. 12). But the apparently simple question of how facts are generated already leads to severe challenges that to date have not been uncontroversially resolved. In any event, once there are facts, they can be causally ordered, and experiments can be causally described. Most issues related to determinism, causation, prediction and retrodiction in science are based on sequential time.

By contrast, non-sequential time is not, or only rarely, addressed in the sciences at all. Chapter 13 lays out some ideas about how it might acquire significance in both mental and physical contexts. On the mental side, the experienced duration of the present is a key issue, which has been much studied in psychology, cognitive science, and the phenomenological tradition of philosophy. From the perspective of physics, related phenomena have been occasionally referred to by the notion of temporal nonlocality or temporal entanglement. Very recent developments linked to the investigation of such phenomena are sketched at the end of the section.

Although the monograph is arranged linearly by necessity, the thoughtful reader will notice that particular key topics re-appear in different contexts, leading to a gradual strengthening of their significance. These repetitive themes are partial Boolean algebras, the notions of complementarity and entanglement, and the contrast between different kinds of knowledge processing. First introduced formally and by examples, they arise again in the way they are used in quantum physics, and eventually are exploited for the discussion of the complementary modes of sequential and non-sequential time. In this way, the subtle layers of "Knowledge and Time" unveil themselves step by step, like the petals of a flower opening up slowly to finally reveal its full beauty.

Given the mathematical sophistication of much of the text, it is clear that the book is not exactly an "easy read". However, it offers such innovative and far-reaching concepts, so much stimulation and inspiration, and so many promising ideas deserving further studies that it truly would have been a shame to leave all this publicly unrecognized. So here it is, and—as always—there are some acknowledgments to be made for the support and encouragement received during the editorial work.

Major thanks go to Margarita Primas, who carefully checked and approved all corrections and editorial changes implemented in the manuscript, and who was of substantial help in the preparation of the bibliography and the index. All additions that clearly go beyond the original text are marked as editor's notes in the text. Mostly, this concerns updates on certain topics with recent literature, or simply translations of non-English quotations into English.

Thomas Filk double-checked Sect. 7.3 on local quantum physics and provided advice on how to redesign parts of the formalism to make sure that the ideas are easier to follow by readers used to standard textbook presentations. And, finally, Angela Lahee at Springer and some referees she consulted for the manuscript accompanied the publication process with professional courtesy and with the insight that such a monograph requires to be broadly accessible.

Collegium Helveticum
University and ETH Zurich
Zürich, Switzerland
August 2016

Harald Atmanspacher

Contents

Part I
Knowledge

Chapter 1
Introduction

It is impossible to study this wonderful theory
without feeling as if the mathematical equations had
an independent life and an intelligence of their own,
as if they were wiser than ourselves,
indeed wiser than their discoverer,
as if they gave forth more than he had put into them.

Heinrich Hertz (1890, pp. 11f)

1.1 A Full Understanding of Time Requires a Non-Boolean Framework

Many areas of present-day science require an unambiguous Boolean language in which empirical statements are truth-definite. In this way we obtain Boolean domains in which the laws of classical logic apply and allow us to successfully predict and control the behavior of systems.

However, we cannot conclude that every valid and useful scientific description can be formulated in terms of a language governed by the laws of Boolean logic. Modern physics proves that the description of matter requires a theory with a non-Boolean logical structure, with the consequence that any description of a universe of discourse including the material world needs to be non-Boolean. To cover the full range of our capacities of insight, non-Boolean descriptions are compulsory. Moreover, I will argue in this monograph that a comprehensive understanding of the notion of time is impossible within the framework of Boolean logic alone.

Much of the discussion to follow will be framed in terms of the theory of *partial Boolean algebras* (cf. Primas 2007). The corresponding mathematical tools are familiar in modern quantum theory, but here I will abstain from any physics-specific deliberations. A partial Boolean algebra is a family of Boolean algebras whose operations coincide on overlaps so that it is *locally Boolean, yet globally non-Boolean*. In an algebraic codification, the projection operators of a non-commutative

H. Primas, *Knowledge and Time*,
DOI 10.1007/978-3-319-47370-3_1

W^*-algebra form a partial Boolean algebra. Every commutative subalgebra corresponds to an empirically accessible Boolean context. The set of all Boolean contexts generates the non-Boolean universe of discourse.

The family of Boolean contexts plays a privileged role since a sound theory should be able to describe intersubjectively communicable empirical propositions. Therefore the overall non-Boolean framework must have a locally Boolean logical structure. This guarantees that aspects of reality can be perceived by projections onto empirically accessible Boolean reference frames.

The requirement that empirical propositions have to be truth-definite can only be achieved by deliberately suppressing irrelevant features. However, what is relevant and what is irrelevant is not determined by the laws of nature. It has to be set by convention, by interest, by evolutionary history, by our cognitive apparatus, or by the measuring instruments used. *Objectivity* in the modern sense is enforced by the requirement that an empirically meaningful statement has to belong to a Boolean domain of discourse.

In order to describe aspects of an overall non-Boolean reality we have to ignore particular global features such that the remainder separates into facts (see Chap. 2). Inevitably, such a description is valid only within the adopted partition of the world, that is, within the *chosen context*. Every locally Boolean description suppresses correlations between Boolean domains. But if we insist on unambiguously communicable empirical propositions in terms of two-valued logic, we have to pay the price of this inevitable sacrifice.

1.2 Two Modes of Understanding

The various modes of learning and knowing are closely related to different conceptions of time. Above all, analytic and intuitive modes of knowing are connected with conceptions of time as sequential and non-sequential, respectively. Correspondingly, there are two different types of learners: *sequential learners* who process information in a sequential and analytic manner, and *non-sequential learners* who follow an intuitive and synthetic approach. Moreover, there is ample evidence that knowledge may originate from different styles of information processing. Differences in cognitive styles are well-known in cognitive processes related to memory, language, art instruction (cf. Edwards 1979), and music.[1]

This situation is particularly clear in teaching and understanding mathematics (for more details see Chap. 5). There are two principal ways of conceptualizing mathematical ideas and procedures: the analytic sequential way and the intuitive synthetic way. Bernhard Riemann and Karl Weierstrass—the greatest 19th century German mathematicians after Gauss—are the major pioneers of modern complex

[1]In his *Schulwerk*, Carl Orff developed a method of music education based on learning by doing which involves experiencing music through the use of percussion instruments. He referred to the ancient Chinese proverb "tell me, I forget ... show me, I remember... involve me, I understand".

analysis. Yet their thinking styles were strikingly different. They developed complex analysis along two fundamentally different and original ways.[2]

Henri Poincaré (1898, p. 7) compared the method of Riemann with that of Weierstrass as follows:

> The method of Riemann is primarily a method of discovery, that of Weierstrass is primarily a method of proof.

Riemann's work was visionary and based on intuitive reasoning, but his insights were often free of proof. In contrast to Riemann, Weierstrass insisted on maximal rigor. Felix Klein (1926, p. 246, author's translation) described this dissimilarity lucidly:

> Riemann is the man of brilliant intuition. His comprehensive genius surpasses all his contemporaries. Where his interest is awakened, he starts afresh, without allowing himself to be swayed by tradition and without accepting any constraint by systematization. ... Weierstrass is primarily a logician, he proceeds slowly, systematically, step by step. Where he works, he strives for completion.

Analytic and intuitive reasoning serve complementary functions and can work together even though one of them may partially inhibit the other. In general, a combination of sequential and non-sequential thinking is conducive for insights which analytic or intuitive reasoning alone are incapable of producing. According to Michael Wertheimer, a founder of Gestalt theory, the crucial point for any deeper understanding is whether the insight that has been acquired can be transferred to different but structurally similar situations (Wertheimer 1985, p. 23):

> Learning has occurred when the learner has understood, has developed insight into the actual nature of the problem situation, and can act in ways that show that the relevant features of the task have indeed been grasped. If understanding has actually occurred, it is only a minor step, and one that presents little difficulty, to transfer the learning to a new situation to which it also applies. Thus, one test of whether learning has really happened is to check whether what has been learned will generalize to a related task—if all that has transpired is sheer memorizing or mechanical associating, the learner will be unable to recognize the similarity between a task that has already been mastered and a new one which, while it may be superficially quite different, requires the same insight to solve it that also worked in the earlier task.

1.3 Two Modes of Learning

In traditional teaching techniques new concepts are introduced in a step-by-step manner which satisfies the needs of sequential learning where each step follows logically from the previous one. This approach is ineffective for non-sequential learners who do not go meticulously through small steps but learn in large jumps. They put things together in novel ways and strive for the big picture, even though they may miss technical details. They typically think in mental images and symbols rather than in formal language and may have difficulty explaining how they did it.

---

[2]For a history of complex analysis, the theory of complex functions, see Bottazzini (2002).

Clearly both modes of learning are important. Everybody utilizes both logical and intuitive modes of reasoning to some degree but most of us have a preferred thinking style. However, this predisposition is not absolute so that the balance among and the coordination of the two modes varies.

In his book *How to Solve It*, the mathematician Georg Polya (1945, p. 207) said:

> The intelligent reader of a mathematical book desires two things: First, to see that the present step of the argument is correct. Second, to see the purpose of the present step.
> The intelligent listener to a mathematical lecture has the same wishes. If he cannot see that the present step of the argument is correct and even suspects that it is, possibly, incorrect, he may protest and ask a question. If he cannot see any purpose in the present step, nor suspect any reason for it, he usually cannot even formulate a clear objection, he does not protest, he is just dismayed and bored, and loses the thread of the argument.

And he added in his *Patterns of Plausible Inference* (Polya 1954, p. 167):

> You go through a lengthy mathematical demonstration. The demonstration is supposed to be decomposed into steps each of which you can check perfectly, and the final conclusion must be correct if each step is correct. Yet you may make mistakes like everybody else. After having checked the successive steps quite carefully, can you trust the final conclusion? Not more, and perhaps less, than the final result of a long computation.
> In fact, a mathematician who has checked the details of a demonstration step by step and has found each step in order may be still dissatisfied. He needs something more to satisfy himself than the correctness of each detail. What?
> He wants to *understand* the demonstration. After having struggled through the proof step by step, he takes still more trouble: he reviews, reworks, reformulates, and rearranges the steps till he succeeds in grouping the details into an understandable whole. Only then does he start trusting the proof.

1.4 Two Notions of Time

> What can be shown, cannot be said.
>
> *Ludwig Wittgenstein (1969, 4.1212)*

1.4.1 Chronos and Kairos

The ancient Greeks divided the notion of time into two complementary manifestations: $\kappa\rho o\nu o\varsigma$ (kronos) and $\kappa\alpha\iota\rho o\varsigma$ (kairos).[3] In Greek mythology, *Kronos* was the son of Gaia and Uranus and was overthrown by his son Zeus and imprisoned in Tartarus. *Kairos*, the youngest son of Zeus, was the personification of opportunity, luck and favorable moments.

---

[3]This division corresponds to the symmetry breakdown of a featureless fundamental time into two complementary modes of sequential time and non-sequential time. Compare Chap. 8 and Sect. 11.4.

In *Physics* (Book IV, Chap. 11, 219b), Aristotle (1983) defined *chronos*[4] as the "number of changes with respect to the before and the after", which is an expression of sequential time: the future passing through the present and so becoming the past. Chronos time, therefore, is neither an independent substance nor identical with motion. It is composed of a past that is gone, a present that exists, and a future that has to come yet. Chronos time is quantitative and can be measured with clocks. It relates to sequential logic and analytic thinking.

Kairos time is qualitative and denotes the sense of doing the right thing at the right time.[5] Kairos time refers to a realm outside of measurable clock time. Kairos time is meaningful and pregnant with possibility, opportunity, and wisdom. In the pedagogical tradition Kairos refers to the crucial moment of transformative insight and encompasses the occasion itself. For Friedrich Copei (1950, p. 17, editor's translation) Kairos refers to

> the "fruitful moments" in an educational process—those peculiar moments in which a new insight awakens in us in a flash, an idea seizes us, and we suddenly see a situation in an entirely new light.

1.4.2 Time Is Neither Material Nor Mental

In Books 10 and 11 of his *Confessions* Augustinus of Hippo (354–430) links the analysis of time to that of memory. He notes that past time no longer exists and future time does not yet exist:[6]

> Yet I say with confidence, that I know that if nothing passed away, there would not be past time; and if nothing were coming, there would not be future time; and if nothing were, there would not be present time.

> But even now it is manifest and clear that there are neither times future nor times past. Thus it is not properly said that "there are three times, past, present, and future". Perhaps it might

[4]According to Panofsky (1939, p. 73), "the Greek expression for time, Chronos, was very similar to the name of Kronos (the Roman Saturn), oldest and most formidable of the gods. ... the fortuitous similarity between the words Chronos and Kronos was adduced as proof of the actual identity of the two concepts which really had some features in common. According to Plutarch, who happens to be the earliest author to state this identity in writing, Kronos means Time in the same way as Hera means Air and Hephaistos Fire."

[5]The depth, complexity and untranslatability of the Greek concept of *Kairos* has been discussed by Ramo (1999) and by Sipiora (2002). For example, Kairos is central to Plato's and Aristotle's conceptions of rhetoric, and to Pythagorean philosophy. Kairos is a significant concept in the Old and New Testaments, for instance it is the term used for time in Paul's Epistle to the Ephesians (5:16). Kairos is also crucial in Hippocratic medicine (*Corpus Hippocraticum*). The *Intermediate Greek-English Lexicon* by Henry George Liddell and Robert Scott (http://www.perseus.tufts.edu/hopper/resolveform?redirect=true) defines Kairos as "exact or critical time, season, opportunity".

[6]*Augustini Confessiones*, Liber XI, Caput XIV, §17; Liber XI, Caput XX, §26. English translation from Augustine (1955).

> be said rightly that "there are three times: a time present of things past; a time present of things present; and a time present of things future". For these three do coexist somehow in the soul, for otherwise I could not see them. The time present of things past is memory; the time present of things present is direct experience; the time present of things future is expectation.

Expressed in modern language, Augustine asserts that for there to be time, there must be some material events ("things") happening. From this Augustine concludes that time resides in the passage from future to past events and that we can be aware of time only while it is passing (Augustine 1955, Book XI, §16).

For Augustine *memory* plays a pivotal role in the perception of time, it not only retains recollections of past experiences, but also holds present realities in the mind and anticipates the future:[7]

> For we, as it were, see some of our thoughts about future things more clearly and more certainly, as though they were very near; and we do this when we are able and insofar as we are able, by our memory which seems to pertain not to the future but to the past. This can be proved by those sayings and hymns which we render from memory in the proper sequence. For unless we foresaw in our thought what follows, we certainty could not utter them. And yet it is memory, not foresight, that enables us to foresee them.

In his *Correspondence with Clarke* (compare Alexander 1956), Leibniz postulated that space and time are ideal systems of relations, and he rejected the idea that space and time are substance-like as absurd. In his relational theory of space and time he defended the view that time is nothing else than the order of successive events. Leibniz's concept of the ideality of time implies that time cannot belong to the material domain. Since time is neither a material nor a self-subsistent entity, a genuine time operator does *not* belong to the algebra of material observables and therefore *commutes* with the material Hamilton operator of physics.[8]

The fact that there exists no material time operator valid for the whole material domain does not imply that there are no such operators for specific temporal phenomena for specific material systems. Although there are dynamical physical systems which produce highly reproducible periodic phenomena, *there is no universal material clock*. (For more details compare Sect. 8.3.1.)

---

[7] *Augustini de Trinitate*, Liber XV, §13. English translation from Augustine (2002, p. 181).

[8] The popular idea in physics that time and energy are canonically conjugate observables is based on a category error since the Hamilton operator and time observables cannot belong to the same factorial algebra. The error is probably due to the wrong assumption that for a derivation of the time-energy inequality $\Delta t \cdot \Delta E \geq \hbar/2$ a commutation relation of the type $[T, H] = TH - HT = i\hbar$ is indispensable, where both the time operator T and the Hamilton operator H refer to the *material* domain (compare Busch 2008 and Hilgevoord 2005).

Without any arguments or references, Heisenberg (1927, p. 177) referred to the equation $Et - tE = h/(2\pi i)$ as "familiar". The only two earlier references known to me are a letter by Pauli to Heisenberg of January 31, 1926 (Hermann et al. 1979, p. 284), where he posits the relation $Wt - tW = h/(2\pi i)$, and Dirac's (1926b, p. 410) publication on relativistic quantum theory in which he introduced the commutation relation $[t, W] = 1$, where $t = -ix_4/c$ and $W = -icp_4$.

1.4.3 Sequential and Non-Sequential Processing

According to von Neumann (1958, p. 8), there are two dichotomous modes of information processing: sequential and non-sequential processing. A sequential process consist of a sequence of operations each of which take place one at a time. In non-sequential processing multiple operations can be performed simultaneously. The term "parallel processing" used in computer science refers to a method of executing several instructions simultaneously by rapid massive *serial* multiprocessing, i.e. executing many serial processes simultaneously. This differs from true non-sequential processing, although it is easily misunderstood as such. On the other hand, quantum computations which rely on the quantum-theoretical superposition principle represent true non-sequential processing (Margolus 1990; Deutsch and Jozsa 1992).

There is some evidence that the distinction between sequential and parallel processing is related to the left and right cerebral hemispheres of the human brain. It has been claimed that the left hemisphere employs analytic sequential procedures, while the right hemisphere processes (visual) stimuli in a parallel fashion (cf. Cohen 1973). Although lateral differences seem to be measurable, both hemispheres actually contribute to sequential and non-sequential processing. The popular generalization that the right hemisphere is responsible for logical sequential processing while the left hemisphere accounts for creative thinking has to be appreciated very carefully (see, e.g., White 1969; Edwards 1979).

1.5 Parts of the Universe of Discourse

While classical science has been based on an analysis of nature in terms of interacting but independently existing objects, an entirely different view emerges in quantum physics. If we take quantum mechanics not only as a bunch of highly successful pragmatic working rules, but as our best candidate for a fundamental theory of the material world, then we are compelled to admit that independently existing individual objects do not exist (with the possible exception of the universe as a whole). This entails a radically holistic worldview. Any analysis of nature must begin with a decomposition of the distinction-free universe of discourse into appropriately discriminable parts.

Quantum theory states that nature is undivided. It is the first and only logically consistent, empirically well-confirmed and mathematically formalized holistic theory. The undividedness of nature is described by so-called Einstein-Podolsky-Rosen (EPR) correlations (Einstein et al. 1935), which exist even in the absence of any interactions. According to quantum theory, subsystems of the world are inextricably entangled so that it is impossible to describe them by so-called pure states. Quantum theory describes the material world as a whole, in fact as *a whole which is not made of parts*. This result has been verified beyond any reasonable doubt in many beautiful experiments within recent years.

If we abstract from the ubiquitous EPR correlations between any subsystem of a mind-independent reality with its environment, we recover the multitude of empirically observed phenomena. The observed empirical world is the result of our decision which EPR correlations to disregard so that parts manifest themselves as *contextually* emergent subsystems.[9] Since there are mutually exclusive ways to disregard EPR correlations, there are mutually exclusive scientific descriptions of empirical reality.

[9]In algebraic quantum mechanics a context with its associated suppression of EPR correlations can be realized by an appropriate GNS-construction. For more details see Sects. 6.5.3.7, 6.5.3.8, and Appendix A.3.

Chapter 2
Boolean Descriptions

2.1 On the Necessity of Distinctions

You'll soon know it all, as planned,
When you've learnt the science of reduction,
And everything's proper classification.

Johann Wolfgang von Goethe (2003, p. 78)

According to the British mathematician George Spencer Brown, the most fundamental of all cognitive operations is to draw distinctions. In his seminal book *Laws of Form* (Spencer Brown 1969), he developed a binary arithmetic based upon distinctions and principles of symbolic logic.[10] He begins his book with the key statement (Spencer Brown 1969, p. 1):

> We take as given the idea of distinction and the idea of indication, and that we cannot make an indication without drawing a distinction.

To make a distinction is to divide the universe of discourse into two parts (be it mathematical, cognitive, linguistic, perceptual, etc.). Every distinction has a purpose but no *a priori* meaning: "There can be no distinction without motive, and there can be no motive unless contents are seen to differ in value" (Spencer Brown 1969, p. 1). As Suzanne Langer (1978, p. 273) observed, "our world 'divides into facts' because we so divide it". Distinctions create frames of reference, necessary for any kind of description.

The search for regularities is the principal concern of all scientific inquiry. Regularities can be found if and only if we suppress irrelevant features. What is relevant and what is irrelevant is not determined by some law of nature but by convention—or by our interests, by our cognitive faculties, by evolution, or by the pattern recognition devices used by experimentalists.

All concepts of empirical science refer to observations obtained by pattern recognition protocols which ignore many features and concentrate on those which we

---

[10] Banaschewski (1977) showed that Spencer Brown's concept of a primary algebra is exactly the theory of join and addition (i.e., symmetric difference) of Boolean algebras.

H. Primas, *Knowledge and Time*,
DOI 10.1007/978-3-319-47370-3_2

consider as relevant. A pattern is something that somebody recognizes as a pattern. What is considered as relevant, and what as irrelevant, is not written down in some first principles but depends in a crucial way on the abstractions chosen by someone. Since the world is neither intrinsically divided into subsystems nor are there unprejudiced sense data or unbiased observations, observable patterns of the world do not exist by themselves. Any recognition of patterns and resulting classifications are based on the primitive act of *making a distinction*. Distinctions eliminate irrelevant features. It follows that *every classification is context-dependent and unavoidably entails a loss of information*.

2.2 Boolean Logic

The concept of a Boolean algebra has its roots in the algebra of logic, introduced by George Boole (1847, 1848) as a study of the laws of thought. He tried to explain how the mind processes thought (Boole 1849, p. 13):

> The object of Logic as a Science is to explain the laws of those mental operations by which ordinary Reasoning is conducted.

Boole's logic is based on a binary approach—the yes-no, true-false approach. His classic *The Laws of Thought* (Boole 1854) begin with these words:

> The design of the following treatise is to investigate the fundamental laws of those operations of the mind by which reasoning is performed; to give expression to them in the symbolic language of a Calculus, and upon this foundation to establish the science of Logic and construct its method.

Boole's idea that logic is the study of the laws of thought was taken over by Frege when he stated that the task of logic can be seen "as the investigation of the mind".[11] Similarly, Hilbert assumed that there is a correspondence between human thinking and the formal system of proof theory:[12]

> Logic is a discipline which expresses the structure of all our thought.

Boolean logic, originally developed by George Boole in the late 1830s, is a two-valued logic, based on the *law of the excluded middle*, asserting that every proposition is either true or false (Boole 1847, 1848, 1854). Alessandro Padoa emphasized that logic is a purely formal system which allows us to make inferences without specifying a particular interpretation:[13]

---

[11] Frege (1918/1919, p. 74, editor's translation): "Neither logic nor mathematics has the task to inquire into the souls and conscious contents of individual human beings. One might rather consider it as their task to study the mind, not the minds."

[12] Hilbert (1931, p. 125, editor's translation): "Through the tertium non datur logic acquires full harmony; its theorems attain such simplicity and the system of its concepts such completeness as it ought to conform to the significance of a discipline expressing the structure of all our thinking."

[13] See Padoa (1901, pp. 309–365). English translation quoted from van Heijenoort (1967, p. 121).

> …for what is necessary to the logical development of a deductive theory *is not the empirical knowledge of the properties of things*, but the *formal knowledge of relations between symbols*.

The importance of logic as a formal symbolic (not natural) language lies in the exclusion of ambiguities and implicit assumptions.

The propositional calculus of Boolean logic (also called classical logic) deals with propositions which can be studied irrespective of their truth. Here, a proposition is understood as an unambiguous sentence that is either true or false. It is then possible to examine how propositions are combined by means of sentential connectives.

- If A is a proposition, then the *negation* "not A", denoted by $A^\perp$, is also a proposition.
- If A, B are two propositions, then the *conjunction* "A and B" is also a proposition, denoted by $A \wedge B$.
- The proposition "A or B, or both" is called the *disjunction* (the *inclusive or*) and is denoted by $A \vee B$.
- The relation "A implies B" is written as $A \leq B$.
- If proposition A is logically equivalent to proposition B, one writes $A = B$.

All false propositions are considered equal and are identified by the falsehood symbol $\mathbb{0}$. The universally true proposition is denoted by the symbol $\mathbb{1}$. Classical logic is a *two-valued* logic which accepts the doctrine of the "law of the excluded middle" that every proposition is either true or false—*tertium non datur*.

A *Boolean algebra* is a set $\mathcal{B}$ of propositions $A, B, \ldots$ with an operation $^\perp$ of rank one, with two operations $\wedge$ and $\vee$ of rank two, and with two distinguished elements $\mathbb{0}$ and $\mathbb{1}$, such that the following axioms hold:

$$\text{(i)} \quad (A^\perp)^\perp = A \text{ for all } A \in \mathcal{B}, \tag{2.1a}$$

$$\text{(ii)} \quad \wedge \text{ and } \vee \text{ are each associative and commutative,} \tag{2.1b}$$

$$\text{(iii)} \quad \wedge \text{ and } \vee \text{ are distributive with respect to one another,} \tag{2.1c}$$

$$\text{(iv)} \quad (A \vee B)^\perp = A^\perp \wedge B^\perp, \; (A \wedge B)^\perp = A^\perp \vee B^\perp, \tag{2.1d}$$

$$\text{(v)} \quad A \vee A = A \wedge A = A, \tag{2.1e}$$

$$\text{(vi)} \quad \mathbb{1} \wedge A = A, \quad \mathbb{0} \vee A = A. \tag{2.1f}$$

In Boolean (classical) logic the following axioms hold:

$$(A^\perp)^\perp = A \quad \text{(law of double negation)} \tag{2.2a}$$

$$A \leq B \text{ if and only if } B^\perp \leq A^\perp \quad \text{(law of contraposition)} \tag{2.2b}$$

$$(A \vee B)^\perp = A^\perp \wedge B^\perp \tag{2.2c}$$

$$\text{and} \quad (A \wedge B)^\perp = A^\perp \vee B^\perp \quad \text{(de Morgan's laws)} \tag{2.2d}$$

$$(A \vee B) \wedge C = (A \wedge C) \vee (B \wedge C) \tag{2.2e}$$

$$\text{and} \quad (A \wedge B) \vee C = (A \vee C) \wedge (B \vee C) \quad \text{(distributive laws)} \tag{2.2f}$$

$$A \wedge A^{\perp} = \mathbb{0} \quad \text{(law of contradiction)} \tag{2.2g}$$

$$A \vee A^{\perp} = \mathbb{1} \quad \text{(law of excluded middle)} \tag{2.2h}$$

2.3 On the Biological Basis of Binary Distinctions

We have found a strange footprint on the shores of the unknown.
We have devised profound theories, one after another,
to account for its origins. At last, we have succeeded in
reconstructing the creature that made the footprint.
And lo! It is our own.

Arthur Stanley Eddington (1920, p. 201)

In contemporary science the results of experiments have to be expressed in terms of an unambiguous *bivalent language* in the sense that every statement about an empirical fact has to be *either true or false*. Since binary yes/no or true/false distinctions have become a dominating principle of scientific reasoning, two-valued classical Boolean logic has a special status in our system of knowledge. But its outstanding simplicity proves in no way that it is given *a priori*. Rather, it is plausible that our innate preference for two-valued logic has a biological basis. Our apparently self-evident so-called "laws of thought" may be related to the neurobiological architecture of our brains.

The central nervous system is in an essential way based on the all-or-none character of neural events. As a result of biological evolution, animal and human neural networks operate with *binary choices*, presumably developed in the struggle for survival. Strictly binary decisions eliminate vagueness by reducing the available information. In this way conflicting responses in the brain are avoided so that simple and fast decisions become possible.

Practically all biological systems use a threshold logic to achieve clear yes-no responses (compare McCulloch and Pitts 1943; Platt 1956; von Neumann 1951, p. 22). It seems that built-in threshold gates ("tertium-non-datur" devices) have evolutionary advantages. They lead to Boolean circuits based on AND, OR and NOT gates. Warren McCulloch and Walter Pitts (1943) showed that anything that can be defined completely and unambiguously in a finite number of words can be realized by a Boolean neural network.

John Platt (1956) suggested that our preference for the "laws of thought" of two-valued logic are a natural consequence of their evolutionary advantage for survival. A neural network realizes a threshold logic to achieve strictly binary decisions which suppress conflicting stimuli so that ambiguous responses are avoided and clear-cut instructions for action become feasible. As Platt (1956, p. 195) remarked, "survival selects the reproducible". On a similar vein, Szent-Györgi (1962, p. 11) stated:

> Primarily the human brain is an organ of survival. It was built by nature to search for food, shelter, and the like, to gain advantage—before addressing itself to the pursuit of truth.

Neural threshold gates explain why we preferably perceive and interpret the world filtered through a Boolean frame of reference.

2.4 Boolean Worldviews

2.4.1 Boolean Frames of Reference

Scientific knowledge is partly based on sense perception, partly on theoretical ideas, partly constructed by social processes, and partly informed by historical preconceptions. Therefore, the established body of scientific knowledge at any given time depends also on history and culture, and determines a system of beliefs. We cannot avoid that scientific knowledge is always influenced by extra-scientific factors. A *conceptual scheme* by which we define reality and within which we think and interpret our experiences is called a *worldview*. According to Feyerabend (1994, p. 152), a worldview is

> a collection of beliefs, attitudes and assumptions that involves the whole person, not only the intellect, has some kind of coherence and universality and imposes itself with a power far greater than the power of facts and fact-related theories.

From the very beginning, humankind has asked for and developed knowledge that can be considered as trustworthy. Knowledge always rests on assumptions and choices. It is never exhaustive, so we need to distinguish what is relevant from what is irrelevant in a certain situation. The idea that science is the only legitimate source of beliefs about reality has been criticized by Michael Polanyi (1958, p. 217):

> ...the scientists actually establish the current meaning of the term "science", determine what should be accepted as science, and establish also the current meaning of the term "scientist" and decide that they themselves and those designated by themselves as their successors should be recognized as such.

Moreover, Polanyi (1958, pp. 274f) argued that even among "scientists" intersubjective agreement is not sufficient to avoid serious errors or mistaken beliefs:

> Ordinary people were convinced of the fall of a meteorite, when an incandescent mass struck the earth with a crash of thunder a few yards away, and they tended to attach supernatural significance to it. The scientific committees of the French Academy disliked this interpretation so much that they managed, during the whole of the eighteenth century, to explain the facts away to their own satisfaction. ... We regard these acts of scepticism as unreasonable and indeed preposterous today, for we no longer consider the falling of meteorites ... to be incompatible with the scientific worldview. But other doubts, which we now sustain as reasonable on the grounds of our own scientific worldview, have once more only our beliefs in this view to warrant them. Some of these doubts may turn out one day to have been as wanton, as bigoted and dogmatic as those of which we have now been cured.

Even in modern science we have to take into account that knowledge is always influenced by biases and ideologies which exert a powerful influence—knowledge and belief cannot strictly be delineated. For these reasons William Cobern (1996, p. 584) disapproved the notion of a "scientific worldview" as misleading since a worldview

> provides a non-rational foundation for thought, emotion, and behavior. A worldview provides a person with presuppositions about what the world is really like and what constitutes valid and important knowledge about the world.

We call a domain, in which all that is knowable in principle is also simultaneously knowable, a *Boolean frame of reference* or a *Boolean context*. Niels Bohr (1949, p. 209) characterized the privileged role of Boolean, or classical, descriptions as follows:

> *However far the phenomena transcend the scope of classical physical explanation, the account of all evidence must be expressed in classical terms.* The argument is simply that by the word "experiment" we refer to a situation where we can tell others what we have done and what we have learned and that, therefore, the account of the experimental arrangement and the results of the observations must be expressed in unambiguous language with suitable application of the terminology of classical physics.

This requirement actually reflects scientific practice: the outcome of every experiment ever performed in physics, chemistry, biology or psychology can be described in a language based on classical Boolean logic.[14] For the description of an observed result of an experiment Bohr proposed to use (Bohr 1948, p. 317)

> the word phenomenon to refer exclusively to observations obtained under specified circumstances, including an account of the whole experiment.

2.4.2 Incompatible Boolean Descriptions

The fact that every outcome of a *single* experiment allows a description in terms of classical Boolean logic does not imply that the family of all feasible experiments can be combined into a single Boolean frame of reference. We adopt the following terminology:

> Two Boolean descriptions are said to be *compatible* if they can be embedded into a single Boolean reference frame.
> Two Boolean descriptions are said to be *incompatible* if they cannot be embedded into a single Boolean reference frame.

Since there are many mutually incompatible Boolean contexts, there are always many different mutually incompatible descriptions of one and the same universe of

[14] According to Bohr (1934, p. 53), our "interpretation of the experimental material rests essentially upon the classical concepts". Bohr's emphasis on the special role of classical physics is somewhat misleading. By a "classical description" Bohr means a description in terms of ordinary language. Important for Bohr's arguments is only that *facts* have to be described in a Boolean language, but not necessarily in terms of classical physics.

discourse. For this situation, Bohr coined the notion of *complementarity* to describe reality using incompatible classical concepts. Although no formal definition of complementarity can be found in his publications,[15] its spirit is closely related to that of incompatibility:[16]

> Two Boolean descriptions are said to be *complementary* if they cannot be embedded into a single Boolean description.

The idea that truths cannot contradict each other characterizes a Boolean worldview which, from a modern point of view, cannot adequately express our knowledge about the structure of the world. The binary "either-or" thinking of the Boolean worldview is a prejudice that prevents us from new ways of thinking. The development of quantum physics throughout the 20th century has shown that an all-encompassing description of the universe does not have a Boolean logical structure with the law of the excluded middle. Consequently, *the strictly Boolean worldview of classical science has to be given up*. This implies the necessity to use different, incompatible, sets of concepts for different contexts. This incompatibility does not just express a cognitive, or epistemic, limit, but is a fundamental features of the world.

In order to understand the world properly we need more than one Boolean frame of reference. Since science requires that empirical results must be expressed in terms of a *Boolean language*, an empirically meaningful framework for incompatible concepts requires a *locally Boolean structure* which must be *globally non-Boolean*.

This is possible by connecting mutually incompatible *locally Boolean* descriptions within a globally non-Boolean framework. The resulting structure is called partially Boolean. A partial Boolean algebra is a set of Boolean algebras pasted together such that their operations agree with each other where they overlap. In this way one obtains a manifold of locally Boolean descriptions, within which each Boolean algebra is a local chart of the manifold. This manifold generates a partial Boolean algebra in which all elements are connected. In general, the global structure of a partial Boolean algebra is not Boolean but allows a rigorous definition of the mutual compatibility of the elements of the Boolean manifold. For more details see Sects. 2.5 and 3.4.

A partial Boolean algebra provides a globally non-Boolean description with restricted sentential connectivity,[17] generated by patching together incompatible local Boolean descriptions in a smooth way—as geometric manifolds can be constructed from locally Euclidean spaces. In such an approach one does not introduce new connectives, the meaning of the Boolean logical connectives remains the same. As

---

[15]Editor's note: For the history of the concept of complementarity and its origin in psychology and philosophy see Holton (1970).

[16]Editor's note: In fact, complementarity is a maximal form of incompatibility, as will be discussed later in this monograph (see Sects. 3.4 and 3.5).

[17]Strauss (1936a, 1936b, 1938, 1967, 1970, 1973) defined a logic of complementarity as a *two-valued logic with restricted sentential connectivity* whose propositional calculus is given by a partial Boolean algebra.

a consequence there are no problems to interpret partial Boolean descriptions. We have to acknowledge that we always perceive nature through Boolean filters, and that these filters are usually incompatible. All Boolean descriptions are equally valid in their own way, and all together are required for a complete picture.

2.4.3 Simultaneously Undecidable Propositions

The modern version of partial Boolean algebras was introduced informally by Ernst Specker (1960) in a parable about a princess, her admirers, and prophecy as a logic of propositions that are not simultaneously decidable.[18]

> At the Assyrian School of Prophets in Arba'ilu in the time of King Asarhaddon, there taught a seer from Nineva. He was a distinguished representative of his faculty (eclipses of the sun and moon) and aside from the heavenly bodies, his interest was almost exclusively in his daughter. His teaching success was limited, the subject proved to be dry, and required a previous knowledge of mathematics which was scarcely available. If he didn't find the student interest which he desired in class, he did find it elsewhere in overwhelming measure. His daughter had hardly reached a marriageable age when he was flooded with requests for her hand from students and young graduates. And though he didn't believe that he would always have her by his side, she was in any case still too young and her suitors in no way worthy. In order that they might convince themselves of their worthiness, he promised her to the one who could solve a "prediction problem" which he set. The suitor was taken before a table on which three little boxes stood in a row and was asked to say which boxes contained a gem and which didn't. But no matter how many tried, the task seemed impossible. In accordance with his prediction, each of the suitors was requested by the father to open two boxes which he had marked as both empty or both full. But it always turned out that one contained a gem and the other one didn't, and furthermore the stone was sometimes in the first box and sometimes in the second. But how should it be possible, given three boxes, neither to mark two as empty nor two as full? The daughter would have remained single until her father's death had she not followed the advice of a prophet's son and quickly opened two boxes, one of which was marked full and the other empty. Following the weak protest of her father that he had wanted two *other* boxes opened, she tried to open the third. But this proved impossible whereupon the father grudgingly admitted that the prediction was correct.

The calculus of propositions underlying this parable takes into account that propositions cannot be connected as in classical thinking.[19] The proper algebraic structure for this calculus is that of a partial Boolean algebra.[20]

---

[18]English translation by A. Stairs as *The logic of propositions which are not simultaneously decidable*, reprinted by Hooker (1975, pp. 135–140). A more recent translation is by M.P. Seevinck: *The logic of non-simultaneously decidable propositions*, arxiv.org/abs/1103.4537. Editor's note: Specker presented a slightly revised version on March 27, 2000, in his seminar on *Quantum Logic and Hidden Parameters* at ETH Zurich. This version has not been translated into English so far.

[19]Editor's note: The particular version of Specker's parable is a complication of the usual pairwise incompatibility of measurements in quantum physics. It expresses that compatible measurements are jointly possible pairwise but not triplewise (see Liang et al. 2011, see also Sect. 3.5.4).

[20]Partial Boolean algebras have been used by Franz Kamber (1964) as the propositional calculus for quantum systems before Kochen and Specker (1965a, 1965b) proved that the partial Boolean

2.5 Boolean Classifications

The search for regularities is the point of departure of all scientific inquiry. Classifications and the recognition of patterns are based on the primitive act of *making a distinction* (Spencer Brown 1969). A distinction splits the world into two parts. Since the world is not intrinsically divided into subsystems, there is no unique way to make a distinction. That is, every classification is context-dependent and always entails a loss of information: it eliminates "irrelevant" information.

In the simplest case, a classification refers to the error-free grouping of finitely many objects into classes according to certain attributes they have in common. Any classification creates and isolates phenomena relevant to us. If we consider something else to be relevant, we need a different classification. The feasibility of a classification presupposes that there are distinct individual objects (like material or mathematical objects, electric signals, pictures, ideas) which can be characterized by well-defined attributes. Since there are no *a priori* distinctions, there is no unique way to classify.

Classifications can be based on *equivalence relations* which allow us (for a well-specified purpose) to merge objects with selected distinguishing characteristics together into *equivalence classes*. A binary relation $\sim$ on a nonempty set Ω is called an *equivalence relation* if it is reflexive, symmetric, and transitive:

$$\text{(i)} \quad \textit{reflexive}: \quad \mathcal{A} \sim \mathcal{A} \text{ for all } \mathcal{A} \in \Omega, \tag{2.3a}$$

$$\text{(ii)} \quad \textit{symmetric}: \quad \text{whenever } \mathcal{A} \sim \mathcal{B}, \text{ then } \mathcal{B} \sim \mathcal{A}, \tag{2.3b}$$

$$\text{(iii)} \quad \textit{transitive}: \quad \text{if } \mathcal{A} \sim \mathcal{B} \text{ and } \mathcal{B} \sim \mathcal{C}, \text{ then } \mathcal{A} \sim \mathcal{C}. \tag{2.3c}$$

If $\mathcal{A} \sim \mathcal{B}$, we say that they are equivalent under the equivalence relation $\sim$. If $\mathcal{A} \in \Omega$, then the set of all elements of Ω that are equivalent to $\mathcal{A}$ is called the *equivalence class* $[\mathcal{A}]_\sim$ of $\mathcal{A}$,

$$[\mathcal{A}]_\sim := \{\mathcal{B} \in \Omega \mid \mathcal{B} \sim \mathcal{A}\}. \tag{2.4}$$

Two equivalence classes either coincide or are disjoint:

$$\mathcal{A}, \mathcal{B} \in \Omega \quad \Rightarrow \quad \text{either} \quad [\mathcal{A}] = [\mathcal{B}] \quad \text{or} \quad [\mathcal{A}] \cap [\mathcal{B}] = \emptyset. \tag{2.5}$$

The strongest equivalence relation is $=$, the equality relation, producing the smallest possible equivalence classes.

Regularities can be found if and only if we suppress features that can be declared irrelevant in the context considered. This elimination of irrelevant features is necessary to classify objects according to attributes, and sometimes even to "produce" such attributes. That is, every classification is context-dependent and entails the loss of "irrelevant" information. Every classification can be realized by a battery

---

structure arising in quantum theory cannot be embedded into a Boolean algebra. Compare Sect. 3.4 for more details.

of filters, i.e., devices which make binary decisions based on binary distinctions (Hammer 1969, p. 107):

> A filter is here defined as any device which accepts or passes certain elements in a set and rejects others. Thus a filter produces an ordered dichotomy in a set, the set of elements passed and its complement—those not passed.

The result of every classification can be expressed by a *finite* number of binary decisions. Accordingly, every exhaustive ideal classification can be represented by finitely many filters $\mathbb{F}_1, \mathbb{F}_2, \ldots, \mathbb{F}_n$. Filters can be represented algebraically by classification operators which we denote be $F_1, F_2, \ldots, F_n$ which can be combined as serial connections or as parallel connections.

The *serial connection* of two filters $\mathbb{F}_j$ and $\mathbb{F}_k$ is a filter which accepts objects that are accepted by both filters. By convention the serial connection of two filters characterized by the classification operators F_j and F_k is written *multiplicatively* and represented by the selection operator $F_j \wedge F_k$.

The *parallel connection* of two filters $\mathbb{F}_j$ and $\mathbb{F}_k$ is a filter which accepts the union of objects that are accepted by at least one of the component filters. By convention the parallel connection of two filters characterized by the classification operators F_j and G_k is written *additively* and represented by the selection operator $F_j \vee G_k$.

- A filter is said to be *ideal* if its repeated application has the same effect as its single application, i.e. if the selection operator F_j is idempotent,

$$F_j \wedge F_j = F_j. \tag{2.6a}$$

- Two filters F_j and F_k are called *compatible* if the result of the filter operation does not depend on the order, so that the corresponding selection operators commute,

$$F_j \wedge F_k = F_k \wedge F_j. \tag{2.6b}$$

- A system of n filters $F_1, F_2, \ldots, F_n$ is said to be *complete* if

$$F_1 \vee F_2 \vee \cdots \vee F_n = \mathbb{1}. \tag{2.6c}$$

  where $\mathbb{1}$ is the trivial filter that accepts every object.

Every system of finitely many *compatible* ideal filters generates a finite *Boolean classification* associated with the idempotent projectors $F_1, F_2, \ldots, F_n$. A Boolean classification is said to be *ideal* if it is exhaustive, reproducible and mutually exclusive:

$$\text{(i)} \quad \textit{exhaustive}: \quad F_1 \vee F_2 \vee \cdots \vee F_n = \mathbb{1}, \tag{2.7a}$$

$$\text{(ii)} \quad \textit{reproducible}: \quad F_j \wedge F_j = F_j, \tag{2.7b}$$

$$\text{(iii)} \quad \textit{mutually exclusive}: \quad F_j \wedge F_k = \mathbb{0} \text{ for } j \neq k. \tag{2.7c}$$

A *partial ordering* on the set of the projections of all ideal filters of a given context is defined by:

$$F_j \leq F_k \quad \text{if and only if} \quad F_j \wedge F_k = F_k \wedge F_j = F_j, \tag{2.8}$$

The zero operator $\mathbb{O}$ is the smallest projector, and the identity operator $\mathbb{1}$ is the largest projector. Every ideal Boolean classification generates a *Boolean algebra* $\{\mathcal{B}, \mathbb{1}, \mathbb{O}, \wedge, \vee, ^{\perp}\}$ of projections $F_1, F_2, \ldots, F_n$ with $F_j^2 := F_j \wedge F_j = F_j$, where the orthocomplementation $^{\perp}$ is defined by

$$F_j^{\perp} := \mathbb{1} - F_j. \tag{2.9}$$

The Boolean algebra $\{\mathcal{B}, \mathbb{1}, \mathbb{O}, \wedge, \vee, ^{\perp}\}$ defines a linear space $\{\mathfrak{C}, \mathbb{1}, \mathbb{O}, \cdot, +, ^*, \mathbb{C}\}$ over $\mathbb{C}$ which is a *commutative *-algebra* over $\mathbb{C}$ under the multiplication $\cdot$ and the addition $+$,

$$F_j F_k := F_j \cdot F_k := F_j \wedge F_k, \tag{2.10a}$$

$$F_j + F_k := F_j \vee F_k + F_j \wedge F_k. \tag{2.10b}$$

It is generated by the projections $F_1, F_2, \ldots, F_n$,

$$\mathfrak{C} := \left\{ \sum_{j=1}^{n} c_j F_j \,\middle|\, c_j \in \mathbb{C} \right\}. \tag{2.10c}$$

The involution $C \to C^*$ of $\mathfrak{C}$ is characterized by $(C, C_1, C_2 \in \mathfrak{C}, c \in \mathbb{C})$:

$$\left(C^*\right)^* = C, \qquad (cC)^* = c^* C^*, \tag{2.10d}$$

$$(C_1 + C_2)^* = C_1^* + C_2^*, \qquad (C_1 C_2)^* = C_1^* C_2^*. \tag{2.10e}$$

Chapter 3
Non-Boolean Descriptions

As far as the laws of mathematics refer to reality, they are not certain;
and as far as they are certain, they do not refer to reality.

Albert Einstein (1921)

To portray means to leave out.
You know, I portrayed you more similar than you are.

Max Liebermann (2005)[21]

3.1 Vagueness

3.1.1 Eliminable Vagueness

The theory of sets introduced by *Georg Cantor* is based on the *principle of bivalence*, characterized by the values 0 (for no membership) and 1 (for membership). This principle provides the foundation of two-valued or bivalent logic. This means that every declarative sentence expressing a proposition has exactly one truth value, either true or false:

Every proposition is either true or false.

Boolean logic satisfies this principle and is therefore called a two-valued logic.

Based on the work of Boole and Cantor, *Friedrich Ludwig Gottlob Frege* developed formal logic. He affirmed that mathematical entities are well-defined and have objective properties regardless of humans thinking about them. In his *Foundations of Mathematics* he claimed that any vagueness is a defect of natural language and should be excluded from the language of science. He insisted to adopt an artificial language in which each predicate was complete and precise (Frege 1952, p. 159):

---

[21] Editor's translation of the German original: "Zeichnen heisst Weglassen. Wissen Sie, ich habe Sie ähnlicher gemalt als Sie sind."

H. Primas, *Knowledge and Time*,
DOI 10.1007/978-3-319-47370-3_3

> A definition of a concept (of a possible predicate) must be complete; it must unambiguously determine, as regards any object, whether or not it falls under the concept (whether or not the predicate is truly ascribable to it). Thus there must not be any object as regards which the definition leaves in doubt whether it falls under the concept; though for us human beings, with our defective knowledge, the question may not always be decidable. We may express this metaphorically as follows: the concept must have a sharp boundary. If we represent concepts in extension by areas on a plane, this is admittedly a picture that may be used only with caution, but here it can do us good service. To a concept without sharp boundary there would correspond an area that had not a sharp boundary-line all round, but in places just vaguely faded away into the background. This would not really be an area at all; and likewise a concept that is not sharply defined is wrongly termed a concept. Such quasi-conceptual constructions cannot be recognized as concepts by logic; it is impossible to lay down precise laws for them. The law of excluded middle is really just another form of the requirement that the concept should have a sharp boundary.

Yet lack of sharp boundaries is prevalent in our use of natural language. Bertrand Russell (1913) claimed that both natural language and logical connectives are necessarily vague:

> ...we are able to imagine a precise meaning for such words as “or” and “not”. We can, in fact, see precisely what they would mean if our symbolism were precise. All traditional logic habitually assumes that precise symbols are being employed. It is therefore not applicable to this terrestrial life, but only to an imagined celestial existence.

Russell considered vagueness as an eliminable feature of natural language, and not as an objective feature of reality. Excluding explicitly quantum physics, also van Inwagen (2009, p. 6f) claims that there is no intrinsic vagueness in reality. He argues that all material objects are either elementary particles or living organisms (van Inwagen 1990) and concludes that vagueness is a result of “purely linguistic” conventions. He thinks that there is no vagueness in reality but that any vagueness is a result of our “rules for naming”. In Sect. 3.3.3 we will give a counterexample from classical physics which shows that vagueness is not merely an artifact of our language.

3.1.2 Intrinsic Vagueness

All descriptions involve idealizations and abstractions of inessential features—without abstractions there is no science. Such idealizations and abstractions are literally false but unavoidable. If we observe certain aspects of reality, we inevitably have to ignore other aspects so that every recognition of particular features has to be paid for by a loss of other features.

Among all the abstractions of physical science the idea of an isolated system is central. Yet, *on the fundamental level there are no isolated systems.* Every physical system is interacting and entangled with the rest of the world which we call its *environment.* The environment of every material object includes the electromagnetic field which is inextricably intertwined with the object under consideration. Therefore every material object has unsharp boundaries. Since every Boolean description depends on the feasibility of making sharp distinctions, a sharp Boolean classification of material objects, or of facts about them, is impossible as a matter of principle.

Common-sense things are created by abstracting from features that are irrelevant for a particular viewpoint. Different points of view correspond to different abstractions. There exist mutually exclusive abstractions which provide alternative descriptions of the very same system. Such incompatible descriptions are not logically contradictory because they refer to different abstractions and pattern recognition methods.

We must draw a distinction between mathematical idealizations and the real world. Many scientific concepts are inevitably vague so that strict equivalence classes are not realizable. We conclude that context-independent sharp classifications are inherently impossible.

3.2 Non-Classical Logic

The cognitive linguist George Lakoff (1973, p. 458) remarked that

> logicians have, by and large, engaged in the convenient fiction that sentences of natural languages (at least declarative sentences) are either true or false or, at worst, lack a truth value, or have a third value often interpreted as "nonsense". And most contemporary linguists who have thought seriously about semantics, especially formal semantics, have largely shared this fiction, primarily for lack of a sensible alternative.

While Boolean logic appears to be the simplest logical calculus we know of, Boolean "black-and-white" thinking is insufficient to capture both many everyday situations and many scientific concepts. For physics, Fritz Zwicky (1933) reminded us that

> any closed set of truths can be stated only after the adoption of a certain definite set of rules describing the type of thinking which is to be used. As a priori there are many such sets of rules, we cannot hope to embrace any part of nature completely by restricting ourselves to a definite set.

He proposed to use a non-classical logic rejecting the law of the excluded middle:

> No set of two-valued truths can be established with the expectation that this set ultimately will stand the test of experience. Formulations of scientific truth intrinsically must be many-valued.
> If then in physics we are confronted with any statement which has the appearance of a well-established absolute truth, we must question it, guess at more general possibilities in accordance with many-valued logics and consult the experiments. In particular, it may be anticipated that no set of nonstatistical truths will stand the test of experience.

3.2.1 Some Examples

Non-classical propositional logic refers to formal systems which differ in a significant way from the classical Boolean logic developed by Russell and Frege. There are quite a number of different important kinds of non-classical propositional logic:[22]

- *Intuitionistic logic* goes back to Luitzen Egbertus Jan Brouwer (1907). It captures constructive mathematical reasoning, explains truth by provability, and rejects the law of the excluded middle, double negative elimination, and de Morgan's laws (Heyting 1931, 1934; Kolmogorov 1932).
- *Paraconsistent logic* rejects the law of non-contradiction and admits propositions that are both true and false, which is a contradiction in most logics (Priest 2002).
- *Quantum logic:* Garrett Birkhoff and John von Neumann found that in quantum mechanics the logical operations obey all the rules for Boolean algebras except the distributive law. This law is replaced by a weaker law called the modular identity (Birkhoff and von Neumann 1936; Dalla Chiara and Giuntini 2002).
- *Many-valued logic* rejects bivalence. Three-valued logic was originally defined by Jan Lukasiewicz (1920). Post (1921) generalized Łukasiewicz's three-valued logic to many-valued variants.
- *Infinitely-many-valued logic* (with the set of all rationals between 0 and 1 as truth degree set) goes back to Łukasiewicz and Tarski (1930).
- *Vague predicates and de Morgan algebras.* Susan Haak (1996) claims that truth does not come in degrees but, according to a detailed analysis by Nicholas Smith (2008), any adequate account of vagueness must involve *degrees of truth*. Hermann Weyl (1940) proposed to replace the two-valued predicate of classical logic by vague predicates:

> the classical logic of propositions as formalized by G. Frege, and later by Russell and Whitehead in the *Principia Mathematica*, is based on the assumption that a proposition puts a question to some realm of reality whose facts answer with a clear-cut yes or no, according to which the proposition is either true or false. Up to the time of the *Principia Mathematica* everybody believed, or at least hoped, that mathematical propositions were of this nature, leaving no room for indeterminacies.

He suggested to replace the all-or-nothing principle of Boolean logic by a continuous grade $f : X \to [0, 1]$, where $f(x)$ describes the extent to which a given property $x \in X$ is possessed by an object $x \in X$. He defined the following operations for such predicates:

$$f \text{ and } g = \min(f, g) =: f \wedge g, \tag{3.1a}$$

$$f \text{ or } g = \max(f, g) =: f \vee g, \tag{3.1b}$$

$$\text{not } f = 1 - f =: \bar{f}, \tag{3.1c}$$

[22]Compare for example Frink (1938), Sinowjew (1970a, 1970b), Rasiowa (1974), Jammer (1974, Chap. 8), Haak (1996), Gabbay and Woods (2007), Gabbay and Guenthner (1986).

so that they define a de Morgan algebra.[23] De Morgan logic was rediscovered within the framework of fuzzy set theory which has been used to represent vague or ambiguous knowledge, in particular to deal with problems in natural language. Fuzzy sets and the corresponding fuzzy logic will be discussed in more detail in Sect. 3.3.

- *Probabilistic logic:* Karl Menger (1942) introduced the idea of a *probabilistic metric space* by replacing the numerical distance between two points by a distance distribution function. In 1951 he proposed an extension of the classical notion of a set by replacing the element-set relation by the probability of an element belonging to a set. He named these entities "ensembles flous" or "hazy sets" (Menger 1951) and used them to describe unsharp concepts in natural language and for pattern classification.
- *Klaua's many-valued set theory:* Inspired by a discussion following a colloquium talk by Karl Menger in 1960,[24] Dieter Klaua (1965, 1967) proposed an extension of Cantor's classical theory to a set theory in which the elements have degrees of membership. This has been discussed and developed in a many-valued set theory based on Łukasiewicz's logic. Independent of the work of Menger and Klaua, Lotfi Zadeh (1965) introduced the concept of *fuzzy sets* whose elements have degrees of membership, and developed a corresponding *fuzzy logic* (see Sect. 3.3 for more details).
- *Temporal logic* refers to propositions qualified in terms of time. Since traditional Boolean logic is a timeless discipline, it is ill-adapted with statements containing temporal references. Charles Sanders Peirce (1933, Sect. 4.523) suggested to include time within logic:

  > Time has usually been considered by logicians to be what is called an "extralogical" matter. I have never shared this opinion. But I have thought that logic had not yet reached the state of development at which the introduction of temporal modifications of its forms would not result in great confusion; and I am much of that way of thinking yet.

  Arthur Norman Prior (1957) introduced a temporal logic under the name of *tense logic* as a modal logic with modalities referring to time.
- *Linear logic* was introduced by Jean-Yves Girard (1987) as an extension of classical and intuitionistic logic. Girard sees the truth of assumptions as a resource that is consumed when it is used. Linear logic introduces new connectives. As Girard (1995, p. 1) states:

  > Classical and intuitionistic logics deal with stable truths:
  >
  > $$\text{if } A \text{ and } A \Rightarrow B, \text{ then } B, \textit{ but A still holds.}$$
  >
  > This is perfect in mathematics, but wrong in real life since real implication is *causal*. A causal implication cannot be iterated since the conditions are modified after its use;

[23] Every Boolean algebra is a de Morgan algebra, but in a de Morgan algebra the law of the excluded middle and the law of non-contradiction do not necessarily hold.

[24] See Gottwald (2010, p. 2369, footnote 2).

this process of modification of the premises (conditions) is known in physics as *reaction*.

Linear logic can also be used to extend Birkhoff and von Neumann's propositional quantum logic whereby the ill-defined implication of static quantum logic is replaced by a well-defined dynamical implication (Pratt 1992).

3.3 Fuzzy Sets and Fuzzy Logic

I believe that, although much of modern science is based on bivalent logic, eventually most scientific theories will be based at least in part on fuzzy logic. It may, however, take a long time for this to become a reality.

Lotfi Zadeh[25]

Lotfi Zadeh (1965) introduced the concept of "fuzzy sets" as collections of elements with graded membership.[26] Fuzzy sets generalize classical sets, where an element either belongs or does not belong to the set insofar as the indicator functions of classical sets are special cases of the membership functions of fuzzy sets if the latter only assume values 0 or 1. By contrast, fuzzy set theory permits the gradual assessment of the membership of elements in a set; this is described with the aid of a *membership function* valued in the real unit interval [0, 1]. This membership function characterizes a partial membership and can be used to describe fuzzy sets. In contrast to Menger, Zadeh considered fuzzy sets as a class of objects which are intrinsically imprecise, where the fuzziness is in general of *non-probabilistic* nature. *Randomness and fuzziness are independent concepts*.

Fuzzy logic replaces the distinction between true and false with degrees of truth. It rejects the law of the excluded middle, accommodates vague sentences both as inputs to and outputs of the relation of implication. Since many concepts are fuzzy rather than bivalent, bivalent logic is in fundamental conflict with reality. In such cases fuzzy logic may lead to a better description of reality.

In particular, the structure of propositions about quantum systems gives rise to a logic in which the connectives AND and OR are not always defined. This logic has two non-standard ortho-complements. The generalized ortho-complementation of fuzzy set theory is called *Zadeh ortho-complementation*, while the generalized ortho-complementation of Brouwerian lattices is called *Brouwer ortho-complementation*. The propositions about quantum systems define a partial Boolean algebra—a weaker version of a Boolean algebra where the operations meet and join are only defined for pairs of compatible elements. This amounts to a partially ordered set endowed with a binary reflexive and symmetric compatibility relation such that every triple of pairwise compatible elements determines a Boolean algebra.

---

[25]Quoted by Dumitras and Moschytz (2007, p. 195) from an interview with Zadeh.

[26]For details of the history of the concepts "vagueness", "haziness" and "fuzziness" compare Seising (2007).

3.3.1 Fuzzy Sets

In the classical set theory of Cantor, Zermelo and Fraenkel an object is either a member of a set or it is not. In a seminal paper, Zadeh (1965) relaxed this condition and defined a fuzzy set as a class of objects which are intrinsically imprecise so that fuzziness is in general of non-probabilistic nature. The degree of membership of an object in a fuzzy set can be any real number between 0 and 1.

In other words, a fuzzy set is a class X with fuzzy boundaries, characterized by a membership function μ which associates with each element of X a number $\mu \in \mathbb{R}$ in the interval $[0, 1]$. If the range of μ is restricted to $\{0, 1\}$, then μ is the characteristic function of classical set theory.

In classical set theory there is a bijective correspondence between the power set $\mathcal{P}(X)$ of all subsets $A \in X$ of a set X and the set of all characteristic functions of elements $x \in X$:

$$\chi_A(x) = \begin{cases} 0 & \text{for } x \notin A \\ 1 & \text{for } x \in A \end{cases}. \tag{3.2}$$

A fuzzy set $\mathcal{F}(X)$ can be characterized in the same way whereby the Boolean characteristic function $\chi_A : X \to \{0, 1\}$ is replaced by the membership function $\mu_A(x) : X \to [0, 1]$. If we define the following operations in $\mathcal{F}(X)$,

$$\mu_A = \mu_B \quad \Longleftrightarrow \quad \mu_A(x) = \mu_B(x) \quad \text{(equality)}, \tag{3.3a}$$

$$(\mu_{A \vee B})(x) := \max\{\mu_A(x), \mu_B(x)\} \quad \text{(union)}, \tag{3.3b}$$

$$(\mu_{A \wedge B})(x) := \min\{\mu_A(x), \mu_B(x)\} \quad \text{(intersection)}, \tag{3.3c}$$

the union operation in fuzzy set theory corresponds to the OR operation in Boolean algebra while the intersection operation in fuzzy set theory corresponds to the AND operation in Boolean algebra. With the definitions (3.3a)–(3.3c) the set $\mathcal{F}(X)$ is a completely distributive lattice[27] (i.e., a complete lattice in which arbitrary joins distribute over arbitrary meets). One can also define a pseudo-complement operation $A \to \bar{A}$ by

$$\bar{\mu}_A(x) := 1 - \mu_A(x) \quad \text{(pseudo-complement)}, \tag{3.3d}$$

but this operation differs from the lattice theoretical complement $A^{\perp}$ since in general

$$\bar{A} \vee A \neq \mathbb{1}, \qquad \bar{A} \wedge A \neq \mathbb{O},$$

while the lattice theoretical complement is defined by the two relations

$$A^{\perp} \vee A = \mathbb{1}, \qquad A^{\perp} \wedge A = \mathbb{O}.$$

[27]For details compare Negoitã and Ralescu (1975, Chap. 1).

Therefore the lattice $\mathcal{F}(X)$ is in general not a Boolean lattice. The laws of commutativity and associativity (2.1b), distributivity (2.1c), and de Morgan's laws (2.2c), (2.2d) hold in both Boolean and fuzzy logic. However, the Boolean law of contradiction (2.2g) and the Boolean law of the excluded middle (2.2h) are in general *not* valid for fuzzy sets.

3.3.2 Fuzzy Logic and Fuzzy Classifications

Cognitive scientists and philosophers of language know well that meaningful natural language sentences will often be neither true nor false. The traditional Boolean view of psychological concepts was challenged in the 1970s when experimental evidence showed that concept categories are graded and thus cannot be represented adequately by classical sets. The role of prototypes in human cognition was studied by Eleanor Rosch (1973a, 1973b, 1975). An experimental study of human systems of semantic categories showed how categories are constructed around a "natural prototype".

There is a continuous relationship between subjective categories and fuzzy logic so that "it is not unreasonable for different rules to be used under various situations" (Oden 1977a, p. 572; see also Oden 1977b). This approach led to the rejection of the traditional classical view of concepts which have to be replaced by graded categories that are not realized by a Boolean structure.

Each set theory has its corresponding logic whereby each basic set theoretical operation is translated into a logical operation. In fuzzy logic, A and B stand for arbitrary propositions and μ_A is the truth value of A, $0 \leq \mu_A \leq 1$. Relation (3.3b) says that the truth value of the disjunction $\mu_{A \vee B}$ is the maximum of the truth value of A and the truth value of B. Similarly, relation (3.3c) says that the truth value of the conjunction $\mu_{A \wedge B}$ is the minimum of the truth value of A and the truth value of B.[28]

The crisp classes of Boolean classifications are an unrealistic oversimplification of reality, and fuzzy classifications are their natural extension. In a Boolean classification, each object is assigned to exactly one class, so that the membership of objects to classes is mutually exclusive. In contrast, in a fuzzy classification, a set of classes C is assumed. A membership function $\mu_c : X \to [0, 1]$ characterizes the degree $\mu_c(x)$ to which object x belongs to class $c \in C$ (Roubens 1978; Amo et al. 2004).

[28] A detailed discussion of the Brouwer-Zadeh logic can be found in Chap. 15 of the monograph by Dalla Chiara et al. (2004).

3.3.3 Example: Bandpass Filters

3.3.3.1 Ideal Bandpass Filters

Bandpass filters are widely used in communication and signal-processing systems. A bandpass filter is an electrical device that transmits the spectral components of an electric signal around a certain frequency in the pass with little or no attenuation, while it rejects other spectral components.

In the time domain a continuous time-translation invariant linear bandpass filter can be represented by an input-output *response function* $t \to R(t)$ which relates an input $t \mapsto s(t+\tau)$ to the output $t \mapsto r(t+\tau)$ by

$$r(t) = r_0 + \int_{-\infty}^{\infty} dt' R(t-t') s(t'). \tag{3.4}$$

A *non-anticipating* filter cannot respond before the input signal arrives, so that in this case the input-output relation is *retarded*,

$$R(t) = 0 \quad \text{for } t < 0, \tag{3.5a}$$

$$r(t) = r_0 + \int_{-0}^{\infty} dt' R(t-t') s(t'). \tag{3.5b}$$

The Fourier-Laplace transform of a retarded response function is called *transfer function* $\hat{R}(\ell)$ with frequency ℓ:

$$\hat{R}(\ell) = \int_{0}^{\infty} e^{-2\pi i \ell t} R(t) dt. \tag{3.6}$$

An *ideal* bandpass filter would have a perfectly flat passband and would completely reject all frequencies outside the passband so that its transfer function in the frequency domain is given by

$$\hat{R}^{\mathrm{id}}(\ell) = \begin{cases} 0 & \text{for } \ell < \ell_1 \\ 1 & \text{for } \ell_1 \le \ell \le \ell_2 \\ 0 & \text{for } \ell > \ell_2 \end{cases} \tag{3.7}$$

where ℓ_1 is the lower cutoff frequency, ℓ_2 is the upper cutoff frequency, and $B := \ell_2 - \ell_1$ is the width of the passband.

With the rectangle function $x \mapsto \Pi(x)$,

$$\Pi(x) := \begin{cases} 0 & \text{for } |x| > 1/2 \\ 1/2 & \text{for } |x| > 1/2 \\ 1 & \text{for } |x| < 1/2 \end{cases} = \frac{\mathrm{sgn}(t+1/2)}{2} - \frac{\mathrm{sgn}(t-1/2)}{2}, \tag{3.8}$$

where $\mathrm{sgn}(t) = t/|t|$, we can write:

$$\hat{R}^{\mathrm{id}}(\ell) = \Pi(\ell/2B), \tag{3.9}$$

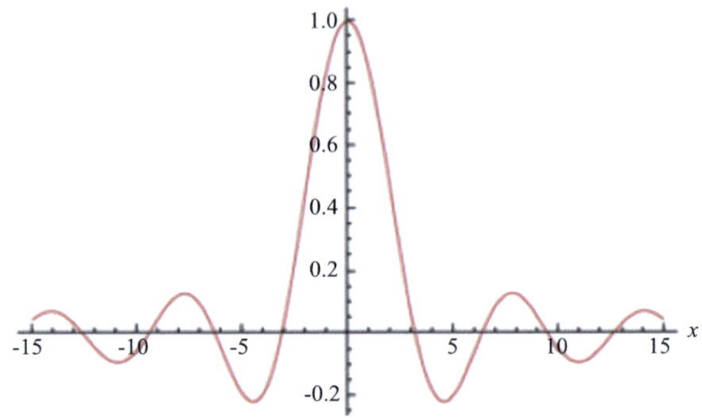

Fig. 3.1 Impulse response function $x \mapsto \sin(x)/x$ for an ideal bandpass filter

and the Fourier transform of $\Pi(\ell)$

$$\int_{-\infty}^{\infty} e^{2\pi i \ell t} \Pi(\ell) d\ell = \frac{\sin(\pi t)}{\pi t} \tag{3.10}$$

yields the ideal response function $t \mapsto R^{\mathrm{id}}(t)$

$$R^{\mathrm{id}}(t) = \int_{-\infty}^{\infty} e^{2\pi i \ell t} \Pi(\ell/2B) d\ell = \frac{\sin(\pi Bt)}{\pi Bt}. \tag{3.11}$$

This response function $t \mapsto R(t)$ equals the impulse response, i.e. the response to an input $t \mapsto \delta(t)$, which for an ideal bandpass filter is given by $t \mapsto \sin(\pi Bt)/(\pi Bt)$. Since this is non-zero before the δ-impulse has been applied (see Fig. 3.1), an ideal bandpass filter is acausal, so that it cannot be realized by a physical device.

More generally, a theorem by Paley and Wiener (1933, 1934) implies that a realizable amplitude characteristic $\ell \mapsto |\hat{R}(\ell)|$ of a causal time-invariant linear system cannot have perfect attenuation. In particular, no realizable filter can have infinite rejection over a finite band of frequencies. This is precisely expressed by the Paley-Wiener theorem:[29]

Paley-Wiener Theorem for Causal Filters
The following two criteria are necessary and sufficient for a square-integrable amplitude characteristic $A(\ell) := \ell \mapsto |\hat{R}(\ell)|$ to be realizable by a non-anticipative linear time-translation invariant dynamical system with a response function $R : \mathbb{R} \to \mathbb{R}$:

$$\text{causality criterion:} \quad R(t) = 0 \quad \text{for} \quad t < 0, \tag{3.12a}$$

$$\text{Paley-Wiener criterion:} \quad \int_{-\infty}^{\infty} \frac{|\ln|\hat{R}(\ell)||}{1+\ell^2} d\ell < \infty. \tag{3.12b}$$

[29]Compare Valley and Wallman (1948, App. A, pp. 721–727). The condition of square-integrability in the original formulation by Paley and Wiener is not essential. It can be replaced by much weaker conditions, for example by the requirement that the amplitude function is a tempered generalized function. Compare Pfaffelhuber (1971), compare also Sect. 12.4.2.7.

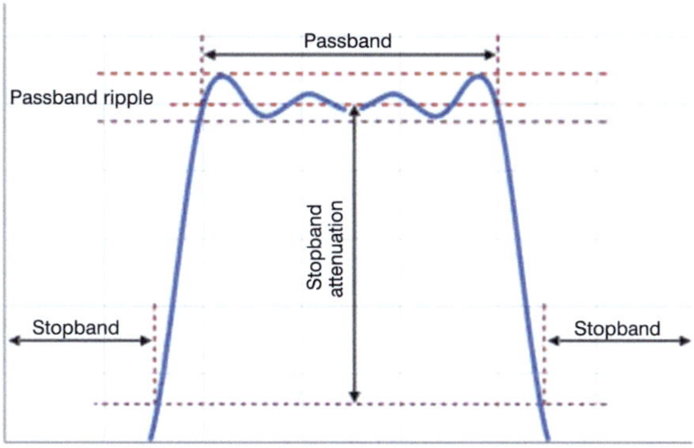

Fig. 3.2 The frequency response of a fuzzy low-pass filter

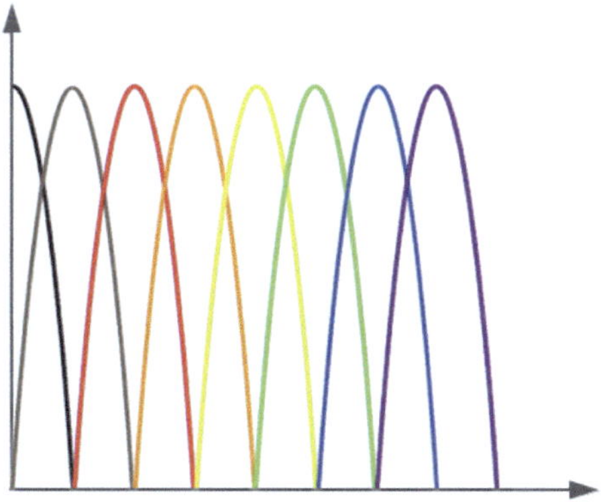

Fig. 3.3 The frequency response of a fuzzy filter bank

The Paley-Wiener theorem implies that an ideal bandpass filter cannot be realized. According to Valley and Wallman (1948, p. 723), "*this is not just a practical difficulty but a theoretical impossibility*".

3.3.3.2 Fuzzy Bandpass Filters

Ideal bandpass filters are acausal, so they cannot be realized in real time since they have an infinite group delay. Every realizable bandpass filter has unsharp boundaries: signals outside the desired frequency band are not rejected but only attenuated (the so-called *filter roll-off*). Real filters can only approximate ideal bandpass filters. In every practical bandpass filter there is a gradual transition between the passband and the upper and lower stopbands, as shown in Fig. 3.2.

The impossibility of realizing ideal electric filters whose frequency passbands have exactly defined threshold frequencies leads to a paradigmatic example for the usefulness of fuzzy classifications. Realizable electric filters have unsharp boundaries, so that any filter bank (that is, an array of bandpass filters) has overlapping bands. We cannot construct a filter bank which allows an ideal classification with respect to frequency. But a *fuzzy classification*, depicted in Fig. 3.3, is possible.

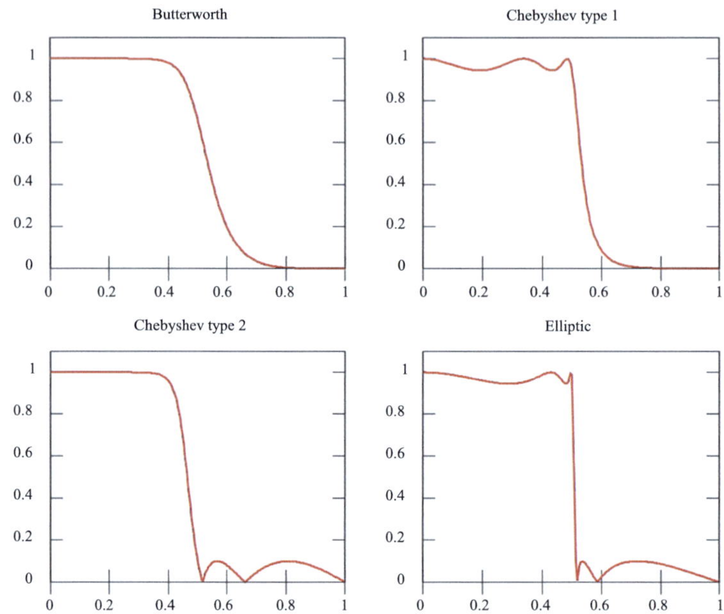

Fig. 3.4 The frequency response $\ell \mapsto |\hat{R}(\ell)|$ of typical low-pass filters (taken from commons.wikimedia.org/wiki/File:Electronic_linear_filters.svg)

3.3.3.3 Butterworth Filters

An approximate bandpass filter can always reduce the roll-off, but only at the expense of pass-band or stop-band ripple. The most popular approximations to an ideal low-pass filter are Butterworth filters, Chebyshev filters and elliptic filters. Their frequency responses are shown in Fig. 3.4.

Among these filters, *Butterworth filters* (Butterworth 1930) are characterized by the flattest pass-band response, at the expense of a less steep transition from pass-band to stop-band. They have a monotonically changing magnitude function $\ell \mapsto |\chi(\ell)|$ which has no ripples in the passband, and rolls off towards zero in the stopband. A Butterworth low-pass filter of order N is characterized by its amplitude response (Fig. 3.5), which is the square of its transfer function $\ell \mapsto \hat{R}_N^{\mathrm{B}}(\ell)$,

$$\left|\hat{R}_N^{\mathrm{B}}(\ell)\right|^2 = \frac{1}{1+(\ell/\ell_{\mathrm{c}})^{2N}}, \tag{3.13}$$

where ℓ_{c} is the filter's cut-off frequency. If we expand the transfer function $\ell \mapsto \hat{R}(\ell)$ in partial fractions, we get

$$\hat{R}_N^{\mathrm{B}}(\ell) = \frac{\chi_0}{\prod_{k=1}^{N}(\ell - \ell_k/\ell_{\mathrm{c}})} = \left|\hat{R}_N^{\mathrm{B}}(\ell)\right| e^{2\pi i \varphi_n^{\mathrm{B}}}(\ell), \tag{3.14}$$

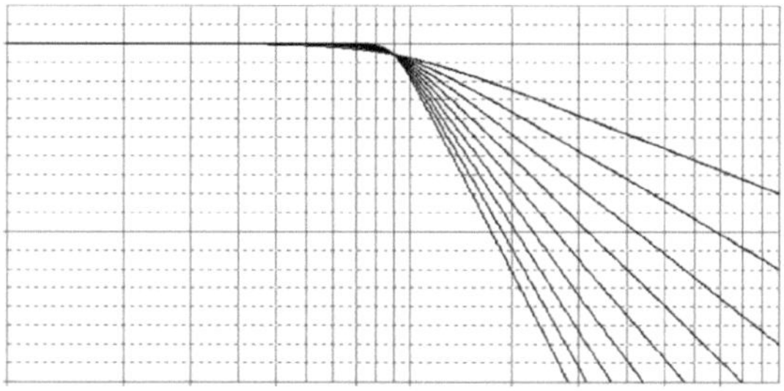

Fig. 3.5 Amplitude response $\ell \to |\hat{R}_N^{\mathrm{B}}(\ell)|^2$ of low-pass Butterworth filters of order $N = 2, \ldots, 10$ (*from top to bottom*; Zumbahlen 2008, p. 605). Note that the amplitude response of a Butterworth filter has no ripples in the passband

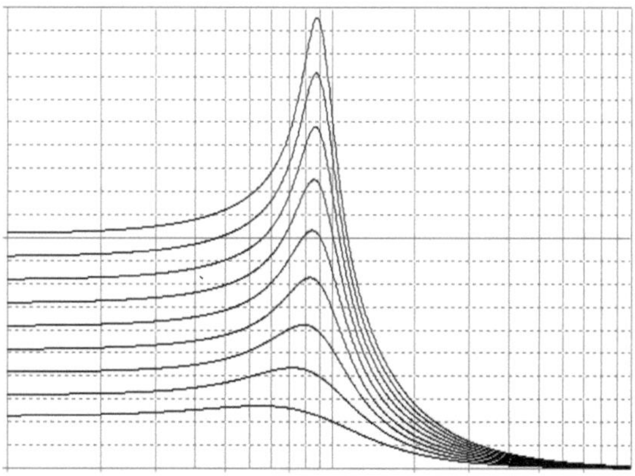

Fig. 3.6 Group delay $\ell \to |\tau_N^{\mathrm{B}}(\ell)|^2$ of low-pass Butterworth filters of order $N = 2, \ldots, 10$ (*from bottom to top*; Zumbahlen 2008, p. 605)

where the poles ℓ_k are given by

$$\ell_k = \ell_c \exp\left\{\pi i \left(\frac{1}{2} + \frac{(1+2k)}{2N}\right)\right\}, \qquad k = 0, 1, 2, \ldots, N-1. \tag{3.15}$$

The group delay τ_N^{B} of an Nth-order Butterworth filter is defined as the derivative of the phase $\ell \mapsto \varphi_N^{\mathrm{B}}(\ell)$ (see Fig. 3.6)

$$\tau_N^{\mathrm{B}}(\ell) = -\frac{1}{2\pi} \frac{d\varphi_N^{\mathrm{B}}(\ell)}{d\ell}. \tag{3.16}$$

The slope of the transition between a pass-band and the upper and lower stop-bands depends on the order of the Butterworth filter. The slope increases by 6.02 dB/oct for each increment of the filter order N. As the order N of a Butterworth filter increases, its amplitude response approaches that of an ideal bandpass filter better and better (see Fig. 3.7) at the expense of an increasing delay. In the limit $N \to \infty$ we get

$$\lim_{N\to\infty} \left|\hat{R}_N^{\mathrm{B}}(\ell)\right| = R^{\mathrm{id}}. \tag{3.17}$$

However, this limit is not realizable since the group delay diverges,

$$\lim_{N\to\infty} \tau_n^{\mathrm{B}}(\ell) = \infty. \tag{3.18}$$

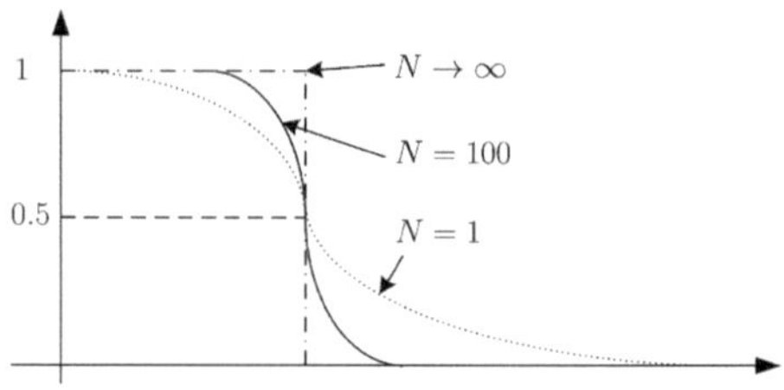

Fig. 3.7 Transition between pass-band and upper stop-band of low-pass Butterworth filters for different orders N

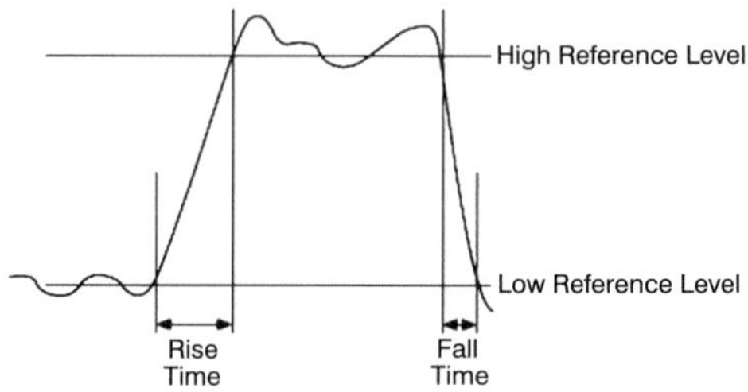

Fig. 3.8 Transition times of a linear bandpass filter

3.3.3.4 The Küpfmüller-Wiener Fuzzyness Principle

There is a relationship between the bandwidth of a filter and the duration of its response which can be summarized as follows:

It is impossible for an amplitude response function $\ell \mapsto |\hat{R}(\ell)|$ and the response function $t \mapsto R(t)$ to have simultaneously sharp boundaries.

The insight that a non-zero function with compact support cannot have a Fourier transform with compact support can be formulated in many different mathematically rigorous ways (compare Havin and Jöricke 1994). It was already used by Sommerfeld (1914) to discuss the coherence length of wave trains. Later Küpfmüller (1924) proved an inequality for settling time and bandwidth of bandpass filters in communication theory.[30]

In engineering applications it is important to know how quickly a system responds to a change in input. There are two types of transition times (see Fig. 3.8):

- the *rise time*, referring to the time required for an output amplitude of a filter to increase from a specified low value to a specified high value (exclusive of overshoot or undershoot),
- the *fall time*, referring to the time required for an output amplitude of a filter to decrease from a specified high value to a specified low value (exclusive of overshoot or undershoot).

The definitions of "rise time" or "fall time" are not unique but depend upon the application. Independent of its precise definition, the transition times are inversely proportional to the bandwidth.

[30] For a modern discussion of relations between bandwidth and duration of acoustic or electric signals in engineering science compare Slepian (1983).

If $t \mapsto s(t)$ is an *approximately* time-limited signal, then it is also *approximately* band-limited. If we use the second moments σ_t^2 as measures of the fuzziness of the duration and the second moments σ_ℓ^2 as measures of the fuzziness in bandwidth,

$$\sigma_t^2 := \int_{-\infty}^{\infty} (t - t_0)^2 |s(t)|^2 dt, \qquad \sigma_\ell^2 := \int_{-\infty}^{\infty} (\ell - \ell_0)^2 |\hat{s}(\ell)|^2 d\ell, \tag{3.19a}$$

$$\text{where } \hat{s}(\ell) := \int_{-\infty}^{\infty} e^{2\pi i \ell t} s(t) dt, \text{ and } t_0, \ell_0 \in \mathbb{R} \text{ are arbitrary}, \tag{3.19b}$$

then we get for the time-bandwidth product of a filter the following fuzziness principle,

$$\sigma_t \sigma_\ell \geq \frac{1}{4\pi} \|s\|^2, \tag{3.20}$$

first derived by Karl Küpfmüller (1924) and Norbert Wiener.[31] The equality in (3.20) holds if and only if $t \mapsto s(t)$ is proportional to the Gaussian function $t \mapsto g(t)$:

$$t \mapsto g(t) = e^{2\pi i a(t-b)} e^{-\pi (t-a)^2/b} \quad \text{for } a, b \in \mathbb{R} \text{ and } c > 0. \tag{3.21}$$

The Küpfmüller-Wiener uncertainty principle (3.20) describes an inverse relationship between bandwidth and duration: the narrower the bandwidth the greater the rise/fall time of a signal. It also implies that the number of telegraph signals that can be transmitted over a single line is directly proportional to its bandwidth.

3.4 Partial Boolean Descriptions

3.4.1 Partial Boolean Algebras

A partial algebraic structure is a set with certain partial operations which are defined only for certain elements of this set. A *partial Boolean algebra* is the union of a family of Boolean algebras which is closed under intersection. The connectives in the partial Boolean logic of a partial Boolean algebra are locally those of any Boolean subalgebra, but only the conjunction is globally defined for the whole algebra. In addition, a partial Boolean algebra is endowed with a binary reflexive and symmetric relation, interpreted as a compatibility relation, such that every maximal subset of pairs obeying that relation is a Boolean algebra. The following *axiomatic characterization* is due to Kochen and Specker (1965a, 1965b):

[31] The mathematically rigorous proof was given by Norbert Wiener in a seminar on generalized harmonic analysis at the University of Göttingen in 1924 (Wiener 1956, pp. 105f).

Definition: Partial Boolean Algebra
A *partial Boolean algebra* $\{\mathcal{P}, \leftrightarrow, \vee, \wedge, ^{\perp}, \mathbb{O}, \mathbb{1}\}$ is an algebraic structure where $\mathcal{P}$ is a set with two distinguished elements $\mathbb{O}$ and $\mathbb{1}$. The binary partial relation $\leftrightarrow$ is symmetric and reflexive (but not transitive) and is called *compatibility*.
The *disjunction* $\vee$ and *conjunction* $\wedge$ are two *partial* binary relations defined only for pairs of compatible elements in $\mathcal{P}$. They have domain $\leftrightarrow$ and range $\mathcal{P}$. The *negation* $^{\perp}$ is a unary relation defined for every element in $\mathcal{P}$. Apart from the partial definability of the logical relations, a partial Boolean algebra satisfies all the relations of a Boolean algebra. In particular, in any partial Boolean algebra the following relations hold for every $F, G, H \in \mathcal{P}$:

- $F \leftrightarrow \mathbb{O}$, $F \leftrightarrow \mathbb{1}$, $F \leftrightarrow F$ for every $F \in \mathcal{P}$,
- if $F \leftrightarrow G$, then $G \leftrightarrow F$,
- if $F \leftrightarrow G$, then $F \leftrightarrow G^{\perp}$,
- if $F \leftrightarrow G$, then $G \vee F$ and $G \wedge F$ are defined,
- if $F \leftrightarrow G \leftrightarrow H \leftrightarrow F$, then F, G, H together with $\mathbb{O}, \mathbb{1}$ generate a Boolean algebra relative to operations $\vee$, $\wedge$ and $^{\perp}$,
- if $F \leftrightarrow G \leftrightarrow H \leftrightarrow F$, then $F \vee G \leftrightarrow H$.

Every Boolean algebra $\{\mathcal{B}, \vee, \wedge, ^{\perp}, \mathbb{O}, \mathbb{1}\}$ is a partial Boolean algebra where all pairs of elements are compatible. Janusz Czelakowski (1978) showed that every partial Boolean algebra can be extended to a partial *-algebra (i.e., to a partial algebra equipped with an involution) so that the set of all idempotents of a partial algebra forms a partial Boolean algebra.

Since every Boolean description is a projection of the totality of reality into a Boolean context, it entails a loss of information: it eliminates "irrelevant" information. A veracious description of nature can be found only by expanding the frame of discourse to include *all mutually incompatible descriptions*, where each of them reveals only a part of the attributes ascribed to reality.

3.4.2 Logical Structures in Partial Boolean Algebras

Every predicate X of a logic of incompatible descriptions (i.e. a logic of complementarity) can be realized by a projection operator P_X in a linear vector space, where the connectives of compatible propositions are realized by

$$X \longleftrightarrow P_X, \tag{3.22a}$$

$$X \wedge Y \longleftrightarrow P_X P_Y, \tag{3.22b}$$

$$X \vee Y \longleftrightarrow P_X + P_Y - P_X P_Y, \tag{3.22c}$$

$$X^{\perp} \longleftrightarrow \mathbb{1} - P_X. \tag{3.22d}$$

The set of projection operators is a partial algebra. This algebra is partial because $X \wedge Y$ and $X \vee Y$ are defined only if X and Y are compatible, that is, if $X \wedge Y = Y \wedge X \longleftrightarrow P_X P_Y = P_Y P_X$. It follows that the logic of incompatible, i.e. complementary, propositions is not closed under propositional combination. Incompatible propositions can be true simultaneously, while propositional combinations of incompatible propositions do not exist.

Martin Strauss (1967, pp. 84f) developed a general theory of transition probabilities for partial Boolean algebras. Since the meaning of the logical connectives in partial Boolean algebras is the same as in Boolean algebras, there is a natural extension of the calculus of probability in terms of Boolean algebras[32] to partial Boolean algebras. According to Strauss the probability that a physical system acted upon by a stochastic state changer undergoes a transition from state X to state Y must have the form

$$\mathrm{Prob}(X \to Y) = \frac{S(P_X P_Y)}{S(P_Y)}, \tag{3.23}$$

where S is a real-valued additive function

$$S(P_X + P_Y) = S(P_X) + S(P_Y). \tag{3.24}$$

The only function satisfying this condition is the trace tr,

$$\mathrm{Prob}(X \to Y) = \frac{\mathrm{tr}(P_X P_Y)}{\mathrm{tr}(P_Y)}. \tag{3.25}$$

3.4.3 Partial Boolean Classifications

The fact that every *particular* description or every *single* experiment allows a realization in terms of Boolean logic does not imply that the family of all feasible experiments can be combined into a single Boolean context. Every classification is contextual and can only be understood against a background of the purposes for which it has been designed. Different purposes require different classifications which may not be compatible.

In spite of this trivial insight, many biologists tried to develop a *natural classification* for plants and animals—a classification being "in accord with nature" and "sharing an underlying essence". They presupposed that this order exists in nature independent of human perception, and referred to classifications which depend on specific purpose as "subjective" or "artificial". However, as emphasized by the botanist John Gilmour (1989, pp. 89, 100):

> Now one aspect of this process that needs special emphasis is that all classification is primarily utilitarian. It is a tool by the aid of which the human mind can deal effectively with the almost infinite variety of the universe. It is not something inherent in the universe, but is, as it were, a conceptual order imposed on it by man for his own purposes. ... Viewed in this light, "natural" and "artificial" are seen to be purely relative terms.

Every classification is a *projection* of reality into a Boolean reference frame. Since there is no universal Boolean frame of reference, there are incompatible classifications. We say that two Boolean classifications are *compatible* if they can be embedded in a common Boolean classification. If not, they are called *incompatible*.

---

[32] Compare Halmos (1944), Kolmogorov (1948), Loś (1955). A detailed study of the "point-free" approach to classical probability can be found in the monograph by Kappos (1969).

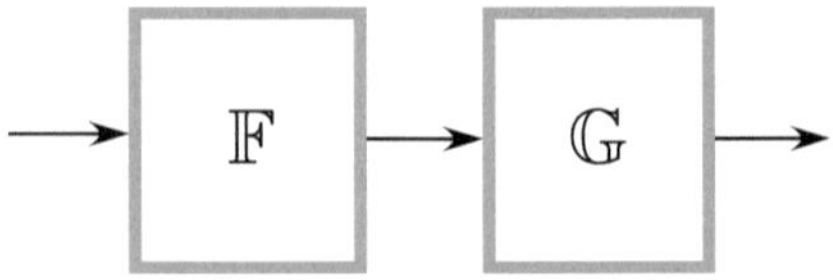

Fig. 3.9 Serial combination of two filters $\mathbb{F}$ and $\mathbb{G}$

Serial (and parallel) connections of two arbitrary filters $\mathbb{F}$ and $\mathbb{G}$ can be characterized algebraically by the corresponding selection operators F and G, respectively. If the serial connection of two filters $\mathbb{F}$ and $\mathbb{G}$ depends on the order of execution, then the two filters are called *incompatible*. The serial combination "first $\mathbb{F}$, then $\mathbb{G}$" (see Fig. 3.9) is usually realized algebraically by the product GF of the corresponding selection operators F and G,

$$\text{“first } \mathbb{F}\text{, then } \mathbb{G}\text{” is realized by the operator } GF. \tag{3.26}$$

This definition is entirely conventional, the opposite convention would be equally valid. If we use the opposite convention, the filters $\mathbb{F}$ and $\mathbb{G}$ can be realized by the *adjoint selection operators* F^* and G^*, respectively. The serial combination "first $\mathbb{F}$, then $\mathbb{G}$" is then realized by the product F^*G^*,

$$\text{“first } \mathbb{F}\text{, then } \mathbb{G}\text{” is realized by the operator } F^*G^*. \tag{3.27}$$

Since no result can depend upon which convention is employed, the underlying mathematical structure possesses a symmetry which can be realized by an involutary automorphism $F \to F^*$, $G \to G^*$, $GF \to F^*G^*$ with (Schwinger 1959, p. 1548)

$$\left(F^*\right)^* = F, \qquad \left(G^*\right)^* = G, \qquad (GF)^* = F^*G^*. \tag{3.28}$$

The involution operation * corresponds to *time reversal*. It reverses the order of the execution of the filtering: GF means that we use the filter G immediately *after* the filter F, while F^*G^* means that the filter G^* is used immediately *before* the filter F^*. Assuming that a filtering operation $\mathbb{F}$ takes only a very short time, we find that the corresponding selection operator F is self-adjoint,

$$F = F^*. \tag{3.29}$$

To discuss incompatible classifications we consider two finite maximal classifications realized by two families of n ideal filters $\{\mathbb{F}_1, \mathbb{F}_2, , \ldots, , \mathbb{F}_n\}$ and $\{\mathbb{G}_1, \mathbb{G}_2, , \ldots, , \mathbb{G}_n\}$, respectively, where F_j is the projector corresponding to filter $\mathbb{F}_j$, and G_j the projector corresponding to filter $\mathbb{G}_j$,

$$F_j F_{j'} = \delta_{j,j'} F_{j'}, \qquad \sum_{j=1}^{n} F_j = \mathbb{1}, \tag{3.30a}$$

$$G_k G_{k'} = \delta_{k,k'} G_{k'}, \qquad \sum_{k=1}^{n} G_k = \mathbb{1}. \tag{3.30b}$$

With these filters we construct the serial combination $\mathbb{F}_j \to \mathbb{G}_k \to \mathbb{F}_j$, realized by the operator $F_j G_k F_j$. If $F_j G_k \neq G_k F_j$, only a fraction of the objects selected by the first filter $\mathbb{F}_j$ is transmitted through the second filter $\mathbb{G}_k$, and the final filter $\mathbb{F}_j$ selects from the output of the objects which are compatible with $\mathbb{F}_j$. The fraction of objects which passes the serial combination is called the *transmission coefficient* p_{jk} of the serial filter $\mathbb{F}_j \to \mathbb{G}_k \to \mathbb{F}_j$:[33]

$$F_j G_k F_j = p_{jk} F_j. \tag{3.31a}$$

The relation $F_j G_k F_j = p_{jk} F_j$ implies $G_k F_j G_k F_j G_k = p_{jk} G_k F_j G_k$. The left hand side equals $p_{kj} p_{kj} G_k$, while the right hand side equals $p_{jk} p_{jk} G_k$, so that

$$G_k F_j G_k = p_{kj} G_k, \qquad p_{jk} = p_{kj}, \tag{3.31b}$$

and

$$\sum_{j=1}^{n} p_{jk} = \sum_{k=1}^{n} p_{jk} = 1, \qquad 0 \leq p_{jk} \leq 1. \tag{3.31c}$$

If the projectors $F_1, F_2, \ldots, F_n$ and $G_1, G_2, \ldots, G_n$ of two maximal classifications satisfy $p_{jk} = p_{kj}$, the disparity of two classifications can be characterized by the Shannon information

$$h(\mathcal{F}, \mathcal{G}) := \sum_{j=1}^{n} \sum_{k=1}^{n} p_{jk} \ln(p_{jk}), \qquad 0 \leq h(\mathcal{F}, \mathcal{G}) \leq n \ln(n). \tag{3.32}$$

The mutual information $h(\mathcal{F}, \mathcal{G})$ measures the amount of information required to get the results of the $\mathcal{F}$-classification given the results of the $\mathcal{G}$-classification. If the classifications $\mathcal{F}$ and $\mathcal{G}$ differ only by the numbering of the projectors, they are considered as equivalent and we get $h(\mathcal{F}, \mathcal{G}) = 0$. The mean information $h(\mathcal{F}, \mathcal{G})$ of the classification $\mathcal{G}$ with respect to the attributes of the classification $\mathcal{F}$ is minimal if all p_{jk} are equal

$$\mathcal{F} \text{ and } \mathcal{G} \text{ are } \begin{cases} \text{compatible if and only if } h(\mathcal{F}, \mathcal{G}) = 0, \\ \text{incompatible if and only if } h(\mathcal{F}, \mathcal{G}) > 0, \\ \text{maximally incompatible if and only if } p_{jk} = 1/n. \end{cases} \tag{3.33}$$

If $\{F_1, F_2, , \ldots, , F_n\}$ and $\{G_1, G_2, , \ldots, , G_n\}$ are the projectors of two *maximally incompatible* maximal classifications, then Eqs. (3.30a), (3.30b), (3.31a)–(3.31c) and (3.33) imply

$$F_j G_k F_j = \frac{1}{n} F_j, \qquad G_k F_j G_k = \frac{1}{n} G_k, \tag{3.34a}$$

[33] In his discussion of quantum theory, Temple (1932, 1934, pp. 28f) introduced the relation $F_j G_k F_j = p_{jk} F_j$ as "law of transitions", where p_{jk} was interpreted as induced "transition probability". An information-theoretical interpretation of this relation was given by Schlögl (1960). In our context we make no use of probabilistic concepts.

$$\sum_{j=1}^{n} F_j G_k F_j = \frac{1}{n}\mathbb{1}, \qquad \sum_{k=1}^{n} G_k F_j G_k = \frac{1}{n}\mathbb{1}. \tag{3.34b}$$

Maximally incompatible classifications are called complementary.

3.4.4 Algebras Generated by Classifications

The compatible projectors $\{F_1, F_2, \ldots, F_n\}$ of a maximal Boolean classification $\mathcal{F}$ generate a commutative *-algebra $\mathcal{U}$,

$$\mathcal{U} := \left\{ \sum_{j=1}^{n} c_j F_j \,\middle|\, c_j \in \mathbb{C} \right\}. \tag{3.35a}$$

If another maximal Boolean classification $\mathcal{G}$ of the same context is represented by the projectors $\{G_1, G_2, \ldots, G_n\}$, they generate the associated commutative *-algebra $\mathcal{V}$

$$\mathcal{V} := \left\{ \sum_{k=1}^{n} d_k G_k \,\middle|\, d_k \in \mathbb{C} \right\}. \tag{3.35b}$$

Let $\mathcal{A}$ be the smallest *-algebra containing both $\mathcal{U}$ and $\mathcal{V}$. If the algebra $\mathcal{A}$ is commutative, then the Boolean classifications $\mathcal{F}$ and $\mathcal{G}$ are compatible. If the classifications $\mathcal{F}$ and $\mathcal{G}$ are incompatible, then $\mathcal{A}$ is a *non-commutative* *-algebra and $\mathcal{U}$ and $\mathcal{V}$ are commutative subalgebras of $\mathcal{A}$. The algebra $\mathcal{A}$ is a *-algebra over the complex numbers, together with an involutory map $^* : \mathcal{A} \to \mathcal{A}$, which fulfills for all $A, B, C \in \mathcal{A}$, $\ell \in \mathbb{C}$, the relations:

- $(A + B)C = AC + BC, \; A(B + C) = AB + AC$,
- $(AB)C = A(BC) = ABC, \; \ell(AB) = (\ell A)B = A(\ell B)$,
- $(A + B)^* = A^* + B^*$ for all $A, B \in \mathcal{A}$,
- $\mathbb{1}A = A\mathbb{1} = A$,
- $(\ell A)^* = \ell^* A^*$ for every ℓ in $\mathbb{C}$ and every $A \in \mathcal{A}$, where ℓ^* stands for the complex conjugation of ℓ,
- $(AB)^* = B^* A^*$ for all $A, B \in \mathcal{A}$,
- $(A^*)^* = A$ for all $A \in \mathcal{A}$.

On the *-algebra $\mathcal{A}$ we can define a linear functional $\tau : \mathcal{A} \to \mathbb{C}$ by

$$\tau(A) := \sum_{j=1}^{n} F_j A F_j, \qquad A \in \mathcal{A}, \tag{3.36}$$

where $F_1, F_2, \ldots, F_n$ are n mutually compatible projectors of a maximal Boolean classification. The functional τ is independent of the maximal Boolean classification chosen. The functional τ has the following properties:

$$\tau\{\ell\mathbb{1}\} = \ell n \quad \text{for all } \ell \in \mathbb{C}, \tag{3.37a}$$

$$\tau\{A + B\} = \tau\{A\} + \tau\{B\} \quad \text{for all } A, B \in \mathcal{A}, \tag{3.37b}$$

$$\tau\{A\} = \tau\{A^*\} \quad \text{for all } A \in \mathcal{A}, \tag{3.37c}$$

$$\tau\{\ell A\} = \ell\tau\{A\} \quad \text{for all } A \in \mathcal{A} \text{ and all } \ell \in \mathbb{C}, \tag{3.37d}$$

$$\tau\{A^* A\} \geq 0 \quad \text{for all } A \in \mathcal{A}, \tag{3.37e}$$

$$\tau\{AB\} = \tau\{BA\} \quad \text{for all } A, B \in \mathcal{A}. \tag{3.37f}$$

The transmission coefficient (3.31a)–(3.31c) of the filter combination $\mathbb{F}_j \to \mathbb{G}_k \to \mathbb{F}_j$ with the projectors $\{F_1, F_2, \ldots, F_n\}$ and $\{G_1, G_2, \ldots, G_n\}$ of two maximally incompatible maximal classifications (3.30a), (3.30b) defines a metric on the algebraic structure of filters,

$$p_{jk} = p_{kj} = \langle F_j | G_k F_j \rangle = \langle G_k | F_j G_k \rangle, \tag{3.38}$$

where the bilinear map

$$\langle A|B \rangle := \mathrm{tr}\{A^* B\} \quad \text{for all } A, B \in \mathcal{A} \tag{3.39}$$

defines an *inner product* on the non-commutative *-algebra $\mathcal{A}$. With this metric the *-algebra $\mathcal{A}$ becomes a *H*-algebra* (Ambrose 1945; Loomis 1953, §27) defined as a Banach *-algebra whose norm $\|\cdot\|_2$ is defined in terms of a scalar product $\langle\cdot|\cdot\rangle$,

$$\|A\|_2^2 = \langle A|A \rangle \quad \text{for all } A \in \mathcal{A}, \tag{3.40a}$$

such that its involution $A \to A^*$ satisfies the condition

$$\langle AB|C \rangle = \langle B|A^* C \rangle = \langle A|CB^* \rangle \quad \text{for all } A, B, C \in \mathcal{A}. \tag{3.40b}$$

3.4.5 Boolean Manifolds

All knowledge of reality is inevitably knowledge from a particular Boolean point of view, but there is no *global* Boolean description of our world. The incomplete and mutually incompatible *locally* Boolean descriptions can be smoothly "glued together"[34] to form a *Boolean manifold.*

A Boolean manifold is a structured family of partially overlapping Boolean algebras that are pasted together such that the Boolean operations agree on their overlaps

[34] An appropriate technique for the unification of mutually incompatible Boolean descriptions has been introduced by Greechie (1968) as "pasting". Compare also Finch (1969).

and locally look like a Boolean algebra. A family $\{\mathcal{B}_\gamma | \gamma \in \Gamma\}$ of Boolean algebras $\{\mathcal{B}_\gamma, \vee_\gamma, \wedge_\gamma, {}^{\perp_\gamma}, \mathbf{0}_\gamma, \mathbf{1}_\gamma\}$ with the same smallest element $\mathbf{0}$ and the same greatest element $\mathbf{1}$ is called a *Boolean manifold* if it satisfies the following conditions (Hardegree and Frazer 1981, p. 63; Lock and Hardegree 1984; Hughes 1985):

$$\text{(i)}\ \mathbf{0}_\alpha = \mathbf{0}_\beta \text{ and } \mathbf{1}_\alpha = \mathbf{1}_\beta \text{ for all } \alpha, \beta \in \Gamma, \tag{3.41a}$$

$$\text{(ii)}\ \text{for all } \alpha, \beta \in \Gamma \text{ there is a } \gamma \in \Gamma, \text{ such that } \mathcal{B}_\alpha \cap \mathcal{B}_\beta = \mathcal{B}_\gamma, \tag{3.41b}$$

$$\text{(iii)}\ \text{if } F, G \in \mathcal{B}_\alpha \cap \mathcal{B}_\beta, \text{ then the Boolean operations in } \mathcal{B}_\alpha \text{ and in } \mathcal{B}_\beta \text{ coincide:}$$

$$F \wedge_\alpha G = F \wedge_\beta G,\ F \vee_\alpha G = F \vee_\beta G, \text{ and } F^{\perp_\alpha} = F^{\perp_\beta}. \tag{3.41c}$$

Two elements F, G in a Boolean manifold are *compatible* if there is a Boolean algebra $\mathcal{B}_\alpha \subset \cup_\gamma \mathcal{B}_\gamma$ such that $F, G \in \mathcal{B}_\alpha$.

Definition: Coherent Boolean Manifolds
A Boolean manifold is said to be *coherent* if it fulfills the condition

$$\begin{array}{l}\text{If } F, G \in \mathcal{B}_\alpha,\ G, H \in \mathcal{B}_\beta,\ H, F \in \mathcal{B}_\gamma \\ \text{then there is a } \delta \in \Gamma \text{ such that } F, G, H \in \mathcal{B}_\delta.\end{array} \tag{3.42}$$

By "pasting together" mutually incompatible Boolean contexts one can arrive at a non-Boolean description in terms of partial Boolean algebras.

3.4.6 A Simple Geometric Analogy

The locally Euclidean geometric structure of the globally non-Euclidean theory of general relativity is an apt analogy for the locally Boolean behavior of globally non-Boolean descriptions. The proper tool for a mathematical formulation of this analogy are *Boolean manifolds*. Algebraically, Boolean contexts play an analogous role as Euclidean spaces play for *geometric manifolds*.

Recall that a *geometric n-manifold* is a topological space that is locally like a n-dimensional Euclidean space (that is, every point has a neighborhood homeomorphic to an open set in $\mathbb{R}^n$) such that the locally Euclidean patches are pasted together continuously. In spite of the fact that manifolds look *locally* like Euclidean spaces, their *global* structure can be unimaginably complex. Examples for geometric 2-manifolds are the plane, the sphere, the torus, the Möbius strip, or the Klein bottle.

As a simple geometric analogy we may consider *cartographic map projections*. Map projections are methods of constructing latitudes and longitudes on maps of the earth. The surface of the earth—here for simplicity idealized as a sphere—is a manifold. Locally the earth seems to be flat, but viewed as a whole it is spherical. If

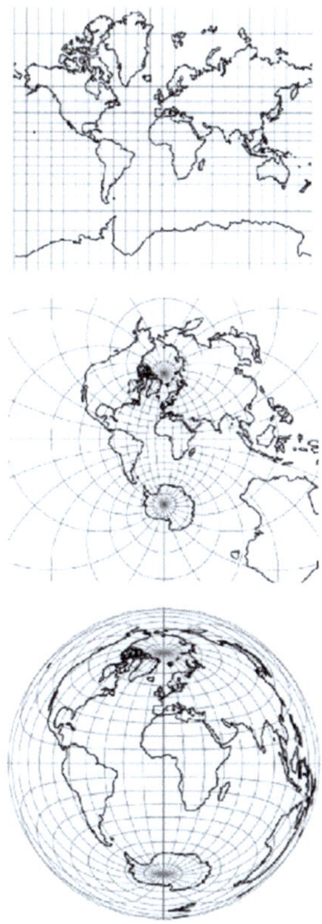

Mercator projection

preserves angles and circles
useful for aeronautical charts

$$x = \lambda$$
$$y = \arctan\{\sin(\phi)\}$$

Stereographic projection

preserves shapes and directions
used in the polar regions

$$x = k\cos(\phi)\sin(\lambda)$$
$$y = k\sin(\phi)$$

Lambert azimuthal projection

preserves areas
used for mapping large regions in middle latitude

$$x = \frac{\sqrt{2}\cos(\phi)\sin(\lambda)}{\sqrt{1+\cos(\phi)\cos(\lambda)}}$$
$$y = \frac{\sin(\phi)}{\sqrt{1+\cos(\phi)\cos(\lambda)}}$$

Fig. 3.10 Examples of cartographic maps (projections taken from http://mathworld.wolfram.com/topics/MapProjections.html)

the surface of a sphere is a two-dimensional manifold $\mathbb{S}^2$, any function defined on $\mathbb{S}^2$ with values on the plane $\mathbb{R}^2$ is called a map projection.

Every projection map $\mathbb{S}^2 \to \mathbb{R}^2$ which transfers a feature (like area, shape, direction, distance, scale) of the surface of the sphere onto the plane $\mathbb{R}^2$ produces some distortions. Some projections preserve relative distances in all directions from the center of the map (equidistant projections), some preserve areas (equal-area projections), while others preserve angles (conformal projections). Therefore, various incompatible or even complementary types of map projections are inescapable. The particular projection chosen for a given map depends on the purpose for which the map is intended. There are literally hundreds of projections used in cartography and in national grid systems (see Fig. 3.10 for examples).[35] The fact that all cartographic maps are valid only "in the small", i.e. locally, is a typical property of the description of geometric manifolds.

[35]Compare the *Map Projection Catalog* accessible at http://mathworld.wolfram.com/topics/MapProjections.html.

3.5 Mathematical Models of Complementarity

3.5.1 Incompatible Classifications Generate Weyl Systems

For a given maximal classification $\mathcal{F}$ with the minimal self-adjoint projectors $\{F_1, F_2, , \ldots, , F_n\}$ with

$$F_j F_{j'} = \delta_{j,j'} F_{j'}, \qquad \sum_{j=1}^{n} F_j = \mathbb{1}, \qquad \tau\{F_j\} = 1, \tag{3.43}$$

there is (up to unitary equivalence) just one maximally incompatible maximal classification. To construct a pair of maximally incompatible classifications we first define two unitary operators $U \in \mathcal{A}$ and $V \in \mathcal{A}$ by

$$U := \sum_{j=1}^{n} e^{2\pi i j/n} F_j, \qquad V := n \sum_{k=1}^{n} F_{k+1} P F_k, \tag{3.44}$$

where P is any one-dimensional projector with

$$\tau\{F_j P\} = 1/n, \quad j = 1, 2, \ldots, n. \tag{3.45}$$

With

$$UV = n \sum_{j=1}^{n} \sum_{k=1}^{n} e^{2\pi i j/n} F_j F_{k+1} P F_k = n e^{2\pi i/n} \sum_{k=1}^{n} e^{2\pi i k/n} F_{k+1} P F_k,$$

$$VU = n \sum_{j=1}^{n} \sum_{k=1}^{n} e^{2\pi i j/n} F_{k+1} P F_k F_j = n e^{2\pi/n} \sum_{k=1}^{n} e^{2\pi i k/n} F_{k+1} P F_k,$$

we obtain

$$UV = e^{2\pi i/n} VU. \tag{3.46}$$

This is a finite Weyl system whose unitary operators U and V fulfill the commutation relation (Weyl 1927, p. 20, 1931, p. 272):

$$U^b V^a = e^{2\pi i ab/n} V^a U^b, \quad a, b = 1, 2, \ldots, n, \quad U^n = V^n = \mathbb{1}. \tag{3.47}$$

The spectral operators $G_1, G_2, \ldots, G_n$ of the unitary operator V

$$V = \sum_{k=1}^{n} e^{-2\pi i k/n} G_k, \tag{3.48}$$

fulfill the condition

$$\langle F_j | G_k \rangle = \tau\left\{F_j^* G_k\right\} = \frac{1}{n} \quad \text{for all } j, k = 1, 2, \ldots n, \tag{3.49}$$

so that they generate a maximal classification $\mathcal{G}$ which is maximally incompatible with the classification $\mathcal{F}$ in the sense of Eq. (3.33).

Schwinger (1960, p. 575) refers to the unitary operators U and V as a *complementary pair of operators*. They generate a unitary operator basis for the *-algebra $\mathcal{A}$,

$$n^{-1}\tau\{U^b V^a\} = \delta_{n,a(\mathrm{mod}\,n)}\delta_{n,b(\mathrm{mod}\,n)}, \quad a,b = 1,2,\ldots,n. \tag{3.50}$$

3.5.2 Complementary Commutative Algebras

Let $\mathcal{F}$ and $\mathcal{G}$ be two finite maximal classifications with the corresponding projectors $\{F_1, F_2, \ldots, F_n\}$, $\{G_1, G_2, \ldots, G_n\}$, and the two maximal commutative *-algebras $\mathcal{U} \in \mathcal{A}$ and $\mathcal{V} \in \mathcal{A}$, respectively. The *-algebras $\mathcal{U} \subset \mathcal{A}$ and $\mathcal{V} \subset \mathcal{A}$ are said to be *complementary* if the underlying classifications are maximally incompatible. The only elements which the complementary algebras $\mathcal{U} \subset \mathcal{A}$ and $\mathcal{V} \subset \mathcal{A}$ have in common are multiples of the identity $\mathbb{1}$,

$$\mathcal{U} \cap \mathcal{V} = \mathbb{C}\mathbb{1}. \tag{3.51}$$

For $\mathcal{U}$ and $\mathcal{V}$ to be complementary it is necessary and sufficient that the traceless subspaces $\mathcal{U}$ and $\mathcal{V}$ be orthogonal.

Definition*: *Complementarity of Commutative Subalgebras (3.52)
Let $\mathcal{U} \subset \mathcal{A}$ and $\mathcal{V} \subset \mathcal{A}$ be two maximal commutative subalgebras of the finite-dimensional algebra $\mathcal{A}$ generated by $\mathcal{U}$ and $\mathcal{V}$. Then the following conditions are equivalent (Parthasarathy 2004; Petz 2007):

- The *-algebras $\mathcal{U}$ and $\mathcal{V}$ are *complementary*.
- If $F \in \mathcal{U}$ and $G \in \mathcal{V}$ are minimal projectors, then $\tau\{FG\} = 1/n$.
- If $A \in \mathcal{U}$ and $B \in \mathcal{V}$ then $\tau\{AB\} = \tau\{A\}\tau\{B\}$.
- The unitary operators $U \in \mathcal{U}$ and $V \in \mathcal{V}$ fulfill $UV = e^{2\pi i/n}VU$.
- The subspaces $\mathcal{U} \ominus \mathbb{C}\mathbb{1}$ and $\mathcal{V} \ominus \mathbb{C}\mathbb{1}$ are orthogonal with respect to the inner product (3.39) of $\mathcal{A}$.

This definition makes it evident that *complementarity is not a physical but a logical notion*. In particular, the general concept of complementarity is independent of Planck's constant (which features as the commutator in the canonical commutation relations in quantum mechanics).

3.5.3 Complementarity in Quantum Mechanics

Any physical system can be interpreted by a W*-algebra $\mathcal{M}$.[36] For classical systems this W*-algebra is commutative while for quantum systems the W*-algebra

[36]For details about the definition and properties of *-algebras as discussed throughout Sect. 3.5.3 see Appendix A.1.

$\mathcal{M}$ is non-commutative. Replacing the commutative algebra of dynamical variables of classical mechanics by a non-commutative algebra is equivalent to replacing the Boolean algebra of two-valued dynamical variables by a non-Boolean algebra. Every set of pairwise commuting projection operators in $\mathcal{M}$ generates a commutative W*-algebra, which corresponds to a *Boolean context* of a particular classical description. Every experimentally accessible property corresponds to a particular commutative W*-subalgebra of $\mathcal{M}$. The non-commutative W*-algebra $\mathcal{M}$ can be obtained by "pasting together" the Boolean algebras of all feasible experimentally accessible Boolean contexts.

The algebraic formalism of quantum mechanics is a generalization of the formalism of finite partial Boolean classifications. In the early days of quantum mechanics the position operator Q and the momentum operator P were characterized by Heisenberg's canonical commutation relation $QP - PQ = i\hbar$. Traditionally the pair (P, Q) is considered as the paradigmatic example for a pair of *complementary observables*. But even if one requires that P and Q are essentially self-adjoint, Heisenberg's commutation relation is insufficient to define the operators P and Q up to unitary equivalence (see Putnam 1967). The correct formulation relates the position operator Q and the momentum operator P to the appropriate kinematical group.

In modern quantum mechanics the canonical self-adjoint operators Q and P are defined as the unbounded generators of two unitary one-parameter Abelian groups $\{U(p)|p \in \mathbb{R}\}$ and $\{V(q)|q \in \mathbb{R}\}$, respectively:

$$U(p) = e^{2\pi i p Q/h}, \qquad V(q) = e^{-2\pi i q P/h}, \quad p, q \in \mathbb{R}, \tag{3.53}$$

where the unitary operators $U(p)$, $V(q)$ fulfill Weyl's canonical commutation relations (Weyl 1927):

$$U(p)V(q) = V(q)U(p)e^{2\pi i pq/h}, \qquad U(p)U(p') = V(q)V(q') = 0. \tag{3.54}$$

The unbounded essentially self-adjoint generators P and Q fulfill the Heisenberg commutation relation on an appropriate domain. According to a theorem by Stone (1930) and von Neumann (1931), Weyl's commutation relation (3.54) defines, up to unitary equivalence and multiplicity, uniquely the self-adjoint canonical operators P and Q.

Let $\mathcal{M}$ be the W*-algebra generated by the two unitary groups $\{U(p)|p \in \mathbb{R}\}$ and $\{V(q)|q \in \mathbb{R}\}$. It follows that $\mathcal{M}$ is a factor of type I_∞, which is a non-commutative W*-algebra with trivial center. The spectral decompositions of the self-adjoint operators P and Q are given by

$$P = \int_{\mathbb{R}} p E_P(dp), \qquad Q = \int_{\mathbb{R}} q E_Q(dq), \quad E_P, E_Q \in \mathcal{M}. \tag{3.55}$$

The spectral measures E_P, E_Q generate two maximal commutative W*-algebras $\mathcal{P}$ and $\mathcal{Q}$,

$$\mathcal{P} := \{E_P(\mathcal{B}) | \mathcal{B} \in \Sigma_{\mathbb{R}}\}'', \qquad \mathcal{Q} := \{E_Q(\mathcal{B}) | \mathcal{B} \in \Sigma_{\mathbb{R}}\}'', \tag{3.56}$$

where $''$ denotes the double commutant with respect to the W*-algebra $\mathcal{M}$, and $\Sigma_{\mathbb{R}}$ is the σ-algebra of the Borel sets in $\mathbb{R}$.

A Boolean algebra of compatible projections in $\mathcal{M}$ is a set of projections $E(\mathcal{B}) \in \mathcal{M}$ which is a Boolean algebra under the operations $\vee$, $\wedge$ and $^{\perp}$,

$$E(\mathcal{B}_j) \vee E(\mathcal{B}_k) := E(\mathcal{B}_j) + E(\mathcal{B}_k) - E(\mathcal{B}_j)E(\mathcal{B}_k), \tag{3.57a}$$

$$E(\mathcal{B}_j) \wedge E(\mathcal{B}_k) := E(\mathcal{B}_j)E(\mathcal{B}_k), \tag{3.57b}$$

$$E^{\perp}(\mathcal{B}) := \mathbb{1} - E(\mathcal{B}), \tag{3.57c}$$

where the operator $\mathbb{1}$ is the greatest element and $\mathbb{O} = \mathbb{1}^{\perp}$ is the smallest element of $E(\mathcal{B})$. The two commutative W*-algebras $\mathcal{P}$ and $\mathcal{Q}$ define two Boolean algebras $\mathcal{P}$ and $\mathcal{Q}$ of projections

$$\mathcal{P} := \{E_P(\mathcal{B}) \big| E_P(\mathcal{B}) \in \mathcal{P}\}, \qquad \mathcal{Q} := \{E_Q(\mathcal{B}) \big| E_Q(\mathcal{B}) \in \mathcal{Q}\}. \tag{3.58}$$

The commutative W*-algebras $\mathcal{P}$ and $\mathcal{Q}$ or, equivalently, the Boolean algebras $\mathcal{P}$ and $\mathcal{Q}$ define two *Boolean contexts*. The characterization of the complementarity of finite commutative algebras in (3.52) allows a generalization to the infinite-dimensional case:

Definition*: *Complementarity in Quantum Mechanics (3.59)
Let $\{U(p) = e^{2\pi i p Q/h} | p \in \mathbb{R}\}$ and $\{V(q) = e^{-2\pi i q P/h} | q \in \mathbb{R}\}$ be unitary one-parameter groups with the essentially self-adjoint generators Q and P, respectively. Then the following relations hold (Accardi 1984; Cassinelli and Varadarajan 2002; Petz 2007):

- The unitary operators $U(p)$ and $V(q)$ fulfill the commutation relation $U(p)V(q) = V(q)U(p)e^{2\pi i p q/h}$, $p, q \in \mathbb{R}$.
- The maximal commutative subalgebras $\mathcal{P} \subset \mathcal{M}$, $\mathcal{Q} \subset \mathcal{M}$ generated by the operators P and Q are complementary.
- If $\mathcal{B}', \mathcal{B}'' \in \Sigma_{\mathbb{R}}$ are two bounded Borel sets in $\mathbb{R}$, then the operator $E_P(\mathcal{B}')E_Q(\mathcal{B}'')$ is of trace class, and fulfills

$$\mathrm{tr}\{E_P(\mathcal{B}')E_Q(\mathcal{B}'')\} = \frac{1}{h}\ell(\mathcal{B}')\ell(\mathcal{B}''),$$

where ℓ is the Lebesgue measure.

Non-commuting observables stand for incompatible contexts which need not be complementary. If $A \in \mathcal{M}$ and $B \in \mathcal{M}$ are two arbitrary self-adjoint operators with the spectral representation $A = \int_{\sigma(A)} a E_A(da)$, $B = \int_{\sigma(B)} b E_B(db)$, $E_A, E_B \in \mathcal{M}$, then the spectral measures E_A, E_B generate two maximal commutative W*-algebras $\mathcal{A}$ and $\mathcal{B}$ in $\mathcal{M}$. If the generating operators A and B commute, $AB = BA$, both commutative W*-algebras $\mathcal{A}$ and $\mathcal{B}$ can be embedded into a single commutative W*-algebra. In this case the corresponding Boolean contexts are *compatible*. If A and B do not commute, $AB \neq BA$, the two Boolean contexts are *incompatible*. Since the commutative algebras A and B may still contain compatible elements, they are not necessarily complementary.

3.5.4 Coexistent Effects

The most general statistical observable of a quantum system characterized by a W*-algebra $\mathcal{M}$ is an affine map $F : \Sigma \to \mathcal{M}$ from the Borel sets $\mathcal{B} \in \Sigma$ into the *positive operator-valued measures* (POVMs)[37] of the observable W*-algebra $\mathcal{M}$ such that

$$\mathcal{B} \mapsto \rho\{F(\mathcal{B})\}, \quad \mathcal{B} \in \Sigma, \tag{3.60}$$

is a real-valued probability measure for all normal states $\rho \in \mathcal{M}_*$. A device which associates to every normal state ρ a probability measure μ on an outcome space (Ω, Σ) can be realized by an *operator-valued probability measure* $F : \Sigma \to \mathcal{M}$, defined as a normalized positive operator-valued measure with the following properties:[38]

- F is *nonnegative*, i.e., $F(\mathcal{B}) \geq 0$ for each $\mathcal{B} \in \Sigma$,
- F is *additive*, i.e., if $\mathcal{B}_1 \cap \mathcal{B}_2 = \emptyset$, then $F(\mathcal{B}_1 \cup \mathcal{B}_2) = F(\mathcal{B}_1) + F(\mathcal{B}_2)$,
- F is *continuous* in the sense that $\sup F(\mathcal{B}_n) = F(\mathcal{B})$ whenever $\{\mathcal{B}_n\}$ is an increasing sequence of sets in Σ whose union $\mathcal{B}$ is also in Σ,
- F is *normalized* in the sense that $F(\Omega) = \mathbf{1}$.

For all Borel sets $\mathcal{B}_1, \mathcal{B}_2 \in \Sigma$, an operator-valued probability measure F satisfies

$$F(\mathcal{B}_1 \cup \mathcal{B}_2) = F(\mathcal{B}_1)F(\mathcal{B}_2). \tag{3.61}$$

The mapping

$$\mu(\mathcal{B}) := \rho\{F(\mathcal{B})\} \tag{3.62}$$

from Σ to the interval $[0, 1]$ defines a real-valued probability measure μ on the measurable space (Ω, Σ). These results show that POVMs are the natural correspondents of the real-valued Kolmogorov probability measures.

An operator-valued probability measure F with $F(\mathcal{B}_1 \cap \mathcal{B}_2) = F(\mathcal{B}_1)F(\mathcal{B}_2)$ for all $\mathcal{B}_1, \mathcal{B}_2 \in \Sigma$ is projection-valued, $F(\mathcal{B}) = F(\mathcal{B})^* = F(\mathcal{B})^2 \in \mathcal{M}$, and therefore called a *projection-valued measure*. The elements of an operator-valued probability measure commute if and only if it is projection-valued.

Naimark's extension theorem (Neumark 1940) implies that every operator-valued probability measure can be realized by a projection of a POVM. This process can be

[37] This fertile view was proposed independently by Ludwig (1964), Holevo (1972a), and Prugovečki (1967). In quantum mechanics, positive operator-valued measures have also been referred to as "measurements" (Holevo 1973a, 1973b) or "effects" (Ludwig 1964, 1976, 1979), or "mixed tests" (Giles 1970), or "observables" (Davies and Lewis 1970; Davies 1970, 1976), or "fuzzy observables" (Prugovečki 1973, 1984; Ali and Emch 1974), or "unsharp observables" (Busch et al. 1991, 1995). We shall avoid these somewhat misleading notions.

[38] As mathematical objects, positive operator-valued measures (POVMs for short) have been introduced independently many times. They are also called generalized spectral measures, semi-spectral measures, generalized resolutions of the identity, or non-orthogonal resolutions of the identity. The basic mathematical investigations go back to Neumark (1943a, 1943b) and Schreiber (1958). For a very clear and useful review compare Berberian (1966).

reversed by "smearing" of a projection-valued measure. Given two operator-valued probability measures $F_1 : \Sigma \to \mathcal{M}$, $F_2 : \Sigma \to \mathcal{M}$, then the convex combination

$$F(\mathcal{B}, \ell) = \ell F_1(\mathcal{B}) + (1 - \ell) F_2(\mathcal{B}), \quad \ell \in (0, 1), \tag{3.63}$$

is also an operator-valued probability measure. If F_1 and F_2 are projection-valued, then their convex combinations are in general not projection-valued.

Every operator-valued measure with *commutative* range can be written as an integral of a projection-valued measure. Let $(\tilde{\Omega}, \tilde{\Sigma})$ and (Ω, Σ) be two measurable spaces. A commutative resolution of the identity F in (Ω, Σ) satisfies the condition

$$F(\mathcal{B}_1)F(\mathcal{B}_2) = F(\mathcal{B}_2)F(\mathcal{B}_1), \quad \mathcal{B}_1, \mathcal{B}_2 \in \Sigma. \tag{3.64}$$

If $\mathcal{B} \to F(\mathcal{B})$, $\mathcal{B} \in \Sigma$, is a commutative resolution of the identity, then there exists a unique projection-valued measure $\widetilde{\mathcal{B}} \to \widetilde{F}(\widetilde{\mathcal{B}})$, $\widetilde{\mathcal{B}} \in \widetilde{\Sigma}$ such that

$$F(\mathcal{B}) = \int_{\widetilde{\Omega}} \ell(x, \mathcal{B}) E(dx), \quad \mathcal{B} \in \Sigma, \tag{3.65}$$

in the weak sense (Holevo 1972b, p. 143).

The positive operators in the range of an operator-valued probability measure are also called *effects* (Ludwig 1964). An effect stands for a particular outcome of a measurement. The set $\mathcal{F}$ of effects has a partial order. For effects $F(\mathcal{B}_1), F(\mathcal{B}_2) \in \mathcal{M}$, the ordering $F(\mathcal{B}_1) \leq F(\mathcal{B}_2)$ means that $\rho\{F(\mathcal{B}_1)\} \leq \rho\{F(\mathcal{B}_2)\}$ for every normal state functional $\rho \in \mathcal{M}_*$.

A set of effects is called *coexistent* if all effects can be measured together (not necessarily simultaneously) by applying a suitable apparatus to a single object system. If a pair of operator-valued probability measures is jointly measurable, then their measurement outcome can be recovered from the measurement outcome of one single operator-valued probability measure:

Definition: Coexistent Effects
Let F_1, and F_2 be two $\mathcal{M}$-valued effects with the value spaces (Ω_1, Σ_1), and (Ω_2, Σ_2). The effects F_1 and F_2 are said to be coexistent if there is an $\mathcal{M}$-valued effect F with the value spaces (Ω, Σ) such that (Busch et al. 1995; Lahti 2003; Ali et al. 2009):

$$\text{range}(F_1) \cup \text{range}(F_2) \subseteq \text{range}(F). \tag{3.66}$$

The following statements are equivalent:

- Two effects F_1 and F_2 are coexistent.
- F_1 and F_2 are functions of an effect G, i.e. there are two measurable functions $f_1 : \Omega \to \Omega_1$ and $f_2 : \Omega \to \Omega_2$ such that for every $\mathcal{B}_1 \in \Sigma_1$ and $\mathcal{B}_2 \in \Sigma_2$ we have $F_1 = f_1^{-1}(\mathcal{B}_1)$ and $F_2 = f_2^{-1}(\mathcal{B}_2)$.
- There exists a joint event G with the outcome space $\Omega_1 \times \Omega_2$ such that $G(\mathcal{B}_1, \Omega_2) = F_1(\mathcal{B}_1)$, $G(\Omega_1, \mathcal{B}_2) = F_2(\mathcal{B}_2)$.
- Two effects with $F_1 \leq F_2$ are coexistent.
- Two effects with $[F_1, F_2]_- = 0$ are coexistent, but the converse is not true: coexistent effects need not commute.

As an example, consider the three Pauli spin-1/2 operators $\sigma_1, \sigma_2, \sigma_3$ which fulfill the commutation relations

$$\sigma_j \sigma_k = \delta_{jk} \sigma_0 + i \sum_{\ell=1}^{3} \varepsilon_{jk\ell} \sigma_\ell ,$$

where $\varepsilon_{jk\ell}$ is the totally antisymmetric unit tensor and σ_0 is the unit operator. These spin observables generate a factorial W*-algebra $\mathcal{M}_2$ of type I_2. We consider the following three effects represented by "fuzzy" (noisy) spin observables with η, $0 \leq \eta \leq 1$, as a noise factor:[39]

$$F_1(\eta) = \frac{1}{2}\{\sigma_0 + \eta \sigma_1\} = \frac{1}{2} \begin{pmatrix} 1 & \eta \\ \eta & 1 \end{pmatrix},$$

$$F_2(\eta) = \frac{1}{2}\{\sigma_0 + \eta \sigma_2\} = \frac{1}{2} \begin{pmatrix} 1 & -\eta i \\ \eta i & 1 \end{pmatrix},$$

$$F_3(\eta) = \frac{1}{2}\{\sigma_0 + \eta \sigma_3\} = \frac{1}{2} \begin{pmatrix} 1+\eta & 0 \\ 0 & 1-\eta \end{pmatrix}.$$

These three effects are

- pairwise consistent if and only if $\eta \leq 1/\sqrt{2}$,
- triplewise consistent if and only if $\eta \leq 1/\sqrt{3}$.

Therefore F_1, F_2, F_3 can be jointly measured pairwise but not jointly measured triplewise if and only if $1/\sqrt{3} \leq \eta \leq 1/\sqrt{2}$. (Compare the illustrative parable by Specker in Sect. 2.4.3.)

[39]For details compare Busch and Lahti (1986), Busch (1986), Decker and Janzing (2005), Spekkens (2005), Heinosaari et al. (2008), Busch and Heinosaari (2008), Stano et al. (2008), Yu et al. (2010), Liang et al. (2011).

Chapter 4
Rationality and Creativity

> It is no use trying to "see through first principles".
> If you see through everything, then everything is transparent.
> But a wholly transparent world is an invisible world.
> To "see through" all things is the same as not to see.
>
> *Clive Staples Lewis (1944, p. 81)*

4.1 Modes of Reasoning

> The elements of consciousness and the operations of the mental faculties exist independently of and antecedent to the existence of language.
>
> *Thomas Henry Huxley (1879, Chap. V)*

4.1.1 The Fascination of Science

The Oxford English Dictionary defines science as "the intellectual and practical activity encompassing the systematic study of the structure and behavior of the physical and natural world through observation and experiment". Another, perhaps more congenial characterization was given by Robert Musil (1995, p. 231) in his novel *The Man Without Qualities*:

> Knowledge is a mode of conduct, a passion.... It is simply not so that the researcher pursues the truth; it pursues him. He suffers it.

Intellectual curiosity and a Faustian striving for knowledge are most important inner compulsions for a full-blooded researcher. The imaginative and creative activities of scientists and artists are basically not very different. To be sure, artists are usually well aware of the non-rational nature of creative processes while many scientists pretend or even believe that they proceed rationally.

Truly creative scientists see themselves as architects in a world of symbols. Nevertheless, some armchair philosopher maintain that theories are deduced by strict

H. Primas, *Knowledge and Time*,
DOI 10.1007/978-3-319-47370-3_4

logical conclusions from empirical facts. Theoreticians with firsthand experience disagree. For instance, Albert Einstein (1936, p. 360) said:

> We now realize, with special clarity, how much in error are those theorists who believe that theory comes inductively from experience.

And some years later he emphasized (Einstein 1954, p. 266):

> It seems that the human mind has first to construct forms independently before we can find them in things. Kepler's marvelous achievement is a particularly fine example of the truth that knowledge cannot spring from experience alone, but only from the comparison of the inventions of the mind with observed fact.

Another popular misconception is that science is more or less equivalent with rational thinking about facts and figures. Although every scientific work depends on rational activity, there are always extrarational elements involved, as the theoretical physicist Markus Fierz (1970, p. 424) expressed in his appreciation of Wolfgang Pauli: "Theoretical physics surely appears quite rational, but it rises from irrational depths."

Inspirations do not follow from empirical facts by purely logical deduction—they "fall into" consciousness from somewhere. The theoretical physicist Wolfgang Pauli (1957, p. 38; English translation in Pauli 1994, p. 129) stated about the genesis of theories:

> Theories come into being through an understanding inspired by empirical material, an *understanding* which we may best regard, following Plato, as a coming into congruence of internal images with external objects and their behavior. The possibility of understanding again demonstrates the presence of typical regulatory arrangements, to which man's inner as well as outer world is subject.

4.1.2 Intuitive Knowledge and Logical Knowledge

From time immemorial humans have distinguished between two *complementary* modes of reasoning. Knowledge obtained through *images* has been referred to as implicit, tacit, non-conscious, or *intuitive knowledge*, while knowledge obtained through *concepts* has been referred to as explicit, conscious, or *logical knowledge*. According to Michael Polanyi (1962, p. 601), "these two kinds of knowing are not only distinct, but also in an important sense mutually exclusive". Explicit knowledge is codifiable and can be transmitted in formal language. Polanyi (1946, 1958, 1966) defined tacit knowledge as inherently non-verbalizable, intuitive, and unarticulated.[40] Intuition and insight are based on parallel processes. Since verbalization

[40]In their influential model of knowledge management, Nonaka and Takeuchi (1995) have greatly misinterpreted Polanyi's concept of "tacit knowledge". Contrary to Polanyi's line of reasoning, they claim that "tacit knowledge could be converted into explicit knowledge that anyone can understand" (Nonaka and Takeuchi 1995, p. 9). Yet, in the view of Polanyi, "tacit knowledge cannot be 'captured', 'translated', or 'converted' but only displayed and manifested in what we do" (Tsoukas 2003, p. 410).

is a sequential process it is difficult to verbalize insight and intuition as this would require a parallel-to-serial conversion.

Modern science focuses on *explicit knowledge*, based on an analytic mode of thinking, on sequential logical reasoning and explicit description. Explicit knowledge is important because it is communicable. Yet, explicit reasoning ignores underlying presuppositions, notwithstanding the fact that any scientific discourse depends on implicit background knowledge which is taken for granted.

Every thoughtful scientist is aware how important *non-verbal processes and skills* are. Moreover, *learning by doing* provides a great deal of *tacit* knowledge which is difficult to communicate. These implicit aspects of scientific work are necessary parts of knowledge in science. Albert Einstein (1931, p. 97) confessed: "I believe in intuition and inspiration. ... At times I feel certain I am right while not knowing the reason." And in a letter to Hadamard (1945, quoted from "a testimonial from Professor Einstein") he pointed out that his creative thought is non-verbal: "The words or the language, as they are written or spoken, do not seem to play any role in my mechanism of thought." Polanyi characterized this "tacit dimension" of personal knowledge, which is not at the forefront of consciousness, by the statement (Polanyi 1962, p. 601): "There are things that we know but cannot tell."

Many ideas about the dual modes of human knowing have been discussed by philosophers and scientists from different times and in dissimilar cultures. They have been developed in different ways which are not necessarily consistent with each other. Since there is more than one reasonable way to draw a borderline between intuitive and logical knowledge, we will first attend to some typical examples from mathematics and philosophy.

Since the early Greeks, much of Western history shows that its notion of reason was predominantly based on analytic thinking, using logic as "the science of the most general laws of truth" (Frege 1979, p. 128). However, this can lead to a deeply incoherent worldview, as Isaiah Berlin discovered in his studies of the *Age of Enlightenment*. In his analysis, there are three basic assumptions causing this incoherence (Berlin 1998, p. 555):

> (a) that to all genuine questions there is one true answer and one only, all others being deviations from the truth and therefore false, and that this applies to questions of conduct and feeling, that is, to practice, as well as to questions of theory or observation—to questions of value no less than to those of fact;
> (b) that the true answers to such questions are in principle knowable;
> (c) that these true answers cannot clash with one another, for one true proposition cannot be incompatible with another; that together these answers must form a harmonious whole: according to some they form a logical system each ingredient of which logically entails and is entailed by all the other elements; according to others the relationship is that of parts to a whole, or, at the very least, of complete compatibility of each element with all the others.

He then argued that these assumptions of analytic reasoning are not the only way to achieve knowledge and contrasted the Enlightenment with later Romantic movements (Berlin 1999, p. 146):

> The notion that there are many values, and that they are incompatible; the whole notion of plurality, of inexhaustibility, of the imperfection of all human answers and arrangements;

> the notion that no single answer which claims to be perfect and true, whether in art or in life, can in principle be perfect or true—all this we owe to the Romantics.

All theoretical work in science depends on philosophical preconceptions. For a long time, natural science followed a yes-or-no logic and was based on causal mechanisms. Since the beginning of the 20th century it became plain that this approach was one-sided. Inasmuch as science is created by human beings, it depends on human faculties of perception and cognition. For instance, Hermann Weyl (2009, p. 204) distinguished constructive mathematics and reflective meta-mathematics as two different ways of knowing:

> In the intellectual life of man we find discernibly separated, on the one hand, a sphere of *action*, of shaping and constructing to which the active artist, scientist, engineer, and statesman are dedicated and which is governed in the field of science by the norm of objectivity; and on the other hand, the sphere of *reflection*, which fulfills itself in insights and judgments and which, as the struggle of gaining insight into the meaning of our actions, is to be considered the proper domain of the philosopher. The danger faced by the work of creation, if not controlled by reflection, is that it outruns reason, goes astray, and hardens into routine; the danger of reflection is that it becomes just noncommittal "talk about it", paralyzing man's creative powers.

In discussion with Hao Wang (1996, p. 235), Kurt Gödel is reported to have said:

> We perceive objects and understand concepts. Understanding is a different kind of perception: it is a step in the direction of reduction to the least cause.

Albert Einstein (1950, p. 13) contrasted the rational with the intuitive mind as follows:

> Time and again the passion for understanding has led to the illusion that man is able to comprehend the objective world rationally, by pure thought, without any empirical foundations—in short, by metaphysics. I believe that every true theorist is a kind of tamed metaphysicist, no matter how pure a "positivist" he may fancy himself. The metaphysicist believes that the logically simple is also the real. The tamed metaphysicist believes that not all that is logically simple is embodied in experienced reality, but that the totality of all sensory experience can be "comprehended" on the basis of a conceptual system built on premises of great simplicity. The skeptic will say that this is a "miracle creed". Admittedly so, but it is a miracle creed which has been borne out to an amazing extent by the development of science.

While the analytic mode of thought depends on sequential logical reasoning and explicit description, the intuitive mode of thought is obtained through *mental images*. It derives from the non-sequential recognition of patterns, creation of form and synthesis. In contrast to rational processes that are fully describable, intuition is direct knowledge neither concluded through reason nor calling for the specification of explicit reasons.[41] Hermann Weyl (1955, p. 7) characterized intuition as

> mind's originary act of "seeing" what is given to him, limited in science to the *Aufweisbare*,[42] but in fact extending far beyond these boundaries.

---

[41] For a detailed study of intuitive processes and their relation to creativity, teaching and problem-solving, see Bastick (1982).

[42] Editor's note: the German term "aufweisbar" may be translated as "evident", that which is "recognizable with certainty".

The *Concise Oxford Dictionary* (Soanes and Stevenson 2008) defines intuition as "the ability to understand something immediately, without the need for conscious reasoning". The French philosopher Henri Bergson characterized intuition as a method of *thinking in temporal duration*, reflecting the continuous undifferentiated flow of life. He argued that any analysis of intuition in terms of logical analysis will fail to grasp the essence of intuition. Albert Einstein once remarked:[43]

> The intuitive mind is a sacred gift, and the rational mind is a faithful servant. We have created a society that honors the servant and has forgotten the gift.

The role of intuition is perhaps the most misunderstood, undervalued and neglected aspect of knowing. It is often abused to qualify an opinion which has no plausible explanation. For example, Mario Bunge (1962, p. 68) claimed:

> Intuition is the collection of odds and ends where we place all the intellectual mechanisms which we do not know how to analyze or even name with precision, or which we are not interested in analyzing or naming.

Yet, it is inappropriate and illegitimate to evaluate the intuitive mode of knowing in terms of the "intellectual mechanisms" of logical analysis. In his autobiography the mathematician Stanislaw Ulam (1976, p. 18) reported how he became familiar with the basics of intuitive understanding in some popular lectures on the theory of relativity:

> I did not really understand any of the details, but I hat a good idea of the main thrust of the theory. Almost like learning a language in childhood, one develops the ability to speak it without knowing anything about grammar. Curiously enough, it is possible even in the exact sciences to have an idea of the gist of something without having a complete understanding of the basics. ... I believe that so-called understanding is not a yes-or-no proposition. But we don't yet have the technique of defining these levels or the depth of the knowledge of reasons.

The analytic mode of thought and the intuitive mode of thought are both necessary but irreducible to one another. Typically, gained knowledge is never the outcome of one of these modes alone. Since intuition and intellect are mutually exclusive, they cannot be judged by the same criteria. Attempts to ignore one at the expense of the other inevitably leads to incomplete or even incoherent conceptions.

In contrast to the highly developed formalisms of sequential logic, we do not have a language which is sufficiently advanced to articulate intuitive ideas. A science of intuitive knowledge has been developed in rudiments at best. But any theory of knowledge that fails to include intuition cannot be remotely adequate.

4.1.3 Verbalization and Mental Pictures

People think differently but all thinking is inevitably tied to perception. In his work *On Memory and Recollection*, Aristotle wrote in his *De Memoria et Reminiscentia*

---

[43]This quote is attributed to Einstein, but the original source is unknown. Quoted from Calaprice (2011, p. 477).

in 348 B.C. (Aristotle 2004, Chap. 1, 450a): “It is impossible even to think without a mental picture.” Likewise, William James (1890, p. 255) observed that insights may occur in the absence of words:

> Great thinkers have vast premonitory glimpses of schemes of relations between terms, which hardly even as verbal images enter the mind, so rapid is the whole process.

Verbal thought is temporally sequential and is expressed through the language-based communication we are familiar with. By contrast, non-verbal conceptualizations refer to non-sequential thinking in terms of descriptive concepts and mental pictures (Davis et al. 1981, p. 186):

> A person thinking in the nonverbal mode is actually thinking with the meaning of the language in the form of mental pictures of the concepts and ideas it contains. Nonverbal thought doesn't require literacy. An illiterate person can communicate without knowing what the symbols look like. ... Literacy, then, is established as the person learns how the symbols look and becomes able to recognize them as representing certain things or concepts.

There cannot be any doubt that there are particular thoughts which are of non-verbal character. In a discussion with Max Wertheimer, Einstein said (quoted by Wertheimer 1946, p. 228) that

> these thoughts did not come in any verbal formulation. I very rarely think in words at all. A thought comes, and I may try to express it in words afterwards.

This kind of thinking is non-verbal and non-visual. Traditional logic just furnishes the means of testing the validity of conclusions already reached. For communicating these conclusions and their tests, verbal and visible tools have to be used. Carl Mitcham (1978, p. 243) reminded us that

> invention causes things to come into existence from ideas, makes world conform to thought; whereas science, by deriving ideas from observation, makes thought conform to existence.

Eugene Ferguson argued that not only science but also engineering is to a large extent a matter of intuition and non-verbal thinking. An engineering education that ignores this will produce engineers who are dangerously ignorant of the many ways in which the real world differs from mathematical models constructed by academic minds (Ferguson 1977, p. 827):

> This scientific age too readily assumes that whatever knowledge may be incorporated in the artifacts of technology must be derived from science. This assumption is a bit of modern folklore that ignores the many nonscientific decisions, both large and small, made by technologists as they design the world we inhabit.
> Many objects of daily use have clearly been influenced by science, but their form and function, their dimensions and appearance, were determined by technologists—craftsmen, designers, inventors, and engineers—using nonscientific modes of thought. ... Many features and qualities of the objects that a technologist thinks about cannot be reduced to unambiguous verbal descriptions; they are dealt with in his mind by a visual, nonverbal process.

And further (Ferguson 1977, p. 834f):

> Whatever may be the mode of thought of scientists, it is certain that technologists who make choices and decisions regarding the design of objects employ a species of nonverbal thought.

> Much of the creative thought of the designers of our technological world is nonverbal, not easily reducible to words; its language is an object or a picture or a visual image in the mind.

The *meaning* of a sentence is given by its semantic content. Consequently, Robert Rosen (1991, p. 191) called sequential processes *syntactic processes* and referred to non-sequential processes as *semantic processes*. Since insight is represented semantically rather than syntactically, and since it is highly problematic to convert non-sequential into sequential knowledge, it is difficult to verbalize insight and intuition sequentially. According to Berne (1949, p. 223):

> To understand intuition, it seems necessary to avoid the belief that in order to know something the individual must be able to put into words what he knows and how he knows it. ...What is intuited is different from what the "intuiter" verbalizes as his intuition.

4.1.4 Hertz's Metaphor Conception of Theories

According to Ernst Cassirer (1910, p. 244), "the relation between theory and experience assumed here has found its clearest expression in recent times in the system of mechanics of Heinrich Hertz". Cassirer (1929, p. 25) stated (editor's translation):

> Heinrich Hertz is the first modern scientist who, in his *Principles of Mechanics* (1894), effected the turn from the "image theory" of physical knowledge to a purely "symbolic theory" in the most decisive way. The basic concepts of natural science no longer appear as mere copies and reproductions of immediate material data; rather, they are introduced as constructive schemes of physical thinking—schemes whose theoretical validity and significance rely on no other condition than that their logical consequences must always accord with the observable data.

Hertz's influential ideas on the symbolic nature of scientific theories first appeared in his popular lectures (at Kiel, 1884) on the constitution of matter (Hertz 1999, pp. 35f, editor's translation):

> It is a common and necessary feature of human intelligence that we can neither conceive of things nor define them conceptually without adding attributes to them that simply do not exist. This applies not only to every thought and imagination of ordinary life, even the sciences do not proceed otherwise. Only philosophy seeks and finds the difference between things that exist and things that we perceive, and also sees the necessity of this difference. ...What we add are therefore not incorrect conceptions but the conditions for such conceptions in general. We cannot simply remove them and replace them with better ones; either we must add them, or we must abstain from all conceptions of this kind.

As a further explanation, Hertz (1999, p. 117, editor's translation) compared paper money (which symbolizes material values) with our mental images (which symbolize the external world):

> I compare matter with paper money, which our intellect spends to regulate its relations to things. Paper money is a symbol of something else, and precisely the fact that it is a symbol marks its value and significance. Its own composition is insensible to whether it contains this or that depth of color, whether it is printed red or blue, large or small. Likewise, the notion of

> matter is a symbol of something else, and the characteristics which human intellect imprints on it are more or less insensible to the services it will provide us with. Nevertheless, these characteristics are not entirely indifferent, and so it is not without interest for a government to produce their paper money to be persistent and difficult to counterfeit. But the essence of both cases lies in the meaning.

In the introduction to his posthumously published *Principles of Mechanics*, Hertz considered it as a principal aim of mature natural science to foresee future experiences. He claimed that our theories of nature are mental images of external reality (Hertz 1894, pp. 1f, editor's translation):

> We create inner virtual images [Scheinbilder] or symbols of external objects in such a way that the logically necessary [denknotwendigen] consequences of the symbols are always pictures of the physically necessary [naturnotwendigen] consequences of the objects considered. To satisfy this requirement, there must be particular correspondences between nature and our mind. Experience teaches us that this requirement can be satisfied, and hence that such correspondences indeed exist. ... The images we are talking about are our conceptions [Vorstellungen] of the things. They correspond to the things in one essential respect, namely in satisfying the mentioned requirement, and any other correspondence is unnecessary for their purpose. In fact, we do not know, and have no means of knowing, whether our conceptions of things correspond to them in anything else than that one fundamental respect.

Hertz judged images by the criteria of *admissibility* [Zulässigkeit], *correctness* [Richtigkeit] and *appropriateness* [Zweckmässigkeit]:

- Images are admissible if they do not contradict logical laws (the laws of thought).
- Images are correct if their essential relations do not contradict the relations between external objects.
- Images are appropriate if they involve the least possible number of dispensable, or idle, relations.

Since Hertz abstained from any claim to the truth of such images, he could speak without contradiction about unobservable entities or idle relations.

Decades later, Pauli extended the role of images in the process of understanding to unconscious activity. He characterized understanding as a laborious process, guided by the unconscious long before conscious contents can be formulated in rational terms.[44] In a letter of January 7, 1948, to Markus Fierz, Pauli (1994, p. 496) wrote:

> When we analyze the preconscious stage of concepts, we invariably find notions consisting of "symbolic" images with a generally strong emotional content. The initial stage of thinking is a pictorial mode of visually perceiving [malendes Schauen] these inner images, whose origin cannot generally and primarily be traced back to sensory perceptions.

---

[44] See Pauli (1952), English translation in Pauli (1994). Compare in particular the first section on pp. 220f.

4.1.5 Ludwik Fleck's Thought Collectives

In his book *Genesis and Development of a Scientific Fact*, Ludwik Fleck (1979) introduced the concepts "thought style" or "style of thought" (Denkstil) and "thought collective" (Denkkollektiv).[45] Fleck (1979, p. 38) defined a

> "thought collective" as a community of persons mutually exchanging ideas or maintaining intellectual interaction ... we will find by implication that it also provides the special "carrier" for the historical development of any field of thought, as well as for the given stock of knowledge and level of culture. This we have designated "thought style".

According to Fleck, a scientific thought collective is primarily characterized by a historically determined thought style shared by all its members, so that a so-called scientific fact (Fleck 1979, p. 95) "always occurs in the context of the history of thought and is always the result of a definite thought style". Moreover (Fleck 1979, p. 27):

> Once a structurally complete and closed system of opinions consisting of many details and relations has been formed, it offers enduring resistance to anything that contradicts it. ... A contradiction to the system is unthinkable. What does not fit into the system remains unseen.

If different thought styles are "incommensurable",[46] there can be no universal intersubjective agreement (*consensus omnium*) about them (Fleck 1986, p. 127):

> The *consensus omnium* is not the touchstone of science, for there is never a consensus omnium, but only the agreement of "our collective", and this also depends on false cognition. Every thought collective considers that people who do not belong to it are incompetent. Practical applicability is not a touchstone, for due to the harmony of illusions even a false view is applicable.

According to Fleck (1935b), there is a deeply inherent and unavoidable context-dependence to all descriptions of nature (quoted from Trenn 1981, p. 247):

> to make an object observable it must first be defined, that is isolated and contrasted against a background or support. ... *There are no observations that are true to nature except those that are also true to culture!*

And in the same vein (quoted from Trenn 1981, p. 240):

> Poincaré could still claim, as sufficient advice, that the research worker with unlimited time at his disposal need only "observe, and observe accurately". The sum total of accurate observations of all phenomena should yield knowledge in its entirety. According to current

---

[45]The German original was published in 1935 (Fleck 1935a) and translated 1979 with a foreword by Thomas Kuhn and a preface by one of the translators, Thaddeus Trenn. According to Kuhn, the U.S. high commissioner for the occupied Western Germany in 1953–1955, James Bryant Conant, reported the reaction of a German associate to his mention of Fleck's book: "How can such a book be? A fact is a fact. It has neither genesis nor development."

[46]Editor's note: The term "incommensurability" is due to Fleck, long before Kuhn (1962) made it popular in his *Structure of Scientific Revolutions*. In the context of the present monograph, incommensurability is best translated into incompatibility. For a discussion of how compatibility, commensurability, and comparability can be systematically related to one another see Atmanspacher and beim Graben (2009).

> ideas this claim is wrong for several reasons. Firstly, observations are not simply additive, because the type of observable phenomenon depends upon the observer; it is therefore impossible to speak of a universal sum of the observations of all phenomena.

All empirical “facts” are conditioned by some thought style, and these thought styles create a *context-dependent* reality (quoted from Trenn 1981, p. 248):

> Definite cultural contexts not only define the observations in a certain positive sense (i.e. make them possible), but also they make certain others impossible (i.e. inhibit and destroy observations favored in different cultural contexts).

Since thought styles can be incommensurable, also “scientific facts” can be incommensurable. Partial Boolean algebras (see Sect. 3.4) provide a well-developed tool for an explicit description of the logical structure of this mutual incommensurability. Fleck (quoted from Trenn 1981, p. 254):

> Thought styles present closed structures and have their own historical development. The traditional faith in the immutable human intellect, which functions only correctly or incorrectly, is no longer tenable.

And (Fleck 1979, p. 93):

> The thought style, developed in this particular way, made possible the perception of many forms as well as the establishment of many applicable facts. But it also rendered the recognition of other forms and other facts impossible.

For Fleck (1979, p. 100) it was crucial that scientific truth is ineluctably dependent on thought style:

> Truth is not “relative” and certainly not “subjective” in the popular sense of the word. It is always, or almost always, completely determined within a thought style. One can never say that the same thought is true for *A* and false for *B*. If *A* and *B* belong to the same thought collective, the thought will be either true or false for both. But if they belong to different thought collectives, it will just not be the same thought! It must either be unclear to, or be understood differently by, one of them. Truth is not a convention, *but rather (1) in historical perspective, an event in the history of thought, (2) in its contemporary context, stylized thought constraint.*

4.2 Neuropsychological Aspects

> The question of brain localization is an extremely delicate one,
> because when you destroy a certain part of the brain you destroy a certain function.
> Yet you do not know whether you have really destroyed the function
> because it is quite possible that you have only destroyed the transmitter
> of that function, as if you have taken away the telephone apparatus
> which does not mean that you have killed its owner.
>
> *Carl Gustav Jung (1976, p. 160)*[47]

The idea that duality is fundamental for the many ways in which humans conceptualize nature and themselves has been taught by philosophers and scientists from numerous different cultures and periods of time. The dominant idea is that there are

[47] Quoted from a letter by Jung to D. Cappon, March 15, 1954.

two distinct modes of knowing, which are embedded in our everyday language and present in our everyday lives. Some key dual distinctions are between thinking and feeling; perception and intuition; objective analysis and narrative comprehension.[48]

The human cerebral cortex is anatomically divided into two cerebral hemispheres that have dissimilar information-processing abilities. Some neuroscientists think that the two modes of knowing can be attributed to this laterality, claiming that the left hemisphere is responsible for rational sequential thinking, explicit knowledge and most language skills, while the right hemisphere maintains a metaphoric, non-sequential, holistic record of events, tacit knowledge and the capacity for gestalt perception.[49]

In the early 1960s the American neurobiologist Roger Sperry and his colleagues conducted experiments on an epileptic patient who had his *corpus callosum*[50] split so that the connection was severed. Sperry (1961) discovered that for such "split-brain" patients each hemisphere functioned independently as if it were a complete brain. According to Sperry the left and right hemispheres perform different tasks. Since the left hemisphere controls the motor functions underlying speech it has also been called the *language dominant hemisphere*. However, the right hemisphere is also actively involved in linguistic information processing, even though in a fundamentally different way. Moreover, it is dominant for other functions such as visuo-spatial processing.

Michael Trimble (2007, pp. 178f) described the importance oft the right hemisphere to modulate language as follows:

> The language of the right hemisphere is that of uncertainty, metaphor. prosody, and emotional tone. It is the language of music. The language of the right hemisphere involves features recognized as poetic that have been used, since the early religious invocations of almost preverbal humankind, to express human feelings. It is not the language of science; it lacks the detail and propositionizing qualities of the more logical, categorizing, left hemisphere-driven speech. The metaphors of the latter are much more codified and fixed; they reveal less flexibility, and they function in an entirely different way. ... Without the right hemisphere, the isolated left hemisphere provides only the dull syntax of precision and seems intolerant of rich metaphor.

Sperry (1984) claimed that the human brain has specialized functions on the right and left hemisphere, and that the two sides can operate practically independently. The *corpus callosum* unifies consciousness and cognition between the two hemispheres. Joseph Bogen, a neurosurgeon who worked with Sperry and performed the first split-brain surgery on a human, proposed the view that the left hemisphere is complemented by the right hemisphere. Bogen distinguished two distinct modes of

---

[48]Editor's note: It seems plausible that dualities like these are related to the ubiquity of Boolean frameworks for mental operations (see Chap. 2, in particular Sect. 2.3).

[49]Compare for example Hellige (1993), Beaumont (2008, Chap. 10), or the popular expositions by Segalowitz (1983), Bruner (1986), or Ornstein (1997).

[50]The brain of placental mammals bears a main fiber tract interconnecting the cortical neurons of the left and right cerebral hemispheres: the *corpus callosum*. It is by far the largest fiber tract in the human brain, consisting of more than 2×10^8 fibers (see Aboitiz et al. 1992, p. 151). Most of the inter-hemispheric communication is mediated across the *corpus callosum*.

operation, which are essentially lateralized in the left and right hemispheres, respectively, and which are assumed to have the same informational capacities (cf. Bogen 1969a, 1969b; Bogen and Bogen 1969).

The right hemisphere deals with simultaneous comparisons and contrasts, while the left hemisphere works sequentially and language-oriented. Bogen argued that the left (often dominant) hemisphere is the seat of the "propositional mind", in contrast to the "appositional mind" which resides in the right (non-dominant) hemisphere. According to Bogen (1969b) the difficulty in characterizing the non-dominant hemisphere

> arises largely from our ignorance—we have barely scratched the surface of a vast unknown. We would do well therefore to choose arbitrarily a word, homologous in structure with the word "propositional" but sufficiently ambiguous to permit provisional use. For example, we can say that the right hemisphere has a highly developed "appositional" capacity. This term implies a capacity for apposing or comparing of perceptions, schemas, engrams, etc., but has in addition the virtue that it implies very little else. If it is correct that the right hemisphere excels in capacities as yet unknown to us, the full meaning of "appositional" will emerge as these capacities are further studied and understood. The word "appositional" has the essential virtue of suggesting a capacity as important as "propositional", reflecting a belief in the importance of right hemisphere function.

Many of the important insights of neuroscience derive from studies of mental disorders. A large part of the left-brain/right-brain research originated from the early detailed studies of split-brain patients who had the *corpus callosum* transected for medical reasons. These studies sometimes provoked unsupported speculations, like the view that split-brain patients are left with just one mind associated with the major hemisphere, or that split-brain patients actually have two minds, or that the split between left and right hemisphere is also present in the intact brain.

In addition, it has become popular to say that the left hemisphere functions logically, analytically and rationally, strongly oriented toward reality and controlling language and speech. In contrast, the right hemisphere has been considered as intuitive, synthetic, involved with imagination, creativity, music, art, and symbolism. Yet neurologically intact subjects show a complex interaction between the two hemispheres, so that many of the results found with split-brain patients cannot be generalized. Joseph Hellige (1993, pp. 335f) summarizes the more recent studies as follows:

> In the intact brain, it is rarely the case that one hemisphere can perform a task normally whereas the other hemisphere is completely unable to perform the task at all. Instead, both hemispheres often have considerable ability to perform a task, even though they may go about it in different ways. That is, hemispheric differences tend to be smaller and more subtle than popularized views have suggested. A possible exception is overt speech, which the normal left hemisphere can typically produce whereas the normal right hemisphere cannot. However, even individuals who are born without the left hemisphere learn to speak without any great effort, although their speech and language may be deficient in subtle ways. In general, then, it seems that each hemisphere provides a neural substrate that is capable of supporting a wide range of cognitive abilities and propensities to some extent. In this sense, having two cerebral hemispheres is akin to having two reasonably complete "brains" whose differences, compared with their many similarities, are likely to start out being subtle.

Bogen assumed that in the normal brain both appositional and propositional modes of information processing are available and that both modes have their own particular strengths. Nevertheless, the localization of these different modes remains a controversial issue.[51] However, from our point of view the lateralization of higher mental functions in the brain is not the real crux of the matter. Rather, the point is that complementary modes of information processing require complementary (Boolean or non-Boolean) domains in which processes are governed by their corresponding logic. Whether these domains are spatially localized or not will be of secondary importance for our discussion.

Most modern cultures, particularly in the West, are predominantly "propositional cultures", considering language and the sequential ordering of thought as superior. Our concepts of intelligence refer primarily to the corresponding analytic logic, disregarding qualitatively different thought styles such as narrative comprehension. This preoccupation with propositional (Boolean) features leads to a severe underestimation of the role of appositional (non-Boolean) aspects. According to Sperry (1973),

> our educational system, as well as science in general, tends to neglect the nonverbal form of intellect. What this comes down to is that modern society discriminates against the right hemisphere.

Similarly, Bogen (1975) underlined that in many societies propositionality has been overemphasized at the expense of appositionality. According to Bogen and Bogen (1969) there is a "lack of appositional development in persons whose only education consists of reading, writing and arithmetic". But appositional intuitive capacities are just as important as propositional rational ones. Both propositional and appositional modes of knowing alone are not only essentially incomplete, they are also not fully translatable into one another.

4.3 Time and the Modes of Knowing

Explicit knowledge depends on *sequential logical reasoning*, on a working *memory* and on an awareness of the present moment as embedded in a temporal sequence from past to future events. The process of sequential knowing is not *time-reversal* invariant, as time reversal would reverse sequential order. *Sequential time* is historical, made up of past and future and requires a memory. In spoken natural languages this is reflected by the tenses of verbs.

If logical propositions are realized by electric networks, (idealized) nervous systems or neuronal networks, then time is a primary resource of fundamental significance. In his investigations of the design of reliable systems with unreliable components, von Neumann (1956, p. 44) stated:

[51] Editor's note: A recent neuroimaging study by Nielsen et al. (2013) found no clear evidence for correlations between personality or cognitive style and the activation strength of left-dominant and right-dominant neural networks.

> There is one important difference between ordinary logic and the automata which represent it. Time never occurs in logic, but every network or nervous system has a definite time lag between the input signal and the output response. A definite temporal sequence is always inherent in the operation of such a real system.

Henri Bergson (1992) distinguished between two modes of time: a sequential outer clock time and an inner durational time. He relates sequential time to conceptual knowledge and durational time to intuitive knowledge. Bergson (1998, p. 343) referred to this situation in terms of *two opposing yet complementary ways of knowing*:[52]

> Not only may we thus complete the intellect and its knowledge of matter by accustoming it to install itself within the moving, but by developing also another faculty, complementary to the intellect, we may open a perspective on the other half of the real. ... To intellect, in short, there will be added intuition.

Sequential time structures with a preferred point of reference are of key importance in scientific experiments where initial conditions are deliberately chosen. Sequential thinking makes use of a logical language appropriate for causal descriptions which conceive of a factual event as the effect of a cause. In sequential thinking, time reversal implies sequence reversal.

The ways in which insight and progress are achieved in science (see Sect. 4.1) demonstrate that their proper conceptual framework cannot be narrowed down to a sequential description of facts alone. Scientific research must take into account both rational sequential logical thinking and extrarational factors such as intuition, aesthetic and moral values. Such extrarational factors are not logical, but they are certainly not nonsensical. Both sequential reason and non-sequential intuition are interdependent rather than opposing aspects of knowing. True insight is a combination of rational knowledge and intuitive wisdom. Intuitive knowledge, poetry and music can only be partly verbalized and are often associated with metaphorical language, mapping different conceptual domains onto one another.

This has been wonderfully expressed by Max Frisch (1977, pp. 12f), according to whom sequential time would be

> simply a magic device which separates out our nature and makes it visible by laying out life, an omnipresence of all possibilities, as a series. That alone is what makes it seem transitory, and that is why we always tend to suppose that time, the one-thing-after-another, is not actual, but apparent: it is an aid to visualization, a means of showing us one after another things that are in fact interlocked, forming a whole which we can no more grasp than we can see the separate colors that constitute light, until its rays are broken down and dissected. Consciousness as the intervening prism which breaks down our life into serial form, and dreams as the other lens which reconstitutes it in its original form.

---

[52]Bergson's texts are often not as clear as one would wish. Many of his contemporaries saw contradictions in Bergson's epistemology, e.g. Cunningham (1916, 1924) or Landes (1924). However, our present point of view concerning complementarity entails that his notion of complementary ways of knowing is perfectly consistent. See also Sect. 13.3.

Editor's note: Chap. 4 in Lonergan's (1992) 800-pages book on "Insight" is devoted to a variety of complementary modes of knowing. The book is not widely known, but definitely to be recommended for an extensive and fine-grained study of human understanding and the nature of insight.

As this quote indicates, many dream contents can be regarded as non-sequential. They acquire sequential structure only if they are retrospectively verbalized as narrative. If we recall or reconstruct a dream in a sequential manner, we may lose important parts of it.

4.4 Modes of Comprehending Art

Claude Lévi-Strauss (1969, p. 18) has always insisted that poetry and music are cultural expressions which cannot be translated into verbal language:

> Since music is a language with some meaning at least for the immense majority of mankind, although only a tiny minority of people are capable of formulating a meaning in it, and since it is the only language with the contradictory attributes of being at once intelligible and untranslatable.

Likewise, a complementarity of the propositional rational capacity and the appositional intuitive capacity has been discussed in literary theory. According to Sayyid Hossain Nasr (1987, p. 91),

> logic and poetry have a common source, the Intellect, and far from being contradictory are essentially complementary. Logic becomes opposed to poetry only if respect for logic becomes transformed into rationalism, and poetry, rather than being a vehicle for the expression of a truly intellectual knowledge, becomes reduced to sentimentalism or a means of expressing individual idiosyncrasies and forms of subjectivism.

Moreover, the idea of *critical pluralism* in literary studies emphasizes that literary works typically (Kraut 2009, editor's addition)

> admit of alternative, equally acceptable interpretations, some of which are incompatible with others; it asserts that if there is a way to get an artwork right then there are many ways.

A truly good work of literature cannot be understood at a single level—it has no exclusive meaning. It acquires its meaning by the interplay of all feasible interpretations. Mutually incompatible but intrinsically coherent partial interpretations of the same literary work are not only justifiable but necessary for an apprehension of genuine art.

A classical example of simultaneous, logically incompatible modes of comprehending poetry is provided by the novella *The Metamorphosis* (German: *Die Verwandlung*) by Franz Kafka (1972). Kafka's intentional use of ambiguous terms posed great problems for translators and inspired most diverse mutually exclusive interpretations.[53] Every interpretation involves a particular viewpoint. Various different interpretations are not necessarily misreadings of a text with one unique meaning (e.g. the author's one). Since no single reading is rationally justifiable as universally valid, we conclude that great works of poetic imagination are not uniquely translatable into one single sequence of analytic thought. The meanings of a good story elude the grasp of Boolean logical thinking.

[53]Stanley Corngold (1973) lists more than 140 interpretations of Kafka's *Metamorphosis* in his *The Commentator's Despair*.

4.5 Modes of Learning

As mentioned above, Bogen (1969b) postulated the existence of two different modes of knowing related to hemispheric specialization. The left hemisphere was thought to be responsible for temporally sequential and analytic knowledge, whereas the right hemisphere was considered to be responsible for temporally non-sequential holistic knowledge (Levy-Agresti and Sperry 1968):

> The data indicate that the mute, minor hemisphere is specialized for Gestalt perception, being primarily a synthesis in dealing with information input. The speaking, major hemisphere, in contrast, seems to operate in a more logical, analytic, computerlike fashion and the findings suggest that a possible reason for cerebral lateralization in man is basic incompatibility of language functions on the one hand and synthetic perceptual functions on the other.

This cerebral asymmetry manifests itself in differences concerning the way in which we communicate and learn. It is popular to distinguish between logical sequential learners and intuitive non-sequential learners (Silverman 2002). Everybody learns sequentially sometimes, and sometimes non-sequentially, but usually there is a preference for one category or the other. The following table indicates some basic differences between sequential and non-sequential learning.

Sequential Learner	**Non-Sequential Learner**
thinks primarily in words	thinks primarily in images
learns step by step	learns part-whole relations
learns analytically	learns by seeing connections
focuses on details	attends to the big picture
learns from instructions / by routine	novel methods of problem solving
repetition reinforces learning	dislikes drill or repetition
is happy with one correct answer	likes unusual solutions to problems

4.6 Complementary Modes of Knowing

Niels Bohr (1928) presented his concept of complementarity[54] in quantum physics at a Congress in Como. Bohr's concept of complementarity provided a basis for describing physical reality in the presence of classically incompatible concepts. Bohr's various characterizations of complementarity show that he understood complementary as a *relation between incompatible classical descriptions*. Bohr (1928, p. 581):

> we are not dealing with contradictory but with complementary pictures of the phenomena, which only together offer a natural generalization of the classical mode of description.

[54]Important antecedents of Bohr's concept of complementarity have been discussed by William James (1890, p. 206), referring to split modes of consciousness, and by Henri Bergson (1998, originally published 1911 in French, pp. 342–345), referring to physical time and our intuition of flowing time.

Bohr (1934, p. 10):

> [We are] forced to adopt a new mode of description designated as *complementary* in the sense that any given application of classical concepts precludes the simultaneous use of other classical concepts which in a different connection are equally necessary for the elucidation of the phenomena.

Bohr (1949, p. 210):

> Evidence obtained under different experimental conditions cannot be comprehended within a single picture, but must be regarded as *complementary* in the sense that only the totality of the phenomena exhausts the possible information about the objects.

Bohr (1961, p. 64):

> It is in this situation that the notion of complementarity is called for to provide a frame wide enough to embrace the account of fundamental regularities of nature which cannot be comprehended within a single picture.

In summary, two Boolean descriptions of a situation are complementary in the sense of Bohr if they are both necessary for a complete description of that situation and at the same time incompatible with each other. Each description is correct, but neither one is "true" in the ultimate sense of the term. No single description is sufficient by itself, all of them are necessary. Only the set of all complementary descriptions can represent material reality as a whole. Bohr's early insistence on the dualistic nature of the wave-and-particle descriptions of matter led to many misunderstandings: *Complementarity is not the same as duality.* Modern quantum physics is not dualistic but holistic.

In a discussion of the idea of complementarity, Pauli (1948, p. 307) explained:

> every gain of knowledge of atomic objects by observations has to be paid for by a loss of other knowledge. ... Which knowledge is obtained and which other knowledge is irrevocably lost is left to the free choice of the experimenter, who may choose between mutually exclusive experimental arrangements. It is this situation which Bohr called "complementarity" and which has changed so radically the principles underlying our description of phenomena by laws of nature and even our ideas of physical reality.

Bohr's extensive writings about complementarity in his later years make it clear that his preeminent concern was to extend the idea of complementarity beyond physics (cf. Kalckar 1985, 1996; Favrholdt 1999). Bohr's examples in this respect are thought-provoking but admittedly vague and prone to misinterpretations. Anyway, the quantum physical examples are just a very special case within the broad range of applications that Bohr had in mind.

A context in which a proposition can be evaluated to see whether it is true or false is called a *Boolean context*. These contexts are defined in terms of the Boolean logical operations AND, OR and NOT. A comprehensive description of nature requires an expansion of the frame of discourse to include complementary descriptions based on incompatible Boolean contexts. *Complementarity refers to the existence of mutually exclusive, maximally incompatible descriptions which cannot be combined in a single description based on a Boolean, two-valued logic.*

Bohr formulated his ideas on complementarity in a purely verbal way, no formal definition of the concept can be found in his publications or letters. In the mid 1930s, the mathematician and logician Martin Strauss was the first to recognize that Bohr's complementarity can be formalized by a *two-valued logic with restricted sentential connectivity*. The propositional calculus of this logic is given by a *partial Boolean algebra* (Strauss 1936a, 1936b, 1938, 1967, 1970, 1972, 1973). In this framework, Strauss (1972, p. 210) characterized complementarity as follows:

> The complementary mode of description is the usage of a language in which, according to its syntactic rules, *no simultaneous use* can be made of certain sentences belonging to that language.

Mutually incompatible locally Boolean descriptions can be unified into a globally non-Boolean structure which allows the contradiction-free use of incompatible descriptions.[55] In such a partial Boolean algebraic structure certain Boolean operation are defined only for certain elements. Since mutually incompatible Boolean domains may partially overlap so that no one of them is contained in any other one, a consistent non-Boolean description must satisfy appropriate compatibility conditions on overlapping domains.

The propositional calculus of the *complementarity logic* introduced by Strauss is given by a partial Boolean algebra. It is defined as a set of Boolean algebras pasted together such that the operations of two or more overlapping Boolean algebras agree with each other (for technical details compare Sect. 3.4). In a partial Boolean algebra no new logical connectives enter,[56] and the concept of "truth" has the usual Boolean meaning (see Clavadetscher-Seeberger 1983). In an overlapping *locally Boolean* structure the neighboring Boolean algebras represent neighboring incommensurable but comparable perspectives.

4.7 Where Do Ideas Come From?

> If an idea springs to mind, this means that it comes from somewhere.
>
> *Paul Feyerabend*[57]

In opposition to metaphysical speculations, the first half of the 19th century witnessed the emergence of the philosophical doctrine of positivism, defending the idea

[55] Such a "pasting together" of mutually incompatible descriptions was introduced by Greechie (1968); see also Finch (1969). Depending on how one chooses the compatibility relation one gets non-Boolean theories based on partial Boolean algebras, Boolean atlases (Domotor 1974) or Boolean manifolds (Hardegree and Frazer 1981).

[56] This is in contrast with the lattice-theoretical approach of quantum logic where the operation of meet for incompatible propositions is formally defined but allows no interpretation as sentential connective. In particular, the logic proposed by Birkhoff and von Neumann (1936) leads to a language containing conjunctions of sentences that are unconnected in complementarity logic. Compare Strauss (1972, pp. 200–202).

[57] German quote, translated by the editor, from the Swiss weekly magazine *Die Weltwoche* of January 10, 1985. p. 25: "Einfall bedeutet, dass er von irgendwoher kommt." The noun "Einfall" may be translated as "an idea that springs immediately to mind". The phrase "dass [der Einfall] von irgendwoher kommt" means "that [the idea] comes from somewhere".

that positive facts derived from observation and measurement are the exclusive core of any scientific work. In his monograph *The Logical Structure of the World and Pseudoproblems in Philosophy*, Rudolf Carnap (1928) proposed a "rational reconstruction" of the concepts of all fields of knowledge (quoted from Carnap 2003, p. 5):

> [Logical positivism] attempts a step-by-step derivation or "construction" of all concepts from certain fundamental concepts, so that a genealogy of concepts results in which each one has its definite place. It is the main thesis of construction theory that all concepts can in this way be derived from a few fundamental concepts and it is in this respect that [logical positivism] differs from most other ontologies.

Knowledge of anything beyond empirical data was considered as impossible. Notably, the later development of logical positivism attempted to reduce propositions about empirical observations to pure logic.

In this respect, the mathematician Hans Hahn (1897–1934), one of the founding members of the Vienna Circle (the nucleus of logical positivism in the early 20th century), was of considerable influence. He was concerned with the status of logic and mathematics within the worldview of strict empiricism. Hahn (1933) claimed that the propositions of mathematics and logic are not derived from experience and emphasized the tautological character of mathematics:

> Our thinking cannot give insight into any sort of reality. It cannot bring us information of any fact in the world. It only refers to the manner in which we speak about the world. All it can do is to transform tautologically what has been said. There is no possibility of piercing through the sensible world disclosed by observation to a "world of true being": any metaphysics is impossible! Impossible, not because the task is too difficult for our human thinking, but because it is meaningless.

In his lecture *The Crisis in Intuition* of 1933 Hahn argued that mathematics had to be purged of intuition (Hahn 1980, p. 101):

> For it is not true, as Kant urged, that intuition is a pure a priori means of knowledge, but rather that it is force of habit rooted in psychological inertia.

Hahn claimed that logic does not say anything about the world and denied that thought can reach beyond what is experienced. According to Hahn the rules of a formal system are our own creation. Logic is not something to be found in the world.

This is a long-standing problem in the psychology of invention (or discovery). What happens in creative work? Today, many would agree that (Einstein 1955, p. 146)[58]

> invention is not a process of logical thought, even though the final result is bound to the structure of logic.

[58]Editor's note: This quote by Einstein expresses succinctly an ongoing theme in the philosophy of science, first addressed by Reichenbach (1938): the distinction between the context of discovery and the context of justification, or rational reconstruction. While the former is usually "not a process of logical thought", the latter "is bound to the structure of logic" in Einstein's words. A recent overview concerning the two contexts is due to Schickore and Steinle (2006). See also Sect. 5.3.

Yet, most contemporary scientists still believe that their conscious rational ego is the master in its own house. But what happens when they experience something that cannot be rationally comprehended? It is impossible to understand the genesis of fundamental physics if one dismisses the inherent human capacity for the formation of innovative concepts. This necessary element in the creation of fundamental theories does always contain archetypal elements which are not observed directly via physical experimentation, and which rational access alone fails to grasp.

In his inaugural address to the *Prussian Academy of Sciences* at July 2, 1914, Albert Einstein described the working principles of the theoretical scientist as follows (Einstein 1997, p. 16):

> [The theoretician's] activity, therefore, has two parts: first, he has to ferret out these principles, and second, he has to develop the conclusions that can be deduced from these principles. His school provides him with excellent tools with which to fulfill the second-named task. ... But the former task, namely to establish these principles which can serve as the basis of his deductions, is one of a completely different kind. Here there is no learnable, systematically applicable method which would lead him to the objective. The researcher must rather eavesdrop on nature to become privy to these general principles, by recognizing in larger sets of experiential facts certain general traits that can then be strictly and precisely formulated.

To "eavesdrop on nature" requires good ideas, but good ideas do not follow from observed facts by inductive inferences—modern artificial-intelligence-based expert systems would be capable of such processes. Science amounts to far more. Many of the deepest insights in the history of natural science were not empirically discovered but arose from an urge to comprehend fundamental theoretical aspects. For example, invariances with respect to symmetry groups of automorphisms, which imply conservation laws, are not directly empirically observable. Yet their significance for fundamental science can hardly be overestimated.

Hermann Weyl (1955, p. 8) remarked that, in spite of the fact that there is no logical way leading from experience to theory, "the decision as to which theories are adopted turns out ultimately to be unambiguous". We speak somewhat helplessly of "intuition", of "ideas" that burst into consciousness. However, we have no clue whatsoever about just how this happens, and about where those intuitions and ideas came from.

Chapter 5
Modes of Knowing in Mathematics

I suppose it is tempting,
if the only tool you have is a hammer,
to treat everything as if it were a nail.

Abraham Maslow (1966, p. 15)

5.1 Truth in Mathematics

Mathematics has often been regarded as the paradigm of rational knowledge, and it has even been claimed that mathematics is a source of indubitable truth. But what does it mean to say that something is true in mathematics? In what sense do mathematical propositions have meaning?

The nature of mathematical truth is a problem discussed since the time of the ancient Greeks. The discovery of non-Euclidean geometries early in the 19th century undermined the idea that Euclidean geometry is the one true geometry. In the 20th century, research in the foundations of mathematics led to pluralist conceptions of arithmetic and set theory. It may therefore be worthwhile to begin this chapter, at least at a cursory level, with the state of affairs concerning truth in mathematics. Traditionally, three mutually incompatible viewpoints can be identified:[59]

- The *Platonist* view takes mathematical objects as entities which exist in a timeless sense, once and for all. For the Platonist every mathematical statement is either *necessarily true* or *necessarily false*. Platonism regards mathematics as a descriptive science, and the concept of mathematical truth as an objective one.
- The *constructivist* view takes the natural numbers as given, while all other mathematical notions are or must be constructed. For the constructivist existence in mathematics means *constructibility*. "Truth" is defined intuitionistically, for in-

[59]An overview of some of the greatly varying views on what truth in mathematics is about can be found in Dales and Olivery (1998).

H. Primas, *Knowledge and Time*,
DOI 10.1007/978-3-319-47370-3_5

stance via effective calculability in the sense of Church, Herbrand, Gödel or Turing.
- The *formalist* view takes a set of mathematical rules and declares the meaning of these rules as non-mathematical. Only the *consistency* of the basic rules is important. Mathematical truths are taken as analytic and hence without content. "The formalist has the freedom to choose any axiom system as long as it is consistent" (Ruitenburg 2001, p. 159).

Concerning these three perspectives, Reuben Hersh (1994, p. 12) formulated the following dilemma:

> How to establish mathematical knowledge as certain, indubitable, free of any possible doubt? Three historically important solutions to that problem were offered—logicism (Platonism), formalism, intuitionism. All three proved unsuccessful.

And David Gale (1989, p. 64) noted:

> On the extreme right are those who reject using the law of the excluded middle in any argument, while at the left we have the Platonists, like the set theorists ... who believe the law holds in great generality.

To fit together these incompatible partial views we have to use a more comprehensive scheme. Already Nicholas of Cusa (1401–1464) opposed the law of the excluded middle of Boolean logic. He saw the weakness of human reason in the supposition of the principle of non-contradiction and insisted on the idea of a *coincidentia oppositorum*, a coincidence of opposites.[60] As we will see in more detail in Sect. 5.8, this concept can be interpreted in terms of the notion of complementarity.

This is intimately related to a *mathematical pluralism* according to which mutually incompatible viewpoints in mathematics may coexist peacefully. One of its key examples is the discovery of independence results for set theory. They imply that there is a plurality of arithmetic systems no one of which can be said to be "truer" than others. For example, George and Velleman (1998) discussed two traditional conceptions of natural numbers, the constructivist "bottom-up version", starting with 0 and closed under the successor operation, and the Platonic "top-down version", defined as the intersection of all sets containing 0 and closed under the successor operation. Both conceptions are used in contemporary mathematics, but they are mutually incompatible.

5.2 On the Origin of the Concept of Integers

Both Henri Poincaré (1904a, Chap. 1) and Luitzen Egbertus Jan Brouwer (1907, p. 3) proposed that the starting point for all mathematics are the natural numbers. They are given to us by the fundamental intuition of *counting*. This ability to grasp something without any abstract reasoning is well demonstrated by the fact that (Heyting 1971, p. 7)

---

[60] *De Docta Ignorantia*; English translation: "Of Learned Ignorance" (Cusanus 1954).

> children in the elementary school understand what the natural numbers are and they accept the fact that the sequence of natural numbers can be indefinitely continued.

Penelope Maddy (1990, p. 31) remarked that "mathematicians are not apt to think that the justification for their claims waits on the activities in the physics labs", and she claims that we do not need physics to justify "$2+2=4$". To justify "$2+2=4$" we have to know what "counting" means. Yet, from a physical point of view the concept of counting is neither as simple nor as obvious as it may seem. The first mathematical idea every child learns is to say "this many" by holding up fingers. This suggests that the concept of number is based on the *operation of counting* which in turn depends on the *existence of distinguishable material objects*.

However, from the point of view of the modern theory of matter this assumption is far from being evident—recall the indistinguishability of electrons in quantum mechanics. In general, two arbitrary subsystems of a quantum system cannot be considered as distinguishable objects. Only in the exceptional case when subsystems are essentially disentangled can they be counted.[61] That is, the concept of numbers cannot be considered to be given *a priori*.[62] Arithmetic concepts depend either on the existence of distinguishable objects in the material world or on mental acts of isolating objects.

5.3 Is Mathematics Discovered or Invented?

Mathematicians have always distinguished between the discovery and the justification of a mathematical idea, between intuitive knowledge and formalistic processing in which symbols and formulas are manipulated. The classical accounts by Henri Poincaré[63] and Jacques Hadamard (1945) discussed anecdotal material and the role of intuition in the creative process.

All great mathematicians appeal to their "intuition", yet this notion of intuition has a variety of meanings. Some of its cognitive aspects are discussed in *The Mathematical Experience* by Davis et al. (1981, pp. 440f):

> ... intuition is not a direct perception of something existing externally and eternally. It is the effect in the mind of certain experiences of activity and manipulation of concrete objects. We have intuition because we have mental representations of mathematical objects. ... Of course, we don't know in what way these representations are held in the mind.
> In the realm of ideas, of mental objects, those ideas whose properties are reproducible are called mathematical objects, and *the study of mental objects with reproducible properties is called mathematics*.

---

[61] In more technical terms: In the set of all pure quantum states of an arbitrary quantum system, the set of all product states has the measure zero.

[62] It is likely that the intuition of counting has a neurobiological basis. There is experimental evidence that some numerical abilities do not have to be learnt, but are present from birth. Butterworth (1999) calls this ability to recognize small cardinalities without counting the *number module*, and according to Dehaene (1997) the *number sense* is based on a neural "accumulator".

[63] In 1908 Henri Poincaré gave a presentation "L'invention mathématique" to the *Institut général psychologique* in Paris, translated as "Mathematical Discovery" in Poincaré (1914).

> To be sure, there is always some possibility for disagreement between intuitions. The process of mutual adjustment to ensure agreement is never finally completed. As new questions are asked, new parts of the structure may come into focus that have never been looked at before.

Albert Einstein (1944, p. 287) believed that mathematics is invented, not discovered by sense experiences:

> the concepts which arise in our thought and in our linguistic expressions are all—when viewed logically—the free creations of thought which can not inductively be gained from sense-experiences. ... Thus, for example, the series of integers is obviously an invention of the human mind, a self-created tool which simplifies the ordering of certain sensory experiences. But there is no way in which this concept could be made to grow, as it were, directly out of sense experiences.

In a different vein, Kurt Gödel (1951, p. 311) criticized the view that "mathematics is only our own creation". He was convinced that mathematics is *a priori* given (Gödel 1961, p. 377) :

> Actually, mathematics, by its nature as an a priori science, always has, in and of itself, an inclination toward the right, and, for this reason, has long withstood the spirit of the time [Zeitgeist] that has ruled since the Renaissance; i.e., the empiricist theory of mathematics, such as the one set forth by Mill, did not find much support.

If we ask whether Abraham Robinson's "non-standard analysis" (Robinson 1966) was an invention or a discovery we may realize that neither Einstein's nor Gödel's remark hits the nail on its head. Non-standard analysis gave Leibniz's calculus of infinitesimals a firm foundation based on a novel system of (non-standard) "hyperreal" numbers—300 years after Leibniz. Surprisingly many textbooks and monographs on non-standard analysis explicitly relate it to both discovery and invention (Arkeryd et al. 1997, p. ix, editor's addition):

> The achievements of [non-standard analysis] can be summarized under the headings (i) *explanation*—giving fresh insight or new approaches to established theories, (ii) *discovery*—leading to new results in many fields, (iii) *invention*—providing new, rich structures that are useful in modeling and representation. ...

5.4 Formalism

In its extreme form, formalism in mathematics says that mathematics is just a game with symbols. The formalist's ultimate goal is a secure foundation for all mathematics that is both complete and consistent. In his *Foundations of Geometry*, David Hilbert (1899) proposed a formal set of axioms as implicit definitions substituting the traditional axioms of Euclid. In contrast to Euclid, Hilbert considered axioms not taken as self-evident truths or as intuitions, but as undefined concepts like "point", "line", "plane". Hilbert's axioms are not about anything in particular but rather about anything that satisfies the axioms. Blumenthal reported a discussion in which Hilbert said: "One should always be able to say, instead of 'points, lines, and planes', 'tables, chairs, and beer mugs' " (quoted from Blumenthal 1935, p. 403).

Gottlob Frege (1948–1925) tried to get rid of any resort to intuition in arithmetic and logic and presented the thesis that arithmetic is a branch of logic in the sense that all arithmetic truth rests only on the laws of logic. Frege argued that all of mathematics can be grounded in logic. In an unpublished and undated manuscript Richard Dedekind (1831–1916) claimed that (quoted from Dugac 1976, p. 315)

> there can be no doubt that there are absolutely no limits to [the] further development [of arithmetic]; and the domain of its application is equally immeasurable, for every thinking man, even if he does not clearly realize it, is a man of numbers, an arithmetician.

Bertrand Russell (1919, p. 208) went even further than Dedekind and Frege and asserted "the identity of logic and mathematics". The discovery of Russell's paradox[64] in 1902 revealed that an axiomatic foundation of set theory requires a more precise development of the underlying logical system.

In the 1920s, Hilbert rejected Russell's logicist approach. He expected that by formalizing mathematics in axiomatic systems and by proving its consistency using finitary methods he could give classical mathematics a philosophically satisfactory grounding. In an address delivered in July 1927, Hilbert explained (translation by van Heijenoort 1967, p. 475):

> For this formula game is carried out according to certain definite rules, in which the *technique of our thinking* is expressed. These rules form a closed system that can he discovered and definitively stated. The fundamental idea of my proof theory is none other than to describe the activity of our understanding, to make a protocol of the rules according to which our thinking actually proceeds. Thinking, it so happens, parallels speaking and we form statements and place them one behind another. If any totality of observations and phenomena deserves to be made the object of a serious and thorough investigation, it is this one—since, after all, it is part of the task of science to liberate us from arbitrariness, sentiment, and habit and to protect us from the subjectivism that already made itself felt in Kronecker's view and, it seems to me, finds its culmination in intuitionism.

Later Hilbert worked on a project to reduce mathematics to a finite set of axioms and rules of inference, deliberately ignoring intuitive ideas and possible applications. Already in 1900, at the *Second International Congress of Mathematicians*, Hilbert had delivered his famous lecture "Mathematical Problems" in which he made clear what his standpoint was (Hilbert 1900, p. 262):

> This conviction of the solvability of every mathematical problem is a powerful incentive to the worker. We hear within us the perpetual call: *There is the problem. Seek its solution. You can find it by pure reason, for in mathematics there is no ignorabimus.*

In 1929, Hilbert (1930, p. 3) proposed a "new grounding of mathematics" (a proof theory) which should eliminate once for all

> the foundational question in mathematics by turning every mathematical statement into a concretely exhibitable and rigorously derivable formula.

---

[64]Russell's paradox arises within naive set theory by considering the set of all sets that are not members of themselves. This paradox showed that the axioms Frege used to formalize his logic were inconsistent.

In his address "Problems of Grounding Mathematics" at the *International Congress of Mathematicians* in Bologna in 1928, Hilbert posed a fundamental problem of mathematical logic, the so-called *decision problem* for first-order logic (cf. Hilbert and Ackermann 1928; Hilbert 1930):

> Is it possible to determine whether or not a given statement pertaining to a field of knowledge is a consequence of the axioms?

Hilbert's decision problem asks for an algorithm that takes as its input a description of a mathematical statement in a formal language and produces as its output either "true" or "false", according to whether the statement is true or false.

Even if the decision problem would have an affirmative answer, Hilbert's program cannot give a satisfactory answer to the question of how axioms are justified or what specific content the chosen axioms may have. Furthermore, any variant of formalism which maintains that human reason does not have mathematical intuition at its disposal is not compatible with the mode of thinking of most working mathematicians.[65]

5.5 Gödel's Theorems

In 1931 Kurt Gödel (1931) proved that the program Hilbert had conceived for proving the consistency of mathematics cannot be successfully accomplished. Gödel's first theorem states that in any consistent formalization of mathematics that is sufficiently strong to axiomatize the natural numbers one can construct a true statement that can be neither proved nor disproved within that system itself. Gödel's second incompleteness theorem states that no consistent system can be used to prove its own consistency.[66] It is impossible to use the axioms of mathematics to prove that any axiomatic system would never lead to contradictions. Yuri Manin (2007, p. 55) summarized the essential content of Gödel's theorems informally:

> By pure deductive reasoning, one cannot even deduce from a finite number of basic principles all true statements about integers that can be formulated in the language of high school algebra.

Heavily influenced by Gödel's earlier work on his incompleteness theorem, both Alonzo Church (1936a, 1936b) and Alan Turing (1936) proved independently that Hilbert's decision problem has no solution.[67] Although Hilbert's program as originally conceived cannot be carried out, it was fruitful and can be reconstructed as a defense of mathematical instrumentalism (cf. Giaquinto 1983; Zach 2007).

---

[65]Compare the stimulating book *Proofs and Refutations* by Imre Lakatos (1976).

[66]Editor's note: A readable and accessible explanation to both scholars and non-specialists of the main ideas and broad implications of Gödel's discovery is due to Nagel and Newman in 1951, 50 years later revised and updated by Douglas Hofstadter (Nagel and Newman 2001).

[67]For a complete and definitive proof see Hilbert and Bernays (1968, 1970).

According to Paul Bernays (1971), Gödel's results and related, but less well known, previous work by the Norwegian logician Thoralf Skolem showed[68]

> that all such strictly formal frameworks for mathematics are unable to represent mathematics in its entirety. In this connection, even for number theory treated axiomatically one has that any strictly formalized axiom system possesses "non-standard models", which do not have the intended structure of the series of numbers. In these facts we see that certain concepts do not admit a fully adequate formalization, such as the concept of finiteness (of a finite number) and the general notion of a predicate.

For a Platonist such as Alain Connes, Gödel's theorem is a perfect example of what should be understood by *mathematical truth* (Connes 2001, p. 42):

> The distinction between truth and provability is one of Gödel's essential results. ... Gödel's theorem is much more powerful than [his undecidability result]. It says that, in a logico-deductive system, there will always be a proposition concerning the natural numbers and only the natural numbers that will be true and impossible to prove through the system of axioms, but will not be undecidable. [Here, to say that it is true means] that it cannot be falsified by a counterexample.

In other words: what is intuitively evident goes beyond mathematical proof. Gödel himself was convinced that mathematical intuition can decide the truth of mathematical propositions and that the classically undecidable problems of mathematics possess determinate truth values. He proposed to search for new axioms of set theory that are strong enough to answer questions left undecided by the standard Zermelo-Fraenkel axioms including the axiom of choice (see Hauser 2006).

5.6 Constructivism

Constructivists reject non-constructive proofs, but there is no consensus about what "constructive" means exactly. There are many forms of constructivism, for example the program of intuitionism founded by Brouwer, the finitism of Hilbert and Bernays, the constructive recursive mathematics of Markov, and Bishop's program of constructive analysis. Bernays (1971) asked:

> Why in mathematics, or at least arithmetic, do we not restrict ourselves to a finite framework? The answer to this is that, in an essential respect, the infinite is more simple than the numerous finite.

This quote stands in sharp contrast to Hilbert's finitism. His former student Hermann Weyl described Hilbert's project as replacing meaningful mathematics by a meaningless game of formulas (Weyl 1928b, p. 87; translation by van Heijenoort 1967, p. 483):

> [Hilbert] succeeded in saving classical mathematics *by a radical reinterpretation of its meaning* without reducing its inventory, namely, by formalizing it, thus transforming it in principle from a system of intuitive results into a game with formulas that proceeds according to fixed rules.

---

[68] See Wang (1996) for a detailed study of the intricate relationship between Skolem's and Gödel's work.

Weyl (1925, p. 30) pointed out that if "mathematics is to remain a serious cultural concern, then some *sense* must be attached to Hilbert's game of formulae". The leading figure to implement this desideratum in the early 20th century was Luitzen Egbertus Jan Brouwer. Emphasizing that mathematics needs to be meaningful and constructive, he introduced the intuitionist position that (see Heyting 1971, p. 3) "mathematics does not have only formal, but also contentual meaning".

Intuitionism is based on the idea that mathematics is a constructive mental activity of humans. In particular, Brouwer did not accept any *a priori* logical principles. He contended that mathematics cannot be grounded on logic, but that logic depends on mathematics:

- Logic does not precede mathematics, but is a result of mathematical activity (Brouwer 1907, p. 73).
- "Intuitionism proceeds independently of the formalization, which can but follow after the mathematical construction" (Heyting 1971, p. 5).
- Non-constructive existence proofs indicate "the presence of a treasure without disclosing its location" (Weyl 1949a, p. 51).

All these arguments notwithstanding, the problem of existential versus constructive aspects of mathematics has no simple yes-or-no answer. This has been stressed by Weyl (1925, p. 32; translation by Mancosu 1998, p. 141):

> ...beside Brouwer's way, one will also have to pursue that of Hilbert; for it is undeniable that there is a theoretical need, simply incomprehensible from the merely phenomenal point of view, with a creative urge directed upon the symbolic representation of the transcendent, which demands to be satisfied.

5.7 Platonism

> I am not a Platonist, but I think platonically.
>
> *Ernst Specker*[69]

A mathematical Platonist believes that mathematical objects exist independently and have existed before anyone discovers them. (Of course, a mathematical Platonist need not adhere to the full body of Plato's philosophy.) Many of the world's most eminent mathematicians are avowed Platonists and have no doubt that mathematical entities (like the numbers 1 or π, or the cardinalities $\aleph_0$ or $\aleph_1$) have an objective existence independent of us. For example, the prominent English mathematician Godfrey Harold Hardy (1940, p. 35) said:

> I believe that mathematical reality lies outside us, that our function is to discover or *observe* it, and that the theorems which we prove, and which we describe grandiloquently as our "creations", are simply the notes of our observations.

[69]Discussion remark by Specker after his lecture "Many Interpretations—One Reality? Did God Really Create the Numbers?" at Zurich on April 21, 1998.

Likewise, Gödel (1995, p. 311) noted that

> mathematical objects and facts (or at least something in them) exist objectively and independently of our mental acts and decisions.

However, Gödel (1933, p. 50) also argued that

> ...our axioms, if interpreted as meaningful statements, necessarily presuppose a kind of Platonism, which cannot satisfy any critical mind and which does not even produce the conviction that they are consistent.

Another Platonist, Alain Connes (2008, pp. 30f), stated:

> I have no doubt that mathematical reality is something which exists, that it exists independently of my own brain trying to see it, and has exactly the same properties of resistance as external reality. When you want to prove something, or when you examine if a proof is correct or not, you feel the same anguish, the same external resistance as you do with external reality. ... There is a fundamental mathematical reality out there, and the mathematician creates tools to understand it.

Of course, Platonists have a problem to explain what it means to access mathematical reality. Is this access the mind's way of interacting directly with the Platonic world? According to Tait (2001, p. 103), "Gödel's later arguments for Platonism were ... based primarily on the failure of alternatives". In his thesis, Alexander Wittenberg (1953), a student of Bernays, stated that

> the usual methods in the foundations of mathematics (Platonist, formalist, Hilbertian) do not enable us to reach a purely objective standpoint devoid of dogmatic components.

And in Wittenberg's response to his critics, we read the truly provocative statement that "all mathematicians would like to be Platonists, if only they were able" (Wittenberg 1956, p. 256).

5.8 Complementarity in Mathematics

The opposition between constructivism and Platonism is an age-old discussion topic in mathematics. Let us illustrate the current situation by way of two extreme examples:

1. *Bishop's constructivism:* Erret Bishop's book *Foundations of Constructive Analysis* contains a fully constructive presentation of large parts of analysis using only constructive proofs which ultimately reduce to finite constructions based only on integers $1, 2, 3, \ldots$. He opposed the idealistic nature of classical mathematics and insisted that (Bishop 1967, p. 409)

   > all mathematics should have numerical meaning. In other words, every mathematical theorem should admit an ultimate interpretation to the effect that certain finite computations within the set of positive integers will give certain results.

2. *Gödel's Platonismus:* In John von Neumann's axiomatics of set theory there is an axiom which captures the intuitive idea that the universe of set theory is maximal (Ulam 1958, p. 13, footnote 5):

> The great interest which this axiom has lies in the fact that it is a maximality principle, somewhat similar to Hilbert's axiom of completeness in geometry. For, roughly speaking, it says that any set which does not, in a certain well defined way, imply an inconsistency exists. Its being a maximum principle also explains the fact that this axiom implies the axiom of choice. I believe that the basic problems of set theory, such as Cantor's continuum problem, will be solved satisfactorily only with the help of stronger axioms of *this* kind, which in a sense are opposite or complementary to the constructivistic interpretation of mathematics.

In a letter to Stanislaw Ulam of November 8, 1957, Gödel related this axiom to the idea that the universe of set theory is maximal in the sense that nothing can be added, and continues (Gödel 2003, p. 295):

> Moreover I believe that only by going farther in this direction, i.e., in the direction opposite to constructivism, will the basic problems of abstract set theory be solved.

As introduced in Sect. 5.1, *mathematical pluralism* is the view that different and mutually incompatible types of mathematics (like classical and constructive mathematics) can coexist. For example, we may accept a mathematical universe in which the *axiom of choice* holds, or we may accept a different mathematical universe without this axiom. Both versions are intrinsically coherent.[70]

With his *principle of tolerance*, Rudolf Carnap (1937, p. xv and p. 52) suggested that alternative logics never genuinely conflict:

> ... let any postulates and any rules of inference be chosen arbitrarily; then this choice, whatever it may be, will determine what meaning is to be assigned to the fundamental logical symbols. By this method, also, the conflict between the divergent points of view on the problem of the foundations of mathematics disappears.
> Everyone is at liberty to build up his own logic, i.e. his own form of language, as he wishes. All that is required of him is that, if he wishes to discuss it, he must state his methods clearly, and give syntactical rules instead of philosophical arguments.

There is no uniquely correct way towards the foundations of mathematics: one must decide the framework of discourse before questions about existence and truth make sense. For example, the quantifier $\exists$ in classical mathematics refers to Platonic existence, while in intuitionistic mathematics $\exists$ refers to the possibility of constructing the relevant quantity (cf. Davies 2005a, 2005b). Mark Balaguer (1998) convincingly defended the view that both mathematical Platonism (i.e. the view that non-spatiotemporal mathematical objects, such as numbers, exist independently of us) and anti-Platonism (understood as mathematical fictionalism, i.e. the view that abstract objects such as numbers do not exist) are perfectly workable philosophies of mathematics.

Paul Bernays (1946, p. 79) pointed out that a comprehensive interpretation of different approaches in the discussion of existential versus constructive aspects in the foundation of mathematics requires *complementary perspectives*:

---

[70]That is, the axiom of choice is logically independent of Zermelo-Fraenkel set theory. In 1940, Gödel proved that the axiom of choice is consistent with Zermelo-Fraenkel set theory, and in 1963 Cohen proved that it is also independent of it.

> ... for the interpretation of mathematics the role of mental activity cannot be fully abstracted from ... the objective, theoretical (existential) standpoint on the one hand and the intuitionistic (constructive) one on the other hand could be regarded as "complementary" aspects. ... Using this sort of distinction we might say that Hilbert's aim in the foundations lies in basing geometry on arithmetic (by means of the metamathematical method), in order then to treat mathematics as geometry; Brouwer's tendency is to restrict mathematics to arithmetic; and Frege's purpose, and that of his followers, amounts to establishing logic as a kind of geometry, on which mathematics is then to be based.

Gödel's theorems imply that every mathematical reasoning contains an intuitive component (compare the quote by Connes in Sect. 5.5). In his unpublished essay "Is Mathematics Syntax of Language?", Gödel (1953, p. 346) wrote:

> ... the scheme of the syntactical program to replace mathematical intuition by rules for the use of symbols fails because this replacing destroys any reason for expecting consistency, which is vital for both pure and applied mathematics, and because for the consistency proof one either needs a mathematical intuition of the same power as for discerning the truth of the mathematical axioms or a knowledge of empirical facts involving an equivalent mathematical content.

In a review of a paper by Carnap (1934) about antinomies in mathematics, Gödel (1935) assessed that mathematics cannot be represented in a single formal system:

> Every formal system is incomplete in two respects: 1. insofar as there are propositions undecidable within it, and 2. insofar as there are notions that cannot be defined within it. ... Thus we are led to conclude that, although everything mathematical is formalizable, it is nonetheless impossible to formalize all of mathematics in a *single* formal system, a fact that intuitionism has asserted all along.

And Turing (1939, pp. 208f, 215f) underscored the necessity of using intuition in mathematical reasoning in the following way:

> Mathematical reasoning may be regarded rather schematically as the exercise of a combination of two faculties, which we may call intuition and ingenuity. The activity of the intuition consists in making spontaneous judgments which are not the result of conscious trains of reasoning. These judgments are often but by no means invariably correct (leaving aside the question what is meant by "correct"). ... In pre-Gödel times it was thought by some that it would probably be possible ... that all the intuitive judgments of mathematics could be replaced by a finite number of these rules. The necessity for intuition would then be entirely eliminated. ... [Gödel's theorem shows] the impossibility of finding a formal logic which wholly eliminates the necessity of using intuition, we naturally turn to "non-constructive" systems of logic with which not all the steps in a proof are mechanical, some being intuitive.

That pluralism does not hamper the development of mathematics is documented by Dieudonné's (1970, p. 145) description of Bourbaki's attitude:

> On foundations we believe in the reality of mathematics, but of course when philosophers attack us with their paradoxes, we rush to hide behind formalism and say: "Mathematics is just a combination of meaningless symbols", and then we bring out chapters 1 and 2 on set theory. Finally we are left in peace to go back to our mathematics and do it as we have always done, with the feeling each mathematician has that he is working with something real.

Every well-founded standpoint in mathematics has an intrinsically consistent realization within a particular Boolean context. Incompatible representations refer to alternative viewpoints and different preconceptions. In spite of that, mutually incompatible standpoints can be embedded into a larger non-Boolean structure. Gödel also emphasized the necessity of more than one language:[71]

> We admit as legitimate mathematics certain reflections on the grammar of a language that concerns the empirical. If one seeks to formalize such a mathematics, then with each formalization there are problems, which one can understand and express in ordinary language, but cannot express in the given formalized language. It follows (Brouwer) that mathematics is inexhaustible: one must always again draw afresh from the "fountain of intuition". There is, therefore, no *characteristica universalis* for the *whole* mathematics, and no decision procedure for the whole mathematics. In each and every *closed language* there are only countably many expressions. The *continuum* appears only in "the whole of mathematics" ... If we have *only one language*, and can only make "elucidations" about it, then these elucidations are inexhaustible, they always require some new intuition again.

5.9 Birds and Frogs in Mathematics

The Hedgehog and the Fox is the title of an essay by Isaiah Berlin, regarding Leo Tolstoy's theory of history (Berlin 1953). Berlin took this title from fragment 201 of Archilochus (7th century B.C.):[72]

> The fox knows many things, but the hedgehog knows one big thing.

Based on this line, he discusses two basic modes of knowing: the way of the fox who knows a thousand small clever things and the way of the hedgehog who knows one great thing. Berlin tries to categorize writers and thinkers as either "foxes" or "hedgehogs". Foxes are fascinated by the infinite variety of things, pursue many divergent ends, and think that life is too complex to be squeezed into a unitary scheme. Berlin's examples include Aristotle, Montaigne, Shakespeare, Goethe, Pushkin, Balzac, Joyce. Hedgehogs have a central vision of reality and tend to see the world through the lens of a single all-embracing idea. Examples include Plato, Dante, Pascal, Hegel, Dostoyevsky, Nietzsche.

Berlin (1953, pp. 3f) saw "a great chasm between those, on one side, who relate everything to a single central vision" and those who seize "upon the essence of a vast variety of experiences and objects ... without ... seeking to fit them into ... any unitary inner vision".

The metaphor of the hedgehog and the fox can also be applied to different styles of mathematics. Freeman Dyson (2009, p. 212) divides mathematicians into two categories: those who study the details of specific examples, and those who seek abstraction, unification and generalization:

---

[71] A diary entry by Carnap from December 12, 1929, recording a conversation with Gödel that day about Brouwer's lectures of the previous year. Quoted from Wang (1987, p. 50).

[72] This fragment is compatible with diverse interpretations. Davenport (1995, p. 54) gives the literal translation "Fox knows many / Hedgehog one / solid trick" and an alternative translation that he claims expresses the true thrust of the original "Fox knows / Eleventythree / Tricks and still / Gets caught: / Hedgehog knows / One but it / Always works".

> Some mathematicians are birds, others are frogs. Birds fly high in the air and survey broad vistas of mathematics out to the far horizon. They delight in concepts that unify our thinking and bring together diverse problems from different parts of the landscape. Frogs live in the mud below and see only the flowers that grow nearby. They delight in the details of particular objects, and they solve problems one at a time. ... Mathematics needs both birds and frogs. Mathematics is rich and beautiful because birds give it broad visions and frogs give it intricate details. Mathematics is both great art and important science, because it combines generality of concepts with depth of structures. It is stupid to claim that birds are better than frogs because they see farther, or that frogs are better than birds because they see deeper. The world of mathematics is both broad and deep, and we need birds and frogs working together to explore it.

Dyson's examples for frogs who solve problem, construct proofs and explore fine details of mathematical structures include Francis Bacon, Abram Samoilovitch Besicovitch, Charles Darwin, Paul Erdös. Typical birds who explore new horizons, link one world of ideas with another, and build theories are René Descartes, Hermann Weyl, collaborators of Nicolas Bourbaki, Alexander Grothendieck. Henri Poincaré, whom we will meet again in the next section, refers to a combination of both modes of knowing (Poincaré 1904a):

> Thanks to these images, one can see at a glance what pure deduction will show only after successive steps.

5.10 Mathematical Intuition

5.10.1 Logic Versus Creativity

Intuition is an indispensable part of mathematical activity. Intuitive knowledge does neither refer to visual perception nor does it rely on a physical model. In mathematics, Henri Poincaré discussed the necessity of two modes of knowing as follows (Poincaré 1904a, pp. 9ff):

> It is impossible to study the works of the great mathematicians, or even those of the lesser, without noticing and distinguishing two opposite tendencies, or rather two entirely different kinds of minds. The one sort are above all preoccupied with logic. ... The other sort are guided by intuition and at the first stroke make quick but sometimes precarious conquests. The two sorts of minds are equally necessary for the progress of science; both the logicians and the intuitionalists have achieved great things that others could not have done. ... Pure logic could never lead us to anything but tautologies; it could create nothing new; not from it alone can any science issue.

Or, more succinctly (Poincaré 1904b, p. 266):

> It is by logic we prove, it is by intuition that we invent. Criticism is good, know how to create is better. ... Logic, therefore, remains sterile unless fertilized by intuition.

A mathematician may say that he *knows* that some proposition is true, but that he does *not see* that is it true. In this situation, "seeing" does not refer to a mental visualization but to a direct recognition of something that he experiences as mind-independent. Even Nicolas Bourbaki (a collective pseudonym for a group of mathematicians striving for a fully rigorous foundation of mathematics on set theory) admitted that (Bourbaki 1949, p. 1, 1950, p. 223):

> Logic, so far as we mathematicians are concerned, is no more and no less than the grammar of the language which we use, a language which had to exist before the grammar could be constructed.
> Every mathematician knows that a proof has not really been "understood'" if one has done nothing more than verifying step by step the correctness of the deductions of which it is composed, and has not tried to gain a clear insight into the ideas which have led to the construction of this particular chain of deductions in preference to every other one.

The logician Sigekatu Kuroda (1951, p. 36) affirmed that

> mathematics is a mental activity independent of logic, and that it rests directly on intuitively clear evidence. Mathematicians can carry out their thinking without having to worry about logical laws.

And the mathematician Andrew Gleason (known for his final solution of Hilbert's fifth problem) emphasized that the *explanatory function* of proofs is of vital importance to mathematicians (quoted from Yandell 2001, p. 150):

> Proofs really aren't there to convince you that something is true, they're there to show you *why* it is true.

Intuitive insight often appears as a whole, "at once without the slightest instant of reflection" (Hadamard 1945, p. 8). A famous example is documented in a letter of September 3, 1805, by Carl Friedrich Gauss to his friend Wilhelm Olbers, where Gauss described his great difficulties to determine the sign of a quadratic sum (quoted from Olbers 1900, pp. 268f):

> Perhaps you remember ... my complaints about a theorem which ... had defied all my attempts to find a satisfactory proof.... for four years a week has seldom passed when I would not have made one or another vain attempt to solve this problem—recently very lively now again. But all brooding, all searching has been in vain, sadly I have had to put down the pen again every time. Finally, a few days ago, I succeeded—not as a result of my tedious efforts but by the grace of God, I might say. Like a sudden flash of lightning, the riddle happened to be solved; I myself would not be able to show the guiding thread between that which I previously knew, that with which I had made the last attempts—and that by which I succeeded.

John von Neumann explained to Pascual Jordan the process of a mathematical discovery as follows (Jordan 1947, p. 89):

> After a certain conjecture has arisen, whose proof is henceforth sought, there appear gradually several initially disjointed fragments of it. It's as if an island should be connected with the mainland by a causeway, but not from the mainland. It grows in high from the see bottom in full extension so that it emerges from the water at separate sites where growth was fastest.

Intuition is usually connected to our ability to understand something immediately, without the intervention of a discursive analytic process of the intellect. Many popular definitions of intuition in cognitive psychology fall short of taking this into account, like the following one by Tilmann Betsch (2008, p. 4):

> Intuition is a process of thinking. The input of this process is mostly provided by knowledge stored in long-term memory that has been primarily acquired via associative learning.

Such an understanding clearly fails to cover the well-known and well-documented cases of genuine scientific insight like those by Gauss, Poincaré, or Srinivasa Ramanujan.

5.10.2 The Intuition of Srinivasa Ramanujan

The Indian mathematician Srinivasa Ramanujan (1887–1920) is widely acknowledged as an all-time mathematical genius. As a teenager he received no formal training in mathematics and had almost no access to mathematical libraries or literature (Berndt and Rankin 2000). Nevertheless, he exhibited unique mathematical creativity and an incomparable intuition which was with certainty not "informed by prior knowledge in long-term memory" (Betsch 2008, p. 4). Ramanujan "knew" many things he had never learned. He recorded many of his mathematical discoveries during 1903–1914 (prior to leaving his home country for England) in his *Notebooks*, mostly without providing proofs.

As John Edensor Littlewood (1929) noted in a review of the *Collected Papers of Srinivasa Ramanujan*:

> The clear-cut idea of what is meant by a proof, nowadays so familiar as to be taken for granted, he perhaps did not possess at all. If a significant piece of reasoning occurred somewhere, and the total mixture of evidence and intuition gave him certainty, he looked no further.

Ramanujan's achievements were only fully understood well after his untimely death in 1920. Narayanaiengar, one of the early editors of the Journal of the Indian Mathematical Society stated (in volume XII) that the first article by Ramanujan [73]

> attracted considerable attention. It is, however, a sad confession to say that the Editor's work in connection with Ramanujan's contribution was by no means light. Ramanujan saw intuitively many things and could not bring himself to explain, and the first article had to be referred back to him no less than three times.

Ramanujan could never explain to his mentor Hardy (who discovered Ramanujan in 1913 and arranged a scholarship for the then 26-year-old in Cambridge) how he arrived at his deep insights. Hardy (1921, p. xxiii) wrote in his obituary notice for Ramanujan:

> There was one great puzzle. What was to be done in the way of teaching him modern mathematics? The limitations of his knowledge were as startling as its profundity. Here was a man who could work out modular equations, and theorems of complex multiplication, to orders unheard of, whose mastery of continued fractions was, on the formal side at any rate, beyond that of any mathematician in the world, who had found for himself the functional equation of the Zeta-function, and the dominant terms of many of the most famous problems

[73] See www.imsc.res.in/~rao/ramanujan/newnow/bernoulliims.htm.

> in the analytic theory of numbers; and he had never heard of a doubly periodic function or of Cauchy's theorem, and had, indeed, but the vaguest idea of what a function of a complex variable was. His ideas as to what constituted a mathematical proof were of the most shadowy description. All his results, new or old, right or wrong, had been arrived at by a process of mingled argument, intuition, and induction, of which he was unable to give any coherent account.

Ramanujan himself insisted that his insights were revealed to him by the family deity, the Hindu Goddess of creativity Namagiri of Namakkal (Kanigel 1991). Hardy doubted that Ramanujan was really serious about this assertion and claimed "that all mathematicians think, at bottom, in the same kind of way, and that Ramanujan was no exception" (Hardy 1921, p. xxviii). Clearly there was a cultural clash between the worldviews of Hardy and Ramanujan. The overemphasis of logical reasoning may have prevented Hardy from recognizing Ramanujan's ways to intuitively see mathematical structures. Considering the many inexplicable discoveries of Ramanujan it is reasonable to admit that (Rajagopalan et al. 1993, p. 99)

> Ramanujan is the best authority to say how he got his intuitions on mathematics, and if he says that it is due to the goddess of Namakkal, it should be accepted.

For nearly three decades of research, Bruce Berndt edited Ramanujan's notebooks (Berndt 1985, 1989, 1991, 1994, 1998; Andrews and Berndt 2005, 2008) and proved each of its 3542 theorems, mainly by methods unknown to Ramanujan. Berndt (1989, p. 5, 1991, p. 1) concluded:

> Ramanujan left us no clues of how he discovered these fascinating theorems. Are these results prototypes for further yet undiscovered theorems? Although we have given proofs, we do not have a firm understanding of how these wonderful theorems fit with the rest of mathematics.
>
> Although we provide proofs here for all of Ramanujan's formulas, in many cases, we have been unable to find the roads that led Ramanujan to his discoveries. It is hoped that others will attempt to discover the pathways that Ramanujan took on his journey through the luxuriant labyrinthine forest of enchanting and alluring formulas.

5.10.3 The Intuition of Kurt Gödel

Both Ramanujan and Gödel where Platonists, but while Ramanujan told us that his ideas came from the goddess Namagiri without any formal proof, Gödel believed in *a priori* truth—as is evident from his unpublished essays and lectures (Gödel 1995). He explained his concept of mathematical intuition as follows (Gödel 1964, pp. 271f):

> It should be noted that mathematical intuition need not be conceived of as a faculty giving an *immediate* knowledge of the objects concerned. Rather it seems that, as in the case of physical experience, we *form* our ideas also of those objects on the basis of something else which is immediately given. ... [the "given" underlying mathematics] may represent an aspect of objective reality, but, as opposed to the sensations, their presence may be due to another kind of relationship between ourselves and reality.
>
> But, despite their remoteness from sense experience, we do have something like a perception of the objects of set theory, as is seen from the fact that the axioms force themselves

> upon us as being true. I don't see any reason why we should have less confidence in this kind of perception, i.e. in mathematical intuition, than in sense-perception.

According to Wang (1987, p. 304), Gödel said:

> Intuition is not proof; it is the opposite of proof. We do not analyze intuition to see a proof but by intuition we see something without a proof.

In contrast to Gödel, other mathematicians such as Kuroda or Gleason denied that mathematical intuition is indubitable or gives *a priori* knowledge. According to Morris Kline (1980, p. 317):

> If intuition is master and logic the servant, the servant has some power over the master. It restricts unbridled intuition. Though intuition does play the major role, it can lead to assertions that are far too general. The proper limiting conditions are imposed by logic. Intuition throws caution to the winds but logic teaches restraint.

In view of all these examples, I assume the following characterization:

Intuition is an immediate insight that can neither be acquired by inference or observation, nor by reason or experience.

Chapter 6
Bottom-Up Approaches in Physics

6.1 Atomistic Physics

The atomism of Leucippus and Democritus has been widely regarded as the metaphysical background to classical physics. Paradigmatic for the classical bottom-up approach is the *atomistic reductionist doctrine* which holds that the physical universe is composed of separate minute indivisible ultimate building blocks. It is well characterized by Isaac Newton's (1730, Quest. 31) *dictum* that

> God in the Beginning form'd Matter in solid, massy, hard, impenetrable moveable Particles.

The reductive method is an approach to understanding the nature of complex compound systems by reducing them to simpler parts and their interactions. In contemporary science atomism is the programmatic preconception that there exist theory-independent objects which are constitutive of everything else. It adheres to Hegel's (1969, p. 167) doctrine that "the whole consists of parts". In 1906, Bertrand Russell formulated the atomistic program with the following words (Russell 1994, p. 146):

> In this world, whatever is complex is composed of related simple things, and analysis is no longer confronted at every step by an endless regress.

Accordingly, all reality is supposed to be *made of* basic building blocks and all changes are due to interactions between these fundamental units. This atomistic ontology has been most influential in the development of all sciences.

The traditional physical worldview of Newtonian mechanics, quantum mechanics and Lagrangian quantum field theory results from attempts to understand how nature works bottom-up, from lower to higher levels of organization. The dynamics of elementary building blocks at the lowest level is governed by a small set of laws which in principle determine everything we can calculate about the universe.

In this approach, parts are *isolated deliberately* from their environment, and wholes are described as assemblages of these contrived parts. Such *bottom-up approaches* are reductive and take for granted that moving from the basic constituents of the universe to higher levels of organization is the proper way of understanding complex systems. Such complex systems are often addressed in terms of a hierarchical structure. In Herbert Simon's (1962) words, a hierarchic system is

H. Primas, *Knowledge and Time*,
DOI 10.1007/978-3-319-47370-3_6

> composed of interrelated subsystems, each of the latter being, in turn, hierarchic in structure until we reach some lowest level of elementary subsystem. In most systems in nature, it is somewhat arbitrary as to where we leave off the partitioning, and what subsystems we take as elementary.

Within these approaches a law is considered as *fundamental* if it cannot be reduced to any lower-level law. Also, physical atomism often assumes that elementary building blocks are ontologically prior to the systems they constitute, and that determination operates exclusively bottom-up: forces generated at a lower level determine what happens at higher levels. However, as Philip Warren Anderson (1972, p. 393) emphasized,

> the reductionist hypothesis does not by any means imply a "constructionist" one: The ability to reduce everything to simple fundamental laws does not imply the ability to start from those laws and reconstruct the universe.

The atomistic doctrine is diametrically opposed to the insights of modern quantum theory. We cannot take for granted that the world is primordially divided into subsystems, that "there are things in themselves"—a position still defended by Bunge (1973, p. 86). According to quantum mechanics, material reality is neither composed of any kind of context-independent building blocks nor are there objects in the sense of an absolute ontology. In quantum field theory, "particles" are small-amplitude excitations of a quantum field. Therefore it is illegitimate to say that the world is made of molecules, atoms, electrons or quarks. Rather, a description based on such theoretical concepts may, in a particular context, be useful or even the best possible one. Matter, as described by the first principles of quantum theory, resembles matter in the Aristotelian sense: It is not a substance, but the capacity to form patterns.

The fact that material reality is not made of elementary systems does, however, not imply that in a well-specified context one cannot introduce *context-dependent building blocks of physical descriptions*. For instance, in molecular chemistry it is certainly reasonable and appropriate to consider nuclei, electrons and the electromagnetic field as basic entities.

6.2 Bottom-Up Quantum Mechanics

6.2.1 Quantum Mechanics Is a Non-Boolean Theory

All classical physical theories have a Boolean logical structure and fulfill the principle of *the mutually independent existence of spatially distant things*.[74] By contrast, quantum mechanics teaches us that the material universe is an inseparable whole

[74]The so-called *separability principle* claims that any two spatially separated things always possess distinct individual states, and that these states completely determine the state of the compound system.

with a non-Boolean logical structure, so that we are compelled to adopt a holistic worldview. As opposed to the classical worldview where all correlations between spacelike separated events can be explained in a local and separable way, the quantum world cannot be decomposed into pieces without loss of information. We are so accustomed to the classical separability of nature as self-evident that we easily forget how artificial this doctrine really is. The non-separability of the material world as predicted by quantum mechanics has been verified in recent decades in numerous beautiful experiments beyond any reasonable doubt. As a consequence, quantum theory has established the priority of inseparable wholes over parts. Therefore, physical atomism and mechanistic reduction cannot anymore be the basis of an acceptable worldview today.

Quantum mechanics is a non-Boolean generalization of classical mechanics, so its underlying logic is fundamentally different from that of classical physics. Quantum theory has produced some of the most accurately confirmed scientific predictions and is the best corroborated theory in the history of science. No noteworthy discrepancies between predictions and experiments are known. Yet, just accepting the logical consistency and the empirical correctness of quantum theory does not imply a true *understanding* of quantum phenomena.

While classical physics is based on propositions that are either true or false, the non-Boolean logic of quantum physics leads to new phenomena such as complementarity, nonlocality and entanglement. Still today many physicists try to understand quantum theory in classical terms and then refer to its non-classical effects as "quantum weirdness". Most popular accounts are misleading and claim that quantum physics is crazy, utterly counter-intuitive and incomprehensible. For non-technical writers the most vexing problem seems to be an explanation of coherent superpositions of quantum states which evades any classical understanding. Even high-impact-factor journals like *Nature* ("the authoritative source for research in science") and *Science* ("the world's leading outlet for scientific news") do not shy away from plenty of silly chatter about the so-called "weird nature of entangled quantum states".[75]

Since our intuition is trained by everyday events and empirical experience, even classical point mechanics is counter-intuitive for someone with no physics background. It is not surprising that scientists trained in classical physics will try to use intuitions from classical physics which are fundamentally undermined by quantum theory. But even more astonishing is the fact that after many decades of most successful applications of quantum theory so many physicists still have a great deal of difficulty understanding the worldview that quantum mechanics implies, entailing

[75]Here are some examples of hair-raising nonsense about "quantum weirdness": "Can a cat be simultaneously dead and alive? ... According to quantum physics ... the cat is 'dead-and-alive' " (*Science* **312**, p. 63, 2006); "In the quantum world the cat is both dead and alive at the same time" (*Nature* **448**, p. xi and p. 784, 2007); "Everyone finds quantum mechanics counterintuitive" (*Nature* **450**, p. 1167, 2007); "Quantum mechanics is weird, there is no doubt about that" (*Nature* **460**, p. 464, 2009); "It is possible for the same particle to be doing two contradictory things simultaneously" (*Nature News*, 17 March 2010).

that quantum theory is in disrepute to be unintuitive. The physics Nobel Laureate Leon Max Lederman (1993, p. 157) even questioned "whether the human brain will ever be prepared for the mysteries of quantum physics". And the physics Nobel Laureate Richard Feynman (1965, p. 129) claimed:

> I think I can safely say that nobody understands quantum mechanics. Nobody knows how it can be like that.

Twenty years later, Feynman (1985, p. 10) wrote:

> The theory of quantum electrodynamics describes Nature as absurd from the point of view of common sense. And it agrees fully with experiment. So I hope you can accept Nature as She is—Absurd.

Feynman does not claim that the formal structure of quantum mechanics is contradictory or inconsistent. He looks for pragmatic, instrumental guidelines for controlling nature, which provide answers to questions of the form (Feynman 1998, p. 17): "If I do this, what will happen?" We may perfectly be at home with quantum mechanics "for performing calculations, but not yet for viewing our world, and our relation to it" (Costa de Beauregard 1983, p. 516).

In particular, features such as nonlocality, non-separability and the possibility of entanglement may appear non-intuitive. Of course, quantum physics appears non-intuitive if we stick to the preconceptions of an obsolete atomistic-mechanistic worldview. For instance, for Henry Margenau (1950, p. 327), "*intuitable* means conceivable or representable in terms of mechanical models", and according to James Cushing (1991, p. 351) "understanding for us is possible for physical processes only when we can tell a causal picture story of those processes". The situation may be similar to our appreciation of modern art. Here is an instructive quote by Gertrude Stein (1938, p. 16) about Picasso:

> A picture may seem extraordinarily strange to you and after some time not only does it not seem strange but it is impossible to find what there was in it that was strange.

A genuine grasp of quantum theory requires an appropriate non-classical conceptual scheme. The primary cause of the apparently strange features of quantum theory is well known for long: *spatially separated and non-interacting parts of the material world are in general holistically correlated.*[76] The query of "*how* correlations of the Bohm-EPR type are produced in spatially separated systems" is not a riddle (as maintained by Cushing 1991, p. 348), but just an ill-posed question: *quantum correlations are not "produced", they are an intrinsic feature of the quantum world*—as perfectly confirmed by many experiments. Correspondingly, a Boolean description is inappropriate for material reality, but this does not imply that a non-Boolean quantum-theoretical description is incomprehensible.

A non-Boolean description of matter yields a natural and convincing explanatory framework for quantum phenomena. It stresses the genuine non-separability of the

[76]More technically: In the framework of traditional Hilbert-space quantum mechanics the set of all pure product states with respect to any decomposition is of measure zero. That is, almost all pure states of composite systems are holistically entangled.

world and tells us why context-independent isolated objects do not make sense except in some approximation. Moreover, it explains the probabilistic nature of experimentally accessible quantum features by the necessity of projecting deterministic non-Boolean processes onto an appropriate Boolean context. On the other hand, this Boolean restriction of the domain of discourse can lead to the appearance of novel properties and new phenomena (for details see Primas 1977, 1981, 1998).

6.2.2 Building Blocks in Bottom-Up Quantum Mechanics

Quantum theory dissolves the historical view based on the atomistic idea that the ultimate nature of reality can be characterized by basic building blocks (also called "atoms" or "elementary particles"). It was Wigner (1939) who first introduced the modern point of view that elementary systems have to be characterized group theoretically. Nowadays, elementarity is always defined with respect to some group. In W^*-algebraic quantum theory (cf. Sect. 6.3.1), an elementary system is defined via an ergodic action of a kinematical group on the algebra of observables.[77]

The "reality" of an object is philosophically addressed by the question of its *ontological* status. It is the question of whether and how it *exists* disregarding any epistemic access to it, e.g. in terms of its observation or description. The "objectivity" of a description does not reside in entities like molecules, atoms, electrons or quarks—these are just manifestations of material reality. The fact that modern quantum theory successfully describes many aspects of the behavior of matter in terms of elementary systems and their interactions does not imply that matter *is composed* of such elementary systems. The context-dependence of objects and the fact that they are entangled with their environment does not mean that realism is refuted by quantum mechanics, but it means that the idea of atomistic context-independent objects is at stake.

As a nonlocal theory, quantum theory does not deal with isolated objects, but with relations and processes. It is important to see that we cannot retain the classical world of objects simply by adding their interconnectedness as a supplementary property. The central problem of understanding quantum theory is not a problem of physics. Rather, we must abandon the prejudices of classical physics and *develop an intuition for non-Boolean ways of thinking*. For instance, we may train our intuitions for non-Boolean thinking by studying the results of modern experiments and the logical structure of quantum theory.

---

[77]A physical system is called *elementary* with respect to a kinematical group $\mathfrak{G}$ if the action of the group is *ergodic*. An action α of a group $\mathfrak{G}$ into the automorphisms group $\mathrm{Aut}\{\mathcal{M}\}$ of a W*-algebra $\mathcal{M}$ is called *ergodic* if $\alpha_{\mathfrak{g}}(M) = M$ for all $\mathfrak{g} \in \mathfrak{G}$ and $M \in \mathcal{M}$ implies that M is a multiple of the identity operator. In the traditional irreducible Hilbert-space representation of all bounded operators $\mathcal{M} = \mathcal{B}(\mathcal{H})$ acting on a Hilbert space $\mathcal{H}$ this implies that a system is elementary if and only if the implementation of the automorphism $\mathfrak{g} \to \alpha_{\mathfrak{g}}$ by a projective unitary representation acts *irreducibly* on the Hilbert space $\mathcal{H}$.

The fact that the *description* of multitudinous aspects of the behavior of matter in terms of elementary systems has been triumphantly successful does not imply that it is comprehensive, and certainly not that matter is *made of* building blocks. In spite of the fact that atomism has been a very successful *mode of description* for a long time, it is not a law of nature, but contradicts the holism suggested by quantum physics, where the notion of ultimate constituents loses its meaning. The so-called "elementary" entities (such as electrons or quarks) stand for *patterns of reality*, not for fundamental building blocks of reality. They are not primordial, but rather secondary and derived, in the same sense as solitons are.[78] A key reason why traditional reductionism fails is that *the presumed lower-level entities do not exist in a theory-independent sense.*

Nevertheless, *for some purposes* it can be convenient, useful and legitimate to introduce theoretically well-defined *fictional* context-dependent building blocks. This amounts to adopting a *contextual ontology* that is legitimate even if we reject an absolute ontology with elementary building blocks of reality. The artificial concept of an isolated system remains a vital element in modern physics since it may give rise to useful mathematical idealizations for natural phenomena. But the often associated metaphysical commitment of an absolute ontology becomes inacceptable. Hilary Putnam proposed to replace it by a *conceptual relativity*: the question of what objects the world consists of only makes sense relative to a theoretical framework (Putnam 1981, pp. 49f):

> I shall refer to it as the *internalist perspective*, because it is characteristic of this view to hold that *what objects does the world consist of?* is a question that it only makes sense to ask *within* a theory or description. Many "internalist" philosophers, though not all, hold further that there is more than one "true" theory or description of the world. "Truth", in an internalist view, is some sort of (idealized) rational acceptability—some sort of ideal coherence of our beliefs with each other and with our experiences *as those are themselves represented experiences in our belief system*—and not correspondence with mind-independent or discourse-independent "states of affairs". There is no God's Eye point of view that we can know or usefully imagine; there are only the various points of view of actual persons reflecting various interests and purposes that their descriptions and theories subserve.

6.2.3 Traditional Hilbert-Space Quantum Mechanics

The first landmark monographs on quantum mechanics were Hermann Weyl's (1928a) profound *Theory of Groups and Quantum Mechanics* and Paul Adrien Maurice Dirac's (1930a) brilliant and influential *Principles of Quantum Mechanics*. The first complete mathematical formulation of quantum mechanics for systems with finitely many degrees of freedom was von Neumann's (1932a) pioneering book

[78]Solitons (discovered by Scott Russell 1844) are localized steady two-dimensional waves of elevation that maintain their shape while propagating at the surface over a flat bed. They are important in our understanding of water waves.

Mathematical Foundations of Quantum Mechanics, in which he introduced the language of Hilbert spaces and the spectral theory of normal operators.

In subsequent years, both Dirac's bra-ket formalism and von Neumann's Hilbert-space theory of the early 1930s have often been adopted in an uncritical manner. Von Neumann's *ad hoc* requirement that to each observable there corresponds a self-adjoint operator in the Hilbert space of state vectors implies that the algebra of observables is *irreducible*. This means that there are no non-trivial observables which commute with all observables within the algebra, so that this formalism does not include *classical properties of quantum systems*. Furthermore, the irreducibility postulate implies that every unit ray represents a possible physical state, which is not true in general (cf. Streater and Wightman 1964, Sect. 1-1).

The irreducible Hilbert-space representation of traditional quantum mechanics is a *statistical* description and starts with a separable Hilbert space $\mathcal{H}$. The associated algebra $\mathcal{B}(\mathcal{H})$ of all bounded linear operators acting on $\mathcal{H}$ plays the role of the algebra of bounded contextual observables. The states of traditional quantum mechanics refer to the statistical outcomes of experiments. They enjoy the property of σ-additivity and are represented by normalized positive elements of the predual $\mathcal{B}(\mathcal{H})_*$ of the algebra $\mathcal{B}(\mathcal{H})$ of observables. The predual $\mathcal{B}(\mathcal{H})_*$ is isomorphic to the Banach space $\mathcal{B}^1(\mathcal{H})$ of all trace-class operators on $\mathcal{H}$, $\mathcal{B}^1(\mathcal{H}) \cong \mathcal{B}(\mathcal{H})_*$. Therefore, a statistical state ρ of traditional quantum mechanics can be represented by a positive trace-class operator D_ρ of unit trace,

$$\rho \longleftrightarrow D_\rho, \quad D_\rho \in \mathcal{B}^1(\mathcal{H}), \quad D_\rho \geq 0, \quad \mathrm{tr}(D_\rho) = 1, \tag{6.1a}$$

$$\rho(A) = \mathrm{tr}(D_\rho A) \quad \text{for every } A \in \mathcal{B}(\mathcal{H}). \tag{6.1b}$$

The operator D_ρ representing the statistical state ρ is called *density operator* (or, sometimes, *statistical operator*). Pure states are represented by idempotent density operators:

$$\rho \text{ is a pure state functional if and only if } D_\rho = (D_\rho)^2. \tag{6.2}$$

6.2.4 Superselection Rules

John von Neumann (1932a) proposed to formulate quantum mechanics in terms of an *irreducible* representation of the canonical commutation relations on a Hilbert space. The idea that irreducibility has a physical meaning is a popular misconception until today. Since irreducibility is a representation-dependent concept, its physical relevance is restricted. This can be seen most clearly in a representation-independent algebraic formulation: von Neumann's irreducibility postulate implies that there are no non-trivial observables which commute with all observables of the system, i.e. in von Neumman's codification the center of the algebra of observables is trivial. In other words, irreducibility implies that in principle there are coherent superpositions for all quantum states: the quantum-mechanical superposition principle in the traditional Hilbert-space representation (see Varadarajan 1968, p. 117).

At the *International Conference on Nuclear Physics and the Physics of Fundamental Particles* at Chicago in 1951, Eugene Wigner shocked the conference by rejecting von Neumann's irreducibility postulate. He proposed rules prohibiting the preparation of quantum states that exhibit coherence between eigenstates of certain observables (cf. Wightman 1995). Subsequently, Wick et al. (1952) introduced so-called *superselection rules* for electric charge, beyond the traditional selection rules.

Later, Wightman (1959) proposed commutative superselection rules to the effect that all observables commute with the electric charge. He introduced the concept of a *superselection rule* between states of different charge so that the Hilbert space decomposes into a direct sum of subspaces ("coherent sectors") labeled by a charge quantum number. Linear combinations of vectors from different sectors do not represent a physical state. The empirically well-established univalence, charge, baryon and symmetrization superselection rules showed that a generalization of von Neumann's formulation is necessary.

Josef Maria Jauch (1960) proved hat the W*-algebra of observables is irreducible if and only if there are no superselection rules. In the framework of the W*-algebraic formulation of quantum mechanics, Jauch and Misra (1961) also showed that commutative superselection rules are always connected with the occurrence of classical observables commuting with *all* observables of the system.

The simplest case of a discrete superselection rule can be represented by a N-fold direct sum of orthogonal subspaces $\mathcal{H}_k$,

$$\mathcal{H} = \mathcal{H}_1 \oplus \mathcal{H}_2 \cdots \oplus \mathcal{H}_N, \tag{6.3}$$

such that under the action of any observable a vector in $\mathcal{H}_k$ is transformed into a vector in $\mathcal{H}_k$. This expresses that the superposition principle holds within every subspace $\mathcal{H}_k$. Such subspaces are called *coherent subspaces* or *sectors*. A superselection rule for two subspaces $\mathcal{H}_k$ and $\mathcal{H}_\ell$ $(k \neq \ell)$ means that they cannot be connected to each other by any observable M,

$$\langle \Psi_k | M | \Psi_\ell \rangle = 0, \quad k \neq \ell, \quad \Psi_j \in \mathcal{H}_k. \tag{6.4}$$

States which are separated by a superselection rule are called *disjoint*. So we can state that *the existence of a superselection rule prohibits the superposition of disjoint states.*

The observables of a quantum system with superselection rules are the self-adjoint operators acting on $\mathcal{H}_k$. The projection operator P_k with the domain $\mathcal{H}$ and the kth subspace $\mathcal{H}_k$ as range commutes with all observables. The corresponding algebra $\mathcal{M}$ of observables,

$$\mathcal{M} = \mathcal{B}(\mathcal{H}_1) \oplus \mathcal{B}(\mathcal{H}_2) \cdots \oplus \mathcal{B}(\mathcal{H}_N), \tag{6.5}$$

is much smaller than the algebra of all bounded operators $\mathcal{B}(\mathcal{H})$ on the Hilbert space $\mathcal{H}$. Its non-trivial center $\mathcal{Z}(\mathcal{M})$,

$$\mathcal{Z}(\mathcal{M}) := \{Z | Z \in \mathcal{M}, ZM = MZ \text{ for every } M \in \mathcal{M}\}, \tag{6.6}$$

is generated by the projection operators $P_1,\ldots,P_N$,

$$\mathcal{Z}(\mathcal{M}) = \{Z|c_1 P_1 + \cdots c_N P_N, c_k \in \mathbb{C}\}. \tag{6.7}$$

The center $\mathcal{Z}(\mathcal{M})$ is a *commutative* W*-subalgebra. It is called trivial if it consists of multiples of the identity element only. A W*-algebra with a trivial center is called a *factor*. Every factor $\mathcal{A}$ can be realized as a W*-algebra $\mathcal{B}(\mathcal{H})$ of all bounded linear operators on some Hilbert space $\mathcal{H}$, $\mathcal{A} \cong \mathcal{B}(\mathcal{H})$. The self-adjoint elements of the center $\mathcal{Z}$ are called *classical observables*. They can be used to label the distinguishable components of the joint system.

A state functional which is dispersion-free with respect to every classical observable is called a *factor state*:

$$\begin{aligned}&\rho \text{ is a factor state functional on } \mathcal{M} \text{ if and only if}\\ &\rho\left(Z^2\right) = \left\{\rho(Z)\right\}^2 \text{ for every classical observable } Z \in \mathcal{Z}(\mathcal{M}).\end{aligned} \tag{6.8}$$

If two state functionals ρ' and ρ'' are disjoint, then there exists a classical observable $Z \in \mathcal{Z}(\mathcal{M})$ such that $\rho'(Z) \neq \rho''(Z)$. Hence disjoint states can be distinguished and classified in a classical manner by the numerical value of an appropriate classical observable.

A natural extension of the concept of a direct sum (6.3) of Hilbert spaces is a direct integral of mutually orthogonal sectors relative to the positive measure μ:[79]

$$\mathcal{H} := \int^{\oplus} \mathcal{H}_\nu d\mu(\nu). \tag{6.9}$$

If $\{P_\nu\}$ are projections corresponding to this direct-integral decomposition,

$$\mathcal{H}_\nu = P_\nu \mathcal{H}, \quad P_\nu P_\mu = \delta_{\nu\mu} P_\nu, \tag{6.10}$$

then every classical observable $Z \in \mathcal{B}(\mathcal{H})$ is of the form $Z = \int \lambda_\nu P_\nu d\mu(\nu)$ and commutes with every observable. Consequently, the corresponding algebra $\mathcal{M}$ of observables is given by the central decomposition

$$\mathcal{M} = \int^{\oplus} \mathcal{B}(\mathcal{H}_\nu) d\mu(\nu). \tag{6.11}$$

Its center $\mathcal{Z}(\mathcal{M})$

$$\mathcal{Z}(\mathcal{M}) = \int^{\oplus} P_\nu d\mu(\nu) \tag{6.12}$$

is a commutative algebra generated by the projections $\{P_\nu\}$ which describes the *classical part* of the quantum system. The subsystem with the algebra $\mathcal{Z}(\mathcal{M})$ is

[79]For details and proofs see for example Takesaki (1979, Chap. IV, Sect. 8).

also referred to as a *classical quantum system*. It is an important feature of classical quantum systems that they depend crucially on the value of Planck's constant.

The concepts of superselection sectors and the associated classical observables have become a most successful tool in molecular physics, in solid state physics, and in statistical physics.[80]

6.2.5 Non-Pure States Versus Classical Mixtures

Disjoint states are particularly important for mixtures in quantum mechanics. Bernard d'Espagnat (1965, 1966, 1976) emphasized over and over again how crucial it is to distinguish between Weyl's (1927, p. 12) concept of a (German) "Gemenge" and Landau's (1927) concept of a "reduced state". He proposed the term "proper mixture" for Weyl's "Gemenge", and the term "improper mixture" for a state represented by Landau's reduced density operator.

As all-important as this distinction is, the term "improper mixture" is also most unfortunate since by tracing out irrelevant degrees of freedom to obtain the reduced density operator, no "mixing" operation whatsoever is involved. We shall therefore avoid the expression of a "mixed state" for a reduced density operator and refer to it as a "non-pure state". *As a rule, non-pure quantum states cannot be interpreted as mixtures of factor states.*[81]

The mixtures we know in everyday life have a *unique* decomposition into components. If we mix water and alcohol, then we tacitly presuppose that we can distinguish operationally between the two components. It makes no sense to speak of a mixture if its components cannot be distinguished. Therefore a proper mixture is characterized by the fact that it is possible to label every component such that different components can be identified. Since such a label must be determinable together with any other property of the component, it must be given by a *classical observable*.

A classical mixture is an *ensemble* of N distinguishable components. Quantum theoretically it can be described by N *mutually disjoint factor states* with probabilities $p_1, \ldots, p_N$. In the Hilbert-space formalism such a classical mixture can be represented on the N-fold direct sum (6.3) by a density operator D with the unique *central decomposition into mutually disjoint* density operators $D_1, \ldots, D_N$:

$$D = p_1 D_1 \oplus p_2 D_2 \oplus \cdots \oplus p_N D_N, \qquad D_\nu \in \mathcal{B}(\mathcal{H}_\nu). \tag{6.13}$$

[80]Compare for example the reviews by Araki (1987) and by Wightman and Glance (1989), and the monographs by Strocchi (1985, 2008).

[81]In many elementary texts and even in handbooks (e.g. Münster 1959, p. 319) one finds the misleading claim that the referent of a thermal equilibrium state is a proper mixture of energy eigenstates. Fortunately, there are also competent texts, like that by Landau and Lifschitz (1980, p. 18), with appropriate *caveats*: "...it would be quite incorrect to suppose that the description by means of the density matrix signifies that the subsystem can be in various states with various probabilities and that the averaging is over these probabilities. Such a treatment would be in conflict with the basic principles of quantum mechanics."

The associated algebra $\mathcal{M} = \mathcal{B}(\mathcal{H}_1) \oplus \mathcal{B}(\mathcal{H}_2) \oplus \cdots \oplus \mathcal{B}(\mathcal{H}_N)$ has the non-trivial center $\mathcal{Z}(\mathcal{M})$ generated by the projection operators P_ν, defined by $\mathcal{H}_\nu = P_\nu \mathcal{H}$, which label distinguishable classical subsystems. The components of a (proper) classical mixture can then be identified via the labeling classical observable $Z := P_1 \oplus P_2 \oplus \cdots \oplus P_N$ by

$$Z D_\nu = \lambda_\nu D_\nu, \quad \nu = 1, 2, \ldots, N. \tag{6.14}$$

Therefore one can interpret the density operator (6.13) as the state of a classical system that is actually in one of the states D_ν. An observer does not know in which of these states the system actually is, he only knows their probabilities p_ν. So the classically mixed state expresses the observer's ignorance about the actual state.

It is a popular error in interpretations of quantum mechanics that an "improper" mixture can also be interpreted as an observer's ignorance about the state of the system in terms of a classical probability distribution. Therefore we emphasize that *only classical (proper) mixtures* allow an *ignorance interpretation*. A non-pure factor state never represents a proper mixture and ignorance arguments are illegitimate. Since in traditional quantum mechanics without superselection rules all states are factor states, there are no classical mixtures in this theory. *A non-pure quantum state can be interpreted as a classical mixture if and only if its components are disjoint states* (cf. also Primas 2000).

6.2.6 *Measurements of the First Kind*

In the framework of the old quantum theory, Bohr (1913) suggested that light and matter interact in such a way that an atom jumps instantaneously between stationary states and thereby absorbs or emits a quantum of light. During the further development of quantum mechanics, Dirac (1926a, p. 666) proposed to replace this quantum jump postulate by the following interpretation:

> We may regard an eigenfunction ψ_n as being associated with definite numerical values for some of the constants of integration of the system. Thus, if we find constants of integration $a, b, \ldots$ such that
>
> $$a\psi_n = a_n \psi_n, \qquad b\psi_n = b_n \psi_n, \qquad \ldots,$$
>
> where $a_n, b_n, \ldots$ are numerical constants, we can say that ψ_n represents a state of the system in which $a, b, \ldots$ have the numerical values $a_n, b_n, \ldots$

The question of whether "quantum jumps", later called "reductions of the wave function" or "collapses of the wave function", occur objectively in reality or are just a useful metaphor to describe the phenomena, has been a contentious subject of debate since the early days of Bohr's model of the hydrogen atom.

John von Neumann (1932a) formulated quantum mechanics as a statistical theory. He tried to unify the epistemic (sometimes referred to as subjective) aspects of the Copenhagen view with a theory of the entire physical universe, including not only measuring devices, but also human bodies and brains. Starting out from the (utterly unrealistic!) presupposition that the result of an experiment is a real number, von Neumann (1927, p. 271) introduced idealized measurements with the intuition that they yield the same numerical outcome when they are immediately repeated.

It should be evident that no single measurement gives a numerical value which can be attributed to a physical observable.[82] The requirement that experiments have to be reproducible does not refer to single experiments but to *equivalence classes* of experimental data. Therefore von Neumann's repeatability axiom refers to the Boolean classification of the *statistical* decision method necessary to arrive at "experimental facts", and not to the Boolean structure of the projection-valued measures associated with self-adjoint operators.

In spite of the fact that no laboratory measurement is instantaneous, or exactly repeatable, Pauli (1933, p. 152) called such operations *measurements of the first kind*. Unfortunately, in the jargon of quantum mechanics, measurements of the first kind are often simply called "measurements". This misuse of a well-established engineering terminology has led to many confusions in the discussion of the interpretation of quantum theory. *No laboratory measurement is of the first kind.*

In the framework of the irreducible Hilbert-space representation of traditional quantum mechanics, a maximally informative state can be represented by a normalized vector in a separable complex Hilbert space $\mathcal{H}$, called *state vector* $\varXi$. In von Neumann's schematic model, the time evolution $\varXi \to \varXi_\tau$ is assumed to be *deterministic, reversible* (invariant under time-reversal) and given by a *unitary group* $\{V(\tau)|\tau \in \mathbb{R}\}$,

$$\varXi_\tau = V(\tau)\varXi, \quad \varXi_\tau \in \mathcal{H},\ \tau \in \mathbb{R}. \tag{6.15}$$

In von Neumann's model, also the measuring apparatus is treated as a quantum mechanical system, so that the Hilbert space $\mathcal{H}$ is given by the tensor product of the Hilbert space $\mathcal{H}_{obj}$ of the object system and the Hilbert space $\mathcal{H}_{app}$ of the apparatus (together with the overall environment).

$$\mathcal{H} = \mathcal{H}_{\text{obj}} \otimes \mathcal{H}_{\text{app}}. \tag{6.16}$$

If the state of the universe of discourse is given by the state vector $\varXi$, this state vector provides an exhaustive description of the universe of discourse. With respect to the tensorization $\mathcal{H}_{\text{obj}} \otimes \mathcal{H}_{\text{app}}$, the state vector $\varXi \in \mathcal{H}_{\text{obj}} \otimes \mathcal{H}_{\text{app}}$ has a canonical Schmidt decomposition (6.58) into product states,

$$\varXi(\tau) = \sum_k c_k(\tau)\Psi_k(\tau) \otimes \Phi_k(\tau), \quad \Psi_k(\tau) \in \mathcal{H}_{\text{obj}},\ \Phi_k(\tau) \in \mathcal{H}_{\text{app}}. \tag{6.17}$$

In the simple case of a first-kind measurement of a self-adjoint operator A with a purely discrete non-degenerate spectrum, let the eigenvalues of A be $a_1, a_2, \ldots$. The

---

[82] In quantum mechanics, the experimentally relevant sample space is always uncountable, even for the simplest case of a two-dimensional Hilbert space. For a more realistic discussion of measurements on a spin-1/2 system compare Primas (1990b, pp. 53f).

corresponding eigenvectors $\alpha_k \in \mathcal{H}_{\text{app}}$ form an orthonormal basis for $\mathcal{H}_{\text{app}}$, so that we can represent $\Phi_k(\tau)$ as

$$\Phi_k(\tau) = \sum_n d_{k,n}(\tau)\alpha_n(\tau), \tag{6.18}$$

so that

$$\Xi(\tau) = \sum_k \sum_n c_k(\tau)d_{k,n}(\tau)\Psi_k(\tau) \otimes \alpha_k = \sum_k \kappa_k(\tau)\Psi_k(\tau) \otimes \alpha_k, \tag{6.19}$$

where

$$\kappa_k(\tau) := \sum_n c_k(\tau)d_{k,n}(\tau). \tag{6.20}$$

Von Neumann's (1927, Chap. VI) version of a "quantum jump" refers to a first-kind measurement of an observable A which induces an *instantaneous, indeterministic, and irreversible* transition from state Ξ to the state vector $\Psi_\ell \otimes \alpha_\ell$, leading to the directly observable value a_ℓ of the measuring apparatus:

$$\Xi \xrightarrow[\text{of the first kind}]{\text{measurement}} \Psi_\ell \otimes \alpha_\ell \quad \text{with probability } p_\ell = \left|\kappa_\ell(\tau)\right|^2. \tag{6.21}$$

The probabilities $p_1, p_2, \ldots$ are *conditional* with respect to the fact that the measuring apparatus has irreversibly registered one of the eigenvalues $a_1, a_2, \ldots$. Since the operation (6.21) gives the same result when it is immediately repeated, von Neumann associated it with a repeatable instantaneous measurement of the first kind. This version of a quantum jump is usually called von Neumann's *projection postulate*.

Tacitly referring to an individual description, it has been claimed that a measurement of the first kind transforms an arbitrary initial state vector Ξ with probability p_ℓ to the state vector $\Psi_\ell(\tau) \otimes \alpha_\ell(\tau)$,

$$\Xi(\tau) \longrightarrow \begin{cases} \Psi_1(\tau) \otimes \alpha_1(\tau) & \text{with probability}\, p_1, \\ \Psi_2(\tau) \otimes \alpha_2(\tau) & \text{with probability}\, p_2, \\ \cdots \end{cases} \tag{6.22}$$

The operation (6.22) has been denoted as the *collapse postulate*, or the *reduction of the wave packet*.[83] Since the state vector is projected onto an eigenspace of the measured observable, the map $\alpha \to \alpha_j$ is also referred to as the *projection postulate*. The "reduction" (or "collapse") is not a physical process but an auxiliary tool of a reduced description which does not include the measuring tools. In the words of Pauli (1933 p. 148, editor's translation):[84]

[83] This way of speaking was introduced by Heisenberg in a letter to Pauli of February 23, 1927 (see Hermann et al. 1979, p. 379).

[84] Note that the English translation in Pauli (1980, p. 71) is misleading since it misses the essential point of the German phrase "Setzung einer physikalischen Tatsache".

> Positing a physical fact [die Setzung einer physikalischen Tatsache] by means of measuring tools (observer or registration apparatus) regarded as separate from the system is, from the viewpoint of the mathematical formalism which directly describes only possibilities (probabilities), a particular, indeterminate act that is unpredictable by natural laws and must be accounted for retrospectively by the reduction of wave packets.

If the vectors $\Psi_1(\tau)\otimes\alpha_1(\tau)$, $\Psi_2(\tau)\otimes\alpha_2(\tau)\ldots$ would be mutually disjoint (and not only orthogonal, cf. Appendix A.4) the projection postulate would imply that the expectation value of the observable A be given by

$$\mathcal{E}(A)=\sum_k p_k a_k = \langle \Xi | A\otimes \mathbb{1} | \Xi\rangle, \tag{6.23}$$

where Ξ is the state vector of the universe of discourse immediately before the experiment begins. However, because in traditional quantum mechanics there are no disjoint states, this formulation of the measuring process is ill-defined.

Moreover, the operation (6.21) is a *nonlinear* and *stochastic* map which transforms an arbitrary state vector of the universe of discourse into one of the product states $\Psi_\ell\otimes\alpha_\ell$, where α_ℓ is an eigenvector of the observable A. Since this operation cannot be derived from the linear Schrödinger equation, "the 'reduction of the wave packet' enters quantum mechanics as a *deus ex machina*, without any relation to the other laws of this theory" (Wigner 1964a, p. 251). These problems have led to a long controversy, lasting for over half a century. They are the main topics of the alleged "measurement problem of quantum mechanics".[85]

The instantaneous character of quantum jumps or reductions of the wave function is unphysical and leads to deep difficulties. It precludes a dynamical explanation of the irreversibility of the outcomes of every measurement: Measured facts cannot be undone. And a realistic model for measurements has to take into account that every measurement is extended over a non-zero time interval: Measurement takes time.

Moreover it is hard to see how Wigner's (1964a, p. 250) claim that a "measurement is not completed until its result enters our consciousness" can be squared with empirical facts.[86] It is reasonable to assume that any measurement is a dynami-

---

[85] Speculations about direct relations of quantum mechanics with human free will and human consciousness, and that the consciousness of an observer causes the wave function to collapse go back to von Neumann (1932a) and to London and Bauer (1939). Wigner (1962, p. 297) claimed that the reduction of the wave packet in the quantum-mechanical measuring process occurs through an act of consciousness on the part of a human observer at the interface between mind and brain (for a recent critical overview see Atmanspacher 2015). This speculation is without any empirical support. Quite to the contrary, there are good reasons to assume that this world of ours existed long before there were any beings with consciousness. In cosmological or biological evolution there are objective occurrences, encodings and registrations which are independent of the existence of beings with consciousness.

[86] Consider as an example the NASA spacecraft mission Pathfinder that landed a base station with a roving probe called Sojourner on Mars in 1997. This robotic Mars rover *automatically* investigated nearby rocks with its alpha proton X-ray spectrometer which used a radioisotope source to bombard the rock surface with alpha particles, and then took an energy spectrum of the X-rays, protons and

cal process which is governed by the deterministic evolution laws of quantum mechanics whereby the consciousness of a human observer plays no role whatsoever. Wigner's (1952, p. 102) objection that quantum theory gives no rules for how a measurement of an observable can be performed disagrees with engineering physics. It is true that there is no universal relationship between observables and measuring instruments, but in simple cases (like in magnetic resonance experiments) quantum mechanics gives the full information necessary for the *construction* of an appropriate measurement instrument.

6.2.7 The Expectation-Value Postulate

The projection postulate is restricted to observables with a purely discrete spectrum. A less restrictive description of the results of measurements is based on the so-called expectation-value postulate. Other than the projection postulate, the expectation-value postulate applies to observables with an arbitrary spectrum but says nothing about the state after the measurement.[87] Nevertheless, the expectation-value postulate is often sufficient for a comparison of quantum-mechanical predictions with experimental results.

The Expectation-Value Postulate for First-Kind Measurements
Let A be a self-adjoint observable with arbitrary spectrum Λ, let Σ be the σ-algebra of Borel sets of Λ and let E_A be the spectral measure of A. The conditional probability that a predictive measurement of A gives a value in the Borel set $\mathcal{B} \in \Sigma$ is given by the Kolmogorov probability measure $\mu : \Sigma \to \mathbb{R}^+$ defined by $\mu(\mathcal{B}) = \operatorname{tr}\{DE(\mathcal{B})\}$, where D is the density operator describing the initial (pure or non-pure) state of the object system immediately before the measurement.

Conditioned by a measurement of the first kind, the expectation-value postulate implies that the observable A is a random observable on the Kolmogorov probability space (Ω, Σ, μ). The expectation value $\mathcal{E}\{f(A)\}$ of an arbitrary measurable function f of this random variable is given by

$$\mathcal{E}\{f(A)\} = \int_\Omega f(\omega)\mu(d\omega). \tag{6.24}$$

---

alpha particles that return. The robot *automatically* collected and coded the data, and sent them to NASA's control center in Pasadena, where they were stored in the usual way computer data are stored. This data can be *automatically* duplicated arbitrarily often and sent to many scientists. Up to this point, no human observer is involved. According to well-established engineering wisdom, the crucial measurements (perhaps not the further data processing) are completed in the spectrometer on the Sojourner and stored in its local computer memory.

[87]The expectation-value postulate was first formulated in a slightly weaker form by Pauli (1927, footnote 1 on p. 83). Note that the expectation-value postulate gives the correct conditional probabilities but does not identify their referents.

All measurements of the first kind can be expressed in terms of the expectation-value postulate. *The projection postulate is not needed for the statistical description of an experiment.*[88]

The discussion of measurements of the first kind is based on the dogma that physical observables need to be valuated by point functions, and not—as both mathematical formalism and experimental practice strongly suggest—by set functions. Although the projection postulate is compatible with the expectation-value postulate, it is a much too strong requirement. In contrast to the projection postulate, the expectation-value postulate says nothing about the state after the measurement. Nevertheless, the expectation-value postulate is often sufficient for a comparison of quantum-mechanical predictions with experimental results. Other than the projection postulate, the expectation-value postulate is empirically well-confirmed.

6.2.8 Operator-Valued Measures

Von Neumann's theory was based on the problematic concept of a measurement of the first kind which assumes that after a measurement the system is in an eigenstate of the measured observable. It is doubtful whether such measurements can be realized experimentally. Moreover, since two non-commuting observables with discrete and non-degenerate spectra cannot have a common eigenstate, it has been claimed that in general non-commuting observables cannot be measured simultaneously.

But this idealization is not necessary since a theory of realistic measurements is possible on the basis of the expectation-value postulate, which is not restricted to repeatable measurements of the first kind. The engineering theory of quantum measurements and quantum communication channels (compare, e.g., the work of Alexander Holevo) is highly developed and in a much better shape than many rather naive philosophical discussions of measurement as projection.

In this framework many questions like the problem of joint measurements of non-commuting observables (Arthurs and Kelly 1965; Gordon and Louisell 1966; She and Heffner 1966), the problem of observables with continuous spectra (Davies 1970, 1976; Davies and Lewis 1970; Ozawa 1984), the problem of sequential quantum measurements (Benioff 1972a, 1972b, 1972c, 1973) and the problem of the position measurement for photons (Jauch and Piron 1967; Amrein 1969; Neumann 1972; Ali and Emch 1974; Kraus 1977) have been solved since long. Their solution is not only based on an operational axiomatic foundation (Ludwig 1964, 1970, 1983, 1985a, 1985b, 1987) but also in the best engineering tradition in terms of optimal estimation and detection theory (Helstrom 1976; Holevo 1978, 1982).

Every laboratory experiment, in its final stage, needs to be framed in terms of a Boolean language. Therefore, any experimental investigation of a quantum system can be described by a *channel* which transfers information about the state of the

[88]This conclusion was often emphasized by Henry Margenau (1936, p. 242, 1958, p. 31, 1963a, p. 6, 1963b, p. 476).

quantum object system to an output system with a *Boolean* laboratory instrument. The outcome space of such a finite-precision statistical experiment is a measurable space (Ω, Σ), where Ω is a non-empty set of outcome events and Σ is a σ-algebra of subsets of Ω.

As introduced in Sect. 3.5.4, the most general statistical observable of a quantum system characterized by the observable algebra $\mathcal{B}(\mathcal{H})$ is an affine map $F : \Sigma \to \mathcal{M}$ from Borel sets $\mathcal{B} \in \Sigma$ onto *positive-operator-valued measures* (POVMs) of the observable algebra $\mathcal{B}(\mathcal{H})$ such that

$$\mathcal{B} \mapsto \rho\{F(\mathcal{B})\}, \quad \mathcal{B} \in \Sigma, \tag{6.25}$$

is a real-valued probability measure for all normal states $\rho \in \mathcal{B}(\mathcal{H})_*$. A device which associates to every normal state ρ a probability measure μ on a outcome space (Ω, Σ) can be realized by an *operator-valued probability measure* $F : \Sigma \to \mathcal{B}(\mathcal{H})$, defined as a POVM with the properties defined in Sect. 3.5.4. The map Λ,

$$\Lambda(\rho) := f_\rho := \frac{d\mu}{d\nu}, \quad \mu(\mathcal{B}) := \rho\{F(\mathcal{B})\}, \quad \mathcal{B} \in \Sigma, \tag{6.26}$$

defines a POV channel. Provided one can construct (in terms of hardware) such a channel, it can be used for gaining information about the state of the object system. The state of the classical output system of a POV channel $F : \Sigma \to \mathcal{B}(\mathcal{H})$ can be determined by the familiar methods of statistical decision theory. If it is given by the probability measure μ_{exp}, then the *expectation-value postulate* says that

$$\mu_{\mathrm{exp}}(\mathcal{B}) := \rho\{F(\mathcal{B})\}, \quad \mathcal{B} \in \Sigma. \tag{6.27}$$

The expectation-value postulate *presupposes* the existence of POV channels; it does not warrant the laboratory realization of such channels. It only gives the *conditional* probability for an event $\mathcal{B} \in \Sigma$ registered on a Boolean output device.

In general, the quantity measured in an experiment is not a self-adjoint operator but a POVM whose expectation value is a real-valued set function which can be determined by statistical experiments. A POV channel transfers information about the *state* of the object system, and not about "observables". Only in the rather special case where the probability-operator measure F is *projection-valued*, one can use F as a spectral measure for an observable. Every operator $A \in \mathcal{B}(\mathcal{H})$ commuting with this spectral measure F can be realized as

$$A = \int_\Omega \alpha(\omega) F(d\omega), \tag{6.28}$$

where $\alpha \in \mathbf{L}^\infty(\Omega, \Sigma, \nu)$. In this case, the expectation value of the operator A is given by

$$\mathcal{E}(A) = \int_\Omega \alpha(\omega)\mu_{\mathrm{exp}}(\omega), \tag{6.29}$$

so that $\mathcal{E}(A)$ equals the expectation value of the random variable α on the measurable space (Ω, Σ) with respect to the probability measure μ_{exp}.

It may be tempting, but is not admissible to assume that a communication channel provides information about "properties" (or "observables") of the object system. The statistical description of quantum systems refers to results of experiments as registered on laboratory instruments, not to properties of quantum systems.

6.3 Decomposition of a Universe of Discourse

6.3.1 Decompositions by Factorizations of W*-Algebras

Traditional Hilbert-space quantum mechanics has a simple rule to describe a system in terms of group-theoretically defined bare elementary systems (cf. Jauch 1968, Sects. 11-7 and 11-8). *The Hilbert space of a joint system decomposable into two different quantum systems with the Hilbert spaces $\mathcal{H}_1$ and $\mathcal{H}_2$ is given by the tensor product $\mathcal{H}_1 \otimes \mathcal{H}_2$.* The corresponding algebra of observables is given by the tensor product of $\mathcal{B}(\mathcal{H}_1)$ and $\mathcal{B}(\mathcal{H}_2)$ as the algebra of all bounded operators acting on the Hilbert space $\mathcal{H}_1 \otimes \mathcal{H}_2$,

$$\mathcal{B}(\mathcal{H}_1)\,\overline{\otimes}\,\mathcal{B}(\mathcal{H}_2) := \mathcal{B}(\mathcal{H}_1 \otimes \mathcal{H}_2). \tag{6.30}$$

Since the W*-algebras $\mathcal{B}(\mathcal{H}_k)$ $(k = 1, 2)$ are factors (i.e. algebras whose centers are scalar multiples of the identity), the decomposition (6.30) is called a *tensor product factorization*. Murray and von Neumann (1936, Chap. III) discovered that not all factors are given by an algebra of bounded operators on some Hilbert space. They defined a W*-algebraic factorization as follows: Let $\mathcal{M} := \mathcal{M}_1 \vee \mathcal{M}_2$ be the W*-algebra generated by the W*-algebras $\mathcal{M}_1$ and $\mathcal{M}_2$, i.e. the smallest W*-algebra containing both $\mathcal{M}_1$ and $\mathcal{M}_2$. Then $(\mathcal{M}_1, \mathcal{M}_2)$ is called a *factorization* of $\mathcal{M}$ if every element of $\mathcal{M}_1$ commutes with every element of $\mathcal{M}_2$, and every element of $\mathcal{M}_2$ commutes with every element of $\mathcal{M}_1$,

$$\mathcal{M}_1 \subset \mathcal{M}_2' \quad \text{and} \quad \mathcal{M}_2 \subset \mathcal{M}_1', \tag{6.31}$$

where $\mathcal{M}_k'$ the relative commutant of $\mathcal{M}_k$ in $\mathcal{M}$,

$$\mathcal{M}_k' := \{M \in \mathcal{M} \mid M M_k = M_k M,\, M_k \in \mathcal{M}_k\}. \tag{6.32}$$

Recall that a W*-subalgebra $\mathcal{M}_k \subset \mathcal{M}$ is called a *factor* if its relative center $\mathcal{M}_k \cap \mathcal{M}_k'$ is trivial (that is, if it consists of multiples of the identity). $\mathcal{M}_k$ is a factor if and only if $(\mathcal{M}_k, \mathcal{M}_k')$ is a factorization. The factors $\mathcal{M}_1$ and $\mathcal{M}_2$ of a factorization $(\mathcal{M}_1, \mathcal{M}_2)$ are called *coupled factors* if $\mathcal{M}_1 = \mathcal{M}_2'$, or equivalently if $\mathcal{M}_2 = \mathcal{M}_1'$. The canonical example for coupled factors is the tensor product factorization with $\mathcal{M}_1 \vee \mathcal{M}_2 = \mathcal{M}_1 \,\overline{\otimes}\, \mathcal{M}_2$, where the W*-tensor product is defined via the preduals.[89]

---

[89]A W*-algebra $\mathcal{M}$ is defined as a C*-algebra which is the dual of a Banach space called the predual $\mathcal{M}_*$, $(\mathcal{M}_*)^* = \mathcal{M}$. The W*-tensor product of two W*-algebras $\mathcal{M}$ and $\mathcal{N}$ is defined as the dual of the Banach space $\mathcal{M}_* \otimes \mathcal{N}_*$, so that $\mathcal{M}_1 \,\overline{\otimes}\, \mathcal{M}_2 := (\mathcal{M}_1 \,\overline{\otimes}\, \mathcal{M}_2)_*^*$. For details see Sakai (1971, p. 1 and p. 67), see also Appendix A.1.

The W*-tensor product is a factorization, but not every factorization is a tensor product factorization.

The conditions under which a W*-factor can be realized as the tensor product of its two subfactors have been considered by Takesaki (1958): *For a W*-factorization to be a W*-tensor product factorization it is necessary that the subsystems are statistically independent.*[90]

Statistical Independence of W*-Algebras (6.33)
Two W*-algebras $\mathcal{M}_1, \mathcal{M}_2 \in \mathcal{M}_1 \vee \mathcal{M}_2$ are called W*-*statistically independent* if for every pair of normal state functionals $\rho_1 \in (\mathcal{M}_1)_*$ and $\rho_2 \in (\mathcal{M}_2)_*$ there exists a normal state functional $\rho \in \mathcal{M}_1 \vee \mathcal{M}_2$ such that $\rho(M_1) = \rho_1(M_1)$ for every $M_1 \in \mathcal{M}_1$ and $\rho(M_2) = \rho_2(M_2)$ for every $M_2 \in \mathcal{M}_2$.

This means that the partial states ρ_1 and ρ_2 are uncoupled so that there are no correlations between the W*-algebras $\mathcal{M}_1, \mathcal{M}_2$, and that any two states which can be independently prescribed on the constituting algebras can be extended to a common uncorrelated state on the joint algebra.

When Is a Factorization a Tensor-Product Factorization? (6.34)
Let $(\mathcal{M}_1, \mathcal{M}_2)$ be a pair of commuting and W*-statistically independent W*-algebras $\mathcal{M}_1, \mathcal{M}_2 \in \mathcal{M}_1 \vee \mathcal{M}_2$. Then $\mathcal{M}_1 \vee \mathcal{M}_2$ is a tensor-product factorization $\mathcal{M}_1 \overline{\otimes} \mathcal{M}_2$ if and only if $\mathcal{M}_1$ and $\mathcal{M}_2$ are type-1 algebras.

6.3.2 Renormalization

The quantities appearing in our most fundamental theories are usually *bare particles* and *bare fields*—hypothetical group-theoretical constructs manifesting a high degree of symmetry without any direct operational meaning. The usual procedure to formulate a Hamiltonian in terms of bare elementary systems (like bare electrons and the bare electromagnetic field) is a provisional solution at best. Wolfgang Pauli (1956, p. 266) repeatedly denounced this procedure:[91]

> It troubles me that it is logically possible to write down Maxwell's equations in vacuum without introducing a charge; for if there were no charge, one could not measure the field. Yet one can mathematically write down the field without charge, and in this sense it seems to me that somehow the classical notion of a field must always remain sheer mathematics.

Since bare quantities have no independent existence, and since they are never directly observable, the fundamental theory needs to be rephrased in terms of more

---

[90]The notion of statistical independence is logically independent of the requirement that the algebras $\mathcal{M}_1$ and $\mathcal{M}_2$ commute (see Ekstein 1969; Napiórkowski 1972; Kruszynski and Napiórkowski 1973). The fact that W*-algebras are completed in the weak*-operator topology while C*-algebras are completed in the operator norm topology implies that the corresponding independence concepts are in general different. Often W*-independence implies C*-independence but not *vice versa*.

[91]Editor's translation; compare also Pauli (1921, p. 775), and his letter to Eddington of September 20, 1923, published in Hermann et al. (1979, p. 117, footnote).

directly accessible quantities—for instance, *dressed particles and dressed fields*. Such a reformulation is also known as *renormalization*.

Renormalization came into fashion in connection with divergence difficulties in quantum electrodynamics, but nevertheless—according to Källen (1962, p. 181)—

> the process of renormalization has nothing to do with the appearance of infinities in the theory. Rather, it comes because we want to replace the "mathematical" quantities in the theory by their physical counterparts, so that the predictions of the theory can be compared with experiments in a more direct way.

For instance, there have been many attempts to formulate statistical (classical or quantum) many-particle problems in such a way as to replace the bare propagators and the bare interaction potentials by more directly observable quantities such as charge densities or Green functions. The efforts toward such renormalized formulations were motivated by the search for more convergent expansions where the bare Coulomb interaction is replaced with an effective interaction.

This complex of ideas goes back to the classical virial expansions in terms of the one-particle density. In the context of quantum electrodynamics, Feynman and Schwinger discovered that the Green functions for fields interacting which each other can be expressed in terms of Green functions for uncoupled fields with an external source. Therefore, the elimination of the interaction potentials is possible in terms of time-dependent Green functions.[92]

Furthermore, Lee and Yang (1959) formulated the grand partition function in terms of the average occupation numbers in momentum space, and established that the partition function is stationary under variations of these one-body functions. Later de Dominicis and coworkers (de Dominicis 1962a, 1962b; de Dominicis and Englert 1967; de Dominicis and Martin 1964a, 1964b) showed that the thermostatic and the dynamical properties of an interacting quantum system with many degrees of freedom can be expressed solely in terms of renormalized propagators ("mass renormalization") and renormalized interactions ("charge renormalization"). These renormalization procedures amount to Legendre transformations from the potential in the Hamiltonian to the distribution functions.

6.3.3 Bare Elementary Systems and Dressing

It is one of the most fundamental preconceptions of all science that one does not have to consider the entire universe as a whole but that one can advance by its compartmentalization. However, quantum mechanics describes a whole, in fact a *whole which is not made up of parts*. While traditional quantum mechanics has the simple rule (6.30) to describe the composition of quantum systems, the inverse problem is much more delicate: *Given the algebra of a quantum universe of discourse, do*

---

[92]Compare the basic paper by Martin and Schwinger (1959) and the text by Kadanoff and Baym (1962).

we have reasonable criteria to find physically relevant subsystems? In the Hilbert-space formulation of quantum mechanics, we start with the Hilbert space $\mathcal{H}$ of the universe of discourse and look for a suitable tensorization $\mathcal{H} = \mathcal{H}_1 \otimes \mathcal{H}_2$, such that the Hilbert spaces $\mathcal{H}_1$ and $\mathcal{H}_2$ represent such physically relevant subsystems.[93]

In traditional quantum mechanics one usually starts with the tensorization induced by group-theoretically defined bare elementary systems. They are *defined* as ergodic representations of a presupposed kinematical group (like the Galilei or Lorentz group), disregarding interactions mediated by gauge fields. Since bare elementary systems are auxiliary constructs without any independent existence, this decomposition has no manifest operational meaning. Systems are called *bare* since they are idealized as objects separated from the associated gauge fields and without environment.

The standard discussion of correlation phenomena in terms of bare systems can be quite misleading. Since an electrically charged system is always surrounded by a cloud of soft photons, a description of charged systems as group-theoretically elementary systems is bound to lead to difficulties. A more reasonable way to discuss complex systems is to use a re-tensorization which reduces the correlations between the redefined subsystems. By self-interactions and interactions with the rest of the world bare elementary systems acquire a very complicated structure. This process is referred to as *dressing*.

There is no operationally relevant context-independent decomposition of the universe of discourse into subsystems. To find an appropriate tensor-product structure is essentially a problem of *pattern detection and recognition*. It amounts to find a transformation which changes the tensorization into bare elementary systems to a tensorization into operationally relevant subsystems, generated by *contextually dressed* objects including environmental effects and clouds of excitations.

For example, the state representing an electron as actually observed in the laboratory can be thought of as consisting of a virtual bare electron, dressed by a cloud of virtual bare photons and virtual bare electron-positron pairs. The presence of such a dynamically coupled cloud not only modifies the properties of the bare elementary system (like its mass) and its bare environment, but also its dynamics. The renormalization replaces the fictitious bare quantities by their empirically accessible counterparts, so that theoretical predictions can be compared with experimental results.

The concept of an elementary system dressed by environmental polarizations or a cloud of field excitations with which the elementary system itself is interacting is an old concept in physics. Well-known examples are the quasiparticle concepts of solid state physics (phonons, magnons, polarons, exitons; compare Pushkarov 1991), or the concept of a *composite fermion* in the theory of the fractional quantum Hall effect (see Jain 2000). Boson-like electron pairs dressed by their surrounding cloud of phonons, incorporating a large part of the electron-phonon interaction, are

[93] As early as 1960, Robert Rosen stated that "this problem seems to be completely ignored in the literature" (Rosen 1960). Rosen's perspective was a system theoretical outlook on biological genetics in a quantum framework.

fundamental in the theory of superconductivity (see McKenna and Blatt 1962). In quantum optics, the concept of a dressed atom is used in the sense of a compound object made of a bare atom and a cloud of photons which the atom continuously absorbs and re-emits (see Cohen-Tannoudji 1968). Only states dressed by the electromagnetic field are observable, therefore experimentally relevant transition probabilities have to be calculated for dressed states rather than bare states (see Heitler 1954, p. 15).

Example: A Classical Dressed System

The paradigmatic example of a dressed system is a *solvated ion*, a purely classical concept in the statistical theory of electrolytes by Debye and Hückel (1923), where each ion is considered as surrounded by a stationary atmosphere of ions of opposite charge. The interaction that takes place when an ion is introduced into a solvent is called the solvation of the ion. Viewing the ion as a point charge and the solvent molecules as electric dipoles, the main effect is that ions orient dipoles.

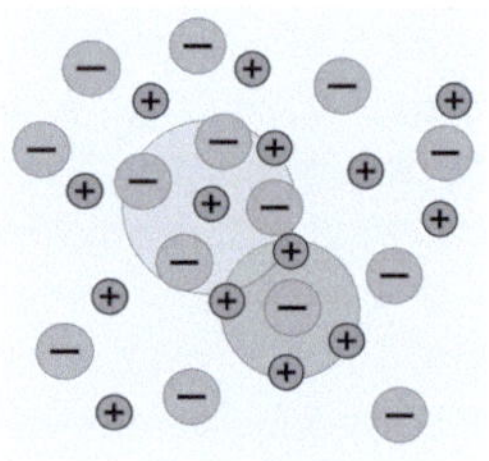

The ion-dipole forces trap and orient a certain number of solvent molecules in the immediate vicinity of the ion. Such solvated ions cease to associate with other water molecules. Due to the solvation, the Coulomb field of the bare ion is screened by the cloud of solvent molecules, so that the solvated ions interact only very weakly with the environment. Hence, at a sufficient distance from the ion the normal structure of the solvent is practically undisturbed.

When the solvated ion moves in an external electric field, the sheath of immobilized solvent molecules moves with the ion. To evaluate the *effective mass* m_{eff} of a solvated ion is a difficult matter since it depends on many parameters of the environment (like the viscosity and the temperature of the solution). The motion of solvated ions can be approximated by the phenomenological Stokes equation in which the dynamical entity is not the bare ion but the solvated ion. In other words, in an appropriate context, the solvated ion behaves like a well-isolated extended particle with an effective mass m_{eff}.

While the potential of a bare point charge q at distance r in a medium of dielectric constant D is given by

$$\phi(r) = \frac{q}{4\pi D\varepsilon_0}\frac{1}{r},$$

the mean potential of the *dressed ion* with charge q is, according to Debye and Hückel, given by

$$\phi(r) = \frac{q}{4\pi D\varepsilon_0}\frac{e^{-r/\lambda}}{r},$$

where λ is the Debye shielding length,

$$\lambda = \sqrt{\frac{D\varepsilon_0 kT}{\sum_k n_k a_k^2}},$$

giving the mean thickness of the ionic atmosphere. Here, n_k is the number of ions of charge q_k per volume, and $a_k = q_k/e$ is the integer charge number relating the charge q_k to the elementary charge e. In a distance essentially larger than the shielding length λ, the bare charge of the ion is effectively screened by this ionic atmosphere.

6.3.4 Renormalization in Classical Electrodynamics

6.3.4.1 The Classical Electromagnetic Field

Modern fundamental physics is largely based on symmetries, both continuous and discrete. According to Emmy Noether's (1918) celebrated theorem, any continuous symmetry of a non-dissipative system gives rise to conservation laws. For example the source-free Maxwell equations in vacuum, the basic framework of electromagnetism, are governed by transformations of the Poincaré group (affine transformations of space-time coordinates). The analysis of the Maxwell equations with respect to Lorentz transformations was an important step in the formulation of the theory of special relativity.

Symmetries entail conservation laws, including (among others) the conservation of energy, linear momentum and angular momentum. However, this basic group-theoretical structure is not sufficient for an operational description of nature. In a polarizable medium the group-theoretical bare fields have to be renormalized in order to include back-reaction effects . Also the group-theoretical bare sources of the electromagnetic field have to be replaced by effective sources which include polarization effects. The resulting dressed fields and dressed particles have a complicated structure. For this reason, realistic particles cannot be considered as bare objects.

James Clerk Maxwell's (1873) *Treatise on Electricity and Magnetism* is generally acknowledged as one of the most significant achievements of physics in the 19th century. However, this work is difficult to read for a modern scientist. It also does not contain the four famous "Maxwell equations". Moreover, it is based on the dated idea that electric and magnetic fields stand for tensions in a physical medium, called ether. Twelve years after the publication of Maxwell's *Treatise*, Oliver Heaviside published in 1885 a completely reformulated set of Maxwell's equations without potentials in the much simpler language of vector calculus (Heaviside 1893; cf. Hunt 1991 and Nahin 1988, Chap. 7).

Later Hermann Minkowski (1908) based the equations governing the electromagnetic field on Einstein's (1905) new way of comprehending the concepts of time and covariance with respect to the Lorentz transformations. Minkowski's differential equations do not make any *a priori* assumptions about the nature of the constitutive equations which characterize the influence of the medium and materials. All information about media and materials is relegated to constitutive relationships.

In the most general setting electromagnetism is described by two pairs of real-valued vector fields. As we will discuss below, the first pair $(\mathbf{E}, \mathbf{B})$ relates to forces, and the second pair $(\mathbf{D}, \mathbf{H})$ relates to sources. Therefore $\mathbf{E}$ is called the *electric* and

B the *magnetic field strength*, while **D** is called the *electric* and **H** the *magnetic excitation*. The Maxwell-Heaviside equations separate into two groups: the Maxwell-Faraday equations involving the field strengths **E** and **B**, and the Maxwell-Ampère equations involving the excitations **D** and **H**. In SI-units (which are free of any properties of particular media),[94] the Maxwell equations of classical electromagnetism are given by:

$$\nabla\times\mathbf{E}(\mathbf{x},t)+\dot{\mathbf{B}}(\mathbf{x},t)=\mathbf{0} \quad \text{Faraday's law,} \tag{6.35a}$$

$$\nabla\cdot\mathbf{B}(\mathbf{x},t)=0 \quad \text{no magnetic monopoles,} \tag{6.35b}$$

$$\nabla\times\mathbf{H}(\mathbf{x},t)-\dot{\mathbf{D}}(\mathbf{x},t)=\mathbf{J}_{\mathrm{el}}(\mathbf{x},t) \quad \text{Ørsted-Ampère's law,} \tag{6.35c}$$

$$\nabla\cdot\mathbf{D}(\mathbf{x},t)=\rho_{\mathrm{el}}(\mathbf{x},t) \quad \text{Gauss' law.} \tag{6.35d}$$

The real-valued vector field $\mathbf{J}_{\mathrm{el}}$ and the scalar field ρ_{el} represent the sources of the electromagnetic field.

The Maxwell equations (6.35a)–(6.35d) do not yet define an electromagnetic theory, since **E**, **D**, **B**, **H** and $\mathbf{J}_{\mathrm{el}}$ are not yet specified vector fields $\mathbb{R}^4 \to \mathbb{R}^2$, and ρ_{el} is not yet a specified scalar field $\mathbb{R}^4 \to \mathbb{R}$. The Maxwell equations have to be supplemented by a force law and by constitutive laws, describing the influence of matter. Equations (6.35c) and (6.35d) imply the continuity equation

$$\nabla\cdot\mathbf{J}_{\mathrm{el}}(\mathbf{x},t)+\dot{\rho}_{\mathrm{el}}(\mathbf{x},t)=0, \tag{6.36a}$$

which is the differential formulation of the conservation law

$$\frac{d}{dt}\iiint_V \rho_{\mathrm{el}}(\mathbf{x},t)d^3\mathbf{x}=\oiint_S \mathbf{J}_{\mathrm{el}}(\mathbf{x},t)\cdot d\mathbf{s}, \tag{6.36b}$$

where V is the volume enclosed by the closed surface S. This relation expresses the *conservation of electric charge*. Equation (6.35c) implies that the total charge $\iiint_V \rho_{\mathrm{el}}(\mathbf{x},t)d^3\mathbf{x}$ contained within the region V can only increase on account of the flux of charge through the unit area $\oiint_S \mathbf{J}_{\mathrm{el}}(\mathbf{x},t)\cdot d\mathbf{s}$. Accordingly, we have to interpret ρ_{el} as the *bare electric charge density*, and $\mathbf{J}_{\mathrm{el}}$ as the *bare electric current density*. Clearly, a necessary condition for solutions of the Maxwell equations (6.35a)–(6.35d) to exist is that the equation of continuity be satisfied.

Accepting the conservation of electric charge as a basic postulate, the fundamental laws of electromagnetism are the two Maxwell equations (6.35a) and (6.35c) and the continuity equation (6.36a), (6.36b), while (6.35b) and (6.35d) can be considered as initial conditions. Taking the divergence of (6.35a) and (6.35c) and using the

[94] Units in SI (Système International) are based on the international MKSA system. In contrast to the older CGS-systems, the SI-system contains factors of 4π only where they are due to geometrical reasons, and the vacuum velocity of light only appears when it is a consequence of Lorentz relativity.

continuity equation, we get

$$\frac{\partial}{\partial t}\nabla \cdot \mathbf{B}(\mathbf{x}, t) = 0, \qquad \frac{\partial}{\partial t}\{\nabla \cdot \mathbf{D}(\mathbf{x}, t) - \rho_{\mathrm{el}}(\mathbf{x}, t)\} = 0. \tag{6.37}$$

Accordingly, the Maxwell equations (6.35b) and (6.35d) just establish the initial conditions for (6.37).

6.3.4.2 The Physical Meaning of the Field Strength

The four Maxwell equations (6.35a)–(6.35d) divide in a natural manner into two pairs: the two homogeneous equations (6.35a) and (6.35b) containing **E** and **B**, and the two inhomogeneous equations (6.35c) and (6.35d) containing **D** and **H** and the sources ($\mathbf{J}_{\mathrm{el}}, \rho_{\mathrm{el}}$). Under a Lorentz transformation, the equations of each pair transform separately so that (**E**, **B**) and (**D**,**H**) transform as a six-vector (i.e. as an antisymmetric tensor of second order), while ($\mathbf{J}_{\mathrm{el}}, \rho_{\mathrm{el}}$) transforms as a four-vector.[95]

According to Gustav Mie (1912, p. 516, 1941, Sect. 78; cf. also Sommerfeld 1949, p. 2) one has to distinguish between intensity magnitudes (how strong?) and quantity magnitudes (how much?). We adopt the view introduced by Mie that **E** and **B** are *intensive* quantities, characterized by

$$[\mathbf{E}] = \frac{\text{force}}{\text{charge}}, \qquad [\mathbf{B}] = \frac{\text{force}}{\text{pole strength}}, \tag{6.38}$$

while **D** and **H** are *extensive* quantities, characterized by (cf. Pohl 1949)

$$[\mathbf{D}] = \frac{\text{charge}}{\text{area}}, \qquad [\mathbf{H}] = \frac{\text{pole strength}}{\text{area}}. \tag{6.39}$$

The intensive quantities **E** and **B** are the primary electromagnetic fields which cause electromagnetic phenomena and can be expressed operationally by the *Lorentz force density* **f** exerted by the field (**E**, **B**) on the bare charge density ρ_{el} and the bare current density $\mathbf{J}_{\mathrm{el}}$,

$$\mathbf{f}(\mathbf{x}, t) = \rho_{\mathrm{el}}(\mathbf{x}, t)\mathbf{E}(\mathbf{x}, t) + \mathbf{J}_{\mathrm{el}}(\mathbf{x}, t)\times\mathbf{B}(\mathbf{x}, t). \tag{6.40}$$

This operational definition characterizes **E** and **B** as force fields—and hence as intensive quantities. Note that the Lorentz force law is not derivable from the Maxwell equations but has to be added in order to give them physical content.

[95] **E** is also known as the electric intensity and **H** as the magnetic intensity, while **D** is also known as the electric displacement or electric induction, and **B** the magnetic induction. This traditional terminology is conceptually misleading. Heaviside was the first to consider the fields **E** and **B** as primary since they represent the physical state (see Nahin 1988, p. 109). Nowadays we know that (**E**, **B**) as well as (**D**, **H**) transform together under a Lorentz transformation. The problem of how to interpret (**E**, **B**) and (**D**, **H**) is a matter of century-old controversy. Since the risk of confusion is still substantial, we will *not* conform to historical terminology but refer to **E** as the fundamental electric field vector, and to **B** as the fundamental magnetic field vector.

Since the charge density ρ and the current density $\mathbf{J}$ refer to the *sources* of the electromagnetic field, the *extensive* fields $\mathbf{D}$ and $\mathbf{H}$ describe *effects* due to the bare sources and the reaction fields which reflect the influence of matter.[96] From this point of view the fields $\mathbf{D}$ and $\mathbf{H}$ are only auxiliary quantities encoding information about the sources. Therefore we call $\mathbf{D}$ the *electric excitation*, and $\mathbf{H}$ the *magnetic excitation*. Note that both the electric excitation $\mathbf{D}$ and the magnetic excitation $\mathbf{H}$ have direct operational significance: in principle, they can be directly measured (Mie 1941, §80, §287; Pohl 1949).

In Minkowski's electrodynamics, free space is considered as an isotropic unpolarizable medium so that the constitutive relations for free space (also called the vacuum) are linear and given by

$$\mathbf{D}(\mathbf{x},t) = \varepsilon_0 \mathbf{E}(\mathbf{x},t), \qquad \mathbf{H}(\mathbf{x},t) = \mu_0^{-1}\mathbf{B}(\mathbf{x},t), \tag{6.41}$$

where $\varepsilon_0 > 0$ is the permittivity of the vacuum and $\mu_0 > 0$ is the permeability of the vacuum. They are related by the vacuum velocity of light c and the characteristic impedance Z_0 of the vacuum:

$$c := 1/\sqrt{\mu_0 \varepsilon_0} \approx 2.99792458 \times 10^8 \text{ m/s}, \tag{6.42a}$$

$$Z_0 := \sqrt{\mu_0/\varepsilon_0} \approx 376.730313\ \Omega \tag{6.42b}$$

6.3.4.3 Constitutive Equations

In contrast to the Maxwell-Heaviside field equations, the so-called *constitutive laws* endow electrodynamics with its *metric* and *kinematical symmetry*. The constitutive laws reflect the properties of the *medium* and establish a relation between the electromagnetic excitation $(\mathbf{D},\mathbf{H})$ and the field strength $(\mathbf{E},\mathbf{B})$. The concrete form of the constitutive laws is determined by the electromagnetic properties of the medium and its state. For every space-time point $(\mathbf{x},t)$, the vectors $\mathbf{D}(\mathbf{x},t)$ and $\mathbf{H}(\mathbf{x},t)$ may depend nonlinearly on the values of $\mathbf{E}$ and $\mathbf{B}$ at all points of the medium (nonlocal case), and at all times prior to time t (hereditary effects; compare, e.g., Eringen and Maugin 1990a, 1990b). Moreover, in most real materials, the magnetic excitation $\mathbf{H}$ is not a single-valued function of the magnetic field strength $\mathbf{B}$.

The two vector fields $\mathbf{D}(\mathbf{x},t)$ and $\mathbf{H}(\mathbf{x},t)$ satisfy the inhomogeneous equations (6.35c) and (6.35d) containing the sources $\mathbf{J}_{\mathrm{el}}$ and ρ_{el}. Here $\mathbf{J}_{\mathrm{el}}$ is the *bare electrical current density* (also called the *true* current density) and ρ_{el} is the *bare electrical charge density* (also called the *true* charge density). Since (6.35c) and (6.35d) include inhomogeneities accounting for the influence of matter, one has to attribute the corresponding pair $(\mathbf{D},\mathbf{H})$ of fields to the influence of matter.

[96] A careful and thorough discussion of the constitutive equations can be found in the handbook article by Truesdell and Toupin (1960, Chap. G) who call $\mathbf{D}$ the charge potential and $\mathbf{H}$ the current potential.

The concrete form of the constitutive equations is determined by the electromagnetic properties of the medium and its state. The presence of a polarizable medium can be completely taken into account by replacing the charge potential $\mathbf{D}$ with the electric polarization density $\mathbf{P}$, and the current potential $\mathbf{H}$ with the magnetic polarization density $\mathbf{M}$ (also called magnetization; compare, e.g., Stratton 1941, p. 12):

$$\mathbf{P}(\mathbf{x},t) = \mathbf{D}(\mathbf{x},t) - \varepsilon_0 \mathbf{E}(\mathbf{x},t), \qquad \mathbf{M}(\mathbf{x},t) = \mu_0^{-1}\mathbf{B}(\mathbf{x},t) - \mathbf{H}(\mathbf{x},t). \tag{6.43}$$

These polarizations are reaction fields which describe the influence of the medium. They vanish in free space. Eliminating $\mathbf{D}$ and $\mathbf{H}$ from the generalized Maxwell equations (6.35a)–(6.35d), we get

$$\nabla \times \mathbf{E}(\mathbf{x},t) + \dot{\mathbf{B}}(\mathbf{x},t) = \mathbf{0}, \tag{6.44a}$$

$$\nabla \times \mathbf{B}(\mathbf{x},t) - c^{-2}\dot{\mathbf{E}}(\mathbf{x},t) = \mu_0 \mathbf{J}_{\mathrm{el}}^{\mathrm{eff}}(\mathbf{x},t), \tag{6.44b}$$

$$\nabla \cdot \mathbf{B}(\mathbf{x},t) = 0, \tag{6.44c}$$

$$\nabla \cdot \mathbf{E}(\mathbf{x},t) = \varepsilon_0^{-1}\rho_{\mathrm{el}}^{\mathrm{eff}}(\mathbf{x},t), \tag{6.44d}$$

where the renormalized effective electrical charge density $\rho_{\mathrm{el}}^{\mathrm{eff}}$ is the sum of the bare electrical charge density ρ and the induced charge density $-\nabla\mathbf{P}$:

$$\rho_{\mathrm{el}}^{\mathrm{eff}}(\mathbf{x},t) := \rho_{\mathrm{el}}(\mathbf{x},t) - \nabla\mathbf{P}. \tag{6.45}$$

The renormalized effective electrical current density $\mathbf{J}_{\mathrm{el}}^{\mathrm{eff}}$ is the sum of the bare electrical current density $\mathbf{J}$ and the induced current density $\nabla \times \mathbf{M} + \dot{\mathbf{P}}$:

$$\mathbf{J}_{\mathrm{el}}^{\mathrm{eff}}(\mathbf{x},t) = \mathbf{J}_{\mathrm{el}}(\mathbf{x},t) + \nabla \times \mathbf{M}(\mathbf{x},t) + \dot{\mathbf{P}}(\mathbf{x},t). \tag{6.46}$$

Both, the bare and the effective sources obey continuity equations,

$$\nabla \cdot \mathbf{J}_{\mathrm{el}} + \dot{\rho} = 0 \quad , \quad \nabla \cdot \mathbf{J}_{\mathrm{el}} + \dot{\rho}_{\mathrm{el}}^{\mathrm{eff}} = 0,$$

expressing the *local* conservation of the bare and renormalized electrical charges.

6.3.4.4 "Les Bonnes Theories Sont Souples"[97]

For an ideal non-polarizable medium with no charges ($\rho_{\mathrm{el}} = 0$) and no currents ($\mathbf{J}_{\mathrm{el}} = 0$), the general Maxwell equations reduce to the classical Maxwell equations in vacuum:

$$\nabla \cdot \mathbf{E}(\mathbf{x},t) = 0, \qquad \nabla \times \mathbf{E}(\mathbf{x},t) = -\dot{\mathbf{B}}(\mathbf{x},t), \tag{6.47a}$$

[97] Editors note: This quote from Poincaré (1900, p. 252) has been interpreted in the sense that good theories should be sufficiently flexible to absorb new constraints or novel insights.

$$\nabla \cdot \mathbf{B}(\mathbf{x}, t) = 0, \qquad \nabla \times \mathbf{B}(\mathbf{x}, t) = c^{-2}\dot{\mathbf{E}}(\mathbf{x}, t). \tag{6.47b}$$

Without additional constraints, the Maxwell equations (6.47a), (6.47b) are covariant under all transformations of the inhomogeneous general linear group $IGL(4, \mathbb{R})$ of degree four over $\mathbb{R}$ (cf. Fushchich and Nikitin 1987, Sects. 39–40 and App. 2).

Both the Galilei group and the Lorentz group are subgroups of $IGL(4, \mathbb{R})$, so that the Maxwell-Heaviside equations (6.47a), (6.47b) allow both a Lorentz-invariant and a Galilei-invariant electrodynamics. They are distinguished by the supplementary constitutive equations. The vector form of the Maxwell equations implies that they are form-invariant under time-independent translations and rotations.

The Maxwell equations in vacuum are invariant under the following duality transformation (Heaviside 1893),

$$\mathbf{E} \to -c\mathbf{B}, \qquad \mathbf{B} \to c^{-1}\mathbf{E}, \tag{6.48}$$

and they are also invariant under the rotations

$$\left.\begin{aligned} \mathbf{E} \to \mathbf{E}(\theta) &= \cos(\theta)\mathbf{E} - c\sin(\theta)\mathbf{B} \\ \mathbf{B} \to \mathbf{B}(\theta) &= \cos(\theta)\mathbf{B} + c^{-1}\sin(\theta)\mathbf{E} \end{aligned}\right\} \quad 0 \le \vartheta \le \pi/2. \tag{6.49}$$

If sources $\mathbf{J}_{\text{el}}(\mathbf{x}, t)$ and $\rho_{\text{el}}(\mathbf{x}, t)$ are added, the Maxwell equations (6.47a), (6.47b) are no longer covariant under the duality transform (6.49). However, the duality symmetry still holds if one adds both electric and magnetic charges and currents. Corresponding to the definition of the induced electrical charge density $\rho_{\text{el}}^{\text{ind}} := -\nabla\mathbf{P}$ one can define an induced magnetic charge density $\rho_{\text{mag}}^{\text{ind}} := -\nabla\mathbf{M}$ (compare, e.g., Carstoiu 1967). Including magnetic sources, Maxwell's equations assume the dual forms (compare Jackson 1998, Sect. 6.11):

$$\nabla \times \mathbf{E}(\mathbf{x}, t) + \dot{\mathbf{B}}(\mathbf{x}, t) = -\mathbf{J}_{\text{mag}}(\mathbf{x}, t), \tag{6.50a}$$

$$\nabla \times \mathbf{H}(\mathbf{x}, t) - \dot{\mathbf{D}}(\mathbf{x}, t) = +\mathbf{J}_{\text{el}}(\mathbf{x}, t), \tag{6.50b}$$

$$\nabla \cdot \mathbf{B}(\mathbf{x}, t) = \rho_{\text{mag}}(\mathbf{x}, t), \tag{6.50c}$$

$$\nabla \cdot \mathbf{D}(\mathbf{x}, t) = \rho_{\text{el}}(\mathbf{x}, t). \tag{6.50d}$$

These modified vacuum Maxwell equations exhibit a duality symmetry between electric and magnetic quantities

$$\text{electric} \to \text{magnetic}, \qquad \text{magnetic} \to \text{electric}, \tag{6.51}$$

so that they are invariant under the following duality transformations

$$\begin{pmatrix} \mathbf{E} \\ \mathbf{H} \end{pmatrix} = \begin{pmatrix} \cos(\vartheta) & Z_0 \sin(\vartheta) \\ -Z_0^{-1} \sin(\vartheta) & \cos(\vartheta) \end{pmatrix} \begin{pmatrix} \widetilde{\mathbf{E}} \\ \widetilde{\mathbf{H}} \end{pmatrix} \tag{6.52a}$$

$$\begin{pmatrix} \mathbf{J}_{\text{el}} \\ \mathbf{J}_{\text{mag}} \end{pmatrix} = \begin{pmatrix} \cos(\vartheta) & Z_0^{-1} \sin(\vartheta) \\ Z_0 \sin(\vartheta) & \cos(\vartheta) \end{pmatrix} \begin{pmatrix} \widetilde{\mathbf{J}}_{\text{el}} \\ \widetilde{\mathbf{J}}_{\text{mag}} \end{pmatrix} \tag{6.52b}$$

$$\begin{pmatrix} \rho_{\mathrm{el}} \\ \rho_{\mathrm{mag}} \end{pmatrix} = \begin{pmatrix} \cos(\vartheta) & Z_0^{-1}\sin(\vartheta) \\ Z_0\sin(\vartheta) & \cos(\vartheta) \end{pmatrix} \begin{pmatrix} \widetilde{\rho}_{\mathrm{el}} \\ \widetilde{\rho}_{\mathrm{mag}} \end{pmatrix} \tag{6.52c}$$

Maxwell expelled isolated magnetic charges from his equations because no isolated magnetic pole had ever been observed. Later, Pierre Curie (1894b) looked into the possibility of free magnetic charges and found no grounds why they should not exist. Since induced magnetic source densities do not appear in the original Maxwell equations, they are often assumed to vanish. Yet, induced magnetic densities are important for the discussion of magnetostatic problems which are characterized by the relation $\nabla \times \mathbf{H} = 0$.[98] For $\mathbf{H}$ not to vanish in magnetostatics, it is necessary that $\nabla \cdot \mathbf{H} \neq 0$, so that the Maxwell equation $\nabla \cdot \mathbf{B} = 0$ implies $\nabla \cdot \mathbf{H} = -\nabla \mathbf{M} = \rho_{\mathrm{mag}}^{\mathrm{ind}} \neq 0$.

In nature, the duality symmetry seems to be spoiled by the fact that we observe electric charges but not magnetic charges (which are usually called magnetic monopoles). But according to Dirac (1931) the existence of magnetic monopoles would be consistent with quantum mechanics. Arguments based on symmetry and conservation laws indicate that in the presence of magnetic sources the equations for the electromagnetic fields would assume a dual form.

6.3.4.5 Intrinsic, Background and External Fields

In a classical description all electromagnetic fields are real vector-valued functions of space-time. The total field $(\mathbf{E}, \mathbf{B})$ can be partitioned into three parts: an intrinsic self-field $(\mathbf{E}^{\mathrm{int}}, \mathbf{B}^{\mathrm{int}})$ interacting with its internal sources, a source-free background field $(\mathbf{E}^{\mathrm{back}}, \mathbf{B}^{\mathrm{back}})$, and an external field $(\mathbf{E}^{\mathrm{ext}}, \mathbf{B}^{\mathrm{ext}})$ prescribed by the experimenter,

$$\mathbf{E} = \mathbf{E}^{\mathrm{int}} + \mathbf{E}^{\mathrm{back}} + \mathbf{E}^{\mathrm{ext}}, \qquad \mathbf{B} = \mathbf{B}^{\mathrm{int}} + \mathbf{B}^{\mathrm{back}} + \mathbf{B}^{\mathrm{ext}}. \tag{6.53}$$

We speak of an intrinsic electromagnetic field $(\mathbf{E}^{\mathrm{int}}, \mathbf{B}^{\mathrm{int}})$ if the sources of this field are included in the dynamical description. The internal sources and the associated intrinsic field are inseparable and constitute a closed system. In spite of the fact that bare sources without an associated electromagnetic field do not exist, it is hardly avoidable to start with a fictitious separation of the whole system into ontologically non-existing bare sources and bare fields. This can be done most conveniently by specifying a Lagrangian L, consisting of a Lagrangian L^{source} for the bare sources, a Lagrangian L^{emf} for the bare electromagnetic field, and an interaction Lagrangian L^{int} describing the interaction between the sources and the field,

$$L = L^{\mathrm{source}} + L^{\mathrm{emf}} + L^{\mathrm{int}}. \tag{6.54}$$

---

[98] See Sommerfeld (1949, §7 and §12) and Stratton (1941, Chap. 4). Compare also Schwinger et al. (1998, Chaps. 2 and 30).

Traditional classical electrodynamics assumes—usually without mentioning[99]—that there is no field without sources, so that in this case the background parts vanish. All three fields, hence also the total field, are solutions of Maxwell's equation.

However, this boundary condition does not follow from Maxwell's equations. Furthermore, there is no empirical evidence for the correctness of this assumption. Timothy Boyer (1975a, 1975b) showed that classical electrodynamics is in much better agreement with quantum electrodynamics if one chooses as boundary condition a source-free Gaussian random classical electromagnetic field with a Lorentz-invariant spectrum. The background field in this so-called "random electrodynamics" is the classical counterpart of the vacuum zero-point field of quantum electrodynamics. Up to a constant, it is uniquely fixed by the requirement of Lorentz invariance, and can be characterized by the spectral energy density

$$\rho_\nu = \frac{\hbar\nu}{2}\frac{\nu^2}{\pi^2 c^3}, \qquad \nu \geq 0, \tag{6.55}$$

through which Planck's constant $\hbar$ enters this neoclassical theory.

In contrast to this approach, we consider quantum theory as basic, so that the material sources and the electromagnetic field are entangled by Einstein-Podolsky-Rosen (EPR) correlations. In this case, we can speak of material objects if and only if we abstract from the actually existing EPR correlations so that the pure states of the matter-field system factorize. If the initial state of the combined system is pure, then the initial state of the field is also pure. If furthermore the initial state of the field is a pure coherent state, then it remains a coherent state for all later times. The convex set generated by all coherent states forms a simplex which can be taken as the state space for a *classical description of the quantum field*. The classical quantum system generated in this way still depends on Planck's constant, and is formally identical with Boyer's random electrodynamics.

6.3.5 Non-Separability of the Quantum World

Classical physics assumes that the material world consists of individual things with well-defined spatiotemporal location. It is a structural property of classical physics that the states of subsystems determine the state of any system composed of them. This property is called *separability*. Moreover, it is tacitly taken for granted that events in spatially and temporally separated regions are independent in the first place.

Just before 1900, it became clear that classical physics was unable to explain the phenomenon of black-body radiation. The further development of quantum mechanics in the early decades of the 20th century revolutionized our understanding of

[99] A notable exception is Lorentz's explicit formulation of this assumption as a boundary condition, compare Lorentz (1909, note 6 on p. 240).

atoms, molecules, and the stability of matter. The papers by Einstein et al. (1935) and by Schrödinger (1935a, 1935b, 1936) started a second phase of quantum theory which enforced a radical change of the worldview of classical physics. It turned out that, in contrast to classical physics, key features of quantum physics are holistic.

This development rejected the kind of local realism Schrödinger and Einstein had favored. In quantum mechanics, the existence of objects isolated in separation is no longer guaranteed. Subsystems of a quantum system do not exist as individual entities. It became clear that quantum theory was not just a "theory of atoms" but a general, mathematically formulated theory of holistically correlated matter. A universe of discourse is called *holistic* if there are no pre-existing parts so that wholes cannot be explained in terms of such parts and their interactions. Erwin Schrödinger (1935b) coined the term *entanglement* to describe these non-classical holistic correlations:

> When two systems, of which we know the states by their respective representatives, enter into temporary physical interaction due to known forces between them, and when after a time of mutual influence the systems separate again, then they can no longer be described in the same way as before by endowing each of them with a representative of its own. I would not call that *one* but rather *the* characteristic trait of quantum mechanics, the one that enforces its entire departure from classical lines of thought. By the interaction the two representatives ... have become entangled.

Entanglement is not "the greatest mystery in physics" (as the title of a popular book by Aczel (2001) states) but a natural and compelling consequence of the non-Boolean structure of nature. In spite of that, entanglement is one of the most often misunderstood concepts—even in physics, where the related experiments form already an important part of modern technology. If we accept the insight that there are no ultimate building blocks, there is nothing spectacular or controversial about the existence of holistic correlations. Holistic correlations have no classical explanation, in particular not in terms of "forces".

Entanglement arises due to *our decision* to distinguish parts. It appears as counterintuitive only within an atomistic worldview. Since there is no *a priori* given decomposition of the world, *entanglement is always relative to a context.* Along the lines of the outdated reductive-atomistic worldview, it has been claimed that quantum entanglement applies only to some "microphysical level" (whatever that means), or only to particles that have previously interacted. This view is in contradiction to the results of quantum physics.[100] Every observable object has to be described as an *open system.*

In principle, every open quantum system is entangled with its environment, even if the interaction is negligibly small. The state of an open quantum system is in general not a product state of the state of the object system and the state of its environmental. The correlations between entangled systems are called *Einstein-Podolsky-Rosen (EPR) correlations*. A quantum object and its environment constitute a single quantum system rather than two coupled systems to which separate state vectors can

---

[100]Note that there is no simple relationship between "interaction" and "entanglement". Compare the example in Sect. 6.3.7 below.

be assigned. Consequently, there are no context-independent individual objects in quantum theory. Moreover, genuine holistic correlations are not restricted to physical systems. They are independent of Planck's constant of action, they are independent of spatial separations, and do not arise from known physical forces.

If we decide to divide the world into pieces exhibiting an approximate individuality, we have to realize that therewith we destroy holistic properties. Because we would like to understand some aspects of nature without understanding everything, we have to learn how to isolate a subsystem from the rest of the world in an appropriate fashion.

Since bare systems are hypothetical constructs that cannot be measured directly, it does not make much sense to consider decompositions in terms of bare elementary systems (usually, but misleadingly called "elementary particles"). There are no preexisting elementary systems. Laboratory experiments always refer to dressed objects, never to bare systems (see Sect. 6.3.3). A dressed object is an *open system* whose phenomenal existence depends upon its environment.

Dressing not only modifies the properties of bare systems and their environment, but also their dynamics. An important effect of "proper" dressing is that object systems may be less entangled with the new effective environment. In an appropriate context, dressing may entrust individuality. Under suitable circumstances we can describe the universe of discourse in terms of dressed subsystems with respect to a *contextually chosen decomposition*.

To sum it up: The objects of our perceivable reality never derive from tensorizations in terms of bare elementary systems but rather from state-dependent contextual tensorizations. *The resulting decompositions of the world are not given a priori, but are to an essential extent contingent.*

6.3.6 Schmidt Decomposition of Composite Systems

The exhaustive description of an arbitrary type-I W*-algebraic quantum system is given by a *pure* state functional ρ in the predual $\mathcal{M}_*$ of the underlying W*-algebra $\mathcal{M}$. An arbitrary state functional $\rho \in \mathcal{M}_*$ can be represented by a density operator $D \in \mathcal{M}$ with $\mathrm{tr}(D) = 1$. The density operator D of a *pure* state fulfills the relation $D^2 = D$ (idempotency), while a non-pure state is characterized by the condition $D^2 < D$.

In an irreducible representation $\mathcal{B}(\mathcal{H})$ on a Hilbert space $\mathcal{H}$ of a type-I W*-system with the underlying W*-algebra $\mathcal{M}$, a pure state functional $\rho \in \mathcal{M}_*$ can be represented either by an idempotent density operator $D = D^2$ or by a state vector Ψ,

$$D = |\Psi\rangle\langle\Psi|, \quad D \in \mathcal{B}(\mathcal{H}),\ \Psi \in \mathcal{H}. \tag{6.56}$$

In the following we consider a composite system with the Hilbert space $\mathcal{H}_1 \otimes \mathcal{H}_2$. The corresponding W*-algebra $\mathcal{M}$ of observables is given by the W*-tensor product of $\mathcal{B}(\mathcal{H}_1)$ and $\mathcal{B}(\mathcal{H}_2)$ as the algebra of all bounded operators acting on the Hilbert space $\mathcal{H}_1 \otimes \mathcal{H}_2$,

$$\mathcal{M} \cong \mathcal{B}(\mathcal{H}_1)\,\overline{\otimes}\,\mathcal{B}(\mathcal{H}_2) := \mathcal{B}(\mathcal{H}_1 \otimes \mathcal{H}_2). \tag{6.57}$$

The state vector $\Psi \in \mathcal{H}_1 \otimes \mathcal{H}_2$ of an exhaustive description of the corresponding universe of discourse has a canonical decomposition into product vectors $\alpha_k \otimes \beta_k$,

$$\Psi = \sum_k c_k \alpha_k \otimes \beta_k, \quad \|\Psi\|^2 = \sum_k |c_k|^2 = 1, \tag{6.58}$$

where $\alpha_k \in \mathcal{H}_1$ and $\beta_k \in \mathcal{H}_2$ are the orthonormal eigenvectors of the eigenvalue problem first discussed by Erhard Schmidt (1907a, 1907b, 1908) long ago:[101]

$$\langle \Psi | \alpha_k \rangle_{\mathcal{H}_1} = c_k \beta_k, \qquad \langle \Psi | \beta_k \rangle_{\mathcal{H}_2} = c_k \alpha_k, \tag{6.59a}$$

$$\langle \beta_k | \phi_\ell \rangle_{\mathcal{H}_1} = \delta_{k,\ell}, \qquad \langle \alpha_k | \psi_\ell \rangle_{\mathcal{H}_2} = \delta_{k,\ell}. \tag{6.59b}$$

If the complex eigenvalues are not degenerate, the Schmidt decomposition is *unique*.[102]

The density operator D corresponding to the state vector Ψ with the Schmidt decomposition (6.58) is given by

$$D := |\Psi\rangle\langle\Psi| = \sum_k \sum_\ell c_k^* c_\ell |\alpha_k \otimes \beta_k\rangle\langle\alpha_\ell \otimes \beta_\ell|, \tag{6.60}$$

so that the reduced density operators D_1 and D_2 are given by the restrictions

$$D_1 := \mathrm{tr}_{\mathcal{H}_2}\{D\} = \sum_k |c_k|^2 |\alpha_k\rangle\langle\alpha_k|, \tag{6.61a}$$

$$D_2 := \mathrm{tr}_{\mathcal{H}_1}\{D\} = \sum_k |c_k|^2 |\beta_k\rangle\langle\beta_k|. \tag{6.61b}$$

It is remarkable that the absolute values $|c_k|$ of the eigenvalues of the reduced density operator D_1 of the first subsystem are the same as those for the reduced density operator D_2 of the second subsystem. Therefore the unconditional expectations are given by

$$\rho_1(X) = \mathrm{tr}_{\mathcal{H}_1}\{D_1 X\} = \sum_k |c_k|^2 \langle\alpha_k|X|\alpha_k\rangle, \quad X \in \mathcal{B}(\mathcal{H}_1), \tag{6.62a}$$

$$\rho_2(Y) = \mathrm{tr}_{\mathcal{H}_2}\{D_2 Y\} = \sum_k |c_k|^2 \langle\beta_k|Y|\beta_k\rangle, \quad Y \in \mathcal{B}(\mathcal{H}_2). \tag{6.62b}$$

---

[101] A useful review of the properties of Schmidt decompositions of state vectors is due to Coleman (1963, Sect. 3, pp. 671ff).

[102] Nowadays, the original literature on decompositions is not regularly studied any more. Often it is claimed that the expansion coefficients of the decomposition are non-negative real numbers (e.g., Nielsen and Chuang 2000, p. 109), but this applies to the von-Neumann decomposition (von Neumann 1932a, p. 231) and is not true for the Schmidt decomposition. The complex expansion coefficients c_k of the Schmidt decomposition contain much more information than the expansion coefficients $|c_k|$ of the von-Neumann decomposition—information which is crucial for a discussion of the holistic correlations of entangled systems.

Via the eigenvalue problems

$$D_1\alpha_k = |c_k|\alpha_k, \tag{6.63a}$$

$$D_2\beta_k = |c_k|\beta_k, \tag{6.63b}$$

for the reduced density operators D_1 and D_2 of the partial descriptions we then know the two orthonormal systems $\{\alpha_1, \alpha_2, \ldots\}$ and $\{\beta_1, \beta_2, \ldots\}$, as well as the absolute values $|c_1|, |c_2|, \ldots$ of the Schmidt eigenvalues $c_1, c_2, \ldots$. For an entangled system a reconstruction of the pure state of the joint system is impossible since the phases of the Schmidt eigenvalues are not given by the two partial descriptions. For a reconstruction of the state vector Ψ of the joint system we also need the relative phases of the Schmidt eigenvalues, e.g. $\chi_k - \chi_1$, where (without loss of generality) we can put $\chi_1 = 0$, so that

$$\Psi = \sum_{k=1}^{\infty} |c_k| e^{i\chi_k} \alpha_k \otimes \beta_k. \tag{6.64}$$

That is, *the holistic correlations of a composite system are encoded by the family* $\{\chi_k | k = 2, 3, \ldots\}$ *of phase factors.*

Only in the exceptional case of a product vector, $\Psi = \alpha \otimes \beta$, can the density operator D of the composite system be recovered from the reduced density operators D_1 and D_2, $D = D_1 \otimes D_2$. The reduced density operators D_1 and D_2 represent pure states if and only if the state vector Ψ is *not entangled but a product state:*

$$\Psi \text{ is not entangled} \iff D_1 = D_1^2 \iff D_2 = D_2^2. \tag{6.65}$$

If we order the eigenvalues $|c_k|$ of the eigenvalue problems (6.63a), (6.63b) according to their absolute values,

$$1 \geq |c_k| \geq |c_{k+1}| \geq 0, \tag{6.66}$$

we can see that the Schmidt decomposition has a remarkable optimality property. For fixed natural numbers m and n,

$$\inf \left\| \Psi - \sum_{j=1}^{m} \sum_{j'=1}^{m'} \widetilde{c}_{j,j'} \widetilde{\alpha}_j \otimes \widetilde{\beta}_{j'} \right\| = \left\| \Psi - \sum_{j=1}^{n} c_j \alpha_j \otimes \beta_k \right\| = 1 - \sum_{j=1}^{n} \|c_j\|^2, \tag{6.67}$$

where $n = \min(m, m')$ and the infimum has to be taken over all complex numbers $\widetilde{c}_{i,j'}$, and all vectors $\widetilde{\alpha}_j \in \mathcal{H}_1$, $\widetilde{\beta}_{j'} \in \mathcal{H}_2$. Accordingly, the best possible approximation (in the Hilbert space norm) of the state vector $\Psi \in \mathcal{H}$ by a product state vector $\Psi_1 \in \mathcal{H}$ is given by the dominant term $\alpha_1 \otimes \beta_1$ of the Schmidt decomposition (6.58)

$$\Psi_1 = \alpha_1 \otimes \beta_1 \tag{6.68}$$

with

$$\|\Psi - \Psi_1\|^2 = 1 - |c_1|^2, \quad |c_1|^2 = 1 - \langle \Psi | \Psi_1 \rangle \leq 1. \tag{6.69}$$

Therefore, Ψ is a product state if and only if $|c_1|^2 = 1$. If $|c_1|^2 < 1$, then there are holistic correlations between the two subsystems. The number $|c_1|^2$ characterizes the degree of entanglement of the subsystems generated by the factorization $\mathcal{H}_1 \otimes \mathcal{H}_2$. We define a non-negative functional I, called the *integrity* of the subsystems, by Primas (1975, pp. 141ff, 1993, pp. 262f):

$$I = |c_1|^2 = \left|\langle\Psi|\alpha_1 \otimes \beta_1\rangle\right|^2, \quad 0 < I \leq 1. \tag{6.70}$$

Note that the integrity I depends on both the state vector Ψ and the chosen factorization $\mathcal{H}_1 \otimes \mathcal{H}_2$. It can be represented by the Rényi entropy H_∞ as

$$I = e^{-H_\infty} \quad \text{with } H_\infty := \lim_{\alpha\to\infty} H_\alpha, \tag{6.71a}$$

where H_α is the Rényi entropy of order α,[103]

$$H_\alpha := \frac{1}{1-\alpha} \ln \sum\nolimits_k |c_k|^{2\alpha}, \quad 0 < \alpha < \infty. \tag{6.71b}$$

6.3.7 Retensorizations

Since *entanglement is always relative to the tensorization used*, any entanglement of a type-I description can be changed by a retensorization of the universe of discourse. A transformation which changes the tensorization in terms of bare group-theoretically defined elementary systems leads to a new description in terms of contextually dressed objects. A desirable tensorization would be a decomposition into almost uncorrelated subsystems. Examples are well known in elementary quantum mechanics, even though they are rarely presented as retensorizations.

A Simple Example of a Retensorization: The Hydrogen Atom
With respect to the bare Hilbert-space tensorization $\mathcal{H} = \mathcal{H}_\mathrm{A} \otimes \mathcal{H}_\mathrm{B}$, the Hamiltonian of a hydrogen atom with Coulomb potential V is given by

$$H = \frac{1}{M_\mathrm{A}}\mathbf{P}_\mathrm{A}^2 \otimes \mathbf{1}_\mathrm{B} + \frac{1}{M_\mathrm{B}}\mathbf{1}_\mathrm{A} \otimes \mathbf{P}_\mathrm{B}^2 + V(\mathbf{Q}_\mathrm{A} \otimes \mathbf{1}_\mathrm{B} - \mathbf{1}_\mathrm{A} \otimes \mathbf{Q}_\mathrm{B}).$$

Here $\mathbf{Q}_\mathrm{A}$ is the position operator, $\mathbf{P}_\mathrm{A}$ the momentum operator, and M_A the mass of the bare nucleus. Likewise, $\mathbf{Q}_\mathrm{B}$ is the position operator, $\mathbf{P}_\mathrm{B}$ the momentum operator, and M_B the mass of the bare electron. These two bare elementary systems are defined via projective irreducible representations of the Galilei group. The familiar center-of-mass transformation is generated by the unitary operator U,

$$\mathbf{Q}_\mathrm{C} := U\{\mathbf{Q}_\mathrm{A} \otimes \mathbf{1}_\mathrm{B}\}U^* = \mathbf{Q}_\mathrm{A} \otimes \mathbf{1}_\mathrm{B} - \mathbf{1}_\mathrm{A} \otimes \mathbf{Q}_\mathrm{B},$$

$$\mathbf{Q}_\mathrm{D} := U\{\mathbf{1}_\mathrm{A} \otimes \mathbf{Q}_\mathrm{B}\}U^* = \frac{M_\mathrm{A}}{M_\mathrm{A}+M_\mathrm{B}}\mathbf{Q}_\mathrm{A} \otimes \mathbf{1}_\mathrm{B} + \frac{M_\mathrm{B}}{M_\mathrm{A}+M_\mathrm{B}}\mathbf{1}_\mathrm{A} \otimes \mathbf{Q}_\mathrm{B},$$

---

[103] The inequality $e^{-H_\alpha} \leq^{-H_\beta}$ for $\alpha > \beta$ implies $I \geq e^{-H_\alpha} \geq \sum_k |c_k|^2 \ln |c_k|^2$. Note that the well-known Shannon entropy can be recovered by the Rényi entropy H_α in the limit $\alpha \to 1$.

$$\mathbf{P}_{\mathrm{C}} := U\{\mathbf{P}_{\mathrm{A}} \otimes \mathbf{1}_{\mathrm{B}}\}U^* = \frac{M_{\mathrm{B}}}{M_{\mathrm{A}} + M_{\mathrm{B}}}\mathbf{P}_{\mathrm{A}} \otimes \mathbf{1}_{\mathrm{B}} - \frac{M_{\mathrm{A}}}{M_{\mathrm{A}} + M_{\mathrm{B}}}\mathbf{1}_{\mathrm{A}} \otimes \mathbf{P}_{\mathrm{B}},$$

$$\mathbf{P}_{\mathrm{D}} := U\{\mathbf{1}_{\mathrm{A}} \otimes \mathbf{P}_{\mathrm{B}}\}U^* = \mathbf{P}_{\mathrm{A}} \otimes \mathbf{1}_{\mathrm{B}} + \mathbf{1}_{\mathrm{A}} \otimes \mathbf{P}_{\mathrm{B}}.$$

U is a unitary dressing transformation which induces the retensorization $\mathcal{H}_{\mathrm{C}} \otimes \mathcal{H}_{\mathrm{D}} := U\{H_{\mathrm{A}} \otimes \mathcal{H}_{\mathrm{B}}\}$ of the Hilbert space $\mathcal{H}$ and allows the separation of internal and external motions. The transformed Hamiltonian

$$UHU^* = \frac{1}{M_{\mathrm{C}}}\mathbf{P}_{\mathrm{C}}^2 \otimes \mathbf{1}_{\mathrm{D}} + V(\mathbf{Q}_{\mathrm{C}}) \otimes \mathbf{1}_{\mathrm{D}} + \frac{1}{M_{\mathrm{D}}}\mathbf{1}_{\mathrm{C}} \otimes \mathbf{P}_{\mathrm{D}}^2,$$

$$M_{\mathrm{D}} := M_{\mathrm{A}} + M_{\mathrm{B}}, \quad M_{\mathrm{C}} := M_{\mathrm{A}} M_{\mathrm{B}}/(M_{\mathrm{A}} + M_{\mathrm{B}}),$$

shows that the center-of-mass system with the total mass M_{D} and the dressed electron with the renormalized mass M_{C} are no longer interacting.

In this example, the dressing transformation from the bare tensorization $\mathcal{H}_{\mathrm{A}} \otimes \mathcal{H}_{\mathrm{B}}$ to the dressed tensorization $\mathcal{H}_{\mathrm{C}} \otimes \mathcal{H}_{\mathrm{D}}$ eliminates both the interaction and the quantum correlations between the bare systems A and B. In the general case this is not possible. Since holistic quantum correlations between subsystems depend on the chosen tensorization, an appropriate criterion for an optimal choice of a dressing transformation is that the dressed tensor decomposition leads to almost uncorrelated subsystems.

Most theories of chemical, biological and social systems are still limited by classical paradigms presupposing, in an unreflected way, the separability of these systems. Yet, fundamental quantum theory refers to a whole which has no pre-existent intrinsic parts; separable subsystems exist only in exceptional cases.[104] Therefore the problem is not to explain the cause of entanglement between objects, but rather to explain why in a fully holistic system which has no parts it may make sense to speak of empirically discernible subsystems.

Every description of a holistic universe of discourse requires its decomposition by a partition, but even an approximate partition leads to serious problems. For example, the inevitable gravitational and electromagnetic interactions have to be taken into account. In particular, self-interactions of bare elementary systems acquire a very complicated structure.

Decompositions of the quantum world are neither given a priori nor determined by first principles. Even though a fully holistic system cannot be decomposed into non-trivial subsystems so that their states determine the state of the whole system, appropriate circumstances may allow us to decompose a quantum system in many inequivalent ways (like molecules, atoms, electrons, or photons) so that we can get approximate *partial contextual descriptions*. Every partition of the universe of discourse breaks the holistic unity of the world. This leads to a loss of information and entails *correlations between the resulting subsystems*.[105]

---

[104] Recall that for quantum systems without classical observables the set of all product states in the set of all pure quantum states has measure zero.

[105] In spite of the fact that quantum mechanics put an end to atomism, modern science is still to a large extent based on an atomistic ontology. Even today experts who know better seem to be unable to disengage themselves from this old metaphysical idea. For example, Wilczek (2002) claimed

Under appropriate circumstances, which reflect our interests and are biased by our theoretical preconceptions, we can deliberately suppress holistic quantum correlations, and thereby generate *contextual* decompositions which allow *contextual* descriptions of a universe of discourse in terms of *contextual* components, like molecules, atoms, electrons, or photons which arise as *contextually* induced secondary manifestations.

6.4 Symmetry and Symmetry Breakdown

6.4.1 Plato on Symmetry and Beauty

Beauty is that to which the human mind
responds at its deepest and most profound.

Subrahmanyan Chandrasekhar (1990, p. 54)

Plato stated in his *Philebus* the unity of beauty, symmetry and truth. Symmetry as we understand this concept nowadays refers to transformations that leave all relevant structures invariant. The symmetry groups of geometric objects date back to Plato's time. Platonic bodies (Plato did not discover them, but mentioned them in his *Timaeus*, written about 340 BC) are regular polyhedra with faces that are identical regular polygons. They come in five possible variants, each of which has its own symmetry group.

As proven by Euclid (in Book XIII of the *Elements*, around 300 BC), the five Platonic bodies are: the tetrahedron (group A_4), the octahedron (group S_4), the icosahedron (group A_5), the hexahedron (group S_4) and the dodecahedron (group A_5). The symmetry groups of Platonic bodies were mathematically established in the early 19th century. Since the symmetry group of any polyhedron coincides with that of its dual, there are only three symmetry groups associated with the Platonic bodies: the alternating group S_4 of degree 4, the alternating group A_4 of degree 4, and the alternating group A_5 of degree 5.

In many of his dialogues Plato wrote about *beauty*, "but he has no word equivalent to our 'symmetry', and this concept was not then formalised" (Lloyd 2010).[106] The modern group-theoretical view goes back to Évariste Galois who disclosed in his last letter to his friend Auguste Chevalier (in the night before he was shot in a duel in 1832) that the three Platonic symmetry groups are precisely the three exceptional Galois groups.

---

that the deepest and most revolutionary insight arising from quantum theory "is the very basic fact that matter is built up from a vast number of copies of a few fundamental components (such as electrons, quarks, photons and gluons". Later in the same publication he corrected this untenable statement by the remark that these so-called building blocks "arise as secondary manifestations—excitations of the fields".

[106]The Greek terms *summetria* and *summetros* refer to "common measure" and "well-proportioned".

6.4.2 Platonic Symmetries in Modern Science

In the Platonic tradition, contemporary physics searches for beauty, where the leading principle is the belief in an ultimate harmony of nature, based on fundamental symmetries and the associated invariances. Hermann Weyl expressed this faith as follows (Weyl 1952, p. 126):

> Symmetry, as wide or as narrow as you may define its meaning, is one idea by which man through the ages has tried to comprehend and create order, beauty and perfection.
> As far as I see, all a priori statements in physics have their origin in symmetry.

And in discussion with Freeman Dyson, Weyl confessed (Dyson 1956, p. 458):

> My work always tried to unite the true with the beautiful, but when I had to choose one or the other, I usually chose the beautiful.

In 1872 Felix Klein introduced the notion of invariance under a symmetry group in his Erlangen Program and used infinite continuous groups to classify the various geometries. In modern science symmetry is characterized by the invariance of some mathematical object under some group transformation. In particular, modern quantum theory strongly suggests to adopt an ontological Platonic conception of natural laws where symmetries are fundamental but hidden behind the phenomena of observable reality.

Since in a perfectly symmetric situation there are no distinctions, a Platonic universe is featureless and symmetries are not manifest at the level of directly observable phenomena. Observable patterns of the world come into being through symmetry breakings. In contrast to this unbroken wholeness, the world we experience emerges as shadows on the wall of Plato's cave by reduction of symmetry. The conceptual necessity to break symmetries in order to obtain phenomena was clearly recognized more than a century ago by Pierre Curie (1894a, p. 400): "C'est la dissymétrie qui crée le phénomène."

Spontaneous symmetry breakdown means that the fundamental equations have the symmetry, but the stable solutions of these equations do not. Nowadays, the so-called "fundamental symmetries" refer to hypothetical context-free situations. Such fundamental symmetries are not concrete properties of the world but statements about equations that describe the world.

According to Tsung-Dao Lee (1970, p. 2), the root of all symmetry principles lies "in the assumption that it is *impossible* to measure certain basic quantities". For example the *invariance under time translation* $t \to t + \tau$ follows from the assumption that absolute time is not observable. The invariance under time reversal $t \to -t$ follows from the fact that the dynamical laws of fundamental physics do not distinguish between past and future. They become discernible by breaking the time-reversal symmetry.

6.4.3 Theory and Experience

Theory and empirical research are not as closely linked as is often assumed. The first principles of physics are not sufficient for an operational description. For example, fundamental symmetries are never directly accessible by experiments, they can only be retrospectively inferred by contingent symmetry breakings. Albert Einstein (1949, p. 89) described this situation as follows:

> A theory can be tested by experience, but there is no way from experience to the setting up of a theory.

As discussed in Sect. 4.1.1, Wolfgang Pauli maintained the following position (Pauli 1957, p. 38; English translation from Pauli 1994, p. 129):

> I hope that no one still maintains that theories are deduced by strict logical conclusions from laboratory-books, a view which was still quite fashionable in my student days. Theories come into being through an understanding inspired by empirical material, an *understanding* which we may best regard, following *Plato*, as a coming into congruence [zur Deckung kommen] of internal images with external objects and their behavior. The possibility of understanding again demonstrates the presence of typical regulatory arrangements, to which man's inner as well as outer world is subject.

And, even more explicitly, Pauli (1952, p. 112; English translation from Pauli 1994, pp. 220f) advanced that

> from the standpoint of this not purely empiristic conception, which we also accept, there arises the question: What is the nature of the bridge between the sense perceptions and the concepts? All logical thinkers have arrived at the conclusion that pure logic is fundamentally incapable of constructing such a link. It seems most satisfactory to introduce at this point the postulate of a cosmic order independent of our choice and distinct from the world of phenomena. ...the conscious realization of new knowledge, seems thus to be based on a correspondence, a "matching" of inner images pre-existent in the human psyche with external objects and their behavior. ...On this level the place of clear concepts is taken by images with strong emotional content, not thought out but beheld, as it were, while being painted.

6.4.4 Science as a Symbolic Construction

In his apology of Socrates, Plato assumed that any true work of art arises essentially from divine inspiration (Plato 1891, Sec. 22c):

> Then I knew that not by wisdom do poets write poetry, but by a sort of genius and inspiration; they are like diviners or soothsayers who also say many fine things, but do not understand the meaning of them.

In spite of the fact that Plato in his *Republic* defined art as essentially mimetic and that Aristotle considered art as basically an imitation of nature, the function of art has been regarded as an expression of spiritual power for a long time.

History of art in the modern sense had its early precursors in the late medieval ages. Giotto di Bondone (1267–1337) was one of the first Renaissance painters who restored the naturalistic approach of the ancients, claiming that there must always be a likeness between the painting and the object painted. This view was not restricted to theories of artistic creation, but also for theories about the nature of discovery in science. Giotto inaugurated the early scientific method based on observations and first-hand experience. Later Filippo Brunelleschi (1377–1446) invented the single-point geometric optical linear perspective which gave rise to the first objective, systematic exploration of the illusion of depth, allowing for naturalistic styles.

In those days scientific knowledge was used to be highly valued as *neutral, value-free, free from prejudice and subjective influences*. In his famous Leiden address "The Unity of the Physical World-Picture" in 1908, Max Planck (1909; translation from Toulmin 1970, p. 25) still characterized this ideal by stressing the necessity of a "constant world picture" by *the complete liberation of the physical world-picture from the individuality of the creative mind:*

> A constant, unified world-picture is, as I have tried to show, the fixed goal which true natural science, in all its forms, is perpetually approaching; and in physics we may justly claim that our present world-picture, although it shimmers with the most varied colors imparted by the individuality of the researcher, nevertheless contains certain features which can never be effaced by any revolution, either in nature or in the human mind.

However, science and mathematics are human endeavors, and every scientific work depends not only on rational design but on irrational elements as well. In his introduction of the natural numbers, Richard Dedekind asked: "What Are and What Is the Purpose of Numbers?" In the preface to the first edition he wrote (Dedekind 1888, pp. VIIf):

> My answer to the problems propounded in the title of this paper is, then, briefly this: numbers are free creations of the human mind; they serve as a means of apprehending more easily and more sharply the difference of things. ... If we scrutinize closely what is done in counting an aggregate or number of things, we are led to consider the ability of the mind to relate things to things, to let a thing correspond to a thing, an ability without which no thinking is possible.

Later in the same text he explained (Dedekind 1888, Sect. 73, p. 21):

> With reference to this freeing the elements from every other content (abstraction) we are justified in calling numbers a free creation of the human mind.

Albert Einstein often used Dedekind's phrase of "free creations of the human mind" (German: "freie Schöpfungen des menschlichen Geistes"). He described the nature of physics in the following way (Einstein and Infeld 1939, p. 33):

> Physical concepts are free creations of the human mind, and are not, however it may seem, uniquely determined by the external world.

Like Einstein, Hermann Weyl emphasized that physical concepts as symbols "are constructions within a free realm of possibilities, freely created by the human mind" (Weyl 1934; quoted from Weyl 2009, p. 115). Weyl (1949b, p. 398) considered Hertz's symbolic conception of theories an adequate approach to mathematics and natural science. In various passages he expressed the view that science and mathematics are symbolic constructions of humans. Weyl (1985, p. 12):

> the power of science rests on the combination of experiment, i.e., observation under freely chosen conditions, with symbolic construction, and the latter is its mathematical aspect.

Weyl (1946, p. 218):

> the physicist's contemplation is not a purely passive attitude—it is creative construction, but construction in symbols, resembling the creative work of the musician.

Weyl (1949b, pp. 376, 381):

> The real is not given to us but posed as a problem. ...in place of a real spatio-temporal-material being we are only left with a construction in pure symbols. ...Man, if he is a creative spirit, is the architect of the world of symbols.

Weyl (1928b, p. 88):

> The individual assumptions and laws [of theoretical physics] have no meaning that can immediately be grasped; in principle, it is not the propositions of physics taken in isolation, but only the theoretical system as a whole, that can be confronted with experience. What is achieved is not a perceptual insight into particular or general states of affairs or a description that faithfully copies the given, but a theoretical, in the last analysis purely symbolic, construction of the world.

Inspirations do not follow from empirical facts by purely logical deduction—they fall into consciousness from somewhere. The imaginative and creative activities of scientists and artists are less different than this may seem (although artists are usually well-informed about the non-rational aspects of creative processes while many scientists often pretend or are even convinced that they proceed rationally). In February 19, 1949, Wolfgang Pauli wrote to Hermann Levin Goldschmidt (1990, p. 39):

> The symbol is always an abstract expression, be it quantitative or qualitative, be it mathematical-theoretical or emotionally valuated. Only part of the symbol can be expressed by conscious concepts, another part acts on the "unconscious" or "preconscious" state of an individual. The same holds for mathematical symbols, for only he is gifted for mathematics for whom mathematical tokens (in the sense mentioned above) have symbolic power. The symbol always is a tertium uniting opposites, what logic alone cannot "provide".

Pauli considered symbolic and quantitative descriptions in natural science as *complementary*. While symbolic descriptions concern both mental and physical aspects, traditional quantitative descriptions disregard mental activity and the archetypal origin of all ideas about the physical.[107]

Pauli used the concept of a symbol in the sense of Carl Gustav Jung, where a symbol never stands for anything that is known otherwise. That is, Jung makes a categorical distinction between the concept of a "symbol" and the concepts of a "sign", "mark", "metaphor", "allegory", "cipher", or "symptom". Jung (1971, par. 815) described a symbolic expression

> as the best possible formulation of a relatively *unknown* thing, which for that reason cannot be more clearly or characteristically represented.

---

[107]Compare Pauli's outline on "complementary modes of description", reproduced by von Meyenn (2009, pp. 22f).

According to Jung, a symbol does not represent an object drawn from the personal nexus, but is instead a message from the collective unconscious. As such, it is a spontaneous manifestation, though by no means a purely subjective, arbitrary invention. Accordingly, its message is expressed in the language of pre-existing images (archetypes), with their strong emotional overtones. For Jung (1971, par. 817) science also operates with symbols:

> Since every scientific theory contains a hypothesis, and is therefore an anticipatory description of something still essentially unknown, it is a symbol.

6.4.5 Fragile and Robust Descriptions

6.4.5.1 Explicit, Spontaneous and Dynamical Symmetry Breakings

Fundamental symmetries refer *to laws of nature*, expected to be universally valid, but *not to states of the world*. Most fundamental symmetries are *hidden*. Observable patterns come into being only through symmetry breakings. The link between fundamental laws and observable empirical facts is not determined by the first principles of physics. What is indiscernible under one context may not be so under another, so that in every application the fundamental symmetries are reduced to *effective contextual symmetries*.

According to Pierre Curie dissymmetry is caused by dissymmetry (Curie 1894a, p. 401):

> If certain causes produce certain effects, then the elements of symmetry of the causes must be found in the effects. When certain effects reveal a certain dissymmetry, then this dissymmetry must also be present in their causes.

In this formulation all extraneous structure is stripped away. For all practical purposes it is impossible to decouple a system from its surroundings. Symmetry implies degeneracy so that the inevitable presence of (even arbitrarily small) fluctuations can trigger responses that break symmetries and lift the related degeneracies.

- *Explicit symmetry breaking* refers to a situation where the dynamical equations are not manifestly invariant under the symmetry group considered. That is, the dynamics contains a (possibly small) term which is not invariant under the considered symmetry group.
- *Spontaneous symmetry breaking* refers to situations where, given a symmetry of the equations of motion, solutions exist which are not invariant under the action of this symmetry without any explicit asymmetric input. For example the periodic structures of crystals are not invariant under all spatial translations.
- *Dynamical symmetry breaking* refers to situations that are not invariant under the symmetry of a dynamical equation, without any explicit breaking of this symmetry. Examples which are connected with a spontaneous breakdown of a *gauge symmetry* are Bose-Einstein condensates, superfluids, superconductors, ferromagnets, anti-ferromagnets. Spontaneously broken gauge symmetries can

be characterized by a *local order parameter*. For example in a ferromagnet the spontaneous magnetization equals the expectation value of the spin observable associated with some microscopically small region.

In quantum theory the loss of symmetry requires both an infinite extension of the system and the existence of non-symmetric ground states, so that each ground state defines a physically disjoint realization of the system (see Strocchi 2008).

6.4.5.2 Bogoliubov's Constructive Method

Descriptions of physical systems with many mechanical degrees of freedom in terms of Hamiltonian mechanics pose a severe conceptual difficulty: it turns out that the exact description of systems with myriads of degrees of freedom is not robust. That is, if we start with a finite system with N degrees of freedom and let N increase, some expectation values become extremely sensitive with respect to small changes of the model or to small external perturbations. As a consequence, the model deteriorates *qualitatively* for small deviations from a reference model.

The behavior of non-robust Hamiltonian systems can be elucidated by Bogoliubov's (1970) constructive *method of quasi-averages*. One considers an expectation value $\langle A \rangle_{N,V}$ of an observable A of a system with N degrees of freedom and the Hamiltonian $H + \epsilon V$. Here H is the Hamiltonian of the object under study and ϵV ($\epsilon > 0$) is an arbitrarily small external perturbation. The physically correct limit for a very large number of degrees of freedom is then taken to be

$$\lim_{\epsilon \to 0} \lim_{N \to \infty} \langle A \rangle_{N,V}.$$

Usually this limit depends critically on the choice of the perturbation ϵV, and it is different from the unphysical limit

$$\lim_{N \to \infty} \lim_{\epsilon \to 0} \langle A \rangle_{N,V}.$$

Different perturbations lead to different limits, but none of them is *a priori* distinguished. An important consequence of Bogoliubov's method of quasi-averages is the insight that *symmetry breaking leads to robust descriptions*.

6.4.6 Breaking the Holistic Symmetry

Since the validity of all symmetry principles rests
on the theoretical hypothesis of non-observables,
the violation of symmetry arises whenever what was thought
to be a non-observable turns out to be actually an observable.

Tsung-Dao Lee (1981, p. 181)

A radically holistic theory describes unbroken wholeness. How do we come to recognize objects in a holistic world? Recall that a system is called holistic if it cannot be decomposed into non-trivial subsystems such that the states of the subsystems determine the state of the system as a whole. By contrast, a system is called separable if the states of the subsystems of every decomposition determine the state of the composite system. Holistic and separable systems are extreme cases along a spectrum of partially holistic and partially separable systems.

In the framework of the theory of W*-systems described by W*-algebras $\mathcal{A}$ and $\mathcal{B}$ of observables, the following two conditions are equivalent:

- Each state on the tensor product $\mathcal{A}\overline{\otimes}\mathcal{B}$ is decomposable;
- $\mathcal{A}$ or $\mathcal{B}$ is commutative.

In other words, entangled states exist if and only if both $\mathcal{A}$ and $\mathcal{B}$ are non-commutative. As a consequence, separable systems are characterized by *commutative* W*-algebras of observables. An example is classical mechanics with the phase space Γ where the corresponding commutative W*-algebra is given by $\mathbf{L}^{\infty}(\Gamma)$. The W*-algebra of a purely holistic system is a *factor*, i.e. an algebra with a trivial center consisting just of multiples of the identity operator. An example is traditional Hilbert-space quantum mechanics with the algebra $\mathcal{B}(\mathcal{H})$ of all bounded operators on some separable Hilbert space $\mathcal{H}$. The algebra $\mathcal{A}$ of observables of a partially holistic system can be written as a W*-tensor product $\mathcal{M} = \mathcal{A}\overline{\otimes}\mathcal{B}$ where $\mathcal{A}$ is a non-commutative W*-algebra and $\mathcal{B}$ is a commutative W*-algebra.

In order to describe a particular aspect of an intrinsically holistic reality we have to break the holistic non-Boolean symmetry such that the world separates into facts which permit a Boolean description. In every Boolean description of particular Boolean aspects it is unavoidable that certain other Boolean aspects are ignored. Therefore every Boolean description is appropriate only within the adopted partition of the world, that is, within the chosen context.

In a holistic theory "objects" do not exist in an absolute sense, but only in particular contexts, i.e. in the framework of a selected Boolean description. In this sense, observable phenomena are created by abstracting from holistic correlations (Primas 1981, p. 253).

6.4.7 Partial Boolean Algebras

The global structure of a quantum-theoretical description is given by an orthocomplemented lattice. Since every orthomodular partially ordered set can be represented by pasting together Boolean algebras (Rogalevicz 1988), the local structure of any quantum-theoretical description is given by a Boolean manifold (see also Sect. 3.4).

In its algebraic formulation the partial Boolean structure is realized by the set of Boolean algebras of the projectors in the underlying W*-algebra of observables. In this framework a Boolean context is defined by the jointly measurable observables or, more precisely, by a maximal commutative subalgebra of the W*-algebra of the universe of discourse. Though Boolean contexts yield only partial descriptions, they play a distinguished role for the description of the quantum world.

Since a quantum logical interpretation of the lattice of quantum propositions leads to difficulties (Primas 1980, Sects. 4.1–4.4, 1981, Chap. 4.4), we prefer to start with the unproblematic interpretation of Boolean contexts as the means by which the world is perceived. Every Boolean context with its classical two-valued logical structure represents a local chart for quantum events relative to which the results of measurements can be characterized. The projection lattice of a Boolean context is a complete two-valued Boolean lattice which provides a unique interpretation in terms of Kolmogorov's classical probability theory.

Boolean contexts represent the classical aspects of a quantum system. Overlaps between the various Boolean contexts characterize non-classical quantum correlations.

6.5 Bottom-Up Algebraic Quantum Theory

6.5.1 Bottom-Up Emergence in the Natural Sciences

Bottom-up emergence refers to the appearance of higher-level behavior of a system which is not evident from the point of view of a lower-level description. But it is *not correct to say* that emergent phenomena "do not appear as logically consequent on [the microscopic] laws" (Anderson 1995, p. 6653). This statement may make sense within a theory that is tentatively postulated as universally valid, but it is void as a context-independent proposition.

There is a widespread confusion about the meaning of emergence. For instance, *ontological emergence* is sometimes said to refer to the unpredictability of the behavior of complex systems—although the whole point of an *ontology* is to go beyond mere matters of knowledge. This and other issues show that philosophers without serious knowledge of natural science are not always helpful to clarify the situation. For example, the "philosopher of science" Charlie Dunbar Broad claimed that there are "emergent" molecular properties in the sense that they are not predictable by known principles of physics and chemistry (Broad 1925, p. 59):

> We will now pass to the case of chemical composition. Oxygen has certain properties and Hydrogen has certain other properties. They combine to form water, and the proportions in which they do this are fixed. Nothing that we know about Oxygen by itself or in its combinations with anything but Hydrogen would give us the least reason to suppose that it would combine with Hydrogen at all. Nothing that we know about Hydrogen by itself or in its combinations with anything but Oxygen would give us the least reason to expect that it would combine with Oxygen at all. And most of the chemical and physical properties of water have no known connexion, either quantitative or qualitative, with those of Oxygen and Hydrogen. Here we have a clear instance of a case where, so far as we can tell, the properties of a whole composed of two constituents could not have been predicted from a knowledge of the properties of these constituents taken separately, or from this combined with a knowledge of the properties of other wholes which contain these constituents.

This claim was unreasonable already in 1925, and today such an assertion is preposterous. The quote makes it evident that in natural science emergence is not an ontological notion which takes non-deducibility as the essential criterion. Rather, emergence is always *a context-dependent concept which refers to the adopted description*. Yet such a *contextual emergence* is not "merely an artefact of a particular model" (Silberstein and McGeever 1999, p. 182), but a reality which manifests itself perfectly well in an appropriate topology.[108]

Significant areas of exact natural science can be grasped by a context-independent basic theory of the universe of discourse, but no fundamental theory can describe *context-dependent aspects* of natural science. Introducing a non-ontological theory-based *hierarchical decomposition* of the universe of discourse, the concept of a contextual emergence makes sense. Then, the necessary context is given by the distinction between what we consider to be relevant or irrelevant for a particular experiment or observable. The context is imposed on the fundamental theory by restricting its domain and introducing a new, coarser topology (usually in terms of a partition).[109]

Of course, what is relevant or irrelevant is not given at the level of the world's fundamental features. The new topology is not given by first principles but depends on the abstractions made by the measuring devices or the pattern recognition devices used by the experimentalist. The descriptions at the various levels of a descriptive hierarchy are context-dependent and differ by the topologies chosen to describe what *we* select as *our* topics of investigation. On the one hand, these topologies restrict the workable domain of discourse, but they also "create" new phenomena in the sense of making them discernible. Although this may be surprising for the non-expert, contextually emergent phenomena such as molecular structure, toxicity of chemical substances, rigidity of solids, optical shadow are consequences of their underlying fundamental theory only under particular contextual constraints.

Contextual topologies can generate families of higher-level theories which do not "approximate" the fundamental theory but create new patterns of reality. Such new descriptions can be achieved by singular asymptotic expansions which do not converge in the original topology of the lower-level theory, or by choosing one of the infinitely possible physically inequivalent representations of the lower-level theory, such as in the GNS-construction of algebraic quantum mechanics (Primas 1998; see also Appendix A.3). Usually such higher-level theories cannot be totally ordered, so that they cannot be represented hierarchically. The validity domains of different higher-level theories usually intersect non-trivially, no domain is contained in any other. Therefore different higher-level theories may give rise to different emergent properties, although they are related to the same lower-level theory.

---

[108]Editor's note: Contextual emergence in this sense has been introduced as an interlevel relation in the philosophy of science by Bishop and Atmansapcher (2006). It has been fruitfully applied to numerous examples in the sciences, see Atmanspacher (2009) for an overview.

[109]In algebraic quantum mechanics a particular context can be introduced by imposing a new, contextually selected topology upon the state space of the C*-algebra of intrinsic observables with its intrinsic topology. For more details compare Sect. 6.5.3.8.

6.5.2 *On the Necessity of Considering Infinite Systems*

6.5.2.1 Descriptions Have to Be Robust

A description of a physical phenomenon makes sense only if "small" changes of the description leave the relevant conclusions "almost" the same. More precisely, a model is said to be *structurally stable* if its topological behavior is preserved under small perturbations (cf. Andronov and Pontriagin 1937). A model which is structurally stable and which depends continuously on initial conditions is called *robust*. An innocent use of non-robust mathematical models can easily lead to qualitatively wrong results.

As a rule, non-phenomenological descriptions of physical phenomena using Hamiltonian models with finitely many degrees of freedom are not robust. At the fundamental level there are no isolated finite system. Every physical system is interacting and entangled with the rest of the world, which we call its *environment*. To get a robust Hamiltonian description on the basis of first principles, we have to include the environment of the object system. Since every material system is inextricably coupled to the electromagnetic radiation field, a description of the environment requires a theory of systems with infinitely many degrees of freedom. This task goes beyond the limits of traditional Hilbert-space quantum mechanics which is restricted to finite systems.

There is a dramatic difference between the algebraic description of finite and infinite systems. For finite systems all Hilbert-space representations are unitarily and physically equivalent. For infinite systems there are infinitely many topologically and physically inequivalent representations. They are related to the appearance of superselection rules, spontaneous symmetry breakings, phase transitions and the emergence of classical observables. In a description with finite-dimensional models, this stupendous richness can only be grasped by regularizing non-robust Hamiltonian models, for instance by Bogoliubov's method of quasi-averages (see Sect. 6.4.5.2).

6.5.2.2 Quantum Systems with Infinitely Many Degrees of Freedom

Impressive as the success of traditional Hilbert-space quantum mechanics is, there is much that it does not explain in an obvious and candid manner. For example, in the traditional formulation there are no physical observables which commute with every observable. Therefore traditional quantum mechanics cannot explain many empirically well known classical features of quantum systems. Concrete examples are the chirality of molecules, the knot type of circular DNA molecules, or the temperature of chemical substances. Such empirical facts can be described in an *ad hoc* manner, but they cannot be rigorously derived from the principles of traditional Hilbert-space quantum mechanics alone.

In every scientific investigation we divide the universe into an object system and its environment—essentially the rest of the universe. In his posthumous book *Fundamental Theory*, Arthur Stanley Eddington (1946, p. 12) called attention to the inevitability of considering the environment:

> The environment must never be left out of consideration. It would be idle to develop formulae for the behavior of an atom in conditions which imply that the rest of matter of the universe has been annihilated. ... We can no more contemplate an atom without a physical universe to put it in than we can contemplate a mountain without a planet to stand on.

Nevertheless, traditional molecular theory is restricted to strictly isolated systems. Even in classical science the environment cannot be neglected, but in classical physics one can describe the influence of the environment suitably by external forces. The situation in quantum theory is radically different. The omnipresent EPR correlations imply that electrons, atoms, or molecules are not even defined in isolation. That is, the environment of a molecular system can never be left out of consideration. There is always the possibility that the environment induces symmetry breakings. In fact, our experience shows that symmetry breakings at the molecular and the macroscopic level are by no means exceptions, but represent the generic case.

Example: Chiral Molecules

Molecular chirality results from the breakdown of parity symmetry (i.e. space inversion symmetry) which has dramatic consequences for living organisms. The fact that most biomolecules (like the amino acids, the sugars, lipids, or steroids) are chiral shows that chirality and chiral recognition is a key concept in the understanding of many biomolecular interactions such as drug-receptor interplay.

Enantiomeric molecules can induce different sensory responses. For example, L-glutaminic acid has a meaty flavor, while D-glutaminic acid is almost tasteless, R-carvone has a caraway odor, while S-carvone has a spearmint odor. The reality of this breakdown of the superposition principle of traditional Hilbert-space quantum mechanics at the molecular level is dramatically demonstrated by the Contergan tragedy in the 1960s, which caused many severe birth defects.

Contergan was the trade name of the drug thalidomide (3-phralimido-2,6-dioxopiperidin, $C_{13}H_{10}N_2O_4$) which exists in two different chiral variants. The left-handed stereoisomer of thalidomide is a powerful and perhaps safe tranquilizer, but the right-handed isomer is a disastrous teratogenic agent, causing physiological deformities in the developing embryo and foetus (DeCamp 1989). Ignorance of stereochemical effects led to the approval of the racemic mixture of the two enantiomers as a drug for marketing in the 1960s.

The molecular Hamiltonian of thalidomide is parity invariant so that its ground state is expected to be symmetric with respect to parity. But the really existing ground state is chiral. The racemic substance is a classical mixture of the two enantiomers, including the teratogenic variant. In contrast, the parity invariant ground state would be a coherent superposition of left-handed and right-handed states with altogether different properties. This fact shows impressively how wrong it is to say that "there is no operational procedure that can distinguish between a 'true' mixture and an apparent mixture caused by a correlation" (Zurek 1991, p. 46).

Molecular superselection rules derive from the coupling of a molecule to an environment with infinitely many degrees of freedom. For example, if a hypothetically isolated molecule has an almost degenerate ground state, it is usually unstable against interactions with the infinitely many soft photons of extremely low frequency (the so-called infrared part of the molecular radiation field; Primas 1978).

The inclusion of the radiation field generates inequivalent sectors and thereby induces molecular superselection rules (cf. Pfeifer 1980, 1983a, 1983b; Amann 1988, 1991a, 1991b, 1991c).

Molecular reality with its phenomenologically well-confirmed superselection rules contradicts the irreducibility postulate of traditional Hilbert-space quantum mechanics. Unfortunately, most modern textbooks on elementary quantum mechanics are still based, almost dogmatically, either on Dirac's bra-ket formalism (and its rigorous implementation in terms of rigged Hilbert spaces), or von Neumann's irreducible Hilbert-space formalism. Neither of these approaches includes the essential further developments of the mathematical formalism which are crucial for a proper understanding even of elementary quantum phenomena.

Yet, already in their first paper *On Rings of Operators*, Murray and von Neumann (1936, p. 121) observed that the algebras of observables used in von Neumann's (1932a) monograph "do not exhaust all possibilities". Von Neumann himself soon became unconvinced of his own account of 1932. In 1935 he wrote to Garrett Birkhoff:[110]

> I would like to make a confession which may seem immoral: I do not believe absolutely in Hilbert space any more. . . . Now we [with F.J. Murray, von Neumann's coauthor] begin to believe that it is not the *vectors* which matter . . . [because they] ought to represent the physical *states*, but they do it redundantly, up to a complex factor, only.

In two pioneering papers (Jordan et al. 1934; von Neumann 1936), von Neumann attempted to generalize quantum mechanics in a purely algebraic way, by focusing on observables which are not necessarily realized as operators acting on a Hilbert space.

Traditional Hilbert-space quantum mechanics is restricted to the discussion of isolated systems with finitely many degrees of freedom. Von Neumann had not postulated this approach *ad hoc*; rather he had based it on the *uniqueness theorem by Stone and von Neumann* which says that for finitely many degrees of freedom there is (up to unitary equivalence) only one irreducible Hilbert-space representation of the canonical commutation relations. Years later, a generalization by Mackey (1949) and Loomis (1952) showed that this uniqueness up to unitary equivalence and multiplicity even holds for projective representations of arbitrary locally compact groups.

These uniqueness theorems imply that the traditional Hilbert-space framework precludes classical observables, classically inequivalent states and time-reversal symmetry breakdown (which is necessary for the distinction of past and future). Although an *ad hoc* decomposition of the Hilbert space of state vectors into a direct sum or a direct integral of coherent subspaces can indeed describe superselection rules and classical observables, these new rules had to be put in by hand and were no logical consequence of the formalism.

---

[110] Letter of von Neumann to Birkhoff of November 13, 1935, quoted from von Neumann (2005, p. 59).

It may be tempting to try a generalization of the Lebesgue-Schrödinger representation of the canonical commutation relations over the configuration space $\mathbb{R}^n$ with $n < \infty$. While for $n < \infty$ the Abelian group $\mathbb{R}^n$ is locally compact and its essentially unique translation-invariant Haar measure is given by the Lebesgue measure, the group $\mathbb{R}^\infty$ is not locally compact and there exists no Haar measure on it. More generally, there exists no translation-invariant measure on any infinite-dimensional topological vector space, so that for these cases the uniqueness theorem for Weyl's canonical commutation relations fails.

As a consequence, there are inequivalent representations. Their existence is implicit in a paper by von Neumann (1938) on infinite tensor products. Later, Cameron and Martin (1947) proved the singularity of scale transformations in Wiener space. Their results also imply the existence of uncountably many inequivalent representations of the canonical commutation relations. However, these insights were not appreciated by physicists at the time. Non-Hilbert (or non-Fock) representations were regarded as pathological features of no physical interest.

6.5.2.3 Inequivalent Representations

In the 1950s, it finally became manifest that inequivalent representations are a necessity in any non-trivial relativistic quantum field theory. Van Hove (1952) and Haag (1955) showed by simple models that inequivalent representations can be physically relevant. There exist many quasi-invariant measures which can be used to define perfectly legitimate representations of Weyl's canonical commutation relations. The simplest examples are generated by Gaussian measures which lead to uncountably many unitarily inequivalent irreducible representations. For details compare Gel'fand and Vilenkin (1964, Chap. IV, Sect. 5.4) and Hida (1980, Sect. 5.8).

Most of these inequivalent representations cannot be constructed easily but fortunately they can be dealt with in a mathematically rigorous fashion within the framework of algebraic quantum mechanics. Inequivalent representations of the canonical commutation relations are indispensable, not only for quantum field theory and for the description of phase transitions, but also in solid state physics, molecular physics and chemistry.

If we adopt a Hamiltonian description with finitely many degrees of freedom, then we have only one natural topology available: the topology induced by the associated finite-dimensional phase space of the canonical commutation relations. In the infinite-dimensional case there are many topologically and physically inequivalent representations of the canonical commutation relations, so that we have to select an appropriate one among them. This choice is context-dependent and reflects our ideas about what is relevant and what is not.

In a mathematical description the distinction between relevant and irrelevant properties is mirrored in the chosen contextual topology. This reflects the topological complexity of phase spaces which are not locally compact, and is by no means a peculiar features of quantum theory. A mathematically equivalent situation is well known in the theory of classical stochastic processes (cf. Hida 1980, Chap. 3.4).

6.5.3 *Algebraic Quantum Theory*

6.5.3.1 C*-Algebraic Quantum Mechanics

A proper discussion of infinite systems exceeds the domain of traditional Hilbert-space quantum mechanics or Fock-space quantum field theory. For example, in these frameworks one cannot derive, in a mathematically sound way, for which kind of molecules there is a strict superselection rule between right- and left-handed enantiomers.

Algebraic quantum mechanics is a full-fledged theory which covers traditional Hilbert-space quantum mechanics, thermodynamics with its phase transitions[111] and classical mechanics as special cases and provides a mathematical explanation for superselection rules and the breakdown of symmetries. It can include the all-important influence of the environment. No real physical system is closed, every material system couples to its environment, which always includes electromagnetic and gravitational fields—systems with infinitely many degrees of freedom. Even weak couplings of small molecular systems with their environment can lead to symmetry breakings and emergent classical observables. In order to understand the influence of the environment on atomic, molecular, mesoscopic, and macroscopic systems as well as their stability, we have to discuss the behavior of open systems coupled to an environment with infinitely many degrees of freedom (Primas 1977, 1981, Sect. 5.6).

In the algebraic codification of physical theories (including classical point mechanics as well as type-I and type-III quantum mechanics as special cases) a physical system is characterized by a C*-algebra $\mathcal{A}$ algebra of observables,[112] where states are realized by a positive linear functional ρ on $\mathcal{A}$,

$$\rho \in \mathcal{A}^*, \qquad \rho(\mathbb{1}) = 1, \qquad \rho\big(A^*A\big) \geq 1 \quad \text{for every } A \in \mathcal{A}. \tag{6.72}$$

The set $\mathcal{S}(\mathcal{A})$ of all states on $\mathcal{A}$ is a convex and weak*-compact subset of the dual $\mathcal{A}^*$ of $\mathcal{A}$, so that by the Krein-Milman theorem the state space $\mathcal{S}(\mathcal{A})$ is the weak*-closed convex hull of the set $\mathcal{P}(\mathcal{A})$ of its extreme points (cf. Bratteli and Robinson 1979, Sect. 2.3.2). The elements of $\mathcal{P}(\mathcal{A})$ are called pure states since they cannot be decomposed into different states. More precisely: A state functional $\rho \in \mathcal{S}(\mathcal{A})$ is pure if and only if $\rho := p\rho_1 + (1-p)\rho_2$ with ρ_1, ρ_2 and $0 < p < 1$ implies $\rho = \rho_1 = \rho_2$.

---

[111]For example, ferromagnetism involves the spontaneous breakdown of the rotation group, crystallization requires the spontaneous breakdown of the translation and rotation group, superfluidity is related to the breakdown of the special Galilei group, while superconductivity is connected with the spontaneously broken gauge group.

[112]Editor's note: Historically, the term "C*-algebra" was introduced by Segal (1947), where "C" was intended to indicate that the algebra is "closed" in the norm topology on $\mathcal{B}(\mathcal{H})$ (see Doran and Belfi 1986, p. 6).

6.5.3.2 W*-Algebraic Quantum Mechanics

An important special kind of C*-algebras are W*-algebras.[113] A C*-algebra is called a W*-algebra if it is a dual space of a Banach space (cf. Sakai 1971, p. 1). That is, for every W*-algebra there exists a unique Banach space $\mathcal{M}_*$ such that $\mathcal{M} = (\mathcal{M}_*)^*$, where $(\mathcal{M}_*)^*$ is the dual Banach space of $\mathcal{M}_*$. The Banach space $\mathcal{M}_*$ is called the *predual* of the W*-algebra $\mathcal{M}$.

W*-algebras are distinguished among all C*-algebras by the property that $\mathcal{M} = \mathcal{M}''$. The so-called *double commutant theorem* implies that a W*-algebra $\mathcal{M}$ has enough projections such that $\mathcal{M}$ is the norm closure of the span of its projections, and that intersections and unions of projections stay within the algebra (cf. Appendix A.1.3). A linear positive functional φ on a W*-algebra $\mathcal{M}$ is said to be *completely additive* if it satisfies $\varphi(\bigvee F_n) = \sum \varphi(F_n)$ for every set $\{F_n\}$ of pairwise orthogonal projections in $\mathcal{M}$, $F_n F_m = 0$ for $n \neq m$.

Complete additivity is equivalent to normality: a linear positive functional φ is said to be *normal* when $\varphi(M_n) \uparrow \varphi(M)$ for each monotonically increasing net $\{M_n\}$ of operators M_n in $\mathcal{M}$ with least upper bound M. Normal state functionals are elements of the predual $\mathcal{M}_* \subset \mathcal{M}^*$ of the W*-algebra $\mathcal{M}$. A normal state on a W*-algebra defines a countably additive *probability measure* over its projection lattice so that normal states are operationally accessible. That is, in a W*-algebraic description *epistemic states* are realized by normal state functionals. The corresponding normal state space is the predual $\mathcal{A}_* \subset \mathcal{A}^*$ of weak* linear state functionals on the W*-algebra $\mathcal{A}$ of observables.

6.5.3.3 Non-Commutative Probability Theory and Gleason's Theorem

For a commutative W*-algebra $\mathcal{M}$ the projection lattice $\mathcal{P}(\mathcal{M})$ is distributive, i.e. a complete Boolean lattice, the point of departure of Kolmogorov's classical probability theory. Non-commutative W*-algebras give rise to a *non-Boolean probability theory*—a generalization on which the statistical description of quantum systems is based. A probability measure $\mu : \mathcal{P}(\mathcal{M}) \to [0, 1]$ on the orthomodular lattice $\mathcal{P}(\mathcal{M})$ is characterized by the following three postulates:

- μ is strictly positive, i.e. $\mu(F) \geq 0$ for every $F \in \mathcal{P}(\mathcal{M})$, and $\mu(F) = 0$ implies $F = 0$,
- μ is normalized, i.e. $\mu(\mathbb{1}) = 1$,
- μ is σ-additive for compatible and disjoint events, i.e. $\mu(\bigvee F_n) = \sum_n F_n$ for every countable sequence $\{F_n\}$ of pairwise orthogonal projections $F_n \in \mathcal{P}(\mathcal{M})$, $F_n \perp F_m$ for $n \neq m$.

[113]Editor's note: Today, W*-algebras are often called von-Neumann algebras, going back to the work of Murray and von Neumann (1936) on "rings of operators". One may speculate that the "W" derives from closure under the "weak" operator topology on $\mathcal{B}(\mathcal{H})$. Note that there is a more recent notion of W-algebras (without *) due to Zamolodchikov (1985) which has nothing to do with W*-algebras.

Given a probability measure μ on the projection lattice $\mathcal{P}(\mathcal{M})$ of a W*-algebra $\mathcal{M}$, one may ask whether there exists a normal state functional $\rho_\mu \in \mathcal{M}_*$ such that μ is the restriction of ρ_μ to $\mathcal{P}(\mathcal{M})$ with $\mu(F) = \rho_\mu(F)$ for every $F \in \mathcal{P}(\mathcal{M})$. It is not at all trivial that a probability measure on $\mathcal{P}(\mathcal{M})$ can be extended to a linear functional on $\mathcal{M}$. For the statistical interpretation of traditional Hilbert-space quantum mechanics, this linear extension is established by a famous theorem by Gleason (1957). In spite of many efforts, the generalization of Gleason's theorem to general W*-algebras remained unsolved for over twenty years. Nowadays this problem is completely solved; compare the excellent review by Maeda (1990).

Generalized Gleason Theorem[114]
Let $\mathcal{M}$ be a W*-algebra without direct summand of type I_2. Every σ-additive probability measure μ on the complete orthomodular lattice $\mathcal{P}(\mathcal{M})$ of projections in a W*-algebra $\mathcal{M}$ can be extended to a normal linear state functional $\rho_\mu \in \mathcal{M}_*$.

6.5.3.4 Central Decomposition

Algebraically, von Neumann's irreducibility postulate for Hilbert-space quantum mechanics implies that the algebra of observables of traditional quantum mechanics is a factor of type I. In a seminal paper, von Neumann (1949) established the decomposition theory for general W*-algebras. For any W*-algebra $\mathcal{M}$ with a separable predual $\mathcal{M}_*$, its center $\mathcal{Z} := \mathcal{M} \cap \mathcal{M}'$ determines the essentially unique (up to measure zero) direct integral decomposition

$$\mathcal{M} = \int^{\oplus} \mathcal{M}_\nu d\mu(\nu), \tag{6.73a}$$

where μ is a central measure. If the measure μ has discrete support, this direct integral decomposition becomes a direct sum decomposition

$$\mathcal{M} = \oplus_k \mathcal{M}_k, \tag{6.73b}$$

where the range of k may be finite or infinite.

Correspondingly, every state functional can be decomposed uniquely into a sum or integral of mutually disjoint factor state functionals. For an arbitrary state functional ρ of the W*-algebra $\mathcal{M}$, such a central decomposition of ρ is given by

$$\rho = \int^{\oplus} \rho_\nu d\mu(\nu), \tag{6.74}$$

[114]Editor's note: Roughly speaking, Gleason's theorem states that a quantum state is completely determined by only knowing the answers to all possible yes/no questions. Excluding factors of type I_2 means to exclude representations in Hilbert spaces of dimension 2. For POVMs, an analog of Gleason's theorem is valid without this restriction (Busch 2003).

where μ is a probability measure. If the measure μ has discrete support, the central decomposition is the direct sum,

$$\rho = \oplus_k \rho_k. \tag{6.75}$$

The central decomposition is the finest unique decomposition of a non-pure state into a classical mixture, which renders possible an ignorance interpretation with respect to this central decomposition into disjoint factor states. However, in general a non-pure factor state functional allows infinitely many physically different decompositions into pure states. In traditional quantum mechanics without superselection rules, where all states are realized by factor state functionals, an ignorance interpretation is illegitimate. The only permissible ignorance interpretation of non-pure quantum states refers to the central decomposition into a classical mixture of mutually disjoint factor state functionals.

6.5.3.5 Pure and Non-Pure State Functionals

If a system with the C*-algebra $\mathcal{A}$ is described by the state functional ρ, then the *reduced state functional* of a subsystem described by the C*-algebra $\mathcal{B} \subset \mathcal{A}$ is given by the restriction $\rho_{\mathcal{B}}$ of ρ to $\mathcal{B}$,

$$\rho_{\mathcal{B}}(B) := \rho(B) \quad \text{for all } B \in \mathcal{B},\ \rho_{\mathcal{B}} \in \mathcal{B}^*. \tag{6.76}$$

As a rule, the restriction of a pure state functional is not pure. Every non-pure state functional ρ can be decomposed in a highly non-unique way into a sum or an integral of pure state functionals ρ_k

$$\rho = \sum_{\kappa} \mu_k \rho_{\kappa} \quad \text{or} \quad \rho = \int d\mu(\kappa) \rho_{\kappa}. \tag{6.77}$$

The fact that a system consisting of an object system and its environment may be in a pure state without the object system being in a pure state is a typical quantum phenomenon reflecting the non-separability of quantum systems. A reduced quantum state functional allows the evaluation of all expectation values of the observables of the object system but gives no information about the holistic correlations between the object system and its environment. Since non-pure factor states cannot be interpreted as mixtures of pure state, it is conceptually misleading to call them "mixed states" (compare Sect. 6.2.5).

6.5.3.6 Superpositions and Superselection Sectors

The superposition principle of quantum mechanics has nothing to do with the linearity of the equations of motion. If one out of two different pure states can produce a new pure state which is different from the two generating pure states, this new state is called a coherent superposition. The superposition principle of quantum mechanics is the postulate that for every two pure states there exists at least one coherent superposition. If this superposition principle is valid, it easily follows that we cannot only construct one new pure state but uncountably infinitely many of them.

In the irreducible Hilbert-space formalism, a pure state can be represented by a ray, and the ray by a state vector. In this representation, the superposition principle takes a mathematically most simple (but perhaps conceptually misleading) form, namely: If Ψ_1 and Ψ_2 are two state vectors representing two different pure states, then any linear combination $\Psi = c_1\Psi_1 + c_2\Psi_2$ with $|c_1|^2 + |c_2|^2 = 1$ represents a coherent superpositions of the two generating states.

A representation-independent formulation is more complicated but very general and conceptually more illuminating. It is based on the concept of intrinsically perturbed states. A perturbed state functional ρ_X is defined as a modification of the state ρ induced by an intrinsic perturbation $X \in \mathcal{A}$ (Haag and Kastler 1964):

Definition: Intrinsically Perturbed States
Let ρ be a state functional on a C*-algebra $\mathcal{A}$, and X an arbitrary element of $\mathcal{A}$. The modification ρ_X of the state ρ induced by an intrinsic perturbation $X \in \mathcal{A}$ is defined by

$$\rho_X(A) := \begin{cases} \dfrac{\rho(X^*AX)}{\rho(X^*X)} & \text{for every } A \in \mathcal{A} \quad \text{if } \rho(X^*X) \neq 0 \\ \rho(A) & \qquad\qquad\qquad\qquad \text{if } \rho(X^*X) = 0 \end{cases} \tag{6.78}$$

With this definition it follows that

- if ρ is a state functional on a C*-algebra $\mathcal{A}$, then ρ_A is also a state functional on $\mathcal{A}$.
- if ρ is a factorial state functional on a C*-algebra $\mathcal{A}$, then ρ_A is also a factorial state functional on $\mathcal{A}$.
- if ρ is a pure state functional on a C*-algebra $\mathcal{A}$, then ρ_A is also a pure state functional on $\mathcal{A}$.
- if $\mathcal{A}$ is a commutative algebra, then $\rho = \rho_X$ for every $X \in \mathcal{A}$.

Definition: Superselection Rules and Sectors
A state functional $\widetilde{\rho}$ is called coherent with the state functional ρ if there exists an element $X \in \mathcal{A}$ such that $\widetilde{\rho} = \rho_X$. A maximal collection of mutually coherent pure states is called a superselection sector (or, briefly, a sector). If two pure states are not related in this way, we say that there is a superselection rule between them.

6.5.3.7 Faces and Folia in the State Space of a C*-Algebra

In many applications, the convex subset of a state space is of special interest. A non-empty convex subset $\mathcal{F}$ of the state space $\mathcal{A}^*$ is called a *face*. Since the elements of the state space are norm-continuous functionals, we consider only faces which are closed in the norm topology. For a given C*-algebra $\mathcal{A}$, there is a one-to-one correspondence between projections in the bidual $\mathcal{A}^{**}$ and the norm closed faces in the state space $\mathcal{A}^*$ (Effros 1963; Prosser 1963). If $\mathcal{F}$ is a norm-closed face in the state space $\mathcal{A}^*$, then there exists a unique projection F in the bidual $\mathcal{A}^{**}$, called the carrier projection of F, such that

$$\mathcal{F} = \{\rho \in \mathcal{A}^* \mid \rho(F) = 1\}, \quad F = F^2 = F^* \in \mathcal{A}^{**}. \tag{6.79}$$

Following Haag et al. (1970) and Sewell (1973), a folium is a norm-closed face $\mathcal{F}$ which is invariant under "quasi-local perturbations" in the sense that:

$$\rho \in \mathcal{F} \quad \text{and} \quad X \in \mathcal{A} \quad \Longrightarrow \quad \rho_X \in \mathcal{F}. \tag{6.80}$$

Theorem: The W*-Algebra Generated by a Folium
Let $\mathcal{F}$ be a folium in $\mathcal{A}^*$. Denote the norm-closed linear span of $\mathcal{F}$ by $[\mathcal{F}]$. Then the dual $[\mathcal{F}]^* \subseteq \mathcal{A}^{**}$ is a W*-algebra.

The concept of a folium generalizes the notion of a superselection sector. The smallest folium containing a particular state functional $\rho \in \mathcal{A}^*$ is denoted by $\mathcal{F}(\rho)$, and is given by

$$\mathcal{F}(\rho) := \{\rho_X | X \in \mathcal{A}\}. \tag{6.81}$$

A state functional ρ is pure if and only if $\mathcal{F}(\rho)$ is the singleton set $\{\rho\}$. It follows that two state functionals in the same folium are equivalent, while state functionals in different folia are disjoint:

- ρ and φ are equivalent: $\rho \sim \varphi \iff \mathcal{F}(\rho) = \mathcal{F}(\varphi)$, (6.82a)
- ρ is contained in φ: $\rho \leq \varphi \iff \mathcal{F}(\rho) \subseteq \mathcal{F}(\varphi)$, (6.82b)
- ρ and φ are disjoint: $\rho \top \varphi \iff \mathcal{F}(\rho) \cap \mathcal{F} = \emptyset$. (6.82c)

Folia arise naturally as the set of *normal* states of W*-representations. If $\mathcal{M}^\pi := \pi(\mathcal{A}'')$ denotes the W*-algebra generated by the representation π, the folium $\mathcal{F}_\pi$ belonging to π is $\mathcal{F}_\pi = \mathcal{M}^\pi_*$. In other words, the W*-algebra $[\mathcal{F}_\pi]$ is W*-isomorphic to the W*-algebra $\mathcal{M}^\pi$. Therefore there is also a one-to-one correspondence between the folia in the state space $\mathcal{M}^\pi = \pi(\mathcal{A})''$ and the generating folium $\mathcal{F}_\pi$. This entails a one-to-one correspondence between the folia in the state space $\mathcal{A}^*$ and equivalence classes of representations of the C*-algebra $\mathcal{A}$. Two representations π and $\bar{\pi}$ of a C*-algebra $\mathcal{A}$ are equivalent if and only if the associated W*-algebras are equivalent,

$$\pi \sim \bar{\pi} \quad \iff \quad \pi(\mathcal{A})'' = \bar{\pi}(\mathcal{A})'' \quad . \tag{6.83}$$

Accordingly, the weak closure $\pi(\mathcal{A})''$ of a representation $\pi(\mathcal{A})$ is a well-motivated extension of the represented C*-algebra $\mathcal{A}$—it explicates an intrinsic structure of $\mathcal{A}$ described by the folium $\mathcal{F}_\pi$.

For infinite systems, the dual $\mathcal{A}^*$ of the C*-algebra $\mathcal{A}$ of context-independent observables is not separable, so that the universal representation $\pi^{\text{uni}}(\mathcal{A})''$, which is isomorphic to the W*-algebra $\mathcal{A}^{**}$, is not a decent algebra of statistical observables appropriate for a statistical description. On the other hand, it is natural to require that the folium $\mathcal{F}$ relevant for a robust statistical description is not too small. It is reasonable to require that $\mathcal{F}$ is weak*-dense in the intrinsic state space $\mathcal{A}^*$. This condition implies that the partially universal representation of $\mathcal{A}$ associated with such a folium is faithful. To such a folium $\mathcal{F} \subset \mathcal{A}^*$ one can associate a *partially universal representation* $(\pi, \mathcal{H})$ of the C*-algebra $\mathcal{A}$,

$$\pi := \bigoplus_{\rho \in \mathcal{F}} \pi_\rho, \qquad \mathcal{H} := \bigoplus_{\rho \in \mathcal{F}} \mathcal{H}_\rho, \tag{6.84}$$

where $(\pi_\rho, \mathcal{H}_\rho)$ denotes the cyclic GNS-representation (according to Gel'fand and Neumark 1943 and Segal 1947) induced by the state functional ρ. Furthermore, there is an associated W*-algebra

$$\mathcal{M}^\pi := \pi(\mathcal{A}'') \subset \mathcal{B}(\mathcal{H}) \tag{6.85}$$

with center

$$\mathcal{Z} = \pi(\mathcal{A}'') \cap \pi(\mathcal{A}'). \tag{6.86}$$

There exists a central projection C_π such that

$$\mathcal{M}_\pi = C_\pi \mathcal{M}^{\text{uni}}, \qquad \mathcal{Z}_\pi = C_\pi \mathcal{Z}^{\text{uni}} \subset \mathcal{Z}^{\text{uni}}, \tag{6.87}$$

where $\mathcal{M}^{\text{uni}}$ is the universal representation with center $\mathcal{Z}^{\text{uni}}$.

6.5.3.8 Emergence in Algebraic Quantum Theory

In algebraic quantum theory, a particular context can be introduced by imposing a contextually selected topology upon the state space of the C*-algebra $\mathcal{A}$ of intrinsic observables in the context-independent description (Primas 1998). This new topology has to be compatible with the algebraically determined, hence context-independent norm topology of $\mathcal{A}$. It can be implemented by a particular reference state, given by a positive linear state functional ρ on the context-independent C*-algebra $\mathcal{A}$ (see also Sect. 6.5.1).

The GNS-construction then provides a context-dependent Hilbert space $\mathcal{H}_\rho$ and an associated faithful representation $\pi_\rho(\mathcal{A})$ of the C*-algebra $\mathcal{A}$ acting on $\mathcal{H}$.[115] The closure of $\pi_\rho(\mathcal{A})$ in the weak topology of the algebra $\mathcal{B}(\mathcal{H}_\rho)$ is a context-dependent W*-algebra $\mathcal{M}_\rho$, called the algebra of *contextual observables* with respect to the contextual topology generated by the reference state functional ρ. The relation $\mathcal{A} \sim \pi_\rho(\mathcal{A}) \subset \mathcal{M}_\rho \subset \mathcal{B}(\mathcal{H}_\rho)$ implies that all *intrinsic observables* reappear in $\mathcal{M}_\rho$, but in addition there are new contextual observables which are not intrinsic (i.e., not elements of $\pi_\rho(\mathcal{A})$).

Observables in the W*-algebra $\mathcal{M}_\rho$ of contextual observables, which are outside the representation of the C*-algebra $\pi_\rho(\mathcal{A})$, are also called *emergent observables*. They represent properties which are novel in the sense that they are absent in the

[115] Compare for example Takesaki (1979), or Bratteli and Robinson (1979).

more fundamental context-independent C*-algebraic description. *The emergence of novelty in contextual descriptions is a compelling fact in algebraic quantum theory.* Examples of *emergent classical observables* are temperature (Takesaki 1970) and chemical potential (Müller-Herold 1980), which arise in a most natural manner from a GNS-construction with respect to canonical Kubo-Martin-Schwinger (KMS) states (Kubo 1957; Martin and Schwinger 1959).

A basic common feature in all cases of emergent behavior is the transition from a context-independent to a contextual description. If the interaction of an object system with its environment is not excluded, there are in general infinitely many *physically inequivalent* representations of the object system. Choosing one of these representations means to select a particular context for the epistemic description. *The simplicity of natural laws manifests itself only in the context-independent C*-algebraic description, while a W*-algebraic representation of the richness of observable phenomena requires the multitude of inequivalent representations.*

6.5.4 Quantum Measurement—The Algebraic Way

6.5.4.1 Disjoint States and the Measurement Problem

We consider the so-called "measuring process" as a dynamical process which is governed by the usual deterministic evolution laws of quantum mechanics. Human observers, or even their consciousness, play no role whatsoever (compare footnote 85 on p. 104).

The model of a measurement of the first kind (Sect. 6.2.6) presupposes an observable $A \in \mathcal{B}(\mathcal{H})$ with purely discrete spectrum. In this model, the result of the measurement is one of the eigenvalues $a_1, a_2, \ldots,$

$$A|\alpha_n\rangle = a_n|\alpha_n\rangle, \quad \alpha_n \in \mathcal{H}. \tag{6.88}$$

If the initial state of the object system is given by the state vector $\Psi \in \mathcal{H}$ with

$$\Psi = \sum_n c_n|\alpha_n\rangle, \quad c_n = \langle\Psi|\alpha_n\rangle, \tag{6.89}$$

then the statistical state after the measurement is given by the density operator

$$D = \sum_n |c_n|^2|\alpha_n\rangle\langle\alpha_n|, \tag{6.90}$$

so that the result of an individual measurement is supposed to be given by some eigenvalue a_n with probability $|c_n|^2$. Yet this argument is invalid for various reasons. If we reject the unfounded indeterministic collapse postulate (6.21), the density operator (6.90) represents a non-pure quantum state, and not a classical mixture.

For a classical mixture it is presupposed that we can *distinguish* operationally between its components: *It must be possible to label every component of a proper mixture so that the components can be distinguished.* Since such a label must be determinable together with any other property of the component, it has to be characterized by the value of a *classical observable*. More precisely: *Non-pure quantum states can be interpreted in terms of a classical mixture of factor states if and only if these factor states are mutually disjoint.* Since in von Neumann's Hilbert-space codification there are no disjoint states, non-pure states cannot be interpreted in terms of an observer's ignorance about an actually pure state of the system. In other words, an "ignorance interpretation" is not feasible in traditional quantum mechanics (see Sects. 6.2.5 and 6.5.3.4).

6.5.4.2 Measurements Produce Facts

Disjoint states are of crucial importance as final states in *any* processes—natural processes or measurement processes—which produce facts. The measurement problem is not, as often asserted, the problem of how a pure statistical state can be transformed into a non-pure state, or how the density operator can become diagonal in a preferred basis. This is a straightforward task—appropriate dynamical linear semigroups and their Hamiltonian dilations can describe such a *decoherence* mechanism. The same is true for the so-called "stochastic unraveling of dynamical semigroups of statistical descriptions": in the framework of traditional quantum mechanics there are always infinitely many stochastic differential equations for pure states which in the statistical average result in one and the same statistical dynamical semigroup.

A proper statistical description of the measurement process requires a dynamics which transforms factor states into a classical mixture of disjoint factor states. Consequently, a quantum system can act as a measurement apparatus only if it can produce disjoint final states.[116] A measurement apparatus has to act as an irreversible causal dynamical system which breaks the time-inversion symmetry. In quantum theory this possibility exists only when the apparatus is described as a system with infinitely many degrees of freedom.

We take the algebra of observables of the system to be measured as a factorial W*-algebra $\mathcal{M}_{\text{sys}}$ of type I. The apparatus is described by a C*-algebra $\mathcal{A}_{\text{app}}$ with a quasi-local structure. The GNS-representation π_ρ with respect to a state functional ρ on $\mathcal{A}_{\text{app}}$ generates the Hilbert space $\mathcal{H}_\rho$, the cyclic vector $\Omega \in \mathcal{H}_\rho$, and the W*-algebra for the apparatus,

$$\mathcal{M}_{\text{app}} := \pi_\rho(\mathcal{A}_{\text{app}})'' , \tag{6.91}$$

so that the W*-algebra of the joint system is:

$$\mathcal{M} = \mathcal{M}_{\text{sys}} \overline{\otimes} \mathcal{M}_{\text{app}} . \tag{6.92}$$

[116] Pauli wrote to Bohr in a letter of March 11, 1955: "As is well known to both of us, it is essential in quantum physics that the apparatus can be described by classical concepts" (cf. von Meyenn 2001, p. 147).

The emergence of classical observables in this algebra is not trivial. A general result due to Klaus Hepp (1972, lemma 2, p. 246) shows that automorphisms preserve the disjointness of states:

> If φ' and φ'' are two disjoint state functionals on a C*-algebra and if α is an automorphism of this C*-algebra, then the transformed state functionals $\varphi' \circ \alpha$ and $\varphi'' \circ \alpha$ are disjoint.

This theorem is not surprising since an automorphism is a symmetry which transforms a description into a fully equivalent description. However, the fact that an automorphic time evolution on any C*-algebra cannot generate new disjoint states is not a "no-go theorem" for the possibility of a theoretical description of measurement-type processes. For automorphic time evolutions Hepp showed that there exist quantum systems with an automorphic dynamics $\{\alpha_t | t \in \mathbb{R}\}$ such that for equivalent initial state functionals φ' and φ'' the asymptotic limits $\varphi' \circ \alpha$ and $\varphi'' \circ \alpha$ exist for $t \to \infty$ and are *disjoint*. Such state functionals are called *asymptotically disjoint.*

It has been objected that processes with asymptotically disjoint final states require an infinite measurement time.[117] This is a misunderstanding: *every measurement in engineering physics is asymptotic* (for details see Primas 1997, 2000). Since there is no scientifically sound way to determine an exact point in time for the death of a biological system, the death of Schrödinger's cat is also an asymptotic process.

6.5.4.3 The Measurement Process Must Be Irreversible

Hepp's crucial contribution did not solve the statistical measurement problem, since it is possible to reverse the measuring process for any finite time by a quasi-local perturbation (Bell 1975). Since facts cannot be undone, a measuring process has to be *irreversible*. The decisive idea of combining the non-anticipative character of laboratory instruments with a dynamics generating asymptotically disjoint final states was worked out by Lockhart and Misra (1986). In the proposed dynamics the broken time-inversion symmetry is given by a weakly contractive semigroup (based on a K-flow) which produces asymptotically disjoint final states in a strictly irreversible manner. The emergence of disjoint states occurs then progressively over finite amounts of time.

Let us assume that the observable $A \in \mathcal{M}_{\text{sys}}$ has a non-degenerate purely discrete spectrum. The key for the proper description of the measurement process is the insight that the measurement process is an irreversible dynamical system which breaks the time-inversion symmetry in such a way that it acts as a causal system.

---

[117] John Bell (1975, pp. 93, 98) insisted "that the 'rigorous reduction' does not occur in physical time but only in an unattainable mathematical limit" and "so long as we do not know exactly when and how it takes over from the Schrödinger equation, we do not have an exact and unambiguous formulation of our most fundamental physical theory". Similarly, Landsman (1995, p. 55) spoke of a "disaster of infinite measurement time" and demanded that we are "supposed to see instantly what the (final) macroscopic state of the pointer is, for otherwise we would not accept it as a measurement device".

This is only possible if the apparatus is a system with infinitely many degrees of freedom. We describe the apparatus by a C*-algebra $\mathcal{A}_{\text{app}}$ with a quasi-local structure. The GNS-representation π_ρ (see above) then generates the W*-algebra $\mathcal{M}_{\text{app}} = \{\pi_\rho(\mathcal{A}_{\text{app}})\}''$, and the W*-algebra $\mathcal{M} = \mathcal{M}_{\text{sys}} \overline{\otimes} \mathcal{M}_{\text{app}}$ of the joint system.

To incorporate irreversible causal behavior one has to select a W*-algebra $\mathcal{M}_0 \subset \mathcal{M}$ of observables which one can actually observe in a causal system. Let $\mathcal{K}_0 = \mathcal{M}_0(\Psi \otimes \Omega) \subset \mathcal{M}_*$ be the corresponding subspace of allowed physical states. The causality requirement means that $\mathcal{K}_0$ has to be closed under the forward time evolution, $U_t\mathcal{K}_0 \subset \mathcal{K}_0$, $t > 0$, where $\mathcal{M}_t = \alpha_t(\mathcal{M}_0)$. In terms of normal conditional expectations, $\mathcal{E}_t = \alpha_t \mathcal{E}_0 \alpha_{-t}$ with $\mathcal{E}_0(\mathcal{M}) = \mathcal{M}_0$, one gets $\mathcal{E}_s \leq \mathcal{E}_t$ for $s \leq t$. The family $\{\mathcal{E}_t | t \in \mathbb{R}\}$ is either constant (Hepp's case, where time-inversion symmetry holds), or increasing as t increases (so that time-inversion symmetry is broken).

The evolution of an observable B is given by $B \rightarrow \mathcal{E}_0\alpha_t(B) := W_t(B)$, and the evolution of a state φ_0 is given by $t \rightarrow W_t^*\varphi_0$, where $\{W_t^*\varphi_0\}(B) = \varphi_0\{W_t(B)\}$. In the case of broken time-inversion symmetry, $\{W_t^* | t \geq 0\}$ is a weakly contractive semigroup on those states which become disjoint (for which a "collapse" occurs) for $t \rightarrow \infty$. This process is gradual, i.e. classical observables in $\mathcal{M}$ emerge progressively as a function of time. The W*-algebra $\mathcal{M}_0$ and the automorphism group α_t give rise to a generalized K-flow in the sense of Emch (1976).

6.5.4.4 Decision Processes

Like any measurement also measurements induced by asymptotically disjoint states have residual errors and therefore require a decision strategy. Decision theory has developed a broad spectrum of techniques for minimizing this error. As an example, we consider the simplest statistical decision test that is commonly used in experimental science.[118]

Decision procedures use attributes of empirical observations to achieve a *Boolean classification* of facts. Since empirical data are invariably contaminated with noise, all experimental observations have to be considered to be subject to random variations. For a formal theory of classifications the cause of these random variations is irrelevant. They may be due to measurement errors, external noise, imperfect experimental procedures, or residual non-Boolean quantum effects.

In the framework of statistical classification theory an observation is considered as a sample value $x = X(\omega)$, $\omega \in \Omega$, of a random element X, defined on a Kolmogorov probability space (Ω, Σ, μ). We consider the special case of a binary decision. Assume that an experimenter knows that an observation $x = X(\omega)$ comes either from a distribution with probability measure μ' or from a distribution with probability measure μ''. A statistical test is a method to assign the observation x to one of the two distributions. The error of a statistical test is defined as the probability

---

[118]For a review of the application of statistical decision theory in engineering science compare for example Chaps. 18–23 in Middleton (1960).

that the decision is false. Independently of how the test is performed, the *minimal error probability* is given by (Rényi 1966, 1967)

$$e_{\min}\big(\mu', \mu''\big) = \inf_{\mathcal{B}\in\Omega}\big\{\mu'(\mathcal{B}) - \mu''(\Omega - \mathcal{B})\big\}. \tag{6.93}$$

The minimal error probability can be estimated using the Hellinger integral $H(\mu', \mu'')$:[119]

$$H\big(\mu', \mu''\big) = \int_\Omega \sqrt{f'(\omega)f''(\omega)}d\mu(\omega), \tag{6.94a}$$

$$f' = d\mu'/d\mu, \qquad f'' = d\mu''/d\mu, \quad 0 \le H\big(\mu', \mu''\big) \le 1, \tag{6.94b}$$

$$1 - \sqrt{1 - H\big(\mu', \mu''\big)^2} \le e_{\min}\big(\mu', \mu''\big) \le H\big(\mu', \mu''\big). \tag{6.94c}$$

Here μ is any dominating measure (for example $\mu = \frac{1}{2}\mu' + \frac{1}{2}\mu''$), so that both μ' and μ'' are absolutely continuous with respect to μ. Since the convergence of the Hellinger integral to zero is equivalent to the convergence of the minimal error probability to zero, we can use the Hellinger integral $H(\mu', \mu'')$ as a criterion for the "nearness" of the two probability measures μ' and μ''. The smaller the Hellinger integral, the easier is it to distinguish the two measures from each other. If the Hellinger integral vanishes, a perfect decision can be made with probability one.

6.5.4.5 Approximate Measurements

For finitely many degrees of freedom n, the W*-algebra $\mathcal{M}_0$ of the joint system is a factor of type I_n with $n < \infty$, so that it can be generated by a unitary Weyl system $\{W(\zeta)|\zeta \in \mathbb{C}^n\}$

$$W(\zeta)W\big(\zeta'\big) = e^{\sum_k \zeta_k^* \zeta_k - \sum_k \zeta_k \zeta_k^*} W\big(\zeta + \zeta'\big), \quad \zeta \in \mathbb{C}^n \tag{6.95a}$$

$$\mathcal{M}_0 = \big\{W(\zeta)|\zeta \in \mathbb{C}^n\big\}''. \tag{6.95b}$$

We define a coherent state vector $|z\rangle \in \mathcal{H}_\rho$ via the state-generating function $z \mapsto \langle z|W(\zeta)|z\rangle$,

$$\langle z|W(\zeta)|z\rangle = e^{-\sum_k |\zeta_k|^2 + \sum_k z_k^* \zeta_k - \sum_k z_k \zeta_k^*}, \quad z_k \in \mathbb{C}. \tag{6.96}$$

Let ρ be a normal state functional for the W*-algebra $\mathcal{M}_0$ for the system with n degrees of freedom. The expectation value of the projector $|z\rangle\langle z|$ with respect to ρ

[119]See Kraft (1955, Lemma 1, p. 127). The Hellinger integral was originally introduced by Hellinger (1909) in his investigation of unitary invariants of self-adjoint operators (cf. also Hahn 1912) and applied to statistical contexts by Bhattacharyya (1943). Kakutani (1948) stressed the fact that the Hellinger integral is an inner product. The Hellinger integral was re-introduced by Matusita (1951) under the name "affinity".

is a positive probability distribution function, called the *Husimi function* $z \mapsto h_\rho(z)$,

$$h_\rho(z) = \rho\{|z\rangle\langle z|\}, \quad h_\rho(z) \geq 0, \; z \in \mathbb{C}^n. \tag{6.97}$$

The positivity properties of Husimi functions can be used to discuss the exact or approximate disjointness of normal state functionals. Consider two normal factor state functionals $\varphi, \phi \in (\mathcal{M}_0)_*$, their associated positive Husimi functions h_φ and h_ψ, and the Husimi function

$$H_{\varphi,\phi} = \int_{\mathbb{C}^n} \sqrt{h_\varphi(z) h_\psi(z)} d^n z. \tag{6.98}$$

Then we obtain the following classification:

$$\varphi \text{ and } \psi \text{ are equal if and only if} \quad H_{\varphi,\phi} = 1, \tag{6.99a}$$

$$\varphi \text{ and } \psi \text{ are classically equivalent if and only if} \quad H_{\varphi,\phi} > 0, \tag{6.99b}$$

$$\varphi \text{ and } \psi \text{ are disjoint if and only if} \quad H_{\varphi,\phi} = 0. \tag{6.99c}$$

This result implies that the Hellinger integral can be used as a criterion for the approximate disjointness of two factor state functionals insofar as $H_{\varphi,\phi}$ measures the degree of disjointness of the two state functionals φ and ψ. The smaller the value of $H_{\varphi,\phi}$, the more the state functionals φ and ψ behave like disjoint states. If $H_{\varphi,\phi} = 0$, the minimal error probability vanishes so that a perfect decision can be made with probability one. In engineering science such error-free decisions are called singular and are rejected as unrealistic idealizations (cf. Root 1963, 1964, 1968). For the quantum-mechanical measurement process this is reflected by the fact that the value $H_{\varphi,\phi} = 0$ can only be approached in the limit of infinitely many degrees of freedom and in the limit of infinite measuring time. Under these ("unrealistic") conditions, arbitrarily precise decisions about disjointness are possible.

In models describing the measurement process statistically with asymptotically disjoint final states, the effective measuring time for concrete experiments can be fairly short. If $\varphi(t) = W_t^* \varphi(0)$ and $\phi(t) = W_t^* \phi(0)$ are two quantum state functionals which for $t \to \infty$ become asymptotically disjoint,

$$\lim_{t\to\infty} \lim_{n\to\infty} \int_{\mathbb{C}^n} \sqrt{h_{\varphi(t)}(z) h_{\psi(t)}(z)} d^n z = 0, \tag{6.100}$$

then the Hellinger integral $\int_{\mathbb{C}^n} [h_{\varphi(t)}(z) h_{\psi(t)}(z)]^{1/2} d^n z$ is arbitrarily small for sufficiently large $t < \infty$ and $n < \infty$. To determine what "sufficiently large" means, one usually sets an appropriate *threshold* adapted to the experimental situation. This is a difficult task because there are many factors that can vary from experiment to experiment. If we choose the threshold level ε for the error probability given by the Hellinger integral, then the necessary number N of degrees of freedom and the minimal measuring time T are determined by:

$$\varepsilon = \int_{\mathbb{C}^N} \sqrt{h_{\varphi(T)}(z) h_{\psi(T)}(z)} d^N z. \tag{6.101}$$

Chapter 7
Top-Down Approaches in Physics

7.1 Limitations of Bottom-Up Approaches

7.1.1 Phenomena Which Cannot Be Easily Understood Bottom-Up

7.1.1.1 Temperature and the Zeroth Law of Thermodynamics

Many philosophers have claimed (e.g., Bunge 1973; Scheibe 1988) that "the classical theory of heat has been reduced to statistical mechanics" (Kemeny and Oppenheim 1956), or that the concept of temperature can be reduced to statistical mechanics by a *mere definition*, namely that of the mean kinetic energy $\frac{3}{2}kT$ of gas molecules, which Nagel (1961) referred to as a *bridge law*. Yet the concept of temperature in phenomenological thermodynamics is incommensurable with that of kinetic theory (Feyerabend 1962, p. 78).

Temperature is neither a mechanical nor a molecular concept, but it is defined via the zeroth law of thermodynamics. This law states: If two systems are in thermal equilibrium with a third system, they are also in thermal equilibrium with each other (Fowler and Guggenheim 1939, p. 56). Systems fulfilling this highly non-trivial transitivity relation "have a property in common, namely the property of being in thermal equilibrium with each other. This property is called *temperature*" (Guggenheim 1949, p. 6; see also Carathéodory 1909, and Born 1921, §2.)

The holistic structure of quantum mechanics implies that the zeroth law cannot be derived from the first principles of quantum theory. The main stumbling block for a bottom-up explanation of the zeroth law is the presupposed separability of thermal systems which is non-trivial within the holistic structure of quantum mechanics (cf. Sect. 7.4.1). Why are there no Einstein-Podolsky-Rosen correlations in thermodynamics?

7.1.1.2 Chemically Pure Substances

Phenomenological thermostatics, as formulated by van der Waals (1927), does not depend on a lower-level description, but is in many respects a much richer theory

H. Primas, *Knowledge and Time*,
DOI 10.1007/978-3-319-47370-3_7

than statistical mechanics. The concept of pure chemical substances plays a central role for a systematic structure of knowledge in chemistry. However, *chemical purity is not a bottom-up concept*. Popular definitions like "substance X is chemically pure if and only if X is composed exclusively of either atoms or molecules of a single species or kind" (Bunge 1985, p. 225) are factually wrong and misleading.

For example, chemically pure liquid water and water vapor are made up of two parts of hydrogen and one part of oxygen, but not of H_2O molecules. Infrared spectra of water vapor show that water is a mixture of many different molecules, among others of monomers (H_2O), dimers $(H_2O)_2$, trimers $(H_2O)_3$ and other polymers $(H_2O)_n$ (cf. Primas 1981, p. 317, 1985).

A proper grasp of chemical purity is provided by *classical thermostatics which does not rely on the molecular view of matter.* That a molecular definition of pure chemical substances is not feasible is discussed at great length in the outstanding monograph on thermostatics by van der Waals (1927, §5, §92, §93). A proper definition of chemical purity can be given in terms of the Gibbs phase rule, which is free from assumptions as to the molecular condition of the substances involved. Gibbs' phase rule cannot be derived from mechanical first principles since no bottom-up theory tells us how many states of aggregations of matter are possible. We do not even know how to characterize quantum-theoretically liquids in terms of molecular interactions and correlations.

7.1.2 Experiments as Human Actions

In the opening lines of his book *The Astonishing Hypothesis*, Francis Crick (1994, p. 3) still emphasized the ideological position of an uncompromising physicalist reductionism:

> You, your joys and your sorrows, your memories and your ambitions, your sense of personal identity and free will, are in fact no more than the behavior of a vast assembly of nerve cells and their associated molecules.

Yet, such a crude explanation certainly has not worked so far, and it becomes more and more questionable that it will work at any future time. There are autonomous top-down actions which are functionally independent of lower-level processes. As George Ellis (2008, p. 76) stated, "top-down causation from the mind to the world underlies all human planning and action". For example, every experimental investigation presupposes a specific design and implementation which is compatible with, but not exclusively determined by bottom-up constructions from physical laws (Ellis 2009, p. 87):

> All scientific experiments are based on purposeful activity and free will, enabling decisions based in abstract analysis that lies beyond the explanatory scope of physical science.

Even more succinctly, Ellis (2008, p. 77) noted that "every physics experiment is an example of intelligent top-down causation".

Bottom-up theories alone cannot give a complete understanding of the world. For example, it is an unresolved problem up to today why and how an experimenter is free to choose the initial conditions in a physical experiment (cf. Sect. 8.1). It was a most consequential insight by Isaac Newton that any description of the world must contain two categorically different elements: *initial conditions* and *laws of nature*. Without initial conditions, no equation of motion could be solved, and without equations of motion, any description would just be a collection of facts.[120]

Furthermore, every stimulus-response experiment requires the distinction of past and future (breakdown of time-inversion symmetry) and the concept of a temporal "now" (breakdown of time-translation symmetry). It is impossible to deduce these requirements on the basis of the first principles of bottom-up physics alone. However, the fact that any description of the behavior of matter solely based on bottom-up principles is fundamentally incomplete does not imply that the first principles of physics are inconsistent or invalid. They are just not sufficient to account for an exhaustive description.

Since no experiment refers to a closed system, the description of any experiment requires an irreversible dynamics. The fundamental laws of bottom-up physics are all time-reversible. Any breakdown of this time-reversal symmetry results in two solutions, one being the time reverse of the other. But there are no valid local bottom-up theories that explain why time unfolds at all, why its proper direction points from past to future, and why the opposite direction seems to be not realized.

Since local time-reversal maps are not completely positive, the time-reversal map of an entangled system is *global* in the sense that it cannot be decomposed into local maps. There is only one known physical solution to the problem of the arrow of time: a top-down action by the universe as a whole, perhaps expressed by boundary conditions at the beginning of space-time, that supports one (forward) solution and rejects the other (Ellis 2002, pp. 2668f, 2008, Example 4.1.5).

7.1.3 Is Physics Causally Closed?

Modern natural science crucially depends on observations and experiments, but in its traditional presentations the role of intentional human actions is reduced to a minimum, and inner experiences are ignored as much as possible. In his famous address *On the Limitations of Nature*, Emil du Bois-Reymond (1872, p. 6) asserted that (editor's translation)

> Insight in natural science ... is the reduction of changes in the world of physical bodies to the motion of atoms which is produced by time-independent central forces or the resolution of natural processes in the mechanics of atoms.

---

[120]Compare Houtappel et al. (1965, p. 596) and Wigner (1964b). The distinction between initial conditions and laws of nature is due to Newton's (1729) *Philosophiae Naturalis Principia Mathematica*, even though in an implicit fashion.

Already in 1925, Moritz Schlick defended the metaphysical doctrine of the *causal closure* of the physical world (quoted from Schlick 1985, pp. 303f):

> Everything that happens in that world is determined solely by physical magnitudes. This principle ... is usually called the "principle that causality in nature is closed". ... Its validity rests on the fact that the natural sciences must banish the sense qualities from their completed conceptualization ... The physicist's world is complete in itself; the world of the psychologist cannot be fitted into it.

The alleged closure (also referred to as completeness) of physics implies that anything that has a physical effect must itself be physical. This is one of the reasons why many philosophers of science have taken *physicalism* for granted. Physicalism started as an ontological doctrine of the *Vienna Circle* of logical positivists and says (roughly) that ultimately everything is explainable by physics.[121] Otto Neurath (1983, p. 56) stated this view as follows:

> Physicalism is the form work in unified science takes in our time. Whatever else is said in pronouncements is either "meaningless" or merely a means to emotion, "poetry".

To define physicalism as the doctrine that all facts, including intention and meaning, are reducible to physical facts, or as the claim that every physically acceptable effect has a physically acceptable cause, is not as clear as it seems. Carnap's (1932) slogan "Physical Language as the Universal Language of Science", and his later superficial explication that physicalist language is a "thing language" (Carnap 1936, pp. 466f) is hardly helpful:

> Let us give the name "physical language" to that language which is used in physics. It contains the thing-language and, in addition, those terms of a scientific methodology which we need for a scientific description of the processes in inorganic matter.

Popular statements such as "physics is causally closed" or "physicalism is the view that all entities, properties, relations, and facts are those which are studied by physics or other physical sciences" are fairly empty if no explanation is given what is meant by "physics". However that may be, there is little agreement on how physicalism should be properly understood in detail. In any case, physics does not have the authority to tell us what exists in nature. In particular, if in a special case we can speak of causal processes, then this refers to our choice of a state space, that is to a *description*, not to the ontology of the world.

The claim that physics is closed and the idea of physicalism indicate a fundamental misunderstanding of physics.[122] There are a number of non-trivial presuppositions of the natural sciences, such as the necessity to distinguish between the

---

[121]The term *physicalism* was introduced into philosophy by Otto Neurath (1932, p. 405, editor's translation): "There are no 'explanations' which would not be physical statements", and Rudolf Carnap (1932, p. 463): "Each matter of fact in science can be explained as a physical matter of fact, i.e. by quantitatively determinable properties of a space-time position". Later, Herbert Feigl (1963, pp. 227f) stated that "physicalism claims that the facts and laws of the natural and the social sciences can all be derived—at least in principle—from the theoretical assumptions of physics".

[122]Editor's note: Although the causal closure of the physical is still defended by influential philosophers of science today (e.g., Spurrett 1999; Papineau 2001), there is an increasing number of cautious and critical voices as well (e.g., Montero 2003; Bishop 2006; Atmanspacher and Filk 2012).

observer and the observed. Physics is based on experiments and fundamental theories. In order to test, or support, a hypothesized cause-effect relationship experimentally, it is indispensable that the experimenter has *the freedom to deliberately choose* (within well-understood limits) a stimulus and then to record the response. Clearly we cannot make measurements on a closed system on which outside influences are disregarded by definition. Yet, there are no known first principles for open systems: the first principles of physics refer to strictly closed and isolated systems where no distinction between cause and effect is possible.

Since all first principles of physics are *invariant under time translation and time reversal*, concepts such as "past", "present", or "future" have no place in fundamental physics. Accordingly there are no physical laws which cover memory and intentions. Memory of past episodes is a necessary precondition of personal identity—in the sense that a person at one time and a person at another time can be said to be the same person.

We conclude that the space-like time of physics does not take into account all aspects of time. In particular, the "arrow of becoming" cannot be consistently integrated into a universally valid picture of physics. Since physics systematically leaves out human intentions, the first principles of physics are not even enough to describe physical experiments or engineering physics exhaustively. The limits of any natural science (even more so if understood in a physicalist spirit) are not the limits of human knowledge.

7.2 Complementarity of Bottom-Up and Top-Down Descriptions

> It seems to me that a systematic foundation of quantum mechanics should be based on the composition and decomposition of systems, to a greater extent than this has been done so far.
>
> *Wolfgang Pauli*[123]

Bottom-up descriptions are based on the idea that the behavior of a system is completely determined by the behavior of its subsystems. Top-down descriptions take into account that the whole is more than the sum of the parts. Quantum theory represents a most radical version of such an approach, due to which parts do not intrinsically exist but arise contextually. The behavior of the contextual parts is determined by the behavior of the system as a whole, so that determination points downward. In bottom-up descriptions, the contextual parts of top-down descriptions appear as "emergent properties" which cannot be reduced to properties of the parts.

Both bottom-up and top-down approaches are indispensable components of scientific research in general. Any description is to some degree constrained by the parts and to some degree constrained by the whole. That is, one has to consider both

[123]From a letter by Pauli to Heisenberg of June 15, 1935 (von Meyenn 1985, p. 404, editor's translation).

upward and downward pathways of determination. A theoretically sound and empirically fruitful method to implement this insight has been proposed as *contextual emergence* (see Sect. 6.5.1, Atmanspacher 2009).

In contextual emergence, lower-level descriptions provide only necessary but not sufficient conditions for higher-level descriptions. The missing sufficient conditions are available as contexts at the higher level, from where they induce their top-down implementation at the lower level and, thus, substitute the lacking sufficient conditions. Avoiding the misleading notion of "downward causation" one might prefer to speak of "downward constraints" in the sense that all processes at the lower level of a hierarchy are *restrained by and act in conformity with* the laws at the higher level. We may say with Haken (1983) that the higher-level dynamical variables "enslave" lower-level dynamical variables.

From another, more general point of view, the interplay between bottom-up and top-down approaches has been expressed by comparing *constructive theories* with *theories of principle* (Einstein 1919, pp. 13):

> There are several kinds of theory in physics. Most of them are constructive. These attempt to build a picture of complex phenomena out of some relatively simple proposition. The kinetic theory of gases, for instance, attempts to refer to molecular movement the mechanical, thermal, and diffusional properties of gases. When we say that we understand a group of natural phenomena, we mean that we have found a constructive theory which embraces them.
> But in addition to this most weighty group of theories, there is another group consisting of what I call theories of principle. These employ the analytic, not the synthetic method. Their starting point and foundation are not hypothetical constituents, but empirically observed general properties of phenomena, principles from which mathematical formula are deduced of such a kind that they apply to every case which presents itself. Thermodynamics, for instance, starting from the fact that perpetual motion never occurs in ordinary experience, attempts to deduce from this, by analytic processes, a theory which will apply in every case.
> The merit of constructive theories is their comprehensiveness, adaptability, and clarity; that of the theories of principle, their logical perfection, and the security of their foundation.
> The theory of relativity is a theory of principle. To understand it, the principles on which it rests must be grasped. But before stating these it is necessary to point out that the theory of relativity is like a house with two separate stories, the special relativity theory and the general theory of relativity.

Arthur Stanley Eddington, Secretary of the Royal Astronomical Society, pointed out in a report on the general theory of relativity (Eddington 1918, preface):

> Whether the theory ultimately proves to be correct or not, it claims attention as one of the most beautiful examples of the power of general mathematical reasoning. The nearest parallel to it is found in the application of the second law of thermodynamics, in which remarkable conclusions are deduced from a single principle without any inquiry into the mechanism of the phenomena.

And Heinrich Hertz's brilliant deduction of Maxwell's equations was not based on directly observed phenomena, but on his "principle of the unity of electric force and that of the unity of magnetic force" (Hertz 1896, p. 374) and the identity of light with electromagnetic waves. Hertz used non-formal criteria like *simplicity* and *beauty* to characterize Maxwell's equations (Hertz 1892, p. 23):

> To the question of what Maxwell's theory is, I know of no shorter or more definite answer than this: Maxwell's theory is the system of Maxwell's equations.

7.3 Local Quantum Physics

There was a most ingenious architect,
who had contrived a new method for building houses,
by beginning at the roof, and working downward to the foundation;
which he justified to me by the like practice
of those two prudent insects, the bee and the spider.

Jonathan Swift (1995, Part I, Chapter V)

Already Roger Joseph Boscovich (1711–1787), a Croatian polymath who worked in Italy and France, rejected atomism as a basis of physics. Later, Michael Faraday (1791–1867) replaced Newton's instant action-at-a-distance between discrete particles with an action-at-a-neighborhood principle and established with his *field concept* the basis for the classical electromagnetic field theory. Einstein changed this action-at-a-neighborhood principle into the Minkowski space-time causality principle in the sense that causes act locally in space-time.

Local quantum field theory is a space-time quantum theory which considers space-time as relational in the philosophical spirit of Leibniz (cf. Sect. 7.3.5.4). Its point of departure are not elementary particles but fields. Quantum field theory addresses systems with an infinite number of degrees of freedom. This is radically different from ordinary quantum mechanics, where physical systems are parametrized by an infinite number of dynamical degrees of freedom. Local quantum field theory is a *holistic top-down theory* that emphasizes the concept of locality, which cannot be properly understood in a bottom-up manner.

The first attempts in this direction dealt with the problem of creating a Lorentz-covariant quantum theory of the electromagnetic field, representing the free electromagnetic field as an infinite set of harmonic oscillators (Jordan and Wigner 1928; Heisenberg and Pauli 1929). The initial step toward such a theory of quantum electrodynamics, which included electric charges and currents and the emission of photons, was the electron theory by Dirac (1928).

Quantum electrodynamics and its successor, quantum chronodynamics (considered as the fundamental theory of strong interactions with its spin $\frac{1}{2}$-fermionic quarks and spin 1-bosonic gluons as fundamental entities), are still bottom-up theories concerned with the ultimate constituents of the universe. They were guided by the idea that a hierarchical structure of matter can be built up in terms of operator-valued quantum fields $x \mapsto \varphi(x)$, which assign to each space-time point x

in Minkowski space-time a field value $\varphi(x)$. This Lagrangian quantum field theory with point-like fields is an extension of quantum mechanics with point-like elementary systems. Both adopt a reductive view in the sense that all behavior of matter can be reduced to a basic set of localized building blocks.

A severe challenge to this idea is the fact that in Lagrangian quantum field theory the concept of a particle is observer-dependent.[124] Moreover, since all generalizations of the particle concept of free quantum fields to interacting systems are inadequate, it has been concluded that a general relativistic quantum field theory cannot be interpreted in terms of particles (Fraser 2008). According to Haag (1992, p. 105),

> in quantum physics just as in classical physics the concept of "field" serves to implement the principle of locality. In particular, a "quantum field" should not be regarded as being more or less synonymous with a "species of particles". ... Free fields are just convenient artifacts.

7.3.1 The Holistic Structure of Local Quantum Physics

In contrast to Lagrangian quantum field theory, the *top-down* description that is achieved by a quantum field theory based on type-III W*-algebras is in a profound manner *holistic*. For a type-III factor there is an association between a unique equilibrium state and the one-parameter group of modular automorphisms of the W*-algebra. The basic concepts of the *modular theory* of W*-algebras were first discovered by the mathematician Minoru Tomita and later developed by Masamichi Takesaki. It is therefore also called the *Tomita-Takesaki theory* (see Appendix A.5).[125]

According to the modular theory of W*-algebras by Tomita and Takesaki, a type-III W*-algebra possesses an intrinsic one-parameter group of automorphisms. It is given by the group of homomorphisms

$$\mathbb{R} \xrightarrow{\delta} \mathrm{Out}(\mathcal{M}) = \mathrm{Aut}(\mathcal{M})/\mathrm{Inn}(\mathcal{M}) \tag{7.1}$$

from the group $\mathbb{R}$ to the group of automorphism classes of a type-III W*-algebra $\mathcal{M}$ modulo inner automorphisms. Alain Connes discovered this fact in 1972 when working on the Tomita-Takesaki theory (Connes 2001, p. 8). This purely mathematical theory yields some deep theorems about the structure of W*-algebras, see Takesaki (1970, 1979), Strătilă and Zsidó (1979), Strătilă (1981), Bratteli and Robinson (1981), Kadison and Ringrose (1983, 1986), Haag (1992), Takesaki (2003).

---

[124]This is demonstrated by the Unruh effect (Unruh 1976). Compare the review by Crispino et al. (2008).

[125]In his fundamental study of the relation between a W*-algebra (also called von Neumann algebra) and its commutant, Tomita (1967) discovered that each left Hilbert algebra is associated with a one-parameter group of automorphisms of the left von Neumann algebra, and that every faithful normal positive functional ρ on a von Neumann algebra $\mathcal{M}$ induces a left Hilbert algebra structure on $\mathcal{M}$ and gives rise to a one-parameter group $\{\sigma_{\rho,s} | s \in \mathbb{R}\}$ of automorphisms of $\mathcal{M}$. His original paper was difficult to read and little notice was taken of it until Takesaki (1970) gave a full account of Tomita's theory.

Quite independently, the same theory was discovered by the physicists Haag et al. (1967) in their study of equilibrium states of systems with infinitely many degrees of freedom in quantum statistical mechanics.[126] They discovered beautiful and unexpected relations between seemingly disparate concepts. Most notably, the equilibrium states of quantum statistical mechanics turned out to be the faithful state functionals whose modular automorphisms are the time translations, while the modular conjugation is related to time reversal. As a consequence, the modular automorphism group of the Tomita-Takesaki theory satisfies the Kubo-Martin-Schwinger (KMS) condition.[127]

In the top-down description of algebraic quantum field theory there are no intrinsic building blocks. We can analyze the universe only if *we divide it* into independent subsystems (cf. Summers 2009). Algebraic quantum field theory fulfills a causal localization principle and uses as its basic structure local nets of algebras which provide an assignment of bounded open regions of Minkowski space-time to C*- or W*-algebras. According to Haag (1992, p. 109), "the intrinsic structure of the [algebraic local quantum field] theory is fully characterized by the *algebraic relations* in the net of *observable algebras*", so that particles are at best derived objects.[128] Local quantum field theory shifts "the localization concept from objects (e.g., particles) to events" (Haag 1999. p. 10) and expresses locality properties in terms of the algebras of observables. It uses mathematical structures different from those used in traditional quantum mechanics (type-III rather than type-I W*-algebras). In particular, new localization (modular localization) and entanglement concepts play a crucial role.

7.3.2 *The Principle of Locality*

The basic feature of local quantum field theory is the principle of locality, which states "that it is meaningful to talk of observables which can be measured in a specific space-time region and that observables in causally disjoint regions are always compatible" (Haag and Kastler 1964, p. 848).

---

[126]Compare also Araki (1968), Bratteli and Robinson (1981).

[127]See Takesaki (1970). A state functional $\rho \in \mathcal{M}_*$ on a W*-algebra $\mathcal{M}$ with a σ-weakly continuous one-parameter group of automorphism α_s of $\mathcal{M}$ satisfies the KMS condition at the parameter value β $(0 < \beta < \infty)$ with respect to $\{\alpha_s | s \in \mathbb{R}\}$ if for any $A, B \in \mathcal{M}$ there exists a complex function $z \mapsto F_{A,B}(z)$ analytic on the strip $\{z \in \mathbb{C} | 0 < \mathrm{Im} z < \beta\}$ and continuous on the closure of this strip such that $F_{A,B}(s) = \rho\{\alpha_s(A)B\}$, $F_{A,B}(s + i\beta) = \rho\{B\alpha_s(A)\}$ for all $s \in \mathbb{R}$.

Since the basic elements of modular theory were discovered independently by mathematicians and physicists, there is still a trivial but unfortunate discrepancy in the formulation of the KMS condition. Mathematicians define the modular flow with opposite sign than physicists, so that there is a sign change in the definition of the KMS condition.

[128]See also Landsman's review of Haag's book (Landsman 1996, p. 512). Editor's note: For a compact recent introduction into the basic concepts of algebraic quantum field theory with slightly different emphases than in the following sections see Fredenhagen (2015).

Local quantum field theory focuses on the locality properties of the algebras of observables and a version of Einstein's causality for causal propagation. It does not start from quantities associated with points but with local observables associated with open space-time regions. Space-time is represented by a four-dimensional Minkowski space[129] which is equipped with a causal metric structure and the Poincaré group as its symmetry group. As in the Maxwell-Lorentz approach to electrodynamics, the principle of locality states that an object is influenced directly only by its immediate surroundings (cf. Jordan 1927; Schroer 2010c).

Algebraic quantum field theory uses a space-time indexed net of operator algebras which generate an operator-valued field. It is mathematically formulated intrinsically in terms of the Tomita-Takesaki operator theory, without any reference to classical physics or to traditional quantum mechanics. The existence of a causal propagation with finite propagation speed leads to fundamentally different physical phenomena. The causal domain of influence in Minkowski's space-time $\mathcal{M}$ with the Lorentz signature $(-,+,+,+)$ is markedly different from that in four-dimensional Euclidean space with the signature $(+,+,+,+)$. As Jordan (1927, p. 106, editor's translation) stated:

> If the four-dimensional world had four real dimensions (while the general physical laws, that is the differential equations of electromagnetism and gravitation, would remain unchanged), there would, in a sense, be much more than causality: One would be able to derive the physical state of temporally or spatially arbitrarily distant regions of the world from the precise knowledge of a small world region.

Since quantum fields $\varphi(x)$ at a point x in space-time make no good sense, one has to introduce smeared fields $\varphi(f) = \int d^4x\varphi(x)f(x)$, where f is a smooth test function on space-time. The field average $\varphi(f)dx$ is the non-commutative analog of a probability measure on Minkowski space. The collection of all sums of products of smeared fields whose test functions vanish outside some bounded region $\mathcal{O}$ in Minkowski space-time generates an algebra $\mathcal{M}(\mathcal{O})$. Thereby the mapping $x \to \varphi(x)$ of the old Lagrangian quantum field theory is replaced by the map $\mathcal{O} \to \mathcal{M}(\mathcal{O})$, where elements of $\mathcal{M}(\mathcal{O})$ are said to be localized in $\mathcal{O}$. Local quantum field theory considers only information coming from operations that can be performed in the region $\mathcal{O}$ associated with the algebra $\mathcal{M}(\mathcal{O})$.[130] Haag (2010, p. 287) emphasized:

---

[129]The Minkowski space $\mathcal{M}$ is the simplest solution of the Einstein field equations in the absence of matter and provides the natural mathematical background of the special theory of relativity. It is a four-dimensional real vector space equipped with a non-degenerate symmetric bilinear form $g : \mathcal{M} \times \mathcal{M} \to \mathbb{R}$. A point $\boldsymbol{x}$ in Minkowski space $\mathcal{M}$ is called an *event* and can be described by the Cartesian space-time coordinates $\mathbf{x} = (x^0, x^1, x^2, x^3) = (ct, x, y, z)$, where c is the speed of light in vacuum. The Minkowski inner product is a non-degenerate symmetric bilinear form of signature $(-,+,+,+)$, defined by $g(\mathbf{x}, \bar{\mathbf{x}}) = \sum_{\mu,\nu=0}^{3} g_{\mu,\nu}x^\mu\bar{x}^\nu$, where $g_{\mu,\nu}$ represents the geometry of space-time. The Minkowski metric is given by the diagonal matrix $(g_{\mu,\nu})$ with elements $g_{0,0} = -1$, $g_{1,1} = g_{2,2} = g_{3,3} = +1$, so that $g(\boldsymbol{x}, \bar{\boldsymbol{x}}) = -x^0\bar{x}^0 + x^1\bar{x}^1 + x^2\bar{x}^2 + x^3\bar{x}^3$. The Minkowski norm $\|\boldsymbol{x}\|$ of a vector $\boldsymbol{x}$ is $\|\boldsymbol{x}\|^2 = -(x^0)^2 + (x^1)^2 + (x^2)^2 + (x^3)^2, x^0 = ct$, which can be positive, negative or zero.

[130]The relation between the traditional way to formulate quantum field theory in terms of distribution-valued fields and local algebraic quantum field theory is not quite trivial. Since in

The essential message is, however, that all physical consequences of the theory are implied by the assignment of sub-algebras to regions in space-time. No further labeling of individual elements (such as a relation to specific species of particles) is needed.

7.3.3 Minkowski Space-Time

7.3.3.1 Lorentz Transformations

If x^μ, $x^{\mu'}$ are the coordinates of two events in some inertial frame, $\bar{x}^\mu$, $\bar{x}^{\mu'}$ are the coordinates of the same events in another inertial frame, and $dx^\mu = x^{\mu'} - x^\mu$ and $d\bar{x}^\mu = \bar{x}^{\mu'} - \bar{x}^\mu$ are (small) differences, then the requirement of a constant speed of light implies the invariance of proper time ds,

$$ds^2 = \sum_{\mu,\nu=0}^{3} g_{\mu,\nu} dx^\mu dx^\nu = \sum_{\mu,\nu=0}^{3} g_{\mu,\nu} d\bar{x}^\mu d\bar{x}^\nu = d\bar{s}^2. \tag{7.2}$$

The *Lorentz group* $\mathbb{L}$, i.e., the group of Lorentz transformations,

$$\mathbb{L} : x^\mu \to \bar{x}^\mu = \sum_{\alpha=0}^{3} L^\mu_\alpha x^\alpha, \quad L^\mu_\alpha \in \mathbb{R}, \tag{7.3}$$

preserves the invariance of the space-time interval

$$\sum_\mu x^\mu x_\mu = \sum_\mu \bar{x}^\mu \bar{x}_\mu,$$

where $\sum_\mu x^\mu x_\mu = \mathbf{x}\cdot\mathbf{x} - c^2 t^2$. For example, a boost in x-direction that relates two inertial frames in relative uniform motion is given by

$$\begin{pmatrix} \bar{ct} \\ \bar{x} \\ \bar{y} \\ \bar{z} \end{pmatrix} \begin{pmatrix} \cosh\phi & -\sinh\phi & 0 & 0 \\ -\sinh\phi & \cosh\phi & 0 & 0 \\ 0 & 0 & 1 & 0 \\ 0 & 0 & 0 & 1 \end{pmatrix} \begin{pmatrix} ct \\ x \\ y \\ z \end{pmatrix}, \tag{7.4}$$

where $\gamma = \cosh\phi$ is the Lorentz factor.

general fields are not observables, a physical theory is not characterized by the smeared fields $\varphi(f)$ but by the net $\mathcal{O} \to \mathcal{M}(\mathcal{O})$ of W*-algebras $\mathcal{M}(\mathcal{O})$ of local observables. For details compare Roberts (2004).

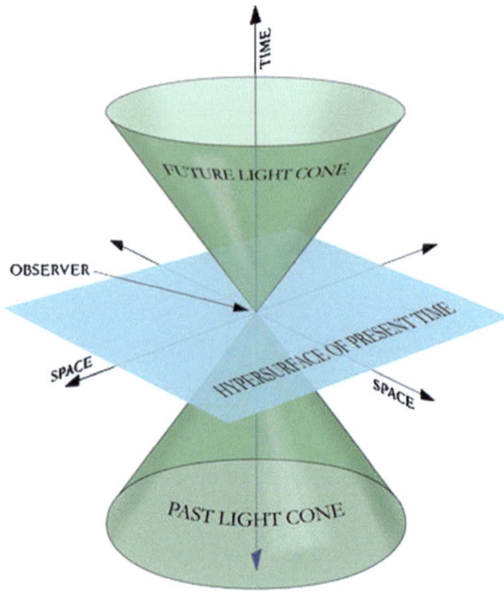

Fig. 7.1 Minkowski space-time: the observer is at the junction of the past light cone (*bottom*) and the future light cone (*top*). The plane is the hypersurface of present time. Causally connected events are "time-like" (vertically) separated, causally unconnected events are "space-like" (horizontally) separated. Credit: *Science Photo Library* at images.sciencesource.com/

7.3.3.2 The Poincaré Group

The *Poincaré group* is the group of all inhomogeneous linear Lorentz transformations,

$$\mathbb{P} : x^\mu \to \bar{x}^\mu = \sum_{\alpha=0}^{3} L^\mu_\alpha x^\alpha + a^\mu, \quad a^\mu \in \mathbb{R}^4. \tag{7.5}$$

It is a ten-dimensional non-compact Lie group given by the semidirect product $\mathbb{L} \times \mathbb{R}^4$ of the Lorentz group $\mathbb{L}$ with the translations. While the Lorentz group $\mathbb{L}$ is the group of isometries of Minkowski space-time $\mathcal{M}$ as a vector space with the (indefinite) inner product $-x^0\bar{x}^0 + x^1\bar{x}^1 + x^2\bar{x}^2 + x^3\bar{x}^3$, the Poincaré group $\mathbb{P}$ is the group of diffeomorphisms of $\mathcal{M}$.

The inhomogeneous linear Lorentz transformations are the *only* non-singular coordinate transformations $x^\mu \to \bar{x}^\mu$ that leave ds^2 invariant (cf., e.g., Weinberg 1972. p. 27). The general Lorentz group $\mathbb{L}$ consists of a proper general Lorentz group $\mathbb{L}_0$ and three discrete transformations P, T and PT, where P and T are space and time reversal, and PT is a so-called full reflection. The Lorentz group may be represented by a semidirect product $\mathbb{L}_0 \rtimes \{1, P, T, PT\}$.

The line element that is invariant under all transformations of the full Poincaré group $\mathbb{P}$, provides a classification of event pairs into three causal types (cf. Fig. 7.1):

- A line element $ds^2 = \sum_{\mu,\nu} g_{\mu,\nu} dx^\mu dx^\nu < 0$ is called *time-like* and characterizes a pair of causally related events.
- A line element $ds^2 = \sum_{\mu,\nu} g_{\mu,\nu} dx^\mu dx^\nu = 0$ ($d\mathbf{x} \neq 0$) is called *light-like* and characterizes a pair of causally related events with signals at speed c.
- A line element $ds^2 = \sum_{\mu,\nu} g_{\mu,\nu} dx^\mu dx^\nu > 0$ is called *space-like* and characterizes a pair of events which are causally unrelated.

An event **x** in a Minkowski space-time $\mathcal{M}$ with a fixed time orientation is said to precede an event $\bar{\mathbf{x}}$ if there exists a future-directed time-like curve extending from **x** to $\bar{\mathbf{x}}$, so that

$$c(t-\bar{t}) < \sqrt{\left(x^1-\bar{x}^1\right)^2+\left(x^2-\bar{x}^2\right)^2+\left(x^3-\bar{x}^3\right)^2}. \tag{7.6}$$

The principle of *relativistic causality* says that a cause and its effect are separated by a time-like interval so that the cause must precede its effect according to all inertial observers. Therefore two points **x** and $\bar{\mathbf{x}}$ from the Minkowski space-time $\mathcal{M}$ are *not* causally related if and only if they are *space-like* separated. The (closed) forward light cone in Minkowski space-time is defined as the set

$$V_+ := \left\{x=\left(x^0,\mathbf{x}\right)\in\mathbb{R}^4 \mid x^0\geq 0,\ \left(x^0\right)^2-|\mathbf{x}|^2\geq 0\right\}. \tag{7.7}$$

Double cones are the open interior of intersections of wedges. Wedge-localized algebras play a distinguished role in algebraic quantum field theory. The standard right wedge region $W_R=\{x\in\mathcal{M}:x_1>|x_0|\}$

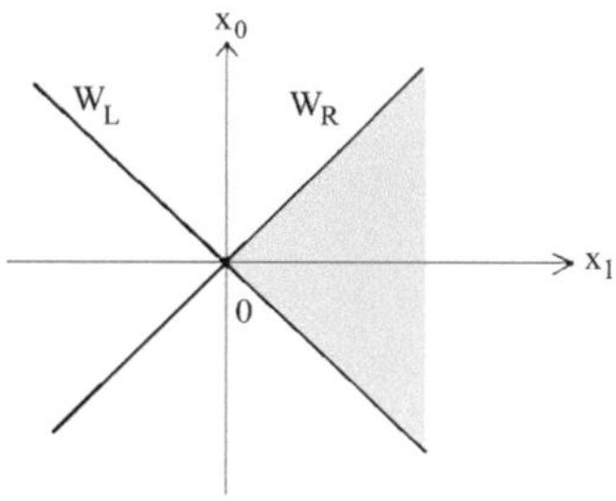

is invariant for boosts relating two reference frames in relative uniform motion along x^1, with speed v given by $\cosh\theta=(1+v^2/c^2)^{-1/2}$. According to Daniela Cadamuro,[131] "wedges are big enough to allow for simple observables being localized in them, but also small enough so that two of them can be space-like separated".

The space-like complement of a wedge in Minkowski space-time coincides with the reflection at the origin of the wedge. This implies that the commutant of the W*-algebra associated with the wedge coincides with the algebra associated with the opposite set. A *double cone W*-algebra* $\mathcal{M}(\mathcal{K})$ is defined as the W*-algebra of the intersection of all wedges containing the set of open double cones $\mathcal{K}$.

The Minkowski metric implies that $\det|L_\mu^\beta|=\pm 1$. Since the transformed coordinates have to be real, we obtain $L_0^0=\pm 1$, so that the Poincaré group $\mathbb{P}$ decomposes into the disjoint union of four connected components $\mathbb{P}_+^\uparrow$, $\mathbb{P}_-^\uparrow$, $\mathbb{P}_+^\downarrow$ and $\mathbb{P}_-^\downarrow$,

$$\mathbb{P}=\mathbb{P}_+^\uparrow\cup\mathbb{P}_-^\uparrow\cup\mathbb{P}_+^\downarrow\cup\mathbb{P}_-^\downarrow, \tag{7.8}$$

[131] http://www.imar.ro/~purice/conferences/2011/EUNCG4/Talks/Cadamuro.pdf.

Here $+/-$ stands for positive/negative determinant and $\uparrow / \downarrow$ for the preservation/reversal of time direction, respectively (cf. Streater and Wightman 1964, p. 10). With the space reversal operation P (which describes the transition from a right-handed to a left-handed spatial frame), the time reversal operation T and the space-time inversion PT,

$$T = \begin{pmatrix} -1 & 0 \\ 0 & +E \end{pmatrix}, \qquad P = \begin{pmatrix} +1 & 0 \\ 0 & -E \end{pmatrix}, \qquad PT = \begin{pmatrix} -1 & 0 \\ 0 & -E \end{pmatrix}, \tag{7.9}$$

where E is the 3×3-unit matrix, we can write:

$$\mathbb{P}_-^\uparrow = P\mathbb{P}_+^\uparrow, \qquad \mathbb{P}_+^\downarrow = PT\mathbb{P}_+^\uparrow, \qquad \mathbb{P}_-^\downarrow = T\mathbb{P}_+^\uparrow. \tag{7.10}$$

Therefore we have:

$\mathbb{P}_+^\uparrow$: $\det L = +1,\ L_0^0 \geq 1$ containing the identity,

$\mathbb{P}_-^\uparrow$: $\det L = -1,\ L_0^0 \geq 1$ containing the parity P,

$\mathbb{P}_+^\downarrow$: $\det L = +1,\ L_0^0 \leq -1$ containing the space-time inversion PT,

$\mathbb{P}_-^\downarrow$: $\det L = -1,\ L_0^0 \leq -1$ containing the time inversion T.

Each of the four components of the decomposition (7.8) are connected, but no Lorentz transformation in one component can be connected to another component. There are four invariant subgroups of the Poincaré group $\mathbb{P}$:

- $\mathbb{P}_+^\uparrow$ is a connected invariant subgroup of the Poincaré group $\mathbb{P}$, called the *proper orthochronous Poincaré group*. The factor group $\mathbb{P}/\mathbb{P}_+^\uparrow$ is isomorphic to the Klein four-group $\{\mathbb{1}, P, T, PT\}$.
- $\mathbb{P}^\uparrow := \mathbb{P}_+^\uparrow \cup \mathbb{P}_-^\uparrow$ is the so-called *orthochronous Poincaré group*. The factor group $\mathbb{P}/\mathbb{P}^\uparrow$ is isomorphic to the discrete group $\{\mathbb{1}, T\}$.
- $\mathbb{P}_+ := \mathbb{P}_+^\uparrow \cup \mathbb{P}_+^\downarrow$ is the so-called *proper Poincaré group*. The factor group $\mathbb{P}/\mathbb{P}_+$ is isomorphic to the discrete group $\{\mathbb{1}, P\}$.
- $\mathbb{P}_0 := \mathbb{P}_+^\uparrow \cup \mathbb{P}_-^\downarrow$ has a factor group $\mathbb{P}/\mathbb{P}_0$ which is isomorphic to the discrete group $\{\mathbb{1}, PT\}$.

Transformations preserving the direction of time, and hence relativistic causality, are called *orthochronous*, they are characterized by $L_0^0 \geq -1$. Since the product of two orthochronous transformations is again orthochronous, the set of all orthochronous transformations of the Poincaré group is called the *orthochronous Poincaré group* $\mathbb{P}^\uparrow$. When orthochronous transformations are composed with the time reversal operation T, general transformations with time reversal are obtained ($L_0^0 \leq -1$), and conversely. We therefore have the following decomposition of $\mathbb{P}$:

$$\mathbb{P} = \mathbb{P}^\uparrow \cup T\mathbb{P}^\uparrow, \qquad \mathbb{P}^\uparrow \cap T\mathbb{P}^\uparrow = \emptyset.$$

7.3.3.3 PCT Symmetry

Another important discrete symmetry is the *charge conjugation* C, which replaces a charged field by its charge-conjugate field or, equivalently, transforms a particle into its corresponding anti-particle. In contrast to the transformations T, P and PT, the discrete symmetry C is not a space-time symmetry. Together with parity P and time reversal T, the charge conjugation C gives rise to the discrete PCT symmetry, which is of far-reaching significance.

A symmetry in algebraic quantum field theory is represented by a Jordan automorphism, i.e. by a linear or anti-linear W*-automorphism. We consider anti-linear automorphisms Θ that implement the PCT-symmetry of a local W*-algebra $\mathcal{M}(\mathcal{O})$, which fulfills for any elements $A, B \in \mathcal{M}(\mathcal{O})$ the relations (compare Borchers and Yngvason 2001)

$$\Theta(AB) = \Theta(A)\Theta(B), \tag{7.11a}$$

$$\Theta(\lambda A) = \lambda^* \Theta(A), \tag{7.11b}$$

$$\Theta\big(\mathcal{M}(\mathcal{O})\big) = \mathcal{M}(-\mathcal{O}), \tag{7.11c}$$

such that Θ transforms the translations in the correct manner, i.e.,

$$\Theta\big\{\alpha_a(A)\big\} = \alpha_{-a}\big\{\Theta(A)\big\}. \tag{7.12}$$

That a field theory is invariant under space reflection P, time reversal T, and charge conjugation C does not imply that these individual symmetries can be unitarily or anti-unitarily implemented by unitary or anti-unitary operators. However, their product can always be implemented by the so-called PCT-operator Θ (Lüders 1954).

The invariance under PCT-transformations is considered as a fundamental law of physics. It can be shown that PCT-invariance holds if the local quantum field theory is Lorentz invariant, fulfills the spectrum condition, and if charge conjugate partners exist. PCT-invariance implies that particles and anti-particles must have the same mass, the same lifetime, and equal but opposite electric charges and magnetic moments. PCT-invariance explains the spin-statistics theorem and the absence of negative energy states.[132]

7.3.4 *Net of Algebras of Local Observables*

Local algebraic quantum theory is a "field theory without fields" (Haag 1970, p. 21) which does not rely on quantizations of any kind. It is an approach to quantum

[132]For more details, precise formulations and proofs compare Pauli (1955), Lüders and Zumino (1958), Streater and Wightman (1964).

physics which does not start with quantizing the fields of a classical Lagrangian but with local algebras of observables associated with bounded regions of space-time that determine what can be measured in a specific space-time region.[133] It does not implement properties of relativistic localization with the help of particles but carries over the idea of a causal propagation of classical waves to quantum theory. The algebraic formulation of local quantum field theory is characterized in terms of its *net of algebras of local observables*, given by the elements of a system of W*-algebras $\mathcal{M}(\mathcal{O})$ of observables. The operators in $\mathcal{M}(\mathcal{O})$ can intuitively be thought of as generating *physical operations* carried out in the open space-time region $\mathcal{O}$ in Minkowski space-time $\mathcal{M}$.

To implement the principle of locality, an ordering of the observables is introduced such that to every region $\mathcal{O} \subset \mathcal{M}$ there is an operator algebra $\mathcal{M}(\mathcal{O})$ whose elements represent physical operations performable within $\mathcal{O}$. The global algebra of all observables generated by the observables localized in $\mathcal{O}$ is taken as the C*-inductive limit

$$\mathcal{M}(\mathcal{M}) = \overline{\bigcup_{\mathcal{O}\in\mathcal{K}} \mathcal{M}(\mathcal{O})}, \tag{7.13}$$

where $\mathcal{K}$ is the set of open double cones in $\mathcal{M}$. The assignment

$$\mathcal{K} \supset \mathcal{O} \mapsto \mathcal{M}(\mathcal{O}) \subset \mathcal{M}(\mathcal{M}) \tag{7.14}$$

contains all the physical information of the theory. It is not assumed that nets can be constructed from point-like objects like fields. The net of W*-algebras indexed by space-time regions can be generated by the family of double cones $\diamond \subseteq \mathcal{M}$ in Minkowski space-time $\mathcal{M}$ which are stable under Poincaré transformations. Such double cones, centered symmetrically around the origin, are also called *diamonds*.

Let ω be a *pure* global state functional of bounded energy (e.g., the Minkowski vacuum state functional) on the global W*-algebra $\mathcal{M}(\mathcal{M})$ associated with the entire Minkowski space-time and let $\mathcal{M}(\mathcal{O})$ be the W*-algebra associated with an open bounded region $\mathcal{O}$ in $\mathcal{M}$. Then $\mathcal{M}(\mathcal{O})$ is a type-III factor and the state functional

$$\omega_{\mathcal{O}} := \omega|_{\mathcal{M}(\mathcal{O})} \tag{7.15}$$

obtained by restricting the vacuum state functional ω to the subalgebra $\mathcal{M}(\mathcal{O})$ is a *non-pure* state functional.[134] Global observables like energy or charge are not elements of $\mathcal{M}(\mathcal{M})$, but they can be defined as weak limit points of local observables in $\mathcal{M}(\mathcal{O})$ (cf. Fredenhagen 1993).

A local net $\mathcal{O} \mapsto \mathcal{M}(\mathcal{O})$ satisfies the following axioms (cf. Haag 1992; Baumgärtel 1995; Araki 1999):

---

[133]Compare the introductions by Horuzhy (1990), Haag (1992), Baumgärtel and Wollenberg (1992), and Araki (1999).

[134]Non-pure state functionals are often called "mixed states" but we shall avoid this misleading term since they do not refer to any kind of (classical) mixture.

- *Isotony:* $\mathcal{O}_1 \subset \mathcal{O}_2$ implies $\mathcal{M}(\mathcal{O}_1) \subset \mathcal{M}(\mathcal{O}_2)$.
- *Local commutativity:* If $\mathcal{O}_1$ and $\mathcal{O}_2$ are space-like separated, then $A_1 A_2 = A_2 A_1$ for all $A_1 \in \mathcal{M}(\mathcal{O}_1)$ and $A_2 \in \mathcal{M}(\mathcal{O}_2)$.
- *Causality:* Space-like localized observables $\mathcal{M}(\mathcal{O}_1) \subset \mathcal{M}(\mathcal{O}_2)'$ for $\mathcal{O}_1 \subset \mathcal{O}_2'$ commute, where $\mathcal{O}_2'$ is the interior of the space-like complement of $\mathcal{O}_2$, and $\mathcal{M}(\mathcal{O})'$ is the commutant of $\mathcal{M}(\mathcal{O})$.
- *Haag duality:* $\mathcal{M}(\mathcal{O}') = \mathcal{M}(\mathcal{O})'$ where $\mathcal{O}' := \{x|(x-y)^2 < 0, y \in \mathcal{O}\}$ is the space-like complement of $\mathcal{O}$. Haag duality implies that a quantum field in a bounded space-time region can never be disentangled from the field in the causal complement.
- *The Poincaré group acts covariantly on the local net:* There exists a continuous representation $\lambda \to \alpha_\lambda$ of the proper orthochronous Poincaré group $\mathbb{P}_+^\uparrow$ by automorphisms α_λ of $\mathcal{M}$ which acts covariantly for all $\lambda \in \mathbb{P}_+^\uparrow$, $\alpha_\lambda \mathcal{M}(\mathcal{O}) \subset \mathcal{M}(\lambda\{\mathcal{O}\})$.
- *Existence of a vacuum state:* There exists a Poincaré covariant state functional ω on $\mathcal{M}(\mathcal{M})$ called a *vacuum state* with $\omega\alpha_\lambda = \omega$.
- *Covariant representation:* The vacuum state functional ω can be represented in a GNS-construction $(\pi_\omega, \mathcal{H}_\omega)$ by a state vector $\Omega \in \mathcal{H}_\omega$. In the induced representation $(\pi, \mathcal{H}_\omega, \Omega, U)$, the joint spectrum of the generators $\boldsymbol{P} = P_0, P_1, P_2, P_3$, of the translations $U(\boldsymbol{x}) = \exp(i\boldsymbol{P}\boldsymbol{x})$ is contained in the forward light-cone: $spec(\boldsymbol{P}) \subset \{\boldsymbol{p} \in \mathbb{R}^4,\ p_0^2 \geq 0,\ p_0^2 - \sum_{k=1}^3 p_k^2 \geq 0\}$.
- *Local algebras are type-III$_1$ factors:* The localized W*-algebra $\mathcal{M}(W_R)$ in a wedge region W_R is a type-III$_1$ factor (Driessler 1977).
- *Type III factors are intrinsically entangled:* If $\mathcal{M}$ is a type-III factor, then $\mathcal{M} \cup \mathcal{M}'$ does not have any product states, so that *all* states are entangled for the pair $(\mathcal{M}, \mathcal{M}')$.

7.3.5 Locality and Entanglement

7.3.5.1 Reeh-Schlieder Theorem

A striking consequence of relativistic local algebraic quantum field theory is the *Reeh-Schlieder theorem* (Reeh and Schlieder 1961), which guarantees that Ω, the GNS-represented state in Hilbert space, is a common cyclic and separating vector for local algebras. It implies that $\mathcal{M}(\mathcal{O})\Omega$ is dense in the vacuum Hilbert space. This implies that any state of a quantum field can be approximated arbitrary closely by applying an operator of a local algebra $\mathcal{M}(\mathcal{O})$ of observables, no matter how small the local region $\mathcal{O}$ is.

Reeh-Schlieder Theorem (7.16)

For any non-empty open region $\mathcal{O}$ in Minkowski space, the vacuum vector Ω (or more generally, any state of bounded energy) is cyclic for the field algebra $\mathcal{M}(\mathcal{O})$.

The Reeh-Schlieder theorem can be proven in the C*-algebraic axiomatic framework for quantum field theory (Haag and Kastler 1964). It is one of the most fundamental results of quantum field theory and provides the key to understanding *top-down descriptions*. It shows that the global Poincaré-invariant vacuum state has a very rich structure: All local observables exhibit fluctuations in the vacuum state and "the entire state vector space of the field could be obtained from measurements in an arbitrarily small region of spacetime" (Segal 1964, p. 1400). Therefore the outcome of any possible measurement procedure will occur with nonvanishing probability in the vacuum state (Redhead 1995, Theorem 2).

The Reeh-Schlieder theorem entails that a state of the form

$$\omega_{\Psi}(M) := \langle \Psi | \pi(M) \Psi \rangle \quad \text{with } \Psi := A\Omega, \tag{7.17}$$

where $A, M \in \mathcal{M}(\mathcal{O})$, and π is any representation of $\mathcal{M}$, can only be regarded as *approximately* localized in $\mathcal{O}$. The approximation is the better the smaller $\|A\|$ is, so that such a state is never *strictly* localized in $\mathcal{O}$.

7.3.5.2 Vacuum Correlations

If the pure vacuum state is restricted to a space-time subregion, it becomes a highly entangled state that cannot be represented by a density operator and that violates Bell's inequalities maximally (Summers and Werner 1985).[135] This non-classical structure of the vacuum implies that for any three causally separated space-time regions there are observables associated with these regions that do not possess a joint distribution in the vacuum state (Landau 1987a, 1987b).

While one may expect that for a localized observable $A \in \mathcal{M}(\mathcal{O})$ the vector $A\Omega$ should be localized in $\mathcal{O}$, this is not the case. The Reeh-Schlieder theorem implies that the states created by the action of elements of the local algebra $\mathcal{M}(\mathcal{O})$ on the vacuum state vector Ω cannot be strictly localized in the region $\mathcal{O}$. Nevertheless, every state can be approximated by states of the kind $A\Omega$, but with increasing distance, creating a unit vector localized outside $\mathcal{O}$ requires operators of ever increasing operator norm so that the long-range effects of the operators of the local algebra will diminish rapidly with distance.

7.3.5.3 Restriction of Ω Yields KMS States

The restriction of the global vacuum state vector Ω, or of any other global finite energy state, to an open submanifold of the Minkowski manifold $\mathcal{M}$, whose boundaries are event horizons (i.e., boundaries in space-time beyond which events cannot

[135] More generally, for every vector state there exist observables localized in complementary wedge-shaped regions in Minkowski space-time that maximally violate Bell's inequalities (Summers and Werner 1987). For further generalizations compare also Summers and Werner (1988, 1995).

affect an outside observer), results in a non-pure KMS state. In other words, a causal localization by restricting the global vacuum state to the local subalgebra is a holistic concept which converts the pure vacuum state Ω to a non-pure KMS state with respect to the modular Hamiltonian $\Delta^{is} = e^{isK}$:

$$\Omega \upharpoonright \mathcal{M}(\mathcal{O}) \text{ is a KMS state,} \tag{7.18}$$

where the modular Hamiltonian is uniquely determined by $A(\mathcal{O})$, and Ω by the modular automorphism.[136] The fact that the restriction $\Omega \upharpoonright \mathcal{M}(\mathcal{O})$ of the pure global vacuum state to a subregion behaves like a stochastic thermal KMS state does not imply that the parameter β of a KMS state is the inverse temperature measured with a thermometer (see Buchholz and Solveen 2013).

7.3.5.4 Modular Localization

It is well known that the spatial localization of particles in terms of position operators in a finite region of space-time is inconsistent with a causality requirement in relativistic quantum physics (Hegerfeldt 1974; Perez and Wilde 1977). But there is a general connection between positive energy representations of the Poincaré group[137] and localization properties of state vectors in the Hilbert space of the representation (Brunetti et al. 2002). This so-called *modular localization* is intrinsic and not associated with a position operator but with the Lorentz boosts corresponding to wedge-like regions in Minkowski space and the PCT-operator.

Every irreducible positive energy representation of the Poincaré group admits dense sets of vectors that are naturally localized in space-like cones in Minkowski space-time. If W is the Poincaré transform of the right wedge $\{x \in \mathcal{M} : x_1 > |x_0|\}$, then there is a one-parameter family $L_W(t)$ of Lorentz boosts leaving W invariant and a reflection j_W mapping W onto the opposite wedge W', the causal complement of W.

From the point of view of traditional quantum theory, which uses W*-algebras of type I, the fact that in quantum field theory all algebras associated with sharp localizations are isomorphic to the unique factor of type III_1 (Buchholz et al. 1987) is rather counter-intuitive. Since every concrete type-III_1 factor is isomorphic to the unique abstract factorial type-III_1 algebra, such wedge-localized algebras are also referred to as *monad* W*-algebras—they are the quantum theoretical counterpart of Leibniz's monads.

---

[136]This result is related to the Unruh effect (Unruh 1976), which states that to a uniformly accelerated observer the vacuum state in Minkowski space-time appears as a modular β_U-KMS equilibrium state, where $\beta_U = 2\pi c/\hbar a$, a is the local acceleration, and c is the speed of light. The modular KMS equilibrium state is often interpreted as a thermal equilibrium state with Unruh temperature $T_U = 1/(k_B \beta_U)$, where k_B is the Boltzmann constant.

[137]Recall that Wigner (1939) identified relativistic particle states with irreducible positive energy representations of the Poincaré group.

The global type-III_1 monads have no internal structure, no pure states and no individuality by themselves. They do not contain operators that annihilate the vacuum but nevertheless generate a dense set of states from the vacuum. If one identifies, with Schroer (2006, p. 36), the type-III_1 algebra

> with the role of *monads* in Leibniz's construct of ideas about what constitutes reality, one finds a perfect match: the physical reality of relativistic local quantum matter in Minkowski space-time originates from the (modular) positioning of a finite number of copies of one abstract monad; no sense of individuality can be attributed to the monad, apart from the fact that those [type-III_1] factors allow inclusions and intersections (which points do not) it is as void of individual structure as a point in geometry (if you have seen one, you know them all).

The global type-III_1 factors act as universal building blocks that create reality by their relations and inclusions like Leibniz's indecomposable monads so that we have (Schroer 2007, p. 284)

> ...the Leibniz viewpoint of physical reality emerging from relations between entities rather than from positions in a pre-assigned absolute "inertia ether".

7.3.5.5 Modular Actions and Lorentz Boosts

The modular theory implies that the modular action is a geometric transformation on a local algebra. Bisognano and Wichmann (1975, 1976) showed that the modular group of every wedge acts like the associated group of Lorentz boosts. In particular, for the wedge region $W_R = \{x \in \mathcal{M} : x_1 > |x_0|\}$ with arbitrary x_0, x_1, the modular group of the corresponding local algebra acts as geometrical transformations on the net like the representation of special Lorentz boosts. The modular conjugation J_W of the wedge W is up to a rotation the physical PCT-conjugation,

$$J_W = \Theta U\{R_W(\pi)\}, \tag{7.19}$$

where Θ denotes the PCT-operator and $U\{R_W(\pi)\}$ is the unitary representation of the rotation by an angle π which leaves the characteristic two-plane of the wedge invariant. It follows that the modular group of a wedge-algebra together with the vacuum vector corresponds to the unitary representation of the group of Lorentz boosts which maps the wedge onto itself. (For more details see Appendix A.5.)

7.3.6 Subsystems in Quantum Physics

7.3.6.1 Subsystems in Type-I Quantum Theories

Traditional quantum mechanics uses a factor $\mathcal{M}$ of type I as its algebra of observables, so that it can be written as the W*-tensor product $\mathcal{M} = \mathcal{M}_1 \overline{\otimes} \mathcal{M}_2$ of two type-I factors $\mathcal{M}_1$ and $\mathcal{M}_2$. The W*-algebra $\mathcal{M}_k$ $(k = 1, 2)$ then represents a subsystem of the system with the W*-algebra $\mathcal{M}$. It is characteristic for traditional quantum mechanics that $\mathcal{M}$ can be written as a tensor product of two type-I factors. In a Hilbert-space representation, the factor $\mathcal{M}$ can be represented by the algebra $\mathcal{B}(\mathcal{H})$ of all bounded operators on some Hilbert space $\mathcal{H}$. Let

$$\mathcal{B} \to E_Q(\mathcal{B}) = \int_{\mathcal{B}} E_Q(dq) \tag{7.20}$$

be the projection-valued spectral measure of the position operator Q,

$$Q = \int_{\mathbb{R}} q E_Q(dq), \quad E_Q \in \mathcal{M}, \tag{7.21}$$

then the conditional probability that a predictive measurement of position Q gives a value in the Borel set $\mathcal{B}$ is given by $\mu(\mathcal{B}) = \rho\{E_Q(\mathcal{B})\}$, where ρ is the state of the system immediately before the measurement. Therefore, localization in type-I quantum mechanics results in a tensor-product decomposition of the Hilbert space $\mathcal{H}$, so that the factor $\mathcal{M} = \mathcal{B}(\mathcal{H})$ decomposes into a tensor product of two factors of type I,

$$\mathcal{B}(\mathcal{H}) = \mathcal{B}\{\mathcal{H}^{\perp}\} \otimes \mathcal{B}\{\mathcal{H}^{\parallel}\}, \tag{7.22}$$

with

$$\mathcal{H}^{\perp} := E_Q(\mathcal{B})\mathcal{H}, \qquad \mathcal{H}^{\parallel} := \{\mathbb{1} - E_Q(\mathcal{B})\}\mathcal{H}. \tag{7.23}$$

7.3.6.2 Subsystems in Type-III Quantum Theories

There is a deep structural difference between the algebras of local quantum field theory and the algebras of traditional quantum mechanics. In local quantum field theories, the local W*-algebras $\mathcal{M}(\mathcal{O})$ are of type III_1, so that they are not isomorphic to an algebra of bounded operators on some Hilbert space. Therefore, it is not possible to define a linear subspace $\mathcal{H}_{\mathcal{O}} \subset \mathcal{H}$, which can be considered to correspond to the state vectors localized in $\mathcal{H}_{\mathcal{O}}$ (for details see Haag 1992, p. 256; Schroer 2010a, 2010b). Although the local W*-algebras $\mathcal{M}(\mathcal{O})$ commute with their causally disjoint $\mathcal{M}(\mathcal{O}')$ (which equals its commutant, $\mathcal{M}(\mathcal{O}') = \mathcal{M}(\mathcal{O})'$), and notwithstanding that the algebras $\mathcal{M}(\mathcal{O})$ and $\mathcal{M}(\mathcal{O}')$ span the total algebra $\mathcal{M}(\mathcal{O}) \vee \mathcal{M}(\mathcal{O}')$, the full algebra cannot be brought into the form of a tensor product (see also Sect. 8.5.6).

A space-time region can be interpreted as an (open) *subsystem* if it is an identifiable component of the universe of discourse that can be screened off from the rest of the universe. In this case, it can be studied experimentally without considering the influence of the rest of the universe, and we can assign a partial state to such an open subsystem. However, the fact that two systems are spatiotemporally distinct does not imply that they are independent. The essential characteristic of all independence notions in local quantum field theory is the so-called *split property* (Summers 2009).

7.3.6.3 Split Inclusion

A basic feature of the net structure of algebraic quantum field theory are *inclusions* $\mathcal{M}_1 \subseteq \mathcal{M}_2$ of W*-algebras $\mathcal{M}_1$ and $\mathcal{M}_2$. Doplicher and Longo (1984) introduced a *standard split inclusion* as the triple $(\mathcal{M}_1, \mathcal{M}_2, \Omega)$, where $\mathcal{M}_1$ and $\mathcal{M}_2$ are W*-algebras on a Hilbert space $\mathcal{H}$ with an intermediate type-I factor $\mathcal{F}$ with

$$\mathcal{M}_1 \subset \mathcal{F} \subset \mathcal{M}_2, \tag{7.24}$$

and $\Omega \in \mathcal{H}$ is a joint cyclic and separating vector for $\mathcal{M}_1$, $\mathcal{M}_2$ and $\mathcal{M}_1 \cup \mathcal{M}_2'$. If $(\mathcal{M}_1, \mathcal{M}_2, \Omega)$ is a standard split inclusion, then so is $(\mathcal{M}_1', \mathcal{M}_2', \Omega)$.[138]

The split property leads to kinds of factorization and localization that are radically different from traditional quantum theory. If either $\mathcal{M}_1$ or $\mathcal{M}_2$ are of type III, and if $(\mathcal{M}_1, \mathcal{M}_2')$ is a split inclusion, then there is a natural *-isomorphism between $\overline{\mathcal{M}_1 \cup \mathcal{M}_2'}$ and the W*-tensor product $\mathcal{M}_1 \overline{\otimes} \mathcal{M}_2'$,

$$\overline{\mathcal{M}_1 \cup \mathcal{M}_2'} \cong \mathcal{M}_1 \overline{\otimes} \mathcal{M}_2'. \tag{7.25}$$

Note that, if $\mathcal{M}_1$ or $\mathcal{M}_2$ are of type I or type II, the W*-algebra $\overline{\mathcal{M}_1 \cup \mathcal{M}_2'}$ generated by $\mathcal{M}_1$ and $\mathcal{M}_2'$ is strictly larger than the tensor product $\mathcal{M}_1 \overline{\otimes} \mathcal{M}_2'$ (Summers 1990, 2009). In a local net $\mathcal{O} \mapsto \mathcal{M}(\mathcal{O})$ the split property refers to any bounded open set $\mathcal{O}_1$ and a slightly larger set $\mathcal{O}_2$ such that there is a type-I factor $\mathcal{F}$ with

$$\mathcal{M}(\mathcal{O}_1) \subset \mathcal{F} \subset \mathcal{M}(\mathcal{O}_2). \tag{7.26}$$

In a Hilbert-space representation, the split property allows a canonical factorization of the Hilbert space $\mathcal{H} = \mathcal{H}_1 \otimes \mathcal{H}_2$ such that $\mathcal{M}(\mathcal{O}_1)$ acts on the factor $\mathcal{H}_1$ and the causal disjoint of the larger algebra $\mathcal{M}(\mathcal{O}_2)$ acts on $\mathcal{H}_2$.

While space-like regions *in contact* are holistically correlated, local algebras with *strictly space-like separated regions* are independent. Here the concept of independence refers to the mutual commutativity of subalgebras. A proper generalization of independence for non-commutative subalgebras is the notion of *statistical independence*, which is logically independent of the requirement of commutativity (de Facio and Taylor 1973).

In order to find traditional quantum mechanical structures inside local type-III quantum field theory, one needs type-I factors inside local hyperfinite type-III factors. A type-I algebra can be thought of as the algebra of a sharply localized core region, augmented by a "halo" in the region where the localization is "fuzzy" (Mund et al. 2006). This fuzziness can be achieved by the split property, which leaves a collar region between the double cone algebra $\mathcal{M}(\mathcal{O})$ and its causal disjoint $\mathcal{M}(\mathcal{O}')$, so that it is possible to construct a type-I tensor factor $\mathcal{F}$ which extends into the collar.

---

[138]The split property is a consequence of the Buchholz-Wichmann nuclearity condition. See Buchholz and Wichmann (1986) and Buchholz et al. (1987).

If the collar region is left out, one obtains a fuzzy inside-outside factorization (see Schroer 2000, p. 282).

There exists a representation of the Poincaré group acting on the algebra $\mathcal{B}(\mathcal{H}_1) \supset \mathcal{M}(\mathcal{O}_1)$ such that the localization of $\mathcal{B}(\mathcal{H}_1)$ is sharp in $\mathcal{M}(\mathcal{O}_1)$ and fuzzy in $\mathcal{M}(\mathcal{O}_2)$. The split property allows to approximate the sharply localized type-III algebra $\mathcal{M}(\mathcal{O})$ by a sequence of type-I algebras with fuzzy boundaries. The following four equivalent statements provide a necessary and sufficient condition for the possibility to prepare an arbitrary state locally (Werner 1987):

(i) *Split property:* There exists a type-I factor $\mathcal{F}$ with $\mathcal{M}(\mathcal{O}_1) \subset \mathcal{F} \subset \mathcal{M}(\mathcal{O}_2)$.
(ii) *Local preparability of a state:* There exists a state ω and a normal positive map T such that $\mathsf{T}(M) = \omega(M)\mathsf{T}(\mathbb{1})$ for all $M \in \mathcal{M}(\mathcal{O})$ and $\mathsf{T}(M') = \mathsf{T}(\mathbb{1})M'$ for $M' \in \mathcal{M}(\mathcal{O}_1)'$.
(iii) *Non-selective local preparability of arbitrary states:* For every normal state ω there is a map T of the form $\mathsf{T}(M) = \sum M_k^*(M)M_k$ with $M_k \in \mathcal{M}(\mathcal{O}_1)$ such that $\mathsf{T}(M) = \omega(M)\mathbb{1}$ for all $M \in \mathcal{M}(\mathcal{O})$.
(iv) *Existence of normal product states:* For two normal states ω_1 and ω_2 there is a state ω such that $\mathsf{T}(M) = \sum M_k^*(M)M_k$ with $M_k \in \mathcal{M}(\mathcal{O}_1)$ such that $\omega(MM') = \omega(M)\omega(M')$ for all $M \in \mathcal{M}(\mathcal{O})$ and $M' \in \mathcal{M}(\mathcal{O}_1')$.

7.3.6.4 Homogeneity of the State Space of a Type-III$_1$ Factor

Connes and Størmer (1978) showed that a factor $\mathcal{M}$ with separable predual is of type III$_1$ if and only if the action of its unitary group on its state space by inner automorphisms is topologically transitive in the norm topology. That is, for any $\varepsilon > 0$ and normal states Φ and Ψ on a factor $\mathcal{M}$ of type III$_1$ there exists a unitary $U \in \mathcal{M}$ such that

$$\|\Phi_U - \Psi\| < \varepsilon \quad \text{where } \Phi_U(M) := \Phi\big(U^*MU\big) \quad \text{for every } M \in \mathcal{M}. \tag{7.27}$$

This result implies that it is impossible to distinguish in any empirical way between the different degrees of entanglement of two states across $(\mathcal{M}, \mathcal{M}')$ (Halvorson 2001, p. 47).

7.4 Intrinsic Time

7.4.1 Transitive Equilibrium States

Following Gibbs, Carathéodory (1909) conceived thermodynamics as a theory of equilibrium states. Before introducing the first and second law, he introduced the concept of “empirical temperature”, which can be constructed from the assumption that thermal equilibrium is transitive, and hence implies an equivalence relation:[139]

> If two systems are both in thermal equilibrium with a third system, then they are in thermal equilibrium with each other.

Fowler and Guggenheim (1939) coined the name “zeroth law of thermodynamics” for this postulate.[140]

Consider an equilibrium system Σ_ℓ completely characterized by a finite set of real state variables $\mathbf{X}_\ell := (X_\ell^{(1)}, X_\ell^{(2)}, \ldots, X_\ell^{(n)})$. If two systems Σ_1 and Σ_2 are in equilibrium, denoted by $\Sigma_j \sim \Sigma_k$, then there exists a functional relation between $\mathbf{X}_j$ and $\mathbf{X}_k$,

$$\Sigma_j \sim \Sigma_k \quad \Longrightarrow \quad f_{j,k}(\mathbf{X}_j, \mathbf{X}_k) = 0. \tag{7.28}$$

If the relation between equilibrium states is transitive, and if any two among the equations $f_{1,2}(\mathbf{X}_1, \mathbf{X}_2) = 0$, $f_{2,3}(\mathbf{X}_2, \mathbf{X}_3) = 0$, $f_{3,1}(\mathbf{X}_3, \mathbf{X}_1) = 0$ are satisfied, then the third is also satisfied. This is only possible if (7.28) can be brought into the form

$$f_{1,2}(\mathbf{X}_1, \mathbf{X}_2) = f_1(\mathbf{X}_1) - f_2(\mathbf{X}_2) = 0, \tag{7.29a}$$

$$f_{2,3}(\mathbf{X}_2, \mathbf{X}_3) = f_2(\mathbf{X}_2) - f_3(\mathbf{X}_3) = 0, \tag{7.29b}$$

$$f_{3,1}(\mathbf{X}_3, \mathbf{X}_1) = f_3(\mathbf{X}_3) - f_1(\mathbf{X}_1) = 0, \tag{7.29c}$$

so that the functions f_1, f_2, f_3 have the same value, implying that $f_1(\mathbf{X}_1) = f_2(\mathbf{X}_2) = f_3(\mathbf{X}_3)$ is a state variable characterizing the fact that the systems Σ_1, Σ_2 and Σ_3 are in equilibrium.[141]

Carathéodory considered simple thermal systems, characterized by the state variables of pressure P and volume V, and called the function $f(P, V)$, which has the same value when the systems are in thermal equilibrium, the *empirical temperature* θ. Of course, the scale of the empirical temperature can be chosen arbitrarily, so that any monotonically strictly increasing function of θ also is an empirical temperature.

---

[139] A binary relation is called an *equivalence relation* if and only if it is reflexive, symmetric and transitive. If $a \sim b$ means “a is equivalent to b”, then the following independent conditions are valid: (i) reflexivity: $a \sim a$; (ii) symmetry: $a \sim b$ implies $b \sim a$; (iii) transitivity: $a \sim b$ and $b \sim c$ imply $a \sim c$. Compare Eq. (2.3a)–(2.3c).

[140] Redlich (1968) objected correctly that this postulate is not a “law”, but he missed the crucial point that the transitivity of a relation between pairs of systems implies the existence of a state function which takes on the same value for all such systems.

[141] Carathéodory stated this result without proof. Many later authors who referred to it did tacitly assume certain uniqueness and continuity properties. For a precise formulation see Whaples (1952) and Lenker (1979).

7.4.2 Intrinsic Modular Time

In the theory of dynamical systems, stationarity is defined in terms of external clock time. Carlo Rovelli (1993a) proposed to reverse this argument and to define an intrinsic time by selecting an appropriate state and to declare this state as stationary. Remarkably, an *intrinsic* definition of equilibrium states is possible by requiring *strong relaxation and stability properties with respect to small local perturbations.* These equilibrium states fulfill the KMS-conditions for modular equilibrium states (Haag et al. 1974).[142]

On the basis of such equilibrium states one can define a one-parameter group of automorphisms (the *modular group*) such that the underlying equilibrium state is stationary with respect to modular automorphisms. Using this result, Connes and Rovelli (1994) introduced a *state-dependent physical time* via the modular flow of an equilibrium state,[143] which gives rise to the associated time evolution. The modular flow of Tomita-Takesaki theory is unique up to an inner automorphism.

Since for arbitrary W*-algebras there are *many* states which are invariant under the action of the modular group (compare Appendix A.5), a so-defined time would be *state-dependent*. While inner automorphisms correspond to internal symmetries, a type-III W*-algebra $\mathcal{M}$ "inherits from its non-commutativity a god-given time evolution $\mathbb{R} \to \mathrm{Out}(\mathcal{M})$" (Connes 2005, p. 16). The co-cycle Radon-Nikodým theorem states that two modular automorphism defined by two states of a W*-algebra are inner equivalent. Since the modular automorphism group $\mathrm{Aut}(\mathcal{D})$ is unique up to inner automorphisms, the quotient group

$$\mathrm{Out}(\mathcal{D}) = \mathrm{Aut}(\mathcal{D})/\mathrm{Inn}(\mathcal{D}) \tag{7.30}$$

of outer automorphisms is independent of the chosen stationary reference state. The action $\mathbb{R} \xrightarrow{\delta} \mathrm{Out}(\mathcal{M})$ is non-trivial if and only if the W*-algebra $\mathcal{M}$ is a type-III factor. In this case it is possible to define an *intrinsic state-independent* canonical time flow by the canonical group homomorphism

$$\mathbb{R} \xrightarrow{\delta} \mathrm{Out}(\mathcal{M}) = \mathrm{Aut}(\mathcal{M})/\mathrm{Inn}(\mathcal{M}) \tag{7.31}$$

from the additive group $\mathbb{R}$ of real numbers to the quotient of the group $\mathrm{Aut}(\mathcal{M})$ of automorphisms of $\mathcal{M}$ with respect to the normal subgroup $\mathrm{Inn}(\mathcal{M})$ of inner automorphisms. By definition $\mathrm{Aut}(\mathcal{M})$ is trivial for inner automorphisms—for factors of type III_1 we have $\mathrm{Aut}(\mathcal{M}) = \mathbb{R}$. Since the modular flow of a type-III algebra is not only state-independent, but depends essentially on the algebra $\mathcal{M}$ itself,

---

[142] For some improvements in the technical clustering conditions compare Bratteli and Robinson (1981, Chap. 5.4). The clustering requirement excludes the mixture of different phases and selects a pure phase.

[143] Recall that a normal state $\rho \in \mathcal{M}_*$ of a W*-algebra $\mathcal{M}$ is faithful if for every $M \in \mathcal{M}$ the relation $\rho(M) = 0$ implies $M = 0$. Every faithful normal state ρ on a W*-algebra $\mathcal{M}$ defines a one-parameter family $\{\sigma_{\rho,s} | s \in \mathbb{R}\}$ of modular automorphisms $\sigma_{\rho,s}$ on $\mathcal{M}$ (compare Appendix A.5).

Connes and Rovelli (1994, p. 2900) considered a type-III W*-algebra in terms of *intrinsically dynamical objects* that "inherit a natural time evolution from their non-commutativity".

7.4.3 Thermal Time?

For finite-dimensional quantum systems, the KMS states are traditionally interpreted as statistical equilibrium states of a Gibbs ensemble of a large number of virtual copies of a system. Following Gibbs, the state of a quantum canonical ensemble in equilibrium with a heat bath at temperature $1/\beta$ is described by a state functional $\rho_\beta = e^{-\beta H}/\text{tr}\{e^{-\beta H}\}$, where H is the Hamiltonian of the system.

Instead of interpreting a Gibbs state functional ρ_β as determined by the Hamiltonian time flow $\tau \to e^{-i\tau H}$, we can characterize this flow by a state ρ_β taken as stationary. As we will argue below, the state of a finite-dimensional quantum system can then be represented by a density operator D_β that determines the dynamics by a Hamiltonian H defined by the logarithm of D_β.

In the thermodynamic limit the Gibbs statistical interpretation with its trace formula $D_\beta = e^{-\beta H}/\text{tr}\{e^{-\beta H}\}$ loses its meaning, but the analytic KMS boundary condition can be extended to infinite-dimensional systems and so provides a connection with the Tomita-Takesaki method for constructing modular automorphisms of arbitrary W*-algebras. Although any thermal equilibrium state generates a corresponding modular flow, the existence of such a flow does in no way imply a *statistical* interpretation of a state that is invariant with respect to the modular automorphism.

The relevance of the Tomita-Takesaki theory (see Appendix A.5) is far greater than just applying to equilibrium states of quantum statistical mechanics. It is of crucial importance for the theory of type-III W*-algebras and for algebraic quantum field theory as a whole. If the W*-algebra $\mathcal{M}$ of a universe of discourse is a type-III W*-algebra, there are no pure normal states in the predual $\mathcal{M}_*$. Therefore, the non-pure states in $\mathcal{M}_*$ are not mixtures of pure states so that *a statistical interpretation is unfeasible* (compare Sect. 6.2.5).

Nevertheless, Rovelli (2001, p. 115) stated (without giving an argument)

> that the temporal aspects of our world have statistical and thermodynamical origin, rather than dynamical. "Time" is ignorance: a reflex of our incomplete knowledge of the state of the world.

Moreover, Rovelli (1993b) even defended the idea

> that the cosmological state of the universe can be described in terms of a statistical state, where a statistical state is understood as a statistical distribution of pure states.

Or (Rovelli 2011, p. 1475):

> When we say that a certain variable is "the time" we are not making a statement concerning the fundamental mechanical structure of reality. Rather, we are making a statement about the statistical distribution we use to describe the macroscopic properties of the system that we describe macroscopically.

Similarly, Connes and Rovelli (1994) adopted a statistical mechanical interpretation of the stationary state of the modular automorphism and claimed "that physical time has a thermodynamical origin".

We do not expect that such an interpretation of type-III modular time in terms of statistical ensembles can be consistent. Thermodynamics is not more fundamental than time. We assume that dynamical information about a global system is encoded in the reference state functional, which does not rely on the notion of time. An intrinsic observer-independent global time is defined via the modular flow on a type-III_1 W*-algebra, without any reference to thermodynamics. This flow is generated by the group $\{\sigma_s | s \in \mathbb{R}\}$ of *outer* automorphisms σ_s of the essentially unique factor $\mathcal{M}$ of type III_1. The self-adjoint generator K of the unitary one-parameter group $s \to \Delta^{is}$ is given by $K = \ln D$,

$$\sigma_s(M) = e^{2\pi i s} M e^{-2\pi i s}, \quad s \in \mathbb{R},\ M \in \mathcal{M}. \tag{7.32}$$

The generator K is called the *modular Hamiltonian* and the flow parameter s is referred to as the global *modular time*, which provides an intrinsic global ordering for top-down descriptions. Modular time s must not be identified with the Hamiltonian time of a bottom-up description of a mechanical system. It is a different flow of completely different origin. A meaningful relation between the Hamiltonian time of a bottom-up description and the global type-III modular time of a top-down description can only be discussed if a joint embedding of these two descriptions is explicitly known.

Part II
Time

Chapter 8
Matter, Mind, and Time

8.1 Limitations of Physical Descriptions

The principles of physics alone cannot provide a complete description of the world and our knowledge about it. For example, intentionality is not a physical concept, but every laboratory experiment is intentional (cf. Sect. 7.1.2). A well-conducted experiment is an intervention in nature that requires artificially produced and deliberately controlled reproducible conditions. In experiments—in contradistinction to mere observations—one prepares systems in initial states, controls some of the variables, and finally measures a particular variable.

The regulative principles of present-day experimental science require the power to create initial conditions, and they stress the facticity of the past and the probabilistic predictability of the future. It is a basic assumption in engineering science in particular that nature can be manipulated and that the initial conditions required by experiments can be prepared by interventions into the world external to the object under investigation. In other words, we assume that the experimenter has a certain freedom of action that is not accounted for by the first principles of physics. Without this freedom of choice, experiments would be pointless.

Even in physics we cannot exclude the subjective dimension of the human condition. Man's free will implies the ability to initialize actions, and it constitutes his essence as a responsible actor. We act under the idea of freedom, but the point here is neither man's sense of personal freedom as a subjective experience nor the question of whether this idea could be an illusion or not. The point is that the framework of *experimental science requires the freedom of action as a constitutive though tacit presupposition.*

Related to intentionality and freedom of action is the concept of the "now", of the present instant. All fundamental physical laws are invariant under time translation and time reversal. Since all instants of physical time are equivalent, the concept of a distinguished now at which distinguished initial conditions for an experiment are prepared cannot be part of those fundamental laws. The concept of nowness has no place in the first principles of physics. In this sense, the first principles of

H. Primas, *Knowledge and Time*,
DOI 10.1007/978-3-319-47370-3_8

physics cannot give a comprehensive description of the empirical fundament of physics.

8.2 What Is Matter?

The concept of matter is far from unproblematic (see Jammer 1961). The most common simple definitions like *res extensa*, or "anything which has mass and occupies space", or "the only substance" are clearly either false or inadequate. With the special theory of relativity of 1905, matter and energy lost their independent and absolute existence and became intimately related notions. From the viewpoint of the first principles of physics, Wolfgang Pauli (1954, pp. 10, 14) gave a modern answer to the question of what matter is in his contribution to the *International Symposium Presented in Honor of the Two-Hundredth Anniversary of Columbia University*:

> Matter has always been and will always be one of the main objects of physics. ... even light has become matter now, due to Einstein's discoveries. It has mass and also weight; it is not different from ordinary matter, it too having both energy and momentum. ...
> Taking the existence of all these transmutations into account, what remains of the old ideas of matter and substance? The answer is *energy*. This is the true substance, that which is conserved; only the form in which it appears is changing.

A fundamental theorem by Emmy Noether (1918) connects conserved quantities and continuous symmetries. Under rather general conditions an inverse of Noether's theorem holds: to every conserved quantity there corresponds a *global continuous one-parameter group*. The constants of motion of such a system are given by the generator of this one-parameter group. In physics the conservation of energy-matter[144] implies that the solutions of the equations of motion of the system are invariant under time translations $t \to t + \tau$ with $\tau \in \mathbb{R}$, where $t \in \mathbb{R}$ refers to an external homogeneous clock time. Since the real numbers $\mathbb{R}$ form an Abelian group under addition,[145] the time translations generate a locally compact commutative one-parameter group, called the *time-translation group* $\{\mathfrak{V}(\tau)|\tau \in \mathbb{R}\}$.

---

[144] A recurrent theme in discussions of the mind-matter problem is the alleged violation of the conservation of physical energy. Yet, the scope of validity of the conservation law of energy is well understood: *it holds if and only if the equations of motion of the system considered do not contain any distinguished moment of time, i.e. no now*. Intentional or other mental influences are not invariant under time translations, so that in this case any argument involving energy conservation is misplaced. Whenever energy is well-defined, it is conserved by definition. The concept of energy is meaningless if time is not homogeneous so that, in this case, it makes no sense to speak of a violation of energy conservation.

[145] Recall that a group $(\mathfrak{G}, \circ)$ is a set $\mathfrak{G}$ on which an associative operation $\circ$ is defined with a unit element e such that $e \circ g = g \circ e = g$ and with an inverse element g' such that $g \circ g' = e$ for all $g \in \mathfrak{G}$.

8.3 What Is Time?

> Space and time taken together constitute the order of possibilities of the entire universe, so that these orders—space and time, that is—relate not to what actually is but also to anything that could be put in its place, just as numbers are indifferent to the things which can be enumerated.
>
> *Gottfried Wilhelm Leibniz (1702)*[146]

8.3.1 *Time Is Not a Physical Observable*

The notion of time is one of oldest and most fundamental concepts in science. And yet there is no general agreement about what time is. Most of the rich *philosophical* literature about time is only marginally helpful to get some ideas about the formal structure of temporal phenomena.

Also, the concepts of *traditional theoretical physics* are not very inspiring. In classical and quantum mechanics, time is a mere ordering parameter—an *external* variable measured by a classical clock that is independent of the dynamics of the system. In the Hamiltonian formalism of classical and quantum mechanics for strictly isolated systems, time can in principle be eliminated without any loss of physical content. Using Jacobi's principle of least action, Hamiltonian dynamics can be formulated in a completely geometric language without a preferred independent time variable.[147]

In other words, time is not an observable. Furthermore, neither classical physics nor special and general relativity nor quantum physics recognizes an asymmetry of temporal directions. Nevertheless, we usually presuppose tacitly a "principle of retarded causality": no effect can precede its cause. But at a fundamental level there is no distinction between past and future. So, at this level, it makes no sense to speak of memory or anticipation, cause and effect.

In addition, concepts like "present", "purpose" or "intention" do not arise in fundamental physics. According to Albert Einstein (quoted by Forsee 1963, p. 81):

> Time and space are modes by which we think and not conditions in which we live.

In physical theories, time is not an intrinsic concept, but retroactively added from the outside. One may think that general relativity is an exception but, according to Kurt Gödel (1995, p. 236),

---

[146]Leibniz's "Reply to the Thoughts on the System of Pre-Established Harmony. Contained in the Second Edition of Mr. Bayle's Critical Dictionary, article Rorarius" of 1702. English translation taken from Loemker (1969, p. 583).

[147]Compare for example Dirac (1933), Lanczos (1949, pp. 186, 199), Synge (1960, Sect. 62, pp. 100–102, Sect. 82–83, pp. 136–139, Sect. 86, pp. 143–144, Sect. 96, pp. 167–170), or Hjalmars (1962).

> ... what remains of time in relativity theory as an objective reality inherent in the things neither has the structure of a linear ordering nor the character of flowing or allowing of change. Something of this kind, however, can hardly be called time.

To experimental physicists "time is what a clock measures".[148] In his famous *Treatise on Analytical Mechanics*, Whittaker (1904, Sect. 20) explained that physical time stands "for any parameter varying continuously with the configuration of the systems considered" and adopted the *convention*

> that the angle through which the earth has rotated on its axis (measured with reference to the fixed stars whose small motions we can for this purpose neglect), in the interval between two events, measures the *time* elapsed between the events in question. This angular measure can be converted into the ordinary measure in terms of mean solar hours, minutes and seconds at the rate of 360 degrees to $24 \times 365\frac{1}{4}/366\frac{1}{4}$ hours.

Yet, relative to atomic clocks, the duration of the earth's revolution around the sun is both irregular and slowing down. Adopting the *hypothesis* that atomic properties are universal, the same at all times and all places, the *International Committee for Weights and Measures* defined in 1967 (*Bureau International des Poids et Mesures* 2006, pp.112f):

> The second is the duration of 9 192 631 770 periods of the radiation corresponding to the transition between the two hyperfine levels of the ground state of the caesium 133 atom.

In 1997 this was amended by the remark that the definition of the SI (Système International) second is based on a caesium atom unperturbed by black body radiation:

> This definition refers to a caesium atom at rest at a temperature of 0 K.

These redefinitions are in accordance with the view of Ernst Mach (1919, p. 224) that the laws of physics simply express correlations between phenomena:

> It is utterly beyond our power to measure the changes of things by time. Quite the contrary, time is an abstraction, at which we arrive by means of the changes of things; made because we are not restricted to any one definite measure, all being interconnected.

We never measure a "true time variable" directly. What we measure as "time" is always a relative evolution between variables. Physics describes only relations between physical processes which can serve as clocks, but cannot define time reductively.

In quantum theories, time is a real-valued parameter as in classical physics. It plays a role as parameter for the Hamiltonian evolution according to the equations of motion, but has no direct physical interpretation. In the very early days of the development of quantum mechanics, Pauli wrote:[149]

---

[148] Compare Einstein (1905, p. 893, editor's translation): "It might seem that all difficulties concerning the definition of time could be overcome if I replace 'time' by 'the position of the pointers of my clock'. Such a definition is indeed sufficient if the issue is to define a time exclusively for the place at which this clock is located. The definition becomes insufficient if the issue is to connect events occurring at different places, or—what is the same—to evaluate events that occur at places distant from the clock."

[149] Letter by Pauli to Bohr of November 17, 1925. Letter 106 in Hermann et al. (1979, p. 260, editor's translation).

> If one reflects where the difficulty lies to make statements about the temporal instants of transition processes from the point of view of the new theory, one soon realizes that time actually does not enter into the new theory at all.

8.3.2 Subjective Time

Awareness of time is a fundamental feature of conscious life. In books 10 and 11 of his *Confessions*, Augustine of Hippo (354–430) links the analysis of time to that of presence. He notes that past time no longer exists and future time does not yet exist and concludes that past and future exist only in the mind's present moment (Augustine 1955, pp. 224, 229):

> What, then, is time? If no one asks me, I know what it is. If I wish to explain it to him who asks me, I do not know. Yet I say with confidence that I know that if nothing passed away, there would be no past time; and if nothing were still coming, there would be no future time; and if there were nothing at all, there would be no present time. ...
> But even now it is manifest and clear that there are neither times future nor times past. Thus it is not properly said that "there are three times, past, present, and future". Perhaps it might be said rightly that "there are three times: a time present of things past; a time present of things present; and a time present of things future". For these three do coexist somehow in the soul, for otherwise I could not see them. The time present of things past is memory; the time present of things present is direct experience; the time present of things future is expectation.

Expressed in modern language, Augustine asserts that for there to be time, there must be some material events happening. From this Augustine concludes that time resides in the passage from future to past events and that we can be aware of time only while it is passing (Augustine 1955, §16).

Moreover, memory plays a pivotal role in the perception of time. Not only does it retain recollections of past experiences, but it also holds present realities in mind and anticipates the future (Augustine 2002, p. 181):

> For we, as it were, see some of our thoughts about future things more clearly and more certainly, as though they were very near; and we do this when we are able and insofar as we are able, by our memory which seems to pertain not to the future but to the past. This can be proved by those sayings and hymns which we render from memory in the proper sequence. For unless we foresaw in our thought what follows, we certainly could not utter them. And yet it is memory, not foresight, that enables us to foresee them.

8.3.3 Newtonian Absolute Time

Our ideas about time have changed dramatically since Isaac Newton proposed a universal, absolute, mathematical time. In his *Philosophiae Naturalis Principia Mathematica* (Latin original of 1687), Newton (1729) assumed that time has an inner nature that exists independently of a spectator's mind and has no relation to sensible things:

> Absolute, true, and mathematical time, in and of itself and of its own nature, without reference to anything external, flows uniformly ...

George Berkeley disagreed with Newton. In his *Treatise Concerning the Principles of Human Knowledge*, Berkeley (1710, Sect. 112t) criticized Newton's doctrine of absolute motion:[150]

> It does not appear to me that there can be any motion other than relative.

Later, in his essay *De Motu*, Berkeley rejected Newton's doctrine of absolute time as meaningless and without any empirical significance, so that it has to be eliminated from science (Berkeley 1721, §63): "It is impossible that absolute motion should be conceived."

In 1883 Ernst Mach published a history of mechanics where he argued strongly against Newton's idea of absolute space and absolute time. Newton claimed that inertial motion was relative to absolute space while Mach argued that inertial motion was relative to the average of all the mass in the universe. Mach advocated the idea that the laws of physics do not describe how things change as time passes, they only express *correlations* between observed quantities (Mach 1866):

> The physical space I have in mind (which already includes time) is therefore nothing but the dependence of the phenomena on one another. A completed physics that knew of this basic dependence would have no need of separate concepts of space and time because these would already be comprised.

Subsequently, Mach (1919, pp. 223f) insisted that absolute time does not exist—rather, we abstract time from motion:

> It would appear as though Newton in the remarks here cited still stood under the influence of the medieval philosophy, as though he had grown unfaithful to his resolves to investigate only actual facts. When we say a thing A changes with time, we mean simply that the conditions that determine a thing A depend on the conditions that determine another thing B. ... [Absolute time] is an idle metaphysical conception.

Similarly, Ludwig Wittgenstein stated in his *Tractatus Logico-Philosophicus* (Wittgenstein 1969, 6.3611):

> We cannot compare a process with "the passage of time"—there is no such thing—but only with another process (such as the working of a chronometer).

8.3.4 Leibnizean Relational Time

In classical mechanics time is considered as an *external parameter* referring to a clock carried by an inertial observer, which is not part of the system considered.

---

[150]Editor's note: The most famous response against Newton's notion of absolute time (and space) was due to Leibniz in 1715/16 (see Sect. 8.3.4). It was phrased in Leibniz's well-known five letters in the *Leibniz-Clarke Correspondence* (Alexander 1956), in which Clarke replied to Leibniz on behalf of Newton. It is not so well known that Berkeley's criticism preceded that of Leibniz by a few years.

Likewise, in traditional quantum mechanics time is taken as an external clock time, very much in the spirit of Newton.

By contrast, Gottfried Wilhelm Leibniz rejected the Newtonian ideas of absolute time, space, and motion. He proposed a relational causal theory of time which is based on the idea that time is absent at the fundamental level. Leibniz held that events are more fundamental than moments and that time is a *relational order* abstracted from sets of connected events (Alexander 1956).

Leibnizean time does not have an existence independent of objects or events. Leibniz's theory of relational time is based on the logical priority of the concept of simultaneity over all spatiotemporal concepts (like *prior* or *posterior*; compare Jammer (2004, p. 1630, 2006, p. 80). In his *Metaphysical Foundations of Mathematics* of 1714, Leibniz (1989) characterized simultaneity as follows (quoted from Loemker 1969, p. 666):

> *If a plurality of states of things is assumed to exist which involves no opposition to each other, they are said to exist simultaneously.* Thus we deny that what occurred last year and this year are simultaneous, for they involve incompatible states of the same thing.
> If one of two states which are not simultaneous involves a reason for the other, the former is held to be *prior*, the latter *posterior*. My earlier state involves a reason for the existence of my later state. And since my prior state, by reason of the connection between all things, involves the prior state of other things as well, it also involves a reason for the later state of those things and is thus prior to them. *Therefore whatever exists is either simultaneous with other existences or prior or posterior.*
> *Time is the order of existence of those things which are not simultaneous.* Thus time is the universal order of changes when we do not take into consideration the particular kinds of change.

8.3.5 Temporal Correlations

Erwin Schrödinger (1931, p. 243) proposed to replace the Platonic concept of time by *correlations*. Later, Wootters (1984) also pointed out that, without loss of observational content, the notion of time can be replaced by correlations between subsystems. More generally, Rovelli (1996, pp. 1648, 1655) noted that

> physics is the theory of the relative information that systems have about each other. ...Quantum mechanics can therefore be viewed as a theory about the states of systems and values of physical quantities relative to other systems.

In fact, all results of classical or quantum physics can be recovered in a timeless description in terms of *correlations and conditional expectations between an atemporal and a temporal domain.*[151] Such a formulation makes no reference to coordinate time. If, for instance in quantum theory, we replace external clock time t by an internal *time operator* T, then this operator must have a simple absolutely

[151] Editor's note: This sounds fairly innocuous at this point, but readers should be aware that the timeless or atemporal nature of the material (and mental) world is of paramount importance for the discussion of matter-mind-time systems which will start with Sect. 8.5 below.

continuous spectrum. Since the extended time operator $\mathbb{1} \otimes T$ commutes with all observables $A \in \mathcal{A}$ of the atemporal domain,

$$[\mathbb{1} \otimes T, A \otimes \mathbb{1}]_- = 0, \tag{8.1}$$

the time operator T acts as a real-valued random observable in the atemporal domain. If the time operator T even commutes with *all* observables of the universe of discourse, then it is a classical observable affiliated to the *center*[152] $\mathcal{Z}(\mathcal{A})$ of the algebra of atemporal observables.

$$\mathbb{1} \otimes T \in \mathcal{Z}(\mathcal{A}). \tag{8.2}$$

In this case the physical content of the theory is not changed and superpositions of states at different times are ruled out.[153]

8.4 What Is a Fact?

> The word fact is one of the most-used words in our language
> and it occurs in almost every book,
> especially those on science and philosophy.
> Yet it is hardly ever defined, and when it is defined,
> the definition is obviously unsatisfactory.
>
> *G. Burniston Brown (1953)*

8.4.1 Good Question!

Webster's Dictionary (Gove 1964, p. 813) defines a *fact* as something that has objective reality and actual existence. In logic, a *fact* refers to a statement of a state of affairs that is true or false *for all time*.

On Ludwig Wittgenstein's (1969) view in his *Tractatus*,

(1) The world is everything that is the case.
(1.1) The world is the totality of facts, not of things.
(1.11) The world is determined by the facts, and by these being all the facts.
(1.12) For the totality of facts determines both what is the case, and also all that is not the case.

Yet (Burniston Brown 1953, p. 154),

---

[152] The *center* $\mathcal{Z}(\mathcal{D})$ of an algebra $\mathcal{D}$ is the set of all elements of $\mathcal{D}$ that commute with all elements of $\mathcal{D}$. If an operator D is said to be affiliated with a W*-algebra $\mathcal{D}$, then $U^*DU = D$ for each unitary operator U in the commutant of $\mathcal{D}$. See also Sect. 6.2.4 and Appendix A.1.3.

[153] Since the extended time operator $\mathbb{1} \otimes T$ commutes with the extended Hamiltonian $H = H \otimes \mathbb{1}$, where H is the Hamiltonian generating the dynamics of *material* systems, the time operator $\mathbb{1} \otimes T$ is *not* canonically conjugate to the Hamiltonian H of the material domain.

> to say that a fact is "whatever is the case", merely transfers the difficulty of discovering what the fact is, to finding what the "case" is, and we are given no instructions as to how this is done.

Giambattista Vico was among the first to adopt a rigorously constructivist epistemology. In his *De Antiquissima*[154] he claimed that the criterion and rule for the truth of a statement is to make it happen. This is the essence of his well-known quote:[155]

> *Verum et ipsum factum convertuntur.*
> The true and the made are convertible.

Vico rejected Descartes' principle that clear and distinct ideas are the source of truth, and he questioned that natural science produces certain knowledge of reality. He proposed that knowledge is constructed by the knower. He pointed out that, whenever we speak of *facts*, we literally refer to something that has been made (*factum* is the past participle of the Latin verb *facere*), so that truth is verified through creation or invention and not, as Descartes claimed, through observation. The human mind can only know what has passed the test of construction. As a natural consequence, Vico did not assume that space and time are *a priori* categories, but regarded them as human constructs.

In the empirical sciences, a fact is something that can be checked and backed up with evidence. In natural science a fact refers to an event in the past which retains its facticity in the future. It may be that the documents which tell us something about facts can be destroyed or changed. Nevertheless, by definition facts themselves cannot be changed by any future influences.

However, the belief that facts are true for all times, that they are independent of what we think or say about them, and that they are perspective-free and context-free (Wittgenstein 1969, 1.21) is in contradiction to the way of thinking and acting in natural science. Every operationally meaningful fact is *contextual*. Today it is well established that all experience also depends on the conceptual structure of theories so that all facts are theory-laden (Hanson 1958). That there is no such thing as "objective" experience and that there are no "hard facts" *per se* has been emphasized by Ernst Cassirer (1957, p. 409):

> Here we find immediate confirmation of Goethe's saying: "the highest thing would be ... to recognize that everything factual is itself theoretical". There is no such thing as a sheer facticity, as an eternal and immutable datum: on the contrary, what we call a fact must always be theoretically oriented in some way, must be seen in reference to a definite conceptual system, which implicitly determines it. The theoretical means of determination are not subsequently added to the sheer fact but enter into the definition of the fact itself.

---

[154]*De Antiquissima Italorum Sapientia*, originally published by Stamperia de' Classici, Naples, 1710. English translation: Vico (1982).

[155]Editor's note: The constructive emphasis of this phrase is also expressed by the proverb "the proof of the pudding is to eat it", which dates back to sources much earlier than Vico.

8.4.2 Facts Are Contextual

Hilary Putnam (1995, p. 295) rejected the external realism of much of philosophical discussion, the thesis that there is a reality totally independent of our perceptions, representations etc., as hopelessly metaphysical. In agreement with the realism of natural science, Hilary Putnam's *pragmatic realism* (also: "internal realism") holds that an object can only be reasonably defined within a framework of some conceptual scheme (Putnam 1981, pp. 50, 52):

> There is no God's eye point of view that we can know or usefully imagine; there are only the various points of view of actual persons reflecting various interests and purposes that their descriptions and theories subserve. ...
> "Objects" do not exist independently of conceptual schemes. we cut up the world into objects when we introduce one or another scheme of description.

According to his position, it makes sense to ask what objects the world consists of *only from within a theory*, and that knowledge of facts presupposes knowledge of values. Truth is not the same as correspondence with reality, but has much more to do with rationality (Putnam 1981, pp. x, 128, 134):

> ... there is an extremely close connection between the notions of *truth* and *rationality*; that, to put it even more crudely, the only criterion for what is a fact is what it is *rational* to accept. ...
> ... factual statements themselves, and the practices of scientific inquiry upon which we rely to decide what is and what is not a fact, presuppose values. ...
> We must have criteria of rational acceptability to even have an empirical world.

Although Putnam does recognize the importance of complementary points of view, he does not attempt to investigate the logical structure of his "internal realism". Putnam's position can be (at least to some degree) appropriately represented as a partial Boolean description. His internal realism requires a preselected conceptual scheme which can be realized by a locally Boolean subdomain of this partial Boolean description. This is consistent with the traditional expression of a fact as a proposition which is either verified (true) or falsified (false). In this context, it is worth noting that in the traditional formulation of quantum theory there are no Boolean subdomains, so that this theory is incapable of describing facts.

8.4.3 Difficulties in Traditional Quantum Mechanics

Yet, this is not the only problem that fundamental quantum theory has with facts. Here are some more non-trivial difficulties:

- In any fundamental time-reversal invariant theory it is not evident what is to be understood by "past" and "future".
- If the time-reversal invariance is broken, then it is not evident why the direction of time should be universal.
- In traditional quantum theory it is not evident how permanent documents can be represented.

Historically, quantum mechanics was conceived as a theory of microscopic systems containing a few electrons, protons, or photons. The early Copenhagen view distinguished between microscopic quantum systems and macroscopic measuring instruments. In view of the fact that there is an essentially continuous transition from the microworld to the macroworld such a sharp distinction is hardly sustainable.

In the older literature there are two differing opinions about how measuring instruments should be described. According to Bohr (1934, p. 53), our "interpretation of the experimental material rests essentially upon the classical concepts". But Bohr's emphasis on the special role of "classical concepts" is somewhat misleading. By classical concepts, Bohr refers to a description in terms of ordinary language. Important for Bohr's arguments is only that facts have to be described in a Boolean language, but not necessarily in terms of classical physics. This requirement reflects actual scientific practice: Every experiment ever performed in physics, chemistry, biology or psychology can be described in the Boolean framework of ordinary language.

While Bohr argued that the description of measuring instruments cannot be included in the realm of traditional quantum mechanics,[156] von Neumann (1932a, Chap. VI) underscored that quantum mechanics is necessary for the description of the measuring process. In a nutshell, the apparent contradiction between these two sensible views is the reason for the fruitlessness of so many discussions about the alleged measuring problem of quantum mechanics. For a long time it was not realized that the two viewpoints are compatible—for example, Wigner (1969, p. 433) still asserted that the requirement of a classical description of measurements "is in conflict with the linearity of the quantum theoretical equations of motion".

8.4.4 Measuring Tools Need Not Be Macroscopic

Wigner's (1963, p. 14) claim "that the macroscopic nature of the apparatus is necessary in principle" is based on a misplaced identification of the terms microscopic/quantal and macroscopic/classical. If Wigner's statement were true, then quantum mechanics would already fail at the molecular level. For example, enzymes act as microscopic measuring instruments, and the genetic code realizes, at a molecular level, a highly reliable classical, irreversible and non-anticipating memory which in fact acts as a measuring instrument of molecular size. Many other examples of extremely small measuring devices can be constructed using modern nanotechnology. The macroscopic nature of most measuring instruments of experimental science is irrelevant for the quantum measurement problem.

The historical distinction between "microscopic" and "macroscopic" has become obsolete since long. There is an essentially continuous transition from atoms to small molecules, to macromolecules, to grains, crystals and macroscopic bodies.

[156]Bohr in a letter to Schrödinger of October 26, 1935, quoted from Kalckar (1996, pp. 510f).

The existence of *classical* features in small molecular quantum systems is an empirically well-known fact in chemistry and molecular biology. The chirality of some biomolecules (like amino acids, sugars, lipids, or steroids), the knot type of circular DNA-molecules, and the temperature of chemical substances are examples of such classical properties.

No quantum mechanical superpositions of enantiomeric states of biomolecules are empirically known. This means that, from a phenomenological point of view, there exists a superselection rule between right- and left-handed isomers. Whatever the explanation in terms of a fundamental theory may be, these phenomena cannot be ignored in phenomenologically valid descriptions of molecular phenomena.

8.4.5 Are There Facts in Quantum Theory?

In non-Boolean theories it is not evident what we should understand by an "event" or by a "fact". Likewise, it is not manifest that we can distinguish different things in our world, and that instruments which can register facts exist at all. Also, we have to bear in mind the possibility that in quantum theory facts in the classical sense turn up only in the limit of long time (which, however, may be quite short from a human perspective). In this case we may speak of asymptotic facts which manifest themselves after sufficiently long but finite time as approximate facts (cf. Primas 1997).

In the early years of quantum mechanics it was not realized that quantum systems are capable of developing Boolean structures. Von Neumann's (1932a) irreducibility postulate that all self-adjoint operators acting on the Hilbert space of state vectors are observables implies that the center of the algebra of observables is trivial, so that Boolean subsystems are impossible. Since experimental science requires a domain of discourse which has a Boolean description, we conclude that the irreducibility postulate is not *universally* valid and that in general the algebra of observables must have a non-trivial center.

8.4.6 Classical Observables

A more comprehensive, mathematically precise codification of the heuristic ideas of early quantum mechanics is *algebraic quantum theory* (see Chaps. 6 and 7). It is based on the theory of topological *-algebras and uses the *same* mathematical language for both classical and quantum systems. It is appropriate for microscopic, mesoscopic and macroscopic systems with finitely or infinitely many degrees of freedom. It includes classical point mechanics, classical statistical mechanics, and traditional quantum mechanics as special cases. In terms of algebraic quantum field theory, it is also capable of addressing fields such as the electromagnetic field.

Algebraic quantum theory rejects von Neumann's irreducibility postulate and allows observables which commute with all physically relevant observables (cf.

Sects. 6.2.3, 6.2.4). Such observables are called *classical observables* because they behave (in an appropriate context) as the dynamical variables of classical point mechanics. In algebraic quantum theory facts are described by classical observables which are compatible with all other observables used in a particular description.

For a suitable description of facts, memories, or of experiments in quantum theory we have to use the Boolean logic associated with the Boolean algebra of all projections of a commutative algebra. In a W*-algebraic description this Boolean algebra is given by the set of all projections in the center of the underlying W*-algebra $\mathcal{M}$ of the universe of discourse. The center $\mathcal{Z}(\mathcal{M})$ is a commutative W*-subalgebra of $\mathcal{M}$ whose elements commute with all elements of $\mathcal{M}$,

$$\mathcal{Z}(\mathcal{M}) := \{\mathcal{Z} | Z \in \mathcal{M}, ZM = MZ \text{for every} M \in \mathcal{M}\}. \tag{8.3}$$

The non-trivial self-adjoint elements of the center are the classical observables of the universe of discourse.

The existence of classical features does not contradict the first principles of quantum mechanics. This is not achieved via the limit $\hbar \to 0$ or by way of the so-called "correspondence principle", but thanks to a description that allows algebras of observables with a non-trivial center.[157] A quantum system is called classical if its algebra of observables is commutative. Note that the dynamical evolution of classical observables still depends on Planck's constant.

In algebraic quantum mechanics it is not necessary to postulate *classical quantum systems*, since they arise as a consequence of a proper mathematical codification. By construction, a classical quantum system does not violate the Heisenberg inequality for non-commuting observables. The behavior of a classical quantum system depends on the value of Planck's constant $\hbar$. Examples for classical observables are the electric charge, the chirality of molecular systems, or the temperature and chemical potential of thermodynamic systems. A single molecule of L-ascorbic acid is an example for a chemically relevant system that is partly quantal and partly classical.

8.5 Matter-Mind-Time Systems

8.5.1 On the Separation of Matter and Mind

At present, an adequate understanding of the relation between mind and matter is not available.[158] A purely materialist understanding of subjective consciousness or intentionality in terms of physical interactions (in the sense of forces) seems to be inconceivable. As a consequence, a straightforward reduction of the mental to the material is premature if not fallacious. Yet, if subjective consciousness or the faculty

[157] Since the fictitious limit $\hbar \to 0$ does not exist in the norm topology, the still widely held view that classical mechanics is the limiting case for a vanishing Planck constant $\hbar$ is untenable.

[158] Compare the recent book *Mind and Cosmos* by Thomas Nagel (2012).

of reason is not reducible to something material, this does not imply that the mental and the material belong to two entirely different realms of reality. Therefore we do not assume that there is a clear-cut separation of mental and material phenomena. In particular, we reject both Cartesian-style substance dualism (claiming that mind and matter are distinct substances) and property dualism (claiming that mental and physical properties are categorically distinct).

Even if we take for granted that there are no physical interactions between mind and matter, we cannot conclude that mental and material phenomena are uncorrelated. As we have argued elsewhere (Primas 2007), a comprehensive description of mind-matter issues requires a non-Boolean approach, so that non-classical correlations are compulsory. In a world with a non-Boolean logical structure, non-classical correlations are inevitable. That is, the well-established but counterintuitive Einstein-Podolsky-Rosen-correlations (see Sect. 6.3.5), or variants thereof, are not restricted to the material domain. In the following, we will discuss these non-classical (also called holistic) correlations with the mathematical tools of algebraic quantum theory. Note that this does in no way imply a physicalist or materialist point of departure.

Two parts of the world are not holistically correlated if and only if at least one part admits in all strictness a Boolean description (Primas 1980, Sect. 5.2, 1981, Sect. 5.6, 1984, Sect. 7). If $\mathcal{A}_1$ and $\mathcal{A}_2$ are W*-algebras, and $\mathcal{A}_1 \overline{\otimes} \mathcal{A}_2$ is their W*-tensor product, then every pure normal state functional on $\mathcal{A}_1 \overline{\otimes} \mathcal{A}_2$ is not entangled if and only if it is a product state (see Sect. 6.3.6). This is the case if and only if either $\mathcal{A}_1$ or $\mathcal{A}_2$ is a commutative W*-algebra (Takesaki 1979, Theorem 4.14, p. 211).

8.5.2 Correlations of Tripartite System

We consider a fictitious tripartite decomposition of the universe of discourse referring to a material, a mental and a temporal domain. This separation is not intended as an ontological commitment referring to the nature of reality, but it is just a methodological assumption which simplifies the distinction between interaction and correlation. To discuss the feasible kinds of entanglement between matter, mind and time we represent the tripartite system by a W*-tensor product

$$\mathcal{D} = \mathcal{A}_{\text{mat}} \overline{\otimes} \mathcal{A}_{\text{men}} \overline{\otimes} \mathcal{T}, \quad \text{where} \tag{8.4}$$

- the W*-algebra $\mathcal{D}$ refers to the total universe of discourse,
- the W*-algebra $\mathcal{A}_{\text{mat}}$ refers to an *atemporal material* domain,
- the W*-algebra $\mathcal{A}_{\text{men}}$ refers to an *atemporal mental* domain,
- the W*-algebra $\mathcal{T}$ refers to a *temporal domain*.

Now we can distinguish the following possibilities for a realization of the W*-algebra $\mathcal{D} = \mathcal{A}_{\text{mat}} \overline{\otimes} \mathcal{A}_{\text{men}} \overline{\otimes} \mathcal{T}$:

(1) $\mathcal{A}_{\text{mat}}$, $\mathcal{A}_{\text{men}}$ and $\mathcal{T}$ are commutative W*-algebras. In this case there are no holistic correlations whatsoever.

(2a) $\mathcal{A}_{\text{mat}}$ and $\mathcal{A}_{\text{men}}$ are commutative W*-algebras and $\mathcal{T}$ is a non-commutative W*-algebras. Holistic correlations exist within the temporal domain.
(2b) $\mathcal{A}_{\text{men}}$ and $\mathcal{T}$ are commutative W*-algebras and $\mathcal{A}_{\text{mat}}$ is a non-commutative W*-algebras. Holistic correlations exist within the material domain.
(2c) $\mathcal{A}_{\text{mat}}$ and $\mathcal{T}$ are commutative W*-algebras and $\mathcal{A}_{\text{men}}$ is a non-commutative W*-algebras. Holistic correlations exist within the mental domain.
(3a) $\mathcal{A}_{\text{mat}}$ is a commutative W*-algebra and $\mathcal{A}_{\text{men}}$ and $\mathcal{T}$ are non-commutative W*-algebras. Holistic correlations exist within the mental and the temporal domain, and between them.
(3b) $\mathcal{A}_{\text{men}}$ is a commutative W*-algebra and $\mathcal{T}$ and $\mathcal{A}_{\text{mat}}$ are non-commutative W*-algebras. Holistic correlations exist within the temporal and the material domain, and between them.
(3c) $\mathcal{T}$ is a commutative W*-algebra and $\mathcal{A}_{\text{mat}}$ and $\mathcal{A}_{\text{men}}$ are non-commutative W*-algebras. Holistic correlations exist within the material and the mental domain, and between them.
(4) $\mathcal{A}_{\text{mat}}$, $\mathcal{A}_{\text{men}}$ and $\mathcal{T}$ are non-commutative W*-algebras. Holistic correlations exist in all three components, and between them. No component is uncorrelated.[159]

8.5.3 *Entanglement of Tripartite Systems*

A state is said to possess genuine tripartite correlations if with respect to every bipartite cut the state is not a product state (Bennett et al. 2011). Tripartite non-Boolean systems can be entangled in many inequivalent ways which cannot be interconverted by local unitary transformations.

Example: Entangled Three-Qubit Pure State
Every pure state vector Ψ of a system consisting of three two-level systems acting in the Hilbert space $\mathcal{H}^{(1)} \otimes \mathcal{H}^{(2)} \otimes \mathcal{H}^{(3)}$, $\mathcal{H}^{(\lambda)} = \mathbb{C}^2$, where the vectors α and β generate an orthonormal basis for $\mathbb{C}^2$, can be written as (Acín et al. 2000, 2001):

$$\Psi = \kappa_1 e^{i\phi} \alpha \otimes \alpha \otimes \alpha + \kappa_2 \beta \otimes \alpha \otimes \alpha + \kappa_3 \alpha \otimes \beta \otimes \alpha + \kappa_4 \alpha \otimes \alpha \otimes \beta + \kappa_5 \beta \otimes \beta \otimes \beta,$$

with $\kappa_j \geq 0$ and $0 \leq \phi < \pi$.
There are two classes of maximal tripartite entanglement (Dür et al. 2000). Every maximally entangled pure state of a system consisting of three two-level systems can be converted by local unitary transformations into one of two standard forms, namely either the so-called *GHZ-state* with the state vector (Greenberger et al. 1989)

$$\Psi_{\text{GHZ}} := \frac{1}{\sqrt{2}} \{\alpha \otimes \alpha \otimes \alpha + \beta \otimes \beta \otimes \beta\}, \tag{8.5}$$

[159]Editor's note: Option (4) replaces what Primas (2003, 2009) referred to as the undivided "one world" (*unus mundus* in the parlance of Pauli and Jung) and lifts it from a verbal speculation to a formal description.

or the so-called *W-state* with the state vector (Coffman et al. 2000)

$$\Psi_{\mathrm{W}} := \frac{1}{\sqrt{3}}\{\alpha \otimes \alpha \otimes \beta + \alpha \otimes \beta \otimes \alpha + \beta \otimes \alpha \otimes \alpha\}. \tag{8.6}$$

The main difference between the GHZ-state and the W-state is that the GHZ state has only tripartite entanglement, while in the W-state there is no genuine tripartite entanglement but the three two-qubit subsystems are maximally entangled. That is, for every $j = 1, 2, 3$ the reduced two-particle density operator

$$\mathrm{tr}_{\mathcal{B}(\mathcal{H}_j)}\big\{|\Psi_{\mathrm{GHZ}}\rangle\langle\Psi_{\mathrm{GHZ}}|\big\} \quad \text{is separable,} \tag{8.7a}$$

$$\text{while} \quad \mathrm{tr}_{\mathcal{B}(\mathcal{H}_j)}\big\{|\Psi_{\mathrm{W}}\rangle\langle\Psi_{\mathrm{W}}|\big\} \quad \text{is entangled.} \tag{8.7b}$$

8.5.4 Borromean-Type Entanglement

To the best of our present knowledge there are no interactions between the material and the temporal domain, between the temporal and the mental domain, and between the mental and the material domain. Nevertheless, we may speculate that the tripartite matter-mind-time system is, as a rule, entangled (option (4) in Sect. 8.5.2).

Some of the structure of possible entanglements of this tripartite system can be elucidated through analogies with the topological connectivity of linked loops in three-dimensional space (Aravind 1997; Kauffman and Lomonaco 2002; Sugita 2006). For instance, a *Borromean link* consists of three mutually disjoint simple closed interlocking curves in three-dimensional space in which no pair of links interlocks.[160]

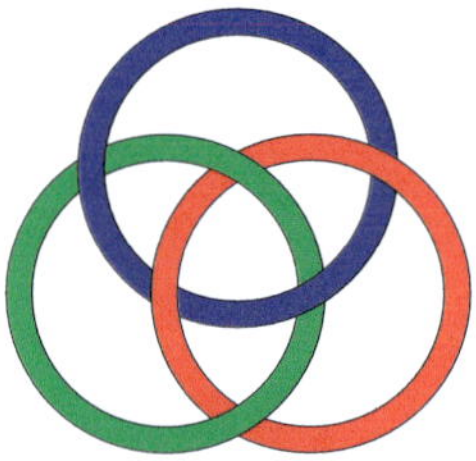

No deformation will separate the rings, but if any one ring is cut, then the three links fall apart. If a state of a tripartite type system is Borromean-like correlated, then there exists a pure triple correlation but there are no pairwise correlations. The structure of such a system is completely different from a pairwise correlated system.

It is worthwhile to speculate about the possibility that the state of matter-mind-time systems described by the tripartite W*-system $\mathcal{A}_{\mathrm{mat}} \overline{\otimes} \mathcal{A}_{\mathrm{men}} \overline{\otimes} \mathcal{T}$ could be

[160]The term Borromean rings derives from the aristocratic Borromeo family, an Italian family from the Renaissance who used the rings in their coat of arms to symbolize the values of collaboration and unity. The Borromean rings have also been used to symbolize the Christian doctrine of Trinity, the unity of Father, Son, and Holy Spirit as three entities in one Godhead.

Borromean-like correlated. In this case, pairwise non-classical correlations between matter and mind, between mind and time, and between time and matter would be absent. The irreducible entanglement between *all three* domains would describe new phenomena which could arise from a *genuine* triple entanglement.[161] However, it is not known whether matter-mind-time systems as outlined here exhibit such Borromean-like triple correlations.

8.5.5 The Universe of Discourse as a Bipartite W-System*

A globally non-Boolean description without Boolean subdomains requires that the representing W*-algebra is a factor.[162] Since the germ of time lies neither in the material nor in the mental domain, we have to distinguish between *atemporal* and *temporal* domains. If we adopt Leibniz's viewpoint (see also Sect. 8.3.5), all time phenomena can be grasped by *correlations* between temporal and atemporal domains. Algebraically we describe the total universe of discourse by a factorial W*-algebra $\mathcal{D}$. It follows that the three composing W*-algebras $\mathcal{A}_{\text{mat}}$ $\mathcal{A}_{\text{men}}$ and $\mathcal{T}$ are also factors.

Since we have no sufficient information about the mental domain, what follows will be restricted to structures that are independent of a hypothetical separation of mind and matter. Since human beings are part of the world, the exclusion of everything mental from the scope of physical science may be premature. Therefore we do not adopt a dualist view, and we do not consider the mental as separate from the non-mental atemporal domain. Algebraically, we describe the atemporal domain as a whole by a factorial W*-algebra $\mathcal{A}$ and the temporal domain by a factorial W*-algebra $\mathcal{T}$. As a consequence, the universe of discourse is represented as a *bipartite system*, realized by a W*-tensor product

$$\mathcal{D} = \mathcal{A} \overline{\otimes} \mathcal{T} \quad \text{with } \mathcal{A} = \mathcal{A}_{\text{mat}} \overline{\otimes} \mathcal{A}_{\text{men}}. \tag{8.8}$$

In this setting, we can basically distinguish four cases:

(1) $\mathcal{A}$ and $\mathcal{T}$ are commutative W*-algebra.
(2) $\mathcal{A}$ is a commutative W*-algebra and $\mathcal{T}$ is a non-commutative W*-algebra.
(3) $\mathcal{A}$ is a non-commutative W*-algebra and $\mathcal{T}$ is a commutative W*-algebra.
(4) $\mathcal{A}$ and $\mathcal{T}$ are non-commutative W*-algebra.

In cases (1), (2) and (3) classical correlations between the atemporal and the temporal domain are possible, but these domains cannot be entangled by non-classical, holistic correlations. In case (4), genuinely holistic correlations are possible in addition. The paradigmatic example for case (1) is classical physics with an external

[161] Editor's note: Compare Specker's parable about the princess of Arba'ilu in Sect. 2.4.3 and the discussion of coexistent effects in Sect. 3.5.4.

[162] A W*-algebra is called a *factor* if its center contains only the scalar multiples of the identity, compare Sect. 6.2.4 and Appendices A.1.3 and A.1.4.

parameter time, where the algebra $\mathcal{A}$ stands for the material system under discussion, while the algebra $\mathcal{T}$ symbolizes time. Case (2) is rarely realized: It corresponds to a classical system driven by a non-Boolean random time. Traditional quantum mechanics with an external parameter time is an example for case (3).

The fact that we use the operator-algebraic formalism of quantum theory *does not imply* that our discussion is restricted to traditional quantum mechanics. Our framework is that of an operator formalism of *harmonic analysis* with its associated non-Boolean projection-valued logic, in which complementary propositions in a very general sense have a natural place.

8.5.6 Type-I Versus Type-III Quantum Theories[163]

Bottom-up approaches in physics (cf. Chap. 6) try to explain the physical world in terms of basic building blocks—often referred to as "elementary particles". In the algebraic bottom-up description the basic building blocks generate a W*-algebra of type I. In this framework the W*-algebra of observables for a composite system is taken as the tensor product of the algebras of observables of the group-theoretically defined elementary systems. The resulting tensor-product structure of the algebra of observables then defines the factorization characterizing the entanglement between the elementary systems.

Top-down quantum theories (cf. Chap. 7) can be formulated in terms of W*-algebras of type III, with properties quite different from type-I algebras.[164] To elucidate this situation, we first recall the classifications of W*-algebras into type I, type II, and type III, introduced in a series of papers by Murray and von Neumann (1936, 1937, 1940, 1943). The finite type-I_n factor is isomorphic to the algebra of $n \times n$ matrices. The infinite type-I_∞ factor is isomorphic to the algebra $\mathcal{B}(\mathcal{H})$ of all bounded operators on a separable infinite-dimensional Hilbert space $\mathcal{H}$. A factor is of type II if it does not contain atoms but contains some non-zero finite projections.

A factor is of type III if it does not contain any non-zero finite projection and therefore no minimal projections. All projections of a type-III factor $\mathcal{M}$ are infinite and equivalent to $\mathbb{1} \in \mathcal{M}$, so that type-III algebras possess no atoms, and hence *there are no pure states for type-III algebras* (see Kadison and Ringrose 1986, Corr. 6.3.5, p. 412.) That is, for every state functional ρ on a type-III algebra there are state functionals ρ_1 and ρ_2, different from ρ, such that

$$\rho(M) = \frac{1}{2}\rho_1(M) + \frac{1}{2}\rho_2(M), \tag{8.9}$$

[163] Compare Appendices A.1.4 and, in particular, A.5 for more background to this subsection.

[164] Compare Yngvason (2005). Besides the factor of type I_2 only the factors of type III are "most non-commutative" in the sense that for each non-trivial projection P there exists a maximally incompatible projection Q (see Raggio and Rieckers 1983, p. 284).

for all M in a type-III W*-factor. But type-III algebras acting on a separable Hilbert space possess a dense set of vectors that are both cyclic and separating (Sakai 1971, Corr. 2.9.28, p. 119), so that all their states can be represented by vector states. For every state functional ρ on a type-III W*-factor $\mathcal{M}$ there is a representation on a Hilbert space $\mathcal{H}$ and a (non-unique) vector $\Psi_\rho \in \mathcal{H}$ such that

$$\rho(M) = \langle \Psi_\rho | M \Psi_\rho \rangle, \tag{8.10}$$

for all M in a type-III W*-factor.

Although it is impossible to interpret non-pure states on a type-III algebra as a mixture, Clifton and Halvorson (2001) referred to such non-pure states as "intrinsically mixed"—which obviously is a misnomer. Let $\mathcal{M} \subset \mathcal{B}(\mathcal{H})$ be a type-III algebra. Then its commutant $\mathcal{M}'$ also is a type-III algebra and we have $(\mathcal{M} \cup \mathcal{M}')'' = \mathcal{B}(\mathcal{H})$. The restriction of any pure state functional of the type-I algebra $(\mathcal{M} \cup \mathcal{M}')'' = \mathcal{B}(\mathcal{H})$ to the type-III subalgebra $\mathcal{M}$ or to the type-III subalgebra $\mathcal{M}'$ results in mixed states. It follows that every state functional on a type-III factor $\mathcal{M}$ is *intrinsically entangled* across $(\mathcal{M}, \mathcal{M}')$ (Halvorson 2001). There are no unentangled states across $(\mathcal{M}, \mathcal{M}')$, hence there are no product states. In particular, *in quantum field theory there are no product states between space-like separated subsystems.*

System-subsystem relations are dramatically different for theories based on W*-algebras of different type. In W*-theories a subsystem is characterized by a W*-subalgebra $\mathcal{M} \subset \mathcal{D}$, but in general spatiotemporal distinctness does not imply the independence of the corresponding subalgebras. An independent subsystem is not merely a spatiotemporally separated portion of the full system, but the influence of the rest of the universe of discourse also has to be screened off. Therefore, we need an appropriate concept of independence to begin with.

In quantum field theory, subsystems are said to be *operationally independent* if they can be localized in space-time regions $\mathcal{O}_1 \subset \mathcal{M}$ or $\mathcal{O}_2 \subset \mathcal{M}$, respectively, of the Minkowsky space $\mathcal{M}$ such that the observables of the W*-algebra $\mathcal{M}(\mathcal{O}_1)$ and the observables of the W*-algebra $\mathcal{M}(\mathcal{O}_2)$ are mutually compatible and can be independently and locally prepared. These requirements are fulfilled if the algebras satisfy the so-called split property (cf. Sect. 7.3.6.3). A pair $(\mathcal{M}_1, \mathcal{M}_2)$ of W*-algebras is said to be split if there exists a type-I factor $\mathcal{F}$ such that $\mathcal{M}_1 \subset \mathcal{F} \subset \mathcal{M}_2$ (see Summers 2009). Clearly, mutually commuting type-I factors are split.

While for type-I factors split inclusions are given by tensor products, split inclusions are highly non-trivial for factors of type III (Connes 1973). In a Hilbert-space representation of type-I systems independent subsystems can be represented by factors $\mathcal{B}(\mathcal{H}_1) \subset \mathcal{B}(\mathcal{H})$ and $\mathcal{B}(\mathcal{H}_2) \subset \mathcal{B}(\mathcal{H})$, acting on the Hilbert-space tensor product $\mathcal{H} = \mathcal{H}_1 \otimes \mathcal{H}_2$, so that the W*-algebra $\mathcal{D}$ of the universe of discourse is given by

$$\mathcal{D} \cong \mathcal{B}(\mathcal{H}) = \mathcal{B}(\mathcal{H}_1) \otimes \mathcal{B}(\mathcal{H}_2). \tag{8.11}$$

This factorization of the observable algebra $\mathcal{D}$ is implied by the existence of minimal projections in $\mathcal{D}$ (i.e., projections which have no proper subprojections in $\mathcal{D}$). Type-III theories are characterized by exactly the opposite: For every non-zero projection P in a type-III factor $\mathcal{D}$ there exists an isometry $W \in \mathcal{D}$ with

$$WW^* = P, \qquad W^*W = \mathbb{1}. \tag{8.12}$$

Moreover, in type-I theories every pure state can be disentangled by an appropriate unitary dressing transformation, while no normal state of a factor of type III can be disentangled.

When Murray and von Neumann introduced their classification of W*-factors, the existence of factors of types II and III was not evident. With the aid of ergodic theory they could construct examples of factors of type II but initially the existence of factors of type III was left undecided. Only in their third paper, Murray and von Neumann (1940) provided examples of factors of type III. In view of this fact it is remarkable that almost three decades later Coleman (1968, p. 77) still noted that

> for any advocate of [an atomistic view of physics] it is obviously disconcerting that there are groups that possess no reducible representations that contain no irreducible subrepresentations

and asked

> whether such type II and type III monsters need actually be taken into account in a reasonable physical theory.

In a seminal paper, Alain Connes (1973), using deep results of the modular theory of W*-algebras, gave a complete classification of factors of type III in terms of a continuum of non-isomorphic factors of type III_λ, with $0 \leq \lambda < 1$. Later, Haagerup (1987) established the uniqueness of the type III_1 factor, see Takesaki (2003, Chap. XII) for a detailed exposition. An example for the physical relevance of type-III W*-algebra is the statistical physics of infinite systems with non-zero density (e.g., the thermodynamic limit of quantum statistical mechanics constructed by the infinite limit of the number of systems while keeping their density finite). In particular, equilibrium states at finite temperatures correspond to direct sums or direct integrals of type-III factors (cf. Bratteli and Robinson 1981, Chap. 5.3)

The modular structure of a W*-algebra encodes essential physical information. Let $\mathcal{M}$ be a factor with a separable predual $\mathcal{M}_*$, $\rho \in \mathcal{M}_*$ a faithful normal state on $\mathcal{M}$, and Δ_ρ the associated modular operator, and denote by $\text{Sp}(\Delta_\rho)$ the spectrum of the self-adjoint operator Δ_ρ. Connes (1973, Sect. III) defined the *modular spectrum* $\mathcal{S}(\mathcal{M})$ as the intersection of the spectra of all possible Δ_ρ,

$$\mathcal{S}(\mathcal{M}) := \bigcap \{ \text{Sp}(\Delta_\rho) \,|\, \rho \in \mathcal{M}_* \}, \tag{8.13}$$

where ρ runs over all faithful normal states on $\mathcal{M}$. The modular spectrum $\mathcal{S}(\mathcal{M})$ is an algebraic invariant of $\mathcal{M}$ and a closed multiplicative subgroup of $\mathbb{R}^+$ and leads to the following classification:[165]

---

[165] For details compare Connes (1994, Sect. V.5), and Takesaki (2003, Chap. XII, §1). See also Sects. 7.3.4 and 7.3.5.

(i) If $\mathcal{M}$ is a factor of type I or II, then $\mathcal{S}(\mathcal{M}) = \{1\}$.
(ii) If $\mathcal{M}$ is of type III_0, then $\mathcal{S}(\mathcal{M}) = \{0, 1\}$.
(iii) $\mathcal{M}$ is of type III_λ, $0 < \lambda < 1$, if and only if $\mathcal{S}(\mathcal{M}) = \{0\} \cup \{\lambda^n | n \in \mathbb{Z}\}$.
(iv) $\mathcal{M}$ is of type III_1, if and only if $\mathcal{S}(\mathcal{M}) = [0, \infty)$.

Type-III_1 factors appear as space-time-indexed local algebras in relativistic quantum field theory in four-dimensional Minkowski space (Baumgärtel 1995). Since the algebras are all unitarily equivalent there is, up to unitary equivalence, only one such algebra.

8.5.7 The Order Parameter of the Universe of Discourse

8.5.7.1 The Order Parameter in Bottom-Up Theories

We adopt Leibniz's view that time is not an entity existing independently of actual events and that time introduces an ordering of events. In the following we depict the universe of discourse by a W*-algebra $\mathcal{D}$ with a separable predual $\mathcal{D}_*$ equipped with a σ-weakly continuous one-parameter group of inner automorphism $\delta_\tau \in \text{Aut}(\mathcal{D})$, where τ is a Leibnizian *order parameter* different from ordinary clock time.

Under the (natural) assumption that there are *no interactions between the atemporal and the temporal domain*, the automorphism $\delta_\tau \in \text{Aut}(\mathcal{A} \overline{\otimes} \mathcal{T})$ of the universe of discourse decomposes into a direct product of the one-parameter automorphisms $\{\alpha_\tau \in \text{Aut}(\mathcal{A})\}$ of the atemporal domain and the one-parameter automorphisms $\{\theta_\tau \in \text{Aut}(\mathcal{T})\}$ of the temporal domain,

$$\delta_\tau = \alpha_\tau \times \theta_\tau, \quad \alpha_\tau \in \text{Aut}(\mathcal{A}), \; \theta_\tau \in \text{Aut}(\mathcal{T}). \tag{8.14}$$

Traditional quantum mechanics is a *bottom-up theory* whose dynamics is generated by a unitary one-parameter group $\{\exp -2\pi i H\tau | \tau \in \mathbb{R}\}$, where the Hamiltonian H is a self-adjoint operator affiliated to a factorial W*-algebra $\mathcal{A}$ of type I of the atemporal domain.[166] More generally, if the W*-algebra $\mathcal{D}$ of the full universe of discourse is a factorial W*-algebra of type I_∞, then the automorphisms δ_τ are inner and can be implemented on $\mathcal{D}$ by a strongly continuous one-parameter group of unitary operators $\exp(-2\pi i G\tau)$ on the factor $\mathcal{D}$,

$$\delta_\tau(D) = e^{2\pi i G\tau} A e^{-2\pi i G\tau}, \quad D \in \mathcal{D}, \; \tau \in \mathbb{R}. \tag{8.15}$$

Here, the generator G replaces H as a self-adjoint operator affiliated with $\mathcal{D}$ (rather than $\mathcal{A}$). Similarly, the automorphism groups $\{\alpha_\tau | \tau \in \mathbb{R}\}$ and $\{\theta_\tau | \tau \in \mathbb{R}\}$ can

[166]Editor's note: Traditional quantum mechanics actually refers to a type-I factor of the atemporal *material* domain. However, here it is used in a wider sense, in which $\mathcal{A}$ covers both the atemporal material and the atemporal mental domain.

be implemented by strongly continuous one-parameter groups of unitary operators $\exp(-2\pi i H\tau)$ and $\exp(-2\pi i \Lambda\tau)$,

$$\alpha_\tau(A) = e^{2\pi i H\tau} A e^{-2\pi i H\tau} := A(\tau), \quad A \in \mathcal{A}, \ \tau \in \mathbb{R}, \tag{8.16a}$$

$$\theta_\tau(B) = e^{2\pi i \Lambda\tau} B e^{-2\pi i \Lambda\tau} := B(\tau), \quad B \in \mathcal{T}, \ \tau \in \mathbb{R}, \tag{8.16b}$$

where the self-adjoint generator Λ of the unitary group $\{\exp -2\pi i \Lambda\tau | \tau \in \mathbb{R}\}$ is affiliated to the W*-algebra $\mathcal{T}$ of the temporal domain and referred to as *frequency operator*. The order parameter τ describes correlations between the atemporal and the temporal domain. It must not be confused with ordinary clock time (for details see Sect. 8.5.9).

8.5.7.2 The Order Parameter in Top-Down Theories

In W*-algebraic top-down descriptions the universe of discourse is described by a general factorial W*-algebra $\mathcal{D}$ of type III. On factors of type III the modular flows induced by different states differ only by an inner automorphism, so that in this case the modular automorphism group is *unique* up to inner automorphisms (Connes 1973). Accordingly, for type-III algebras the quotient $\mathrm{Out}(\mathcal{D})$ of the group of automorphism $\mathrm{Aut}(\mathcal{D})$ of $\mathcal{D}$ by inner automorphisms $\mathrm{Inn}(\mathcal{D})$,

$$\mathrm{Out}(\mathcal{D}) = \mathrm{Aut}(\mathcal{D})/\mathrm{Inn}(\mathcal{D}), \tag{8.17}$$

is an intrinsic one-parameter group. This leads to the possibility to define an intrinsic ordering based on the modular group and thereby to an entirely new notion of time (see Sect. 7.4).

Since we do not assume anything like a "universal time", we choose the automorphism group $\{\theta_\tau | \tau \in \mathbb{R}\}$ such that the resulting temporal correlations are "as simple as possible" in the sense of Einstein (1934, p. 165):

> The basic concepts and laws which are not logically further reducible constitute the indispensable and not rationally deducible part of the theory. It can scarcely be denied that the supreme goal of all theory is to make the irreducible basic elements as simple and as few as possible without having to surrender the adequate representation of a single datum of experience.

Even more specifically with respect to time, Poincaré (1898, p. 6) stated:

> Time should be so defined that the equations of mechanics may be as simple as possible. In other words, there is not one way of measuring time more true than another; that which is generally adopted is only more convenient. Of two watches, we have no right to say that the one goes true, the other wrong; we can only say that it is advantageous to conform to the indications of the first.

8.5.8 The Temporal Weyl System

If the unitary group $\{V(\tau)|\tau \in \mathbb{R}\}$ has a *homogeneous Lebesgue spectrum*, then von Neumann (1932b; see also Cornfeld et al. 1982, p. 457) showed that there exists another continuous one-parameter group $\{U(\lambda)|\lambda \in \mathbb{R}\}$ of unitary operators $U(\lambda)$ related to $\{V(\tau)|\tau \in \mathbb{R}\}$ by Weyl's commutation relation,

$$U(\lambda)V(\tau) = e^{2\pi i\lambda\tau}V(\tau)U(\lambda), \quad \lambda, \tau \in \mathbb{R}. \tag{8.18}$$

By Stone's theorem, the unitary operators $U(\lambda)$ and $V(\tau)$ can be written as

$$U(\lambda) := e^{2\pi i\lambda T}, \qquad V(\tau) := e^{-2\pi i\tau\Lambda}, \quad \lambda, \tau \in \mathbb{R}. \tag{8.19}$$

The self-adjoint generator T of the unitary group $\{U(\lambda)|\lambda \in \mathbb{R}\}$ is called *time operator*, while the self-adjoint generator Λ of the unitary group $\{V(\tau)|\tau \in \mathbb{R}\}$ is called *frequency operator*. Both the time operator T and the frequency operator Λ have *simple absolutely continuous Lebesgue spectra* and are affiliated to the algebra $\mathcal{T}$ of the temporal domain. On an appropriate domain the canonically conjugated operators T and Λ satisfy the commutation relation

$$T\Lambda - \Lambda T = (i/2\pi)\mathbb{1}. \tag{8.20}$$

8.5.9 Evolution in Terms of the Order Parameter τ

8.5.9.1 Interpretation of the Order Parameter τ

The order parameter τ is non-trivially related to the clock time t. Clock time refers to the operational viewpoint that time is what clocks measure (Einstein 1905, p. 893). The evolution $\tau \to T_\tau$ of the time operator T is determined by the one-parameter automorphisms $\alpha_\tau \in \mathrm{Aut}(\mathcal{A})$, while clock time t is measured by a material clock with $C \in \mathcal{A}$, where the evolution $\tau \to C_\tau$ is determined by the one-parameter automorphisms $\theta_\tau \in \mathrm{Aut}(\mathcal{T})$. The relation between the order parameter τ and the sequential clock time t is therefore determined by the two automorphism groups $\mathrm{Aut}(\mathcal{A})$ and $\mathrm{Aut}(\mathcal{T})$.

The unitary group $\{\exp 2\pi i\lambda T|\lambda \in \mathbb{R}\}$ generates a commutative W*-algebra $\mathcal{C} \subset \mathcal{T}$, which can be represented on $\mathbf{L}^2(\mathbb{R})$ as the maximal commutative algebra $\mathbf{L}^\infty(\mathbb{R})$ of all complex-valued essentially bounded Lebesgue-measurable functions on $\mathbb{R}$. Since the time operator $\mathbb{1} \otimes T$ commutes with every atemporal observable $A \otimes \mathbb{1}$ ($A \in \mathcal{A}$), the time operator $\mathbb{1} \otimes T$ behaves as a real-valued random variable[167] with the characteristic function

$$\hat{\mu}(\lambda) := \rho\{\mathbb{1} \otimes e^{2\pi i\lambda T}\} = \int_{\mathbb{R}} e^{2\pi i\lambda t}\mu(dt), \quad \lambda \in \mathbb{R}, \tag{8.21}$$

[167] As far as we know, the concept of a "random time" was first discussed by Wiener and Winter (1958).

where $\rho \in \mathcal{D}_*$ is the normal state functional of the universe of discourse. The characteristic function $\hat{\mu}$ defines the corresponding normalized positive probability density function $\tau \mapsto f(\tau)$ by

$$f(\tau) := \int_{\mathbb{R}} e^{-2\pi i \lambda \tau} \hat{\mu}(\lambda) d\lambda \geq 0. \tag{8.22}$$

The evolution of an atemporal observable $A \in \mathcal{A}$ is determined by the one-parameter automorphisms $\alpha_\tau \in \mathrm{Aut}(\mathcal{A})$,

$$A(\tau) = \alpha_\tau(A) = e^{2\pi i H \tau} A e^{-2\pi i H \tau}, \quad A \in \mathcal{A}, \tag{8.23}$$

where the order parameter τ is a random variable with the probability density function $\tau \mapsto f(\tau)$. The expectation value $\rho\{\mathbb{1} \otimes T\}$ is a real-valued random variable with mean value t and variance σ^2,

$$t := \rho\{\mathbb{1} \otimes T\} = \int_{\mathbb{R}} f(\tau) \tau d\tau, \tag{8.24}$$

$$\sigma^2 := \rho\big\{(\mathbb{1} \otimes T)^2\big\} - \rho\big\{(\mathbb{1} \otimes T)\big\}^2 = \int_{\mathbb{R}} f(\tau)(\tau - t)^2 d\tau. \tag{8.25}$$

The experimentally accessible expectation value of the τ-evolved atemporal observable A depends on the moments $t, \sigma^2, \ldots$ of the probability density f,

$$a\big(t, \sigma^2, \ldots\big) := \mathcal{E}\big\{\rho\big\{A(\tau) \otimes \mathbb{1}\big\}\big\} = \int_{\mathbb{R}} \rho\big\{A(\tau) \otimes \mathbb{1}\big\} f(\tau) d\tau. \tag{8.26}$$

8.5.9.2 Dispersion-Free Time

Traditionally empirical phenomena are described successfully in terms of a dispersion-free clock time that is presupposed both in classical physics and in traditional quantum mechanics. However, according to Schrödinger (1931, p. 247), the assumption of an "ideal clock, which quantum mechanics makes use of through the variable t, is in contradiction with the foundations of quantum mechanics". Schrödinger considered a dispersion-free classical clock time as a serious defect of quantum mechanics because (Schrödinger 1932, p. 293, editor's translation)

> actually the knowledge about the variable t is acquired in the same manner as for any other variable, by observing a physical system, namely a clock. The variable t should then be treated as an observable, which is generally associated with a statistical distribution and not with a single value. An exceptional role for time is not justified.

8.5.9.3 Random Time with Small Dispersion

In a good approximation we can replace the sharp clock time by a random time with small dispersion generated by the time operator $\mathbb{1} \otimes T$. If the probability density

$\tau \mapsto f(\tau)$ has a "quasi-sharp" maximum near the point $\tau = t$ for small values of the variance σ^2, and if $\rho\{A(\tau) \otimes \mathbb{1}\}$ is a smooth function which varies slowly near t, we can obtain an asymptotic expression for the integral (8.26).

Using the method of steepest descent (Jeffreys and Jeffreys 1956, p. 505), we write $f(\tau) = e^{\kappa g(\tau)}$ and evaluate the integral

$$\int_{\mathbb{R}} a(\tau) f(\tau) d\tau = \int_{\mathbb{R}} a(\tau) e^{\kappa g(\tau)} d\tau \tag{8.27}$$

for $\kappa \to \infty$ with the saddle-point method. The value of the integral (8.27) is dominated by the contributions where $\tau \mapsto g(\tau)$ has a local maximum, $dg(\tau)/d\tau = 0$. For each stationary point there will be a significant contribution to the integral. If g has only *one* stationary point at t, (8.27) can be approximated for $\kappa \to \infty$ as

$$\begin{aligned} \int_{\mathbb{R}} a(\tau) f(\tau) d\tau &\approx a(t) e^{\kappa g(t)} \int_{\mathbb{R}} e^{\kappa\{g''(t)(\tau - t)^2/2 + \cdots\}} d\tau \\ &= \sqrt{\frac{2\pi}{\kappa |g''(t)|}} e^{\kappa g(t)} \big\{a(t) + O(1/\kappa)\big\}. \end{aligned} \tag{8.28}$$

Example: Gaussian Probability Function
If we choose the function $\tau \mapsto f(\tau)$ as the Gaussian

$$f(\tau) = \frac{1}{\sqrt{2\pi\sigma^2}} e^{-(\tau - \tau_0)^2/(2\sigma^2)} \tag{8.29}$$

with mean value τ_0 and variance σ^2 we get, for a function $\tau \mapsto a(\tau)$ which is analytic in a neighborhood of $\tau = t$ and bounded for all $\tau \in \mathbb{R}$, the asymptotic expansion

$$\int_{\mathbb{R}} a(\tau) f(\tau) d\tau; = a(t) + \frac{\sigma^2}{2} a''(t) + \frac{\sigma^4}{2^2 2!} a''''(t) + \cdots . \tag{8.30}$$

As an illustration we consider the exactly soluble model of a one-dimensional material harmonic oscillator, where the order parameter τ is given by a Gaussian random time with mean value t and variance σ^2. The self-adjoint Hamiltonian H of a quantum-mechanical one-dimensional harmonic oscillator is given by $H = P^2/2m + m\Omega^2 Q^2/2$ with mass $m > 0$ and frequency $\Omega > 0$. The dynamics of the self-adjoint position operator Q and of the self-adjoint momentum operator P are given by the Heisenberg equations of motion:

$$\frac{dP(\tau)}{d\tau} = \frac{i}{\hbar}\big[H, P(\tau)\big]_-, \qquad \frac{dQ(\tau)}{d\tau} = \frac{i}{\hbar}\big[H, Q(\tau)\big]_-, \qquad \tau \in \mathbb{R}.$$

If the random variable τ is Gaussian with mean value t and variance σ^2, we get for the expectation values $\mathcal{E}\{\rho(P_\tau)\}$ and $\mathcal{E}\{\rho(Q_\tau)\}$

$$\begin{aligned} \mathcal{E}\big\{\rho(P_\tau) \otimes \mathbb{1}\big\} &= \big\{p \cos(\Omega t) - m\Omega q \sin(\Omega t)\big\} e^{-\Omega^2\sigma^2/2}, \\ \mathcal{E}\big\{\rho(Q_\tau) \otimes \mathbb{1}\big\} &= \big\{q \cos(\Omega t) + (m\Omega)^{-1} p \sin(\Omega t)\big\} e^{-\Omega^2\sigma^2/2}, \end{aligned}$$

where

$$p = \rho(P \otimes \mathbb{1}), \qquad q = \rho(Q \otimes \mathbb{1}). \tag{8.31}$$

If the probability density function $\tau \mapsto f(\tau)$ is non-Gaussian but falls off faster than $e^{-\tau^2}$ at infinity, then one may use the simple Gram-Charlier expansion. More generally, the Edgeworth expansion of a non-Gaussian probability density function f is always a true asymptotic expansion. For more details see Johnson and Kotz (1970, p. 17).

Chapter 9
Temporal Groups and Algebras

> It has long been recognized
> that no analysis of natural science,
> whether it be physics or biology,
> is complete unless we possess a proper analysis
> of its appropriate time-concept.
>
> *Norbert Wiener (1948, p. 197)*

9.1 Jordan Automorphisms

In the pioneer days of quantum mechanics, Pascual Jordan (1933, 1934) remarked that the sum $A + B$ of two observables A and B also qualifies as an observable, but that in general the product AB has no conceptual meaning. However, the symmetrized product $\frac{1}{2}(AB + BA)$ is always meaningful since it can be expressed in terms of sums and squares

$$A \diamond B := \frac{1}{2}\{(A+B)^2 - A^2 - B^2\}. \tag{9.1}$$

Nowadays, this commutative ($A \diamond B = B \diamond A$), distributive ($A \diamond \{B + C\} = A \diamond B + A \diamond C$), but non-associative ($A \diamond \{B \diamond C\} \neq \{A \diamond B\} \diamond C$) product is called the *Jordan product*. A map that preserves this product structure is called a *Jordan morphism*. A map α defined on an algebra $\mathcal{A}$ is a *Jordan automorphism* if it is bijective, linear, and satisfies $\alpha(A^2) = \alpha(A)^2$ for every $A \in \mathcal{A}$. Automorphisms and anti-automorphisms of algebras are special cases of Jordan automorphisms.

On a factorial W*-algebra $\mathcal{M}$ of type I every symmetry can be realized either by an inner *-automorphism or by an inner *-anti-automorphism on $\mathcal{M}$. It can be implemented by a unitary or anti-unitary operator which is unique up to a factor of modulus one:

$$\alpha(A) := \begin{cases} UAU^{-1} & \text{if } \alpha \text{ is an automorphism, with } U \text{ unitary,} \\ VAV^{-1} & \text{if } \alpha \text{ is an anti-automorphism, with } V \text{ anti-unitary.} \end{cases} \tag{9.2}$$

H. Primas, *Knowledge and Time*,
DOI 10.1007/978-3-319-47370-3_9

An anti-unitary operator is the combination of an anti-linear and a unitary operator. The complex conjugation operator K is an anti-unitary operator, and the product of two anti-unitary operators V and K is evidently a unitary operator, say U,

$$V = UK, \tag{9.3}$$

so that the most general form of an anti-unitary operator is the product of a unitary operator and the operation of complex conjugation. On a Hilbert space $\mathcal{H}$ an anti-unitary operator V is an anti-linear map $V : \mathcal{H} \to \mathcal{H}$ such that $V^* = V^{-1}$:

$$V(a\Psi + b\Phi) = a^* V\Psi + b^* V\Phi, \langle V\Phi | V\Psi \rangle = \langle \Phi | \Psi \rangle^* = \langle \Psi | \Phi \rangle, \tag{9.4}$$

for all $\Psi, \Phi \in \mathcal{H}$ and $a, b \in \mathbb{C}$. Since in the limit of a vanishing parameter of a one-parameter group the representation has to become the identity map, anti-unitary implementations are possible only for discrete groups.

9.2 Temporal Reversal Groups

9.2.1 The Time-Reversal Group

The time-reversal transformation $T \to -T$ gives rise to a group of order two consisting of two elements $\mathfrak{g}_0$ and $\mathfrak{g}_1$

$$t \xrightarrow{\mathfrak{g}_0} t, \qquad t \xrightarrow{\mathfrak{g}_1} -t, \tag{9.5}$$

with the composition law $\mathfrak{g}_1^2 = \mathfrak{g}_0$. If the corresponding time-reversal symmetry is broken, one gets two realizations, one satisfying the generally accepted rules of retarded forward causation and the other one satisfying the strange rules of advanced backward causation. That is, if it is possible at all to derive the principle of retarded causation ("no effect can precede its cause"), then the very same procedure allows the derivation of processes governed by advanced causation. The decision which of the two possibilities is appropriate can therefore not be derived from the first principles of physics.

In the W*-algebraic description of arbitrary systems, a symmetry is represented by a Jordan automorphism of the underlying W*-algebra $\mathcal{M}$. A Jordan-automorphism ι on a W*-algebra is a linear *-preserving bijection which respects the symmetrized product (cf. Emch 1972, p. 152)

$$\iota(A^*) = \iota(A)^*, \qquad \iota(AB + BA) = \iota(A)\iota(B) + \iota(B)\iota(A), \qquad A, B \in \mathcal{M}. \tag{9.6}$$

Every Jordan automorphism of a W*-algebra is the sum of a W*-automorphism and a W*-anti-automorphism (Kadison 1951, Theorem 10). As an immediate consequence it follows that every Jordan-automorphism of a factor (i.e. a W*-algebra with a trivial center) is either a W*-automorphism or a W*-anti-automorphism. A *W*-automorphism* ι is a Jordan-automorphism which preserves the order of the ordinary product,

$$\iota(AB) = \iota(A)\iota(B), \quad A, B \in \mathcal{M}, \tag{9.7}$$

while a *W*-anti-automorphism* ι reverses the order of the terms in the ordinary product,

$$\iota(AB) = \iota(B)\iota(A), \quad A, B \in \mathcal{M}. \tag{9.8}$$

The product of two anti-automorphisms is an automorphism. There exists only one independent anti-unitary symmetry: time reversal. Any other anti-unitary transformation can be expressed in terms of time reversal.

In the *temporal domain*, the time operator T changes sign under time reversal, $T \to -T$. Since an anti-automorphism transforms a complex number z into its complex conjugate,

$$\iota(z) = z^*, \quad z \in \mathbb{C}. \tag{9.9}$$

we get the following behavior of generators of the affine group (for scale S) and the Weyl-Heisenberg group (for time T and frequency Λ) under time reversal (see Sect. 9.5.2 for more details):

$$\iota(T) = -T, \qquad \iota\left(e^{2\pi i\lambda T}\right) = e^{-2\pi i\lambda T}, \tag{9.10}$$

$$\iota(\Lambda) = \Lambda, \qquad \iota\left(e^{2\pi i\tau\Lambda}\right) = e^{2\pi i\tau\Lambda}, \tag{9.11}$$

$$\iota(S) = -S, \qquad \iota\left(e^{2\pi i\lambda S}\right) = e^{-2\pi i\lambda S}. \tag{9.12}$$

In the *atemporal material domain*, the position operator Q and the momentum operator P as the generators of the material unitary Weyl operators (3.53) are symmetric, but they can be distinguished by their behavior under time reversal. In this case, time reversal is given by *motion reversal* $Q \to Q$ and $\dot{Q} \to -\dot{Q}$, where the velocity $\dot{Q}$ is defined by $\dot{Q} := i\hbar[Q, H]_-$. If the Hamiltonian H is invariant under time reversal, $\iota\{H\} = H$, then we find

$$\iota\{Q\} = Q, \qquad \iota\{P\} = -P. \tag{9.13}$$

9.2.2 The Uniqueness of Time Reversal

The time-reversal symmetry is represented by a group of order two. If the time-reversal symmetry is broken one gets *two* representations, one for retarded causation, the other one for advanced causation. The usual choice of retarded causation (causes precede effects) is not implied by the time-reversal symmetry of fundamental physics.

While in classical physics a breakdown of time-reversal symmetry permits that two non-interacting subsystems have different directions of the time arrow, this is in general impossible in quantum theory. The fact that even non-interacting subsystems of a quantum system are in general entangled implies that in quantum theory the time-reversal operation is *global*.

This can be understood in detail in the algebraic description of physical systems, where time reversal is represented by an involutary anti-automorphism $\iota : \mathcal{M} \to \mathcal{M}$ of the underlying W*-algebra $\mathcal{M}$ (see Bargmann 1964, p. 867),

$$T \to \iota(T) = -T, \tag{9.14}$$

with

$$\iota\big(A^*\big) = \iota(A)^*, \qquad \iota(AB) = \iota(B)\iota(A), \qquad i\big(i(A)\big) = A, \quad A, B \in \mathcal{M}. \tag{9.15}$$

The time-reversal map ι is positive, but for quantum systems (where $\mathcal{M}$ is non-commutative) ι is *not completely positive*.[168] That is, if $(\mathcal{M}_1, \iota_1)$ and $(\mathcal{M}_2, \iota_2)$ describe two non-interacting quantum systems, then the *local* maps $\iota_1 \otimes \mathbb{1}_2$ and $\mathbb{1}_1 \otimes \iota_2$ on the minimal W*-tensor product $\mathcal{M}_1 \otimes \mathcal{M}_2$ are not completely positive, hence they are not time-reversal maps. The time-reversal map for the composite quantum system is given by the *global* positive map

$$(\iota_1 \otimes \mathbb{1}_2)(\mathbb{1}_1 \otimes \iota_2) = \iota_1 \otimes \iota_2. \tag{9.16}$$

This implies that even for *non-interacting* systems the time reversal for the first system is *not* given by $\iota_1 \otimes \mathbb{1}_2$. Hence, it is impossible to define a *local* time-reversal operation for one subsystem only. *The map $\iota_1 \otimes \mathbb{1}_2$ represents the time-reversal operator for the first subsystem alone if and only if the two systems are not entangled.* In an entangled system with broken time-reversal symmetry the direction of the arrow of time has to point to the same direction for all (even non-interacting) subsystems.

As a consequence, the time-reversal operation in the *entangled* bipartite system realized by the W*-tensor product $\mathcal{A} \overline{\otimes} \mathcal{T}$, as introduced in Sect. 8.5.5, is defined only globally. It cannot be decomposed into time-reversal operations acting independently on the atemporal (material or mental) domain $\mathcal{A}$ or on the temporal domain $\mathcal{T}$.

9.2.3 The Frequency-Reversal Group

The frequency-reversal transformation $\Lambda \to -\Lambda$ is a group of order two, which in a W*-algebraic description can be realized by a Jordan automorphisms κ of the underlying W*-algebra $\mathcal{M}$ (cf. (9.15)),

$$\kappa\big(A^*\big) = \kappa(A)^*, \qquad \kappa(AB + BA) = \kappa(A)\kappa(B) + \kappa(B)\kappa(A), \quad A, B \in \mathcal{M}. \tag{9.17}$$

---

[168] A linear map $\iota : \mathcal{A} \to \mathcal{A}$ is said to be *completely positive* if the linear map $\iota \otimes \mathbb{1}_n : \mathcal{A} \otimes \mathcal{B}(\mathbb{C}^n) \to \mathcal{A} \otimes \mathcal{B}(\mathbb{C}^n)$ is positive for all $n \geq 1$. Here $\mathcal{B}(\mathbb{C}^n)$ is the W*-algebra of all complex $n \times n$-matrices, and $\mathbb{1}_n$ is the identity transformation of $\mathcal{B}(\mathbb{C}^n)$ onto itself. Compare Appendix A.2.1.

In the *temporal domain*, the frequency operator Λ changes sign under frequency reversal, $\Lambda \to -\Lambda$, while the time operator is invariant under frequency reversal,

$$\kappa(T) = T, \qquad \kappa(\Lambda) = -\Lambda. \tag{9.18}$$

In the *atemporal material domain*, frequency-reversal can be characterized by the transformation of the position operator $Q \to Q$ and the transformation of the momentum operator $P \to -P$,

$$\kappa(Q) = Q, \qquad \kappa(P) = -P. \tag{9.19}$$

9.3 Temporal Symmetries

> Symmetry denotes that sort of concordance of several parts by which they integrate into a whole.
>
> *Hermann Weyl (1952, p. 3)*

9.3.1 The Basic Temporal Symmetries

Symmetry considerations play an immensely important role in many descriptions of temporal phenomena. A basic symmetry requirement for any science has been stated by Norbert Wiener (1949, p. 11) as follows:

> If we are dealing with any field of science where long-time observations are possible and where experiments can be repeated, it is desirable that our operations be not tied to any specific origin in time. If a certain experiment started at ten o'clock today would give a certain distribution of results by twelve o'clock, then we must expect that if this experiment is carried out under similar conditions at ten o'clock tomorrow, by twelve o'clock we should get the same distribution of results. Without at least an approximate repeatability of experiments, no comparisons of results at different times are possible, and there can be no science. *That is, the operators which come into consideration are invariant under a shift of time.*

Adopting this point of view, a physical law is considered as *fundamental* if it does not change when we change the origin of the time axis.

This situation suggests that the *time-translation group* $\mathfrak{V}$ and its dual group $\mathfrak{U}$ are the cornerstones for a harmonic analysis of temporal phenomena. In addition, the invariance of many temporal phenomena under scale transformations suggests to introduce the dilation group $\mathfrak{D}$. Dennis Gabor (1946) stressed the importance of a holographic time-frequency analysis in the time-frequency phase space and was the first to introduce time-frequency-scale analysis into engineering physics. These time-related engineering concepts can be captured group theoretically. The time-frequency analysis of signal theory and information theory is based on the harmonic analysis of the following five locally compact commutative groups:[169]

(1) the time-translation group $\mathfrak{V} : t \to t + \tau$ with $\tau \in \mathbb{R}$,
(2) the logtime-translation group $\mathfrak{M} : t \to t^{\mu}$ with $\mu > 0$,
(3) the frequency-translation group $\mathfrak{U} : \ell \to \ell + \lambda$ with $\lambda \in \mathbb{R}$,
(4) the logfrequency-translation group $\mathfrak{N} : \ell \to \ell^{\nu}$ with $\nu > 0$,
(5) the dilation group $\mathfrak{D} : t \to e^{\kappa} t, \ell \to e^{-\kappa} \ell$ with $\kappa \in \mathbb{R}$,

where $t \in \mathbb{R}$ stands for time and $\ell \in \mathbb{R}$ for frequency. The *dilation group* $\mathfrak{D}$ describes the scaling operation $t \to e^{\kappa} t, \lambda \to e^{-\kappa} \lambda$ for $\kappa \in \mathbb{R}$. Scale invariance means that there is no universal intrinsic unit of time or frequency.

Besides these five continuous groups there are two additional important discrete groups of order two, namely

(6) the time-reversal group $T \to \iota T = -T$,
(7) the frequency-reversal group $\Lambda \to \kappa \Lambda = -\Lambda$,

where ι and κ are Jordan automorphisms of the underlying W*-algebra.

According to Hermann Weyl (1927, 1928a, 1931), the kinematical structure of a system is determined by its *symmetry group of automorphisms*, which in our case is given by the *temporal symmetry group* $\mathfrak{T}$,

$$\mathfrak{T} := \mathfrak{V} \times \mathfrak{U} \times \mathfrak{D}. \tag{9.20}$$

To assume that the symmetry group underlying temporal phenomena is given by $\mathfrak{T} = \mathfrak{V} \times \mathfrak{U} \times \mathfrak{D}$ does not imply that all observable temporal phenomena are time-translation invariant, frequency-translation invariant and scale invariant. In the next section we will give a simple example of a joint approximative invariance under the three temporal groups $\mathfrak{V}$, $\mathfrak{U}$ and $\mathfrak{D}$. But usually the full symmetry of the temporal group $\mathfrak{T}$ is hidden. Symmetries in the assumed fundamental laws do not imply that all empirically accessible descriptions have the same symmetry. We have to distinguish between the symmetry of the laws and the (often lacking) symmetry of the state of a system.[170] Often, the most stable state of a system fails to have the full symmetry of the underlying temporal symmetry group. This implies that the basic temporal symmetries do not determine the tremendous variety of observable temporal phenomena.

"It is the asymmetry that creates the phenomena" (Curie 1894a, p. 400). There are various possible reasons for the *spontaneous symmetry breaking* that leads to observable phenomena. It is conceivable that a fully symmetric description is not stable, or that a deeper theory possesses less symmetries. In spite of the fact that the underlying temporal symmetry group $\mathfrak{T}$ is usually not manifest, it helps us to discuss fundamental temporal phenomena in a unified framework.

---

[169]Compare for example Folland (1989), Thangavelu (1998), Cohen (1995), Gröchenig (2001), Debnath (2003), Hogan and Lakey (2005).

[170]Editor's note: In other words, the equations of motion for a system often exhibit a symmetry that is broken by their solution, providing the temporal evolution of the system's state.

9.3.2 Operator Realizations of Temporal Groups

The commutative temporal groups $\mathfrak{V}$, $\mathfrak{U}$, $\mathfrak{D}$, can be realized algebraically on the factorial W*-algebra $\mathcal{T}$ of Sect. 8.5.5 by unitary one-parameter groups:

- the time-translation group $\mathfrak{V} : t \rightarrow t + \tau$ by $\{V(\tau) \in \mathcal{T} | \tau \in \mathbb{R}\}$ with the group relation $V(\tau)V(\tau') = V(\tau + \tau')$,
- the frequency-translation group $\mathfrak{U} : \ell \rightarrow \ell + \lambda$ by $\{U(\lambda) \in \mathcal{T} | \lambda \in \mathbb{R}\}$ with the group relation $U(\lambda)U(\lambda') = U(\lambda + \lambda')$,
- the dilation group $\mathfrak{D} : t \rightarrow e^{\kappa} t, \ell \rightarrow e^{-\kappa} \ell$ by $\{D(\kappa) \in \mathcal{T} | \kappa \in \mathbb{R}\}$ with the group relation $D(\kappa)D(\kappa') = D(\kappa + \kappa')$,
- the logtime-translation group $\mathfrak{M} : t \rightarrow t^{\mu}$ by $\{M(\mu) \in \mathcal{T} | \mu \in \mathbb{R}\}$ with the group relation $M(\mu)M(\mu') = M(\mu + \mu')$,
- the logfrequency-translation group $\mathfrak{N} : \ell \rightarrow \ell^{\nu}$ by $\{N(\mu) \in \mathcal{T} | \nu \in \mathbb{R}\}$ with the group relation $N(\nu)N(\mu') = N(\nu + \nu')$.

By Stone's theorem (Stone 1930, 1932), these one-parameter groups can be recast in exponential form

$$V(\tau) := e^{-2\pi i \tau \Lambda}, \tag{9.21a}$$

$$U(\lambda) := e^{+2\pi i \lambda T}, \tag{9.21b}$$

$$D(\kappa) := e^{+2\pi i \kappa \ln S} = S^{2\pi i \kappa}, \tag{9.21c}$$

$$M(\mu) := e^{+2\pi i \mu \ln T} = T^{2\pi i \mu}, \tag{9.21d}$$

$$N(\nu) := e^{-2\pi i \nu \ln \Lambda} = \Lambda^{-2\pi i \nu}. \tag{9.21e}$$

The generators Λ, T, S, $\ln \Lambda$, $\ln T$, and $\ln S$ of these unitary groups are self-adjoint and have a continuous Lebesgue spectrum. We use the following well-established terminology of signal theory for the self-adjoint generators of the three one-parameter subgroups of the affine Weyl-Heisenberg group (Cohen 1995; Hogan and Lakey 2005):

T is called the time operator, (9.22a)

Λ is called the frequency operator, (9.22b)

S is called the scale operator, (9.22c)

$\ln T$ is called the logtime operator, (9.22d)

$\ln \Lambda$ is called the logfrequency operator. (9.22e)

The generator S of the unitary scaling group can be expressed in terms of the self-adjoint operators Λ and T,

$$S = \frac{\Lambda T + T \Lambda}{2}. \tag{9.23}$$

9.3.3 *Symmetries in Music*

> Harmonic analysis, for all its many modern ramifications, has a history going back to Pythagoras and his interest in music and the vibrations of the strings of the lyre.
>
> *Norbert Wiener (1956, p. 105)*

9.3.3.1 Musica est exercitium arithmeticae occultum[171]

There is a tight relation between music and mathematics. Composers have been using symmetry principles for hundreds of years. Any characterization of musical structures relates more or less to the presence of various forms of symmetry. The appearance of symmetry in musical composition sets the beginning of the study of harmony in music. Symmetries in music relate to group theory (Speiser 1952).

Guerino Mazzola (1985), an expert in mathematical music theory, based his discussion of the structure of music on the Yoneda lemma of category theory,[172] which implies that a structure is exhaustively determined as soon as we know how it looks from the "perspective" of *every* other structure. He concluded that for musicians the Yoneda lemma means that "a work of music is known by playing all possible interpretations".[173]

The symmetries of music perception refer to those aspects of music that are preserved under appropriate transformations. Human perception of speech and music is to a certain extent invariant under temporal symmetries. However, exact symmetries are intolerable in music (Speiser 1952, p. 30):

> The nuances are not at all adjuncts of sensitive souls, but the main pillars of playing a piece, since they alone reveal the symmetry content of that piece.

It is an outstanding feature of our mental faculties that we are able to abstract from irrelevant attributes of a signal and be aware only of essentially *invariant* attributes.

9.3.3.2 Symmetries in Music Perception

Symmetry is a fundamental concept in the analysis of music perception. Our ability to recognize a piece of music is quite independent of sound intensity, pitch, and

---

[171] Gottfried Wilhelm Leibniz wrote in a letter to the Russian mathematician Christian Goldbach of April 1712 (quoted after Scagliola 2012, p. 289): "Music is a hidden arithmetic exercise of the soul, which doesn't know that it is counting." Original in Latin: "Musica est exercitium arithmeticae occultum nescientis se numerare animi."

[172] Yoneda's lemma concerns functors from a fixed category to the category of sets. Compare also Mazzola (1985, p. 88).

[173] Mazzola in his lecture *Structure and Music* in Table Ronde Roussel Uclaf 55 (Paris, March 1986) on *Structural Models in Science*.

whether it is played faster or slower. Also the musical quality of a tune is basically the same now as it is at a later time. These invariances refer on the one hand to transformations of the amplitude, and on the other hand to the choice of the time origin and to time and frequency scalings.

In other words, music perception is approximately invariant under affine scaling transformations of time. In this spirit, Georg Picht (1966) stated that music is a *representation of time*:

> We understand music if we know what time is, we learn what time is if we understand music.

In particular, the following *approximate* invariances play a crucial role for the recognition of musical patterns. They refer to the following groups:

- *Time-translation group isomorphic to* $(\mathbb{R}, +)$: It is inessential whether a piece of music is played earlier or later. That is, the relevant information contained in it is independent of the choice of the time origin, and parts of a musical piece can be combined by addition.
- *Logfrequency-translation group isomorphic to* $\mathbb{R}^+$: Human pitch perception is roughly logarithmic with respect to frequency. That is, a note doubled in frequency sounds similar to the original (music perception is octave-cycle invariant). Western musical tradition subdivides the octave into 12 (logarithmically) equally spaced pitch classes, known as the well-tempered scale. Therefore, it makes sense to express acoustic frequencies on a logfrequency axis.
- *Time multiplicative scaling group isomorphic to* $(\mathbb{R}^+\backslash\{0\}, \times)$: In spite of the fact that most tunes have a preferred tempo, we can recognize a piece of music irrespective of its playback speed. This approximate invariance under *time scaling* plays a crucial role for the recognition of musical patterns.
- *Amplitude scaling group* $\mathfrak{D}$: Playing music more quietly or more loudly does not change the musical quality of a tune. Music remains the same no matter at what volume it is played.
- *Loudness scaling group* $\mathfrak{D}$: Perception of sound intensity is approximately logarithmic.
- *Self-similarity:* On the self-similarity of musical processes compare Sect. 10.2.7.

Invariance under scale transformations plays a crucial role for the recognition of patterns in general.

9.3.3.3 Harmonic Analysis and Music

In a seminar at Göttingen in 1925, Norbert Wiener (1956, pp. 105–106) emphasized the importance of harmonic analysis for the development of quantum theory and explained his ideas using the example of a musical score:

> As I have said, my talk concerned harmonic analysis—in other words, the breaking up of complicated motions into sums of simple oscillations. Harmonic analysis, for all its many modern ramifications, has a history going back to Pythagoras and his interest in music

> and the vibrations of the strings of the lyre. There are many ways in which a string can vibrate, but the most elementary and simplest of all is known as the simple harmonic oscillation. The motion of the string of a musical instrument, if indeed it is not simply harmonic, is well known to be the most elementary sort of combination of simple harmonic motions. . . .
> Now, let us see what musical notation really is. The position of a note vertically on the staff gives its pitch or frequency, while the horizontal notation of music divides this pitch in accordance with the time. The time notation contains the indication of the rate of the metronome, the subdivision of sound into whole notes, half notes, quarter notes, etc., the various rests, and much else besides. Thus musical notation at first sight seems to deal with a system in which vibrations can be characterized in two independent ways, namely, according to frequency, and according to duration in time.
> A finer assumption of the nature of musical notation was that things are not as simple as all this. The number of oscillations per second involved in a note, while it is a statement concerning frequency, is also a statement concerning something distributed in time. In fact, the frequency of a note and its timing interact in a very complicated manner.
> Ideally, a simple harmonic motion is something that extends unaltered in time from the remote past to the remote future. In a certain sense it exists *sub specie aeternitatis*. To start and to stop a note involves an alteration of its frequency composition which may be small, but which is very real. A note lasting only over a finite time is to be analyzed as a band of simple harmonic motions, no one of which can be taken as the only simple harmonic motion present. Precision in time means a certain vagueness in pitch, just as precision in pitch involves an indifference to time.

9.3.3.4 The Musical Score

A composer does neither deal with the time representation nor with the frequency representation of his composition, but he produces a representation in the form of a musical score, which is manifestly scale invariant. Consider a harpsichord recording of Prelude 11 of *The Well-Tempered Clavier*, BWV 856, by Johann Sebastian Bach. The output of a record player is an electrical signal $t \mapsto f(t)$. With a graphical representation of $t \mapsto f(t)$ we can hardly recognize the art of Bach.

A quick glance at the function $t \mapsto f(t)$ may give us some information about the rhythmic pattern of the music but hardly about its key. The spectral density $\lambda \mapsto |\hat{f}(\lambda)|^2$ gives us some hints about the key or the orchestration of the work. However, neither the melody nor the rhythm of the music is easily recognizable from the spectral density. As de Bruijn (1967, p. 59) noted:

> . . . the composer does not produce f itself; he does not even define it. He may try to prescribe the exact frequency and the exact time interval of a note (although the uncertainty principle says that he can never be completely successful in this effort), but he does not try to prescribe the phase. The composer does not deal with f; it is only the gramophone company which produces and sells an f. . . . What the composer really does, or thinks he does, or should think he does, is something entirely different from describing either f or $\hat{f}$. Instead, he constructs a function of two variables. The variables are the time and the frequency, the function describes the intensity of the sound. He describes the function by a complicated set

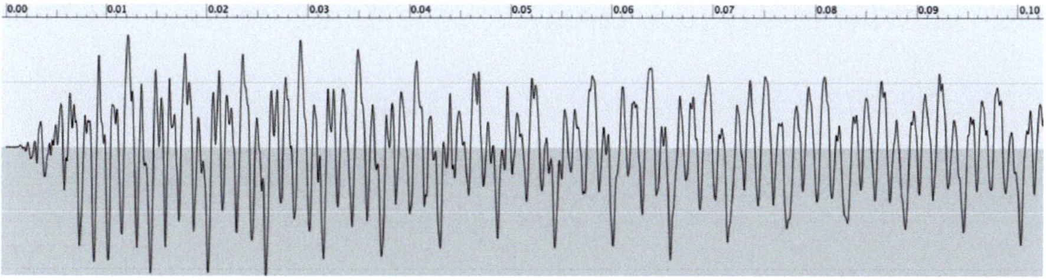

Fig. 9.1 The signal $t \mapsto f(t)$ of a recording of Prelude 11 of *The Well-Tempered Clavier*, BWV 856, for $0 \leq t \leq 0.1$ sec

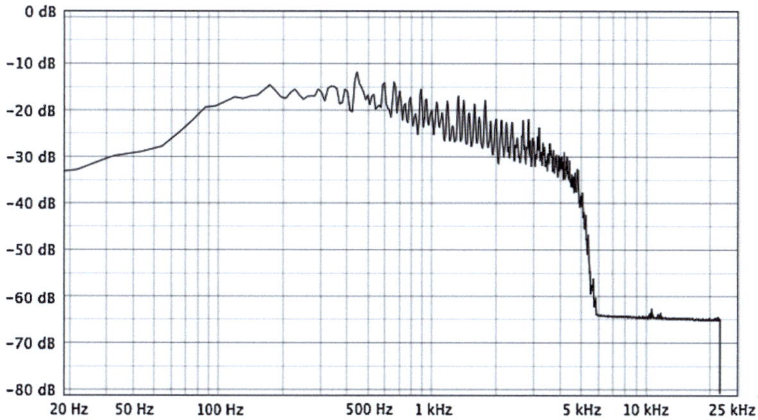

Fig. 9.2 The spectral density $\lambda \mapsto |\hat{f}(\lambda)|^2$ of the time signal represented in Fig. 9.1

Fig. 9.3 Score of the beginning of Prelude 11, BWV 856, by Johann Sebastian Bach

of dots on score paper. His way of describing time is slightly different from what a mathematician would do, but certainly vertical lines denote constant time, and horizontal lines denote constant frequency.

Even though both f and $\hat{f}$ (in Figs. 9.1 and 9.2) contain all information about the music in complementary ways, none of the two representations is adapted to human perception. More informative are representations which yield information about when a note is played and which note is played (as in the score shown in Fig. 9.3). Harmonic analysis combines the relevant features of both f and $\hat{f}$ into a single phase-space function $\mathbb{R} \times \hat{\mathbb{R}} \to \mathbb{R}$ and is called *time-frequency analysis* (Gröchenig 2001; for a short introduction see Dörfler 2001).

9.4 The Weyl-Heisenberg Group

9.4.1 The Phase Space of the Time Domain

Dennis Gabor (1946, p. 429) pointed out that there are two fundamentally distinct approaches to the temporal description of nature: that of time and that of frequency:

> Hitherto communication theory was based an two alternative methods of signal analysis. One is the description of the signal as a function of time; the other is Fourier analysis. Both are idealizations, as the first method operates with sharply defined instants of time, the second with infinite wave-trains of rigorously defined frequencies. But our everyday experiences—especially our auditory sensations—insist on a description in terms of both time and frequency.

The phase space for this approach is parametrized by the two canonical coordinates $\tau \in \mathbb{R}$ and $\lambda \in \mathbb{R}$. The time-translation group $\mathfrak{V}$ characterizes the "configuration space" and the group $\mathfrak{U}$ is the Pontryagin dual $\hat{\mathfrak{V}}$ of $\mathfrak{V}$,

$$\mathfrak{U} := \hat{\mathfrak{V}}. \tag{9.24}$$

In the remainder of this section we introduce some notions that will help us to exploit this duality.

A Lie group $\mathfrak{G}$ is a two-dimensional real smooth manifold, which can itself be thought of as a manifold Γ under the natural group action of left multiplication. A group $\mathfrak{G}$ is said to act transitively on a manifold Γ if every point in Γ can be carried into every point in Γ by some transformation in $\mathfrak{G}$. A *homogeneous space* for a group $\mathfrak{G}$ is a manifold Γ on which $\mathfrak{G}$ acts continuously by symmetry in a transitive way. A *principal homogeneous space* for a group $\mathfrak{G}$ is a homogeneous space Γ for $\mathfrak{G}$ if the stabilizer subgroup of any point is trivial.

For a principal homogeneous space there exists for any $\gamma, \gamma' \in \Gamma$ a unique $\mathfrak{g} \in \mathfrak{G}$ such that $\mathfrak{g} \circ \gamma = \gamma'$, where $\circ$ denotes the right action of $\mathfrak{G}$ on Γ. A principal homogeneous space has a unique analytic structure with the property that $\mathfrak{G}$ acts as a differentiable Lie transformation group of $\mathfrak{G}$ (cf. Helgason 1978, Theorem 4.2, p. 123). A principal homogeneous space is, therefore, a reasonable candidate for a *phase space* on which $\mathfrak{G}$ acts through canonical transformations.

The corresponding symplectic temporal phase space $(\mathfrak{V} \times \mathfrak{U}, \sigma)$ is composed of the two-dimensional smooth manifold of the Lie group $\mathfrak{U} \times \mathfrak{V}$ and the symplectic form

$$\sigma\left(\lambda, \tau \,\middle|\, \lambda', \tau'\right) = \lambda\tau' - \tau\lambda'. \tag{9.25}$$

Every irreducible representation of a commutative group $\mathfrak{G}$ is one-dimensional and of the form $\mathfrak{g} \mapsto \chi(\mathfrak{g})$, where χ is a continuous group homomorphism from $\mathfrak{G}$ with values in the circle group $\mathbb{T}$ satisfying

$$\chi(\mathfrak{g} \circ \mathfrak{g}') = \chi(\mathfrak{g})\chi(\mathfrak{g}'), \quad |\chi(\mathfrak{g})| = 1, \;\; \mathfrak{g}, \mathfrak{g}' \in \mathfrak{G}. \tag{9.26}$$

Such a function χ is called a *character* of $\mathfrak{G}$. The set of all characters of $\mathfrak{G}$ forms a multiplicative commutative group under the ordinary pointwise multiplication. The inverse of a character is its complex conjugate,

$$\chi(\mathfrak{g})^{-1} = \chi(\mathfrak{g})^*. \tag{9.27}$$

The multiplicative group of all continuous characters of an arbitrary topological group $\mathfrak{G}$ is called the *character group* X of $\mathfrak{G}$,

$$X := \{\chi(\mathfrak{g}) \,|\, \mathfrak{g} \in \mathfrak{G}\}. \tag{9.28}$$

The multiplicative character group is a commutative locally compact group which is isomorphic to the *dual group* $\hat{\mathfrak{G}}$ of $\mathfrak{G}$. There is a one-to-one mapping of X onto $\hat{\mathfrak{G}}$,

$$\chi \in X \quad \longleftrightarrow \quad \hat{\mathfrak{g}} \in \hat{\mathfrak{G}}. \tag{9.29}$$

9.4.2 Weyl's Commutation Relations

The Pontryagin duality theorem[174] states that the character group X of a locally compact commutative group $\mathfrak{G}$ is in a canonical way topologically isomorphic to the group $\mathfrak{G}$.

According to a celebrated theorem by Alfred Haar (1933), every locally compact group admits a left-invariant and a right-invariant Borel measure, called the Haar measures, which are unique up to a positive multiplicative constant (see also Appendix A.5.1). If a left-invariant measure is also right-invariant, the group is called unimodular. The Haar measure on the group $\mathfrak{G}$ is denoted by $d\mathfrak{g}$. For $1 \leq p \leq \infty$ we denote by $\mathbf{L}^p(\mathfrak{G}, d\mathfrak{g})$ the Lebesgue space consisting of the set of equivalence classes of complex-valued Lebesgue measurable functions $\mathfrak{g} \mapsto \Phi(\mathfrak{g})$ on the locally compact group $\mathfrak{G}$ such that $\mathfrak{g} \mapsto |\Phi(\mathfrak{g})|^2$ is integrable. We note that the important relation $\{\mathbf{L}^1(\mathfrak{G})\}^* = \mathbf{L}^\infty(\mathfrak{G})$ is valid for every locally compact commutative group $\mathfrak{G}$. This implies that $\mathbf{L}^\infty(\mathfrak{G})$ is a commutative W*-algebra.

Pontryagin's duality theorem makes it possible to construct analogs of classical Fourier theory for arbitrary locally compact commutative groups. Associated with a locally compact commutative group $\mathfrak{G}$ are, up to normalization, a unique Haar measure $d\mathfrak{g}$, a set $\mathbf{L}^1(\mathfrak{G})$ of complex-valued Haar-integrable functions, and the dual group $\hat{\mathfrak{G}}$. The Pontryagin isomorphism is given by the following Fourier transform:

$$\hat{\Phi}(\hat{\mathfrak{g}}) = \int_{\mathfrak{G}} \chi(\mathfrak{g})^* \Phi(\chi) d\mathfrak{g}, \quad \chi \in X, \; \Phi \in \mathbf{L}^1(\mathfrak{G}), \tag{9.30}$$

where $\chi(\mathfrak{g})$ is the value of a character $\chi \in X$, $\hat{\mathfrak{g}} \in \hat{\mathfrak{G}}$, at a point $\mathfrak{g} \in \mathfrak{G}$. If Φ is in $\mathbf{L}^1(\mathfrak{G}) \cap \mathbf{L}^2(\mathfrak{G})$, then $\Phi \mapsto \hat{\Phi}$ is an isometric map onto $\mathbf{L}^2(\mathfrak{G})$, which can be extended to a unitary operator on $\mathbf{L}^2(\mathfrak{G})$ onto $\mathbf{L}^2(\hat{\mathfrak{G}})$. Accordingly, the Fourier transform $\Phi \mapsto \hat{\Phi}$ can be defined for all $\Phi \in \mathbf{L}^2(\mathfrak{G})$.

---

[174] An account of this theorem, also known as Pontryagin-van Kampen duality theorem, along with references to the original papers, can be found in Hewitt and Ross (1970, §24). Compare also Kirillov (1976, p. 168).

If the group $\mathfrak{G}$ is unimodular, then there is a choice of the Haar measure $d\hat{\mathfrak{g}}$ on $\hat{\mathfrak{G}}$, such that the inverse of the Fourier transform is given by

$$\Phi(\mathfrak{g}) = \int_{\hat{\mathfrak{G}}} \chi(\hat{\mathfrak{g}})^* \hat{\Phi}(\hat{g}) d\hat{g}. \tag{9.31}$$

In order to take advantage of the symmetry of $\mathfrak{G}$ and $\hat{\mathfrak{G}}$, we introduce the so-called *bicharacter* $\langle \mathfrak{g}|\hat{\mathfrak{g}}\rangle$ for $(\mathfrak{G}, \hat{\mathfrak{G}})$, defined as a map of $\mathfrak{G} \times \hat{\mathfrak{G}}$ onto the multiplicative group of complex numbers of absolute value 1 which is a character in each argument. That is, the function $\mathfrak{g} \to \langle \mathfrak{g}|\hat{\mathfrak{g}}\rangle$ is a multiplicative character for $\mathfrak{G}$ for fixed $\hat{\mathfrak{g}} \in \hat{\mathfrak{G}}$, and $\hat{\mathfrak{g}} \to \langle \mathfrak{g}|\hat{\mathfrak{g}}\rangle$ is a multiplicative character for $\hat{\mathfrak{G}}$ for fixed $\mathfrak{g} \in \mathfrak{G}$.

Let $\mathfrak{G}$ be an arbitrary additively written self-dual commutative group endowed with an anti-symmetric bicharacter $\langle\cdot|\cdot\rangle$. A bicharacter is called anti-symmetric if $\langle \mathfrak{g}|\mathfrak{g}\rangle = 1$ for all $\mathfrak{g} \in \mathfrak{G}$, which implies that

$$\langle \mathfrak{g}|\mathfrak{g}'\rangle = \frac{1}{\langle \mathfrak{g}'|\mathfrak{g}\rangle}. \tag{9.32}$$

A Weyl system over the phase space $\mathfrak{G} = \mathfrak{V} \times \mathfrak{U}$ is a unitary realization W such that

$$W(\mathfrak{g})W(\mathfrak{g}') = \langle \mathfrak{g}|\mathfrak{g}'\rangle W(\mathfrak{g} + \mathfrak{g}'), \quad \mathfrak{g}, \mathfrak{g}' \in \mathfrak{G}. \tag{9.33}$$

For a Borel subset $\mathcal{B}$ of the time-translation group $\mathfrak{V}$ with the Haar measure μ, the indicator function $\mathcal{B} \mapsto I_{\mathcal{B}}$,

$$I_{\mathcal{B}}(t) = 1 \quad \text{if } t \in \mathcal{B}, \qquad I_{\mathcal{B}}(t) = 0 \quad \text{if } t \notin \mathcal{B}, \tag{9.34}$$

defines a spectral measure E on the Lebesgue space $\mathbf{L}^2(\mathfrak{V}, \mu)$ over the locally compact group $\mathfrak{V}$,

$$\{E(\mathcal{B})\Phi\}(t) = I_{\mathcal{B}}(t)\Phi(t), \quad \Phi \in \mathbf{L}^2(\mathfrak{V}, \mu). \tag{9.35}$$

For the unitary regular representation of $\mathfrak{V}$ on the Hilbert space $\mathbf{L}^2(\mathfrak{V}, \mu)$ we get

$$\{V(\tau)\Phi\}(t) = \Phi(t - \tau), \quad \tau, t \in \mathbb{R} \cong \mathfrak{V}, \ \Phi \in \mathbf{L}^2(\mathfrak{V}, \mu), \tag{9.36}$$

and for the spectral measure:

$$V(\tau)E(\mathcal{B}) = E(\mathcal{B} + \tau)V(\tau). \tag{9.37}$$

Using the spectral measure E, one can define a unitary operator $U(\lambda)$ by

$$U(\lambda) := \int_{\mathfrak{G}} \chi(\tau, \lambda)^* E(d\tau), \quad \chi(\lambda, \cdot) \in \mathfrak{U}, \ \lambda \in \mathbb{R} \cong \mathfrak{U}. \tag{9.38}$$

The group $\{U(\lambda)|\lambda \in \mathbb{R}\}$ is a strongly continuous unitary representation of the dual group $\hat{\mathfrak{V}} = \mathfrak{U}$ (cf. (9.24)). Both unitary groups $\{V(\tau)|\tau \in \mathbb{R}\}$ and $\{U(\lambda)|\lambda \in \mathbb{R}\}$ act on the same Hilbert space $\mathbf{L}^2(\mathfrak{V}, \mu)$ and satisfy Weyl's commutation relation

$$U(\lambda)V(\tau) = \chi(\lambda, \tau)V(\tau)U(\lambda), \quad \lambda \in \mathbb{R} \cong \mathfrak{U}, \ \tau \in \mathbb{R} \cong \mathfrak{V}. \tag{9.39}$$

The character of the time-translation group $\mathfrak{V}$ is given by

$$\chi(\lambda, \tau) = e^{2\pi i \lambda \tau}, \quad \lambda, \tau \in \mathbb{R}. \tag{9.40}$$

so that Weyl's commutation relation can be written as

$$U(\lambda)V(\tau) = e^{2\pi i \lambda \tau} V(\tau)U(\lambda), \quad \lambda, \tau \in \mathbb{R}. \tag{9.41}$$

From a mathematical point of view, the map $(\lambda, \tau) \mapsto \{U(\lambda)V(\tau)\}$ is a unitary projective representation of the commutative group $\mathfrak{V} \times \mathfrak{U}$. The uniqueness theorem by Mackey and Loomis[175] shows that this representation is unique up to unitary equivalence and multiplicity.

9.4.3 Temporal Weyl Operators

According to Eq. (9.41) the operators $U(\lambda)$ and $V(\tau)$ are not closed under multiplication. Weyl's commutation relation (9.41) implies that the operators $cV(\tau)U(\lambda)$ of the family $\{cV(\tau)U(\lambda)|c \in \mathbb{C}, |c| = 1, \tau, \lambda \in \mathbb{R}\}$ form a group under multiplicative composition. It is therefore appropriate to introduce a unitary operator $H(\tau, \lambda, \varphi)$ by

$$H\big(\tau, \lambda, e^{2\pi i \varphi}\big) := e^{2\pi i \varphi} e^{\pi i b \tau} V(\tau)U(\lambda), \tag{9.42}$$

where $\varphi \in \mathbb{R}$ is an auxiliary toral component. Two operators $H\{.\}$ fulfill the commutation relation

$$H\{\tau, \lambda, \varphi\}H\{\tau', \lambda', \varphi'\} = H\big\{\tau + \tau', \lambda + \lambda', \big[\varphi + \varphi' + \big(\lambda\tau' - \lambda'\tau\big)/2\big]\big\}, \tag{9.43}$$

and realize the group $\mathbb{R}^2 \times \mathbb{R}$ with the group law

$$\{\tau, \lambda, \varphi\} \circ \{\tau', \lambda'\varphi'\} = \big\{\tau + \tau', \lambda + \lambda', \varphi + \varphi' + \tfrac{1}{2}\big(\lambda\tau' - \lambda'\tau\big)\big\}. \tag{9.44}$$

The non-commutative group $\mathfrak{H} := \{H(\tau, \lambda, \varphi)|\tau, \lambda, \varphi \in \mathbb{R}\}$ is called the *Heisenberg group*. It is a simply connected and nilpotent unimodular three-dimensional Lie group with the underlying manifold $\mathbb{R}^3$, whose Haar measure coincides with the Lebesgue measure on $\mathbb{R}^3$. The set $\{H(0, 0, \varphi)|\varphi \in \mathbb{R}\}$ is a central closed subgroup $\mathfrak{Z}$ of the Heisenberg group $\mathfrak{H}$. The quotient group $\mathfrak{H}/\mathfrak{Z}$ is called the *Weyl-Heisenberg group* $\mathfrak{W} := \{W(\tau, \lambda)|\tau, \lambda \in \mathbb{R}\}$ with the Haar measure $d\tau d\lambda$ and the unitary *Weyl operators*

$$W(\tau, \lambda) := e^{-2\pi i \varphi} H(\tau, \lambda, \varphi), \tag{9.45}$$

[175] The proof by Mackey (1949) was based on the assumption that the group $\mathfrak{G}$ is separable. Later, Loomis (1952) pointed out that the separability condition is unnecessary.

which can also be written as

$$W(\tau,\lambda) := V(\tau)U(\lambda)e^{\pi i\lambda\tau} = U(\lambda)V(\tau)e^{-\pi i\lambda\tau}, \quad \lambda,\tau \in \mathbb{R}. \tag{9.46}$$

The Weyl operators fulfill Weyl's canonical commutation relation

$$W(\tau,\lambda)W(\tau',\lambda') = e^{\pi i\sigma(\tau,\lambda|\tau',\lambda')}W(\tau+\tau',\lambda+\lambda'), \tag{9.47a}$$

$$= e^{\pi i(\lambda\tau'-\lambda'\tau)}W\{\tau+\tau',\lambda+\lambda'\}. \tag{9.47b}$$

with the symplectic form $\sigma : \mathbb{R}^2 \times \mathbb{R}^2 \to \mathbb{R}$,

$$\sigma(\tau,\lambda \,|\, \tau',\lambda') := \lambda\tau' - \tau\lambda', \quad \lambda,\lambda',\tau,\tau' \in \mathbb{R}. \tag{9.48}$$

The unitary one-parameter groups $\mathfrak{V} = \{V(\tau)|\tau \in \mathbb{R}\}$ and $\mathfrak{U} = \{U(\lambda)|\lambda \in \mathbb{R}\}$ are strongly continuous, so that by Stone's theorem we have

$$U(\lambda) = W(\lambda,0) = e^{2\pi i\lambda T}, \qquad V(\tau) = W(0,\tau) = e^{-2\pi i\tau\Lambda}, \tag{9.49}$$

so that we can write

$$W(\lambda,\tau) = e^{2\pi i(\lambda T - \tau\Lambda)}. \tag{9.50}$$

The generator T is called the *time operator*, while the generator Λ is called the *frequency operator* (cf. Sect. 9.3.2). They are unitarily equivalent unbounded self-adjoint operators with the simple, absolutely continuous spectrum $\mathbb{R}$. The time operator T is the canonical conjugate of the frequency operator Λ. On an appropriate domain, they fulfill the relations

$$e^{2\pi i\tau\Lambda}f(T)e^{-2\pi i\tau\Lambda} = f(T+\tau), \quad \tau \in \mathbb{R}, \tag{9.51}$$

$$e^{-2\pi i\lambda T}g(\Lambda)e^{2\pi i\lambda T} = g(\Lambda+\lambda), \quad \lambda \in \mathbb{R}, \tag{9.52}$$

$$T\Lambda - \Lambda T = (i/2\pi)\mathbb{1}. \tag{9.53}$$

The self-adjoint generators T and Λ are affiliated to the W*-algebra $\mathcal{T}$ of the *temporal domain* and have the spectral representations:

$$T = \int_{-\infty}^{\infty} tF(dt), \qquad \Lambda = \int_{-\infty}^{\infty} \ell E(d\ell), \tag{9.54}$$

where F is the projection-valued spectral measure of the time operator T, and E is the projection-valued spectral measure of the frequency operator Λ. The smallest C*-algebra $\mathcal{T}$ which contains all Weyl operators $W(\tau,\lambda)$ with $\tau,\lambda \in \mathbb{R}$ is called the *temporal Weyl algebra* over the phase space $\mathbb{R}^2$. It is representation-independent and *unique*.[176] The bidual $\mathcal{T}^{**}$ of the Weyl algebra $\mathcal{T}$ is a factor of type I_∞,

$$\mathcal{T} = \mathcal{T}^{**}, \tag{9.55}$$

[176]See Manuceau (1968), Manuceau et al. (1973). Compare also Bratteli and Robinson (1981, Sect. 5.2.2.2, pp. 19–24), Petz (1990, Chap. 2).

hence it is a non-commutative W*-algebra with trivial center. The predual $\mathcal{T}_*$ of the W*-algebra $\mathcal{T}$ is infinite-dimensional and separable. The maximal commutative W*-subalgebras $\mathcal{V}$ and $\mathcal{U}$,

$$\mathcal{V} := \{V(\tau)|\tau \in \mathbb{R}\}'' \subset \mathcal{T}, \tag{9.56a}$$

$$\mathcal{U} := \{U(\lambda)|\lambda \in \mathbb{R}\}'' \subset \mathcal{T}, \tag{9.56b}$$

are complementary in the sense of (3.59).

In contradistinction to the currently standard view that conceives time as the ordered set $(\mathbb{R}, \leq)$ of time points under the standard total order, the generator of the one-parameter group $\{U(\lambda)|\lambda \in \mathbb{R}\}$ is a *generic* time operator exhibiting the (hypothetical) maximal symmetry of time. The spectrum $\sigma(T)$ of the generic time operator T is the *non-directed* line.[177] Richer manifestations of time phenomena like time's arrow arise through symmetry breakings of the basic temporal symmetries. They will be discussed in the following chapters.

Weyl's canonical commutation relation (9.46) implies that the projection-valued spectral measures for the time and frequency operator are related via the unitary *Fourier operator* F

$$\mathsf{F} := e^{-\pi^2 i(T^2+\Lambda^2)/2}, \tag{9.57a}$$

with

$$E(\mathcal{B}) = \mathsf{F}^{-1} F(\mathcal{B})\mathsf{F} \quad \text{for every Borel set } \mathcal{B} \subset \mathbb{R}, \tag{9.57b}$$

so that the frequency operator Λ and the time operator T are related by the Fourier transformation

$$\Lambda = \mathsf{F}^{-1} T \mathsf{F}, \quad T = \mathsf{F} \Lambda \mathsf{F}^{-1}. \tag{9.58}$$

With the aid of the unitary *fractional Fourier operator*

$$\mathsf{F}_\kappa := e^{-i\kappa\pi^2(T^2+\Lambda^2)/2}, \quad \mathsf{F}_1 = \mathsf{F},\ \mathsf{F}_2 = -\mathbb{1}, \tag{9.59}$$

we can construct (Ozaktas et al. 2001) a continuous family $\{\Lambda_\kappa, T_\kappa | 0 \leq \kappa < 4\}$ of fractional canonical operators,

$$\Lambda_\kappa := \mathsf{F}_\kappa^{-1} \Lambda \mathsf{F}_\kappa = \cos(\pi\kappa/2)\Lambda + \sin(\pi\kappa/2)T, \tag{9.60a}$$

$$T_\kappa := \mathsf{F}_\kappa^{-1} T \mathsf{F}_\kappa = \cos(\pi\kappa/2)T - \sin(\pi\kappa/2)\Lambda. \tag{9.60b}$$

The unitary operators $U_\kappa(\lambda) = e^{2\pi i\lambda T_\kappa}$ and $V_\kappa(\tau) = e^{-2\pi i\tau\Lambda_\kappa}$ $(\lambda, \tau \in \mathbb{R})$ generate two families of maximal commutative W*-subalgebras $\mathcal{U}$ and $\mathcal{V}$,

$$\mathcal{U}_\kappa := \{U_\kappa(\lambda)|\lambda \in \mathbb{R}\}'' \subset \mathcal{W}, \quad 0 \leq \kappa < 4, \tag{9.61a}$$

---

[177] A directed line is a line with an arrow on it. An operation which erases the oppositely directed arrows on two directed lines generates a non-directed line.

$$\mathcal{V}_\kappa := \left\{V_\kappa(\tau)|\tau \in \mathbb{R}\right\}'' \subset \mathcal{W}, \quad 0 \leq \kappa < 4. \tag{9.61b}$$

The family $\{(\mathcal{U}_\kappa, \mathcal{V}_\kappa)|0 \leq \kappa < 4\}$ then realizes infinitely many mutually incompatible pairs of complementary Boolean descriptions.

9.4.4 Elementary Weyl Systems

9.4.4.1 Temporal Ergodic Weyl-System

According to Hermann Weyl (1927, 1928a), the kinematics of a physical system is determined by its symmetry group of automorphisms of a W*-algebra $\mathcal{D}$ of observables. A symmetry is a structure-preserving map, i.e., a *-isomorphism of $\mathcal{D}$ onto itself. An automorphic action α of a locally compact, separable group $\mathfrak{G}$ on $\mathcal{D}$ is a homomorphism $\mathfrak{g} \to \alpha_\mathfrak{g}$, such that each $\alpha_\mathfrak{g}$ is an automorphism of $\mathcal{D}$ with $\alpha_{\mathfrak{g}\circ\mathfrak{g}'} = \alpha_\mathfrak{g} \circ \alpha_{\mathfrak{g}'}$, $\alpha_\mathfrak{g}(XY) = \alpha_\mathfrak{g}(X)\alpha_\mathfrak{g}(Y)$, and $\alpha_\mathfrak{g}(X^*) = \alpha_\mathfrak{g}(X)^*$.

In a Hilbert-space representation a pure state is represented by a ray and a connected group by a unitary ray representation,

$$O_\mathfrak{g} O_{\mathfrak{g}'} = e^{i\xi(\mathfrak{g},\mathfrak{g}')} O_{\mathfrak{g}\circ\mathfrak{g}'}, \tag{9.62}$$

where $O_\mathfrak{g}$ is a unitary operator, and the function $\xi : \mathfrak{G} \times \mathfrak{G} \to \mathbb{R}$ is a multiplier exponent. The symmetry is said to be *spontaneously broken* if the automorphism $\alpha_\mathfrak{g}$ of $\mathcal{D}$ is not implemented by unitary operators.

A W*-system is called *elementary* if it is ergodic with respect to the group $\mathfrak{G}$[178] and integrable, i.e., if its irreducible ray representations are square integrable (see Amann 1986, 1987). In the algebraic formulation, the phase space $\mathfrak{V} \times \hat{\mathfrak{V}} = \mathfrak{V} \times \mathfrak{U}$ of the temporal domain introduced in Sect. 9.4.1 is replaced by a W*-algebra $\mathcal{T}$, where an elementary action of the group $\mathfrak{G} = \mathfrak{V} \times \mathfrak{U}$ is given by an *ergodic* automorphic realization $\alpha : \mathfrak{G} \to \mathrm{Aut}(\mathcal{T})$ of $\mathfrak{G}$ in $\mathcal{T}$. In this formulation the temporal domain is given by an ergodic W*-system $(\mathcal{T}, \mathfrak{G}, \alpha)$.

An operator $V \in \mathcal{T}$ is an *eigenoperator*[179] of a realization of the time-translation group $\mathfrak{V}$ by the automorphism group $\{\theta_\tau^\mathcal{U} \in \mathrm{Aut}(\mathcal{T})|\tau \in \mathbb{R}\}$ with the order parameter τ if

$$\theta_\tau^\mathcal{V}(V) = \langle\hat{\lambda}, \tau\rangle V, \tag{9.63}$$

where

$$\tau \mapsto \langle\hat{\lambda}, \tau\rangle = e^{2\pi i\tau\hat{\lambda}}, \quad \tau \in \mathfrak{V} \cong \mathbb{R},\ \hat{\lambda} \in \hat{\mathfrak{V}} \cong \mathbb{R}, \tag{9.64}$$

---

[178] An action α is called *ergodic* if $\alpha_g(X) = X$ for all $g \in \mathcal{G}$ and $X \in \mathcal{T}$ implies that X is a multiple of the identity operator.

[179] If $\mathcal{D}$ is a W*-algebra, $\mathfrak{G}$ a group and $\alpha \in \mathrm{Aut}(\mathcal{D})$ a realization of $\mathfrak{G}$ by automorphisms, then an operator G in $\mathcal{D}$ is an eigenoperator for α if $G \neq 0$ and for each $\mathfrak{g} \in \mathfrak{G}$ there is a complex number $(\hat{\mathfrak{g}}, \mathfrak{g})$ such that $\alpha_\mathfrak{g}(G) = (\hat{\mathfrak{g}}, \mathfrak{g})G$. It follows that $(\hat{\mathfrak{g}}, \mathfrak{g})$ with $\hat{\mathfrak{g}} \in \hat{\mathfrak{G}}$ is a character of $\mathfrak{G}$. Compare Størmer (1974, p. 203).

is a character of the time-translation group $\mathfrak{V}$. The weak closure of the vector space spanned by the eigenoperators V defined by Eq. (9.63) is called the eigenalgebra $\mathcal{V}$ for $\mathfrak{V}$ in $\mathcal{T}$.

Likewise, an eigenoperator U of a realization $\theta_\lambda^{\mathcal{U}} \in \{\mathrm{Aut}(\mathcal{T})|\lambda \in \mathbb{R}\}$ of the dual group $\mathfrak{U} := \hat{\mathfrak{V}}$ by automorphisms is given by

$$\theta_\lambda^{\mathcal{U}}(U) = \langle\hat{\tau}, \lambda\rangle U, \tag{9.65}$$

where

$$\lambda \mapsto \langle\hat{\tau}, \lambda\rangle = e^{-2\pi i \hat{\tau}\lambda}, \quad \lambda \in \mathfrak{U} \cong \mathbb{R},\ \hat{\tau} \in \hat{\mathfrak{U}} \cong \mathbb{R}, \tag{9.66}$$

is a character of the frequency-translation group $\mathfrak{U}$. The weak closure of the vector space spanned by the eigenoperators U defined by Eq. (9.65) is called the eigenalgebra $\mathcal{U}$ for $\mathfrak{U}$ in $\mathcal{T}$.

We choose the W*-algebra $\mathcal{T}$ as the W*-tensor product of the eigenalgebras $\mathcal{V}$ and $\mathcal{U}$,

$$\mathcal{T} = \mathcal{V}\overline{\otimes}\mathcal{U}, \tag{9.67}$$

so that the one-parameter group δ_τ of automorphisms of the temporal domain is given by

$$\delta_\tau = \mathbb{1}^{\mathcal{A}} \times \theta_\tau = \mathbb{1}^{\mathcal{A}} \times \theta_\tau^{\mathcal{U}} \times \theta_\tau^{\mathcal{V}}, \tag{9.68}$$

where $\mathbb{1}^{\mathcal{A}}$ is the trivial identity morphism on the W*-algebra $\mathcal{A}$ of the atemporal domain. The W*-algebra $\mathcal{T} = \mathcal{V}\overline{\otimes}\mathcal{U}$ is the eigenalgebra for the group $\mathfrak{V} \times \mathfrak{U}$, generated by the set of all eigenoperators V, U,

$$\{\theta_\tau^{\mathcal{V}} \times \theta_\lambda^{\mathcal{U}}\}(VU) = \langle\hat{\lambda}, \tau\rangle\langle\hat{\tau}, \lambda\rangle VU = e^{2\pi i(\tau\hat{\lambda} - \hat{\tau}\lambda)} VU. \tag{9.69}$$

It follows that $(\mathcal{T}, \mathfrak{V} \times \mathfrak{U})$ is an ergodic, faithful and integrable W*-system (de Schreye 1983), so that the eigenalgebra $\mathcal{T}$ is the relevant algebra of observables for the temporal domain (Amann 1986, Sect. III.4). The phase space $\mathfrak{V} \times \mathfrak{U}$ admits only two types of integrable ergodic W*-systems (Amann 1986, p. 199), namely:

(i) A classical Boolean system with a maximal commutative W*-eigenalgebra $\mathcal{T}^{\mathrm{cl}} \cong \mathbf{L}^\infty(\mathfrak{V} \times \mathfrak{U}) \cong \mathbf{L}^\infty(\mathbb{R}^2)$ of all complex-valued essentially bounded Lebesgue-measurable functions on $\mathbb{R}^2$ acting on the Hilbert space $\mathbf{L}^2(\mathbb{R}^2)$, which is generated by the two commuting unitary operator groups $\{V^{\mathrm{cl}}(\tau)|\tau \in \mathbb{R}\}$ and $\{U^{\mathrm{cl}}(\lambda)|\lambda \in \mathbb{R}\}$.

(ii) A non-Boolean system with a non-commutative factor $\mathcal{T}$ of type I_∞, which is generated by the two non-commuting unitary operator groups $\{V(\tau)|\tau \in \mathbb{R}\}$ and $\{U(\lambda)|\lambda \in \mathbb{R}\}$ that satisfy the canonical commutation relation $U(\lambda)V(\tau) = e^{2\pi i\lambda\tau} V(\tau)U(\lambda)$.

9.4.4.2 Elementary Boolean Time Systems

The evolution of a classical *Boolean* system with the commutative W*-algebra $\mathcal{T}^{\mathrm{cl}}$ as indicated in point (i) above is determined by the unitary operator group

$\{V^{\mathrm{cl}}(\tau)|\tau\in\mathbb{R}\}$. The unitary operators $\tau\mapsto V^{\mathrm{cl}}(\tau)$ and $\lambda\mapsto U^{\mathrm{cl}}(\lambda)$ can be written as

$$V^{\mathrm{cl}}(\tau)=\exp\bigl(-2\pi i\tau\Lambda^{\mathrm{cl}}\bigr)\in\mathcal{T}^{\mathrm{cl}},\tag{9.70a}$$

$$U^{\mathrm{cl}}(\lambda)=\exp\bigl(2\pi i\lambda T^{\mathrm{cl}}\bigr)\in\mathcal{T}^{\mathrm{cl}}.\tag{9.70b}$$

The self-adjoint operator T^{cl} with the spectrum $\mathrm{sp}(T^{\mathrm{cl}})=\mathbb{R}$ is called the *classical time operator*. Since the classical time operator is affiliated with the center $\mathcal{Z}(\mathcal{D})$, T^{cl} commutes with all operators of the universe of discourse (including the atemporal domain).

The self-adjoint generator Λ^{cl} of the frequency-translation group is called the *classical frequency operator*. It is canonically conjugate to the classical time operator,

$$\bigl\{T^{\mathrm{cl}},\Lambda^{\mathrm{cl}}\bigr\}_{\mathrm{P}}=\mathbb{1}.\tag{9.71}$$

9.4.4.3 Elementary Non-Boolean Time Systems

The evolution of a *non-Boolean* system with the non-commutative W*-algebra $\mathcal{T}$ as indicated in point (ii) above is determined by the unitary operator group $\{V(\tau)|\tau\in\mathbb{R}\}$. By Stone's theorem, the unitary operators $\tau\mapsto V(\tau)$ and $\lambda\mapsto U(\lambda)$ can be written as

$$V(\tau)=\exp(-2\pi i\tau\Lambda)\in\mathcal{T},\tag{9.72a}$$

$$U(\lambda)=\exp(2\pi i\lambda T)\in\mathcal{T}.\tag{9.72b}$$

9.5 The Affine Weyl-Heisenberg Group

9.5.1 Definition of the Affine Weyl-Heisenberg Group

The affine Weyl-Heisenberg time group is an extension of the Weyl-Heisenberg group $\mathfrak{G}$ by the dilation group. It is a unimodular non-commutative group with the elements

$$\bigl\{(\tau,\lambda,\kappa,\varphi)\bigm|\tau,\lambda,\kappa,\varphi\in\mathbb{R}\bigr\}\tag{9.73}$$

and the group law[180]

$$\begin{aligned}&(\tau,\lambda,\kappa,\varphi)\circ\bigl(\tau',\lambda',\kappa',\varphi'\bigr)\\&\quad=\bigl(\tau+e^{-\kappa}\tau',\lambda+e^{\kappa}\lambda',\kappa+\kappa',\varphi+\varphi'+e^{-\kappa}\lambda\tau'\bigr).\end{aligned}\tag{9.74}$$

[180]Compare Grossmann et al. (1986), Torrésani (1991), and Kalisa and Torrésani (1991).

The identity element is given by $(0,0,0,0)$. The affine Weyl-Heisenberg group is topologically isomorphic to $\mathbb{R}^+ \times \mathbb{R}^{2+1}$. Furthermore, it is solvable and unimodular so that the Haar measure $d\kappa d\tau d\lambda d\varphi$ is both left- and right-invariant.

Since the affine Weyl-Heisenberg group is non-compact, all finite-dimensional representations are not unitary. A faithful non-unitary finite-dimensional matrix representation of the affine Weyl-Heisenberg group is given by the following 3×3-matrices

$$(\tau,\lambda,\kappa,\varphi) \sim \begin{pmatrix} 1 & e^{-\kappa}\lambda & \varphi \\ 0 & e^{-\kappa} & \tau \\ 0 & 0 & 1 \end{pmatrix}, \qquad \tau,\lambda,\kappa,\varphi \in \mathbb{R}. \tag{9.75}$$

The matrix multiplication

$$\begin{aligned} &\begin{pmatrix} 1 & e^{-\kappa}\lambda & \varphi \\ 0 & e^{-\kappa} & \tau \\ 0 & 0 & 1 \end{pmatrix} \times \begin{pmatrix} 1 & e^{-\kappa'}\lambda' & \varphi' \\ 0 & e^{-\kappa'} & \tau' \\ 0 & 0 & 1 \end{pmatrix} \\ &\quad = \begin{pmatrix} 1 & e^{-\kappa'}\lambda' + e^{-\kappa-\kappa'}\lambda' & \varphi + \varphi' + e^{-\kappa}\lambda\tau' \\ 0 & e^{-\kappa-\kappa} & e^{\kappa}\tau' + \tau \\ 0 & 0 & 1 \end{pmatrix} \end{aligned} \tag{9.76}$$

realizes the group law (9.74).

In a realization of the affine Weyl-Heisenberg group by unitary operators $W(\tau,\lambda,\kappa,\varphi)$ the group law is given by

$$\begin{aligned} &W(\tau,\lambda,\kappa,\varphi)W(\tau',\lambda',\kappa',\varphi') \\ &\quad = W\left(\tau + e^{-\kappa}\tau', \lambda + e^{\kappa}\lambda', e^{\kappa'}, \varphi + \varphi' + \frac{1}{2}\left[e^{-\kappa}\lambda\tau' - e^{\kappa}\lambda'\tau\right]\right). \end{aligned} \tag{9.77}$$

The unitary operators $W(\tau,\lambda,\kappa,\varphi)$ can be represented by a suitable combination of the unitary time-translation operator $U(\tau)$, the unitary frequency-shift operator $V(\lambda)$ and the unitary dilation operator $D(\kappa)$:

$$W(\tau,\lambda,\kappa,\varphi) = e^{2\pi i\varphi}U(\tau)V(\lambda)D(\kappa), \tag{9.78a}$$

$$= e^{2\pi i\varphi}e^{-2\pi i\lambda\tau}V(\lambda)U(\tau)D(\kappa). \tag{9.78b}$$

9.5.2 The Subgroups of the Affine Weyl-Heisenberg Group

The affine Weyl-Heisenberg group with the elements $\{(\tau,\lambda,\kappa,\varphi)|\tau,\lambda,\kappa,\varphi \in \mathbb{R}\}$ has three non-trivial commutative one-parameter subgroups, two non-commutative two-parameter subgroups, and one non-commutative three-parameter subgroup:

- the time-translation group $\mathfrak{V} = \{V(\tau)|\tau \in \mathbb{R}\}$ with $V(\tau) = (\tau, 0, 1, 0)$ and the group relation $V(\tau)V(\tau') = V(\tau + \tau')$,

- the frequency-translation group $\mathfrak{U} = \{U(\lambda)|\lambda \in \mathbb{R}\}$ with $U(\lambda) = (0, \lambda, 1, 0)$ and the group relation $U(\lambda)U(\lambda') = U(\lambda + \lambda')$,
- the dilation group $\mathfrak{D} := \{D(\kappa)|\kappa \in \mathbb{R}\}$ with $D(\kappa) = (0, 0, \kappa, 0)$ and the group relation $D(\kappa)D(\kappa') = D(\kappa + \kappa')$,
- the affine two-parameter group $\mathfrak{A}_{\mathfrak{U}} := \mathfrak{U} \rtimes \mathfrak{D} = \{A_U(\tau, \kappa)|\tau, \kappa \in \mathbb{R}\}$ with $A_U(\tau, \kappa) = (0, \tau, \kappa, 0)$ and the group relation $A_U(\tau, \kappa)A_U(\tau', \kappa') = A_U(\tau + e^{\kappa}\tau', \kappa + \kappa')$,
- the affine two-parameter group $\mathfrak{A}_{\mathfrak{V}} := \mathfrak{V} \rtimes \mathfrak{D} = \{A_V(\lambda, \kappa)|\lambda, \kappa \in \mathbb{R}\}$, with $A_V(\lambda, \kappa) = (\lambda, 0, \kappa, 0)$ and the group relation $A_V(\lambda, \kappa)A_V(\lambda', \kappa') = A_V(\lambda + e^{-\kappa}\lambda', \kappa + \kappa')$,
- the three-parameter Weyl-Heisenberg group $\mathfrak{W} = \{W(\tau, \lambda, \varphi)|\tau, \lambda, \varphi \in \mathbb{R}\}$ with $W(\tau, \lambda, \varphi) = (\tau, \lambda, 1, \varphi)$ and the group relation $W(\tau, \lambda, \varphi)W(\tau', \lambda', \varphi') = W(\tau + \tau', \lambda + \lambda', \varphi + \varphi' + \lambda\tau')$.

The self-adjoint generators of the unitary groups $\mathfrak{V}$, $\mathfrak{U}$ and $\mathfrak{D}$ are given by

- the frequency operator Λ with $V(\tau) = e^{-2\pi i\tau\Lambda}$, (9.79a)
- the time operator T with $U(\lambda) = e^{2\pi i\lambda T}$, (9.79b)
- the scale operator S with $D(\kappa) = e^{2\pi i\kappa \ln S}$. (9.79c)

The unitary operators $V(\tau)$, $U(\lambda)$, $D(\kappa)$ fulfill the following commutation relations ($\tau, \tau', \lambda, \lambda', \kappa, \kappa' \in \mathbb{R}$):

$$D(\kappa)D\big(\kappa'\big) = D\big(\kappa + \kappa'\big), \tag{9.80a}$$

$$V(\tau)V\big(\tau'\big) = V\big(\tau + \tau'\big), \tag{9.80b}$$

$$U(\lambda)U\big(\lambda'\big) = U\big(\lambda + \lambda'\big), \tag{9.80c}$$

$$D(\kappa)U(\lambda) = U\big(e^{\kappa}\lambda\big)D(\kappa), \tag{9.80d}$$

$$V(\tau)D(\kappa) = D(\kappa)V\big(e^{\kappa}\tau\big), \tag{9.80e}$$

$$U(\lambda)V(\tau) = e^{2\pi i\lambda\tau}V(\tau)U(\lambda). \tag{9.80f}$$

These commutation relations imply the transformation relations ($\lambda, \tau, \kappa \in \mathbb{R}$):

$$e^{-2\pi i\kappa S}Te^{2\pi i\kappa S} = e^{-\kappa}T, \tag{9.81a}$$

$$e^{-2\pi i\kappa S}\Lambda e^{2\pi i\kappa S} = e^{\kappa}\Lambda, \tag{9.81b}$$

$$e^{2\pi i\tau\Lambda}Se^{-2\pi i\tau\Lambda} = S + \tau\Lambda, \tag{9.81c}$$

$$e^{2\pi i\tau\Lambda}Te^{-2\pi i\tau\Lambda} = T + \tau, \tag{9.81d}$$

$$e^{-2\pi i\lambda T}Se^{2\pi i\lambda T} = S + \lambda T, \tag{9.81e}$$

$$e^{-2\pi i\lambda T}\Lambda e^{2\pi i\lambda T} = \Lambda + \lambda. \tag{9.81f}$$

On an appropriate domain, the generators T, Λ, S fulfill the commutation relations:

$$[T, \Lambda]_- = (i/2\pi)\mathbb{1}, \tag{9.82a}$$

$$[S, T]_- = (1/2\pi i)T, \tag{9.82b}$$

$$[S, \Lambda]_- = (i/2\pi)\Lambda. \tag{9.82c}$$

9.6 The Dilation Group: Scaling

The notion of scaling means that a process or structure is invariant with respect to the scale at which it is considered. Such scale-free behavior pervades a wide range of engineering and science applications such as biological systems, network traffic, economics, noise in electronic devices, and it is the defining characteristic of fractal structures and processes. Scaling can be discussed in terms of the action of the one-parameter *dilation group*.

The *dilation group* $\mathfrak{D}$ describes the scaling operation $t \to e^{\kappa}t, \lambda \to e^{-\kappa}\lambda$ for $\kappa, t, \lambda \in \mathbb{R}$. The dilation group is a continuous infinite Abelian group, isomorphic to the commutative *multiplicative* group $\mathbb{R}^+$ of positive numbers. On the temporal algebra $\mathcal{T}$, the dilation group $\mathfrak{D}$ can be realized by a one-parameter group $\mathfrak{D} = \{D(\kappa)|\kappa \in \mathbb{R}\}$ of unitary operators $D(\kappa) \in \mathcal{T}$ with the group relation

$$D(\kappa) \circ D(\kappa') = D(\kappa + \kappa'), \quad \kappa, \kappa' \in \mathbb{R}. \tag{9.83}$$

The self-adjoint generator S of the unitary dilation group $\{D(\kappa)|\kappa \in \mathbb{R}\}$ is called the *scale operator*,

$$D(\kappa) = e^{2\pi i\kappa S}, \quad \kappa \in \mathbb{R}. \tag{9.84}$$

It can be expressed in terms of the self-adjoint frequency operator Λ and the self-adjoint time operator T as

$$S := \frac{1}{2}(\Lambda T + T\Lambda) = \Lambda T + i\mathbb{1}/4\pi = T\Lambda - i\mathbb{1}/4\pi. \tag{9.85}$$

The self-adjoint generator S of the one-parameter dilation group fulfills the affine commutation relation

$$[S, T]_- = (1/2\pi i)T, \qquad [S, \Lambda]_- = (i/2\pi)\Lambda, \tag{9.86}$$

which characterizes the Lie algebra of the affine group.

9.7 Affine Subgroups of the Affine Weyl-Heisenberg Group

9.7.1 Affine Groups

Besides the two-parameter Weyl-Heisenberg group (describing time and frequency translations) the only other two-dimensional simply-connected Lie group is the affine group, which is isomorphic to the group of all linear transformations $x \to ax + b$ ($a > 0, b \in \mathbb{R}$) of the real line $\mathbb{R}$. It is a non-Abelian Lie group with the associative and non-commutative group multiplication $\circ$,

$$(a, b) \circ \left(a', b'\right) := \left(aa', b + ab'\right), \quad a > 0, a' > 0,\ b, b' \in \mathbb{R}, \tag{9.87}$$

where $(1, 0)$ is the neutral element, and $(1/a, -b/a)$ is the inverse of (a, b). This group is isomorphic to the semi-direct product $\mathbb{R}^* \ltimes \mathbb{R}$. The affine group is not compact but locally compact, so that it has a unique left-invariant measure.

If $a > 0$, the transformation is orientation preserving, if $a < 0$ it is orientation reversing. In the following we consider only the orientation-preserving affine group and write $a = e^{\kappa}$ so that the group law can be expressed as:

$$(\kappa, b) \circ \left(\kappa', b'\right) = \left(\kappa + \kappa', b + e^{\kappa} b'\right), \quad \kappa, \kappa', b, b' \in \mathbb{R}. \tag{9.88}$$

The *translation group* $(0, b)$ is an invariant subgroup of the affine group. It describes the translations $x \to x + b$ and is isomorphic to the commutative additive group $\mathbb{R}$ of real numbers. The subgroup (κ, b) is the multiplicatively written *dilation group* $\mathbb{R}^*$, which has representations with characters $x \mapsto s^{ix}$. According to Gelfand and Neumark (1947), the affine group has two non-trivial inequivalent irreducible unitary representations (cf. also Vilenkin 1968, Chap. V, §2). Let $\mathcal{H} = \mathbf{L}^2(\mathbb{R}^*, dx/x)$, then bounded linear maps on $\mathcal{C}_0(\mathbb{R}^*)$,

$$\left\{\pi_{\pm}\left[\mathfrak{g}(a, b)\right]\Phi\right\}(x) = e^{\mp ibx}\,\Phi(ax), \tag{9.89}$$

can be extended to unitary representations of the affine group on the Hilbert space $\mathbf{L}^2(\mathbb{R}^*)$.

9.7.2 The Affine Subgroups $\mathfrak{A}_{\mathfrak{U}}$ and $\mathfrak{A}_{\mathfrak{V}}$

The affine Weyl-Heisenberg group $\mathfrak{G}$ possesses two affine subgroups introduced in Sect. 9.5.2. They are semi-direct products of the dilation group and a translation group:

- the affine group $\mathfrak{A}_{\mathfrak{V}} := \mathfrak{V} \rtimes \mathfrak{D}$ is generated by the time-translation group $\mathfrak{V}$ and the dilation group $\mathfrak{D}$,
- the affine group $\mathfrak{A}_{\mathfrak{U}} := \mathfrak{U} \rtimes \mathfrak{D}$ is generated by the frequency-translation group $\mathfrak{U}$ and the dilation group $\mathfrak{D}$.

The affine groups $\mathfrak{A}_{\mathfrak{U}}$ and $\mathfrak{A}_{\mathfrak{V}}$ implement two mutually exclusive time concepts, which have been called *sequential time* and *non-sequential time* in previous sections (1.4.3, 4.3–4.5). According to Eq. (9.78a), (9.78b), the two non-commuting unitary operators A_U and A_V can be realized by the self-adjoint operators T and Λ,

$$A_U(\lambda,\kappa) = e^{+2\pi i\lambda T} D(\kappa), \tag{9.90a}$$

$$A_V(\tau,\kappa) = e^{-2\pi i\tau\Lambda} D(\kappa). \tag{9.90b}$$

The well-known representation theory of affine groups implies that the W*-algebra $\mathcal{A}_U$ generated by $\{A_U(\lambda,\kappa)|\lambda,\kappa\in\mathbb{R}\}$ and the W*-algebra $\mathcal{A}_V$ generated by $\{A_V(\tau,\kappa)|\tau,\kappa\in\mathbb{R}\}$ are the direct sums of two factors,

$$\mathcal{A}_U = \mathcal{A}_U^- \oplus \mathcal{A}_U^+, \qquad \mathcal{A}_U^\pm := P_U^\pm \mathcal{T} P_U^\pm, \tag{9.91a}$$

$$\mathcal{A}_V = \mathcal{A}_V^- \oplus \mathcal{A}_V^+, \qquad \mathcal{A}_V^\pm := P_V^\pm \mathcal{T} P_V^\pm, \tag{9.91b}$$

where the projections $P_U^\pm$ and $P_V^\pm$ are defined by

$$P_U^\pm := \vartheta(\pm T), \qquad P_V^\pm := \vartheta(\pm\Lambda). \tag{9.92}$$

The projector $P_U^\pm$ commutes with all elements in $\mathcal{A}_U^\pm$, and the projector $P_V^\pm$ commutes with all elements in $\mathcal{A}_V^\pm$, so that they generate the centers $\mathcal{Z}(\mathcal{A}_U)$ and $\mathcal{Z}(\mathcal{A}_V)$ of the W*-algebras $\mathcal{A}_U$ and $\mathcal{A}_V$,

$$\mathcal{Z}(\mathcal{A}_U) = \{P_U^-, P_U^+\}'', \qquad \mathcal{Z}(\mathcal{A}_V) = \{P_V^-, P_V^+\}''. \tag{9.93}$$

This entails that there are *two mutually exclusive superselection rules* which lead to *Boolean classifications of temporal phenomena.*

The projectors $P_U^\pm$ and $P_V^\pm$ have dispersion-free values with respect to pure state functionals in the preduals $(\mathcal{A}_U)_*$ and $(\mathcal{A}_V)_*$, respectively. For this reason they are referred to as *classical observables* with respect to the sequential-time description by the affine group $\mathfrak{A}_{\mathfrak{U}}$ and the non-sequential-time description by the affine group $\mathfrak{A}_{\mathfrak{V}}$, respectively.

Since the actions of the affine group $\mathfrak{A}_{\mathfrak{U}}$ on the W*-algebra $\mathcal{A}_U^\pm$ and of the affine group $\mathfrak{A}_{\mathfrak{V}}$ on the W*-algebra $\mathcal{A}_V^\pm$ are ergodic,[181] the W*-algebras $\mathcal{A}_U^\pm$ and $\mathcal{A}_V^\pm$ are of fundamental significance for the description of temporal phenomena. We denote these two inequivalent ergodic operator realizations by $\{A_U^\pm(\lambda,\kappa)|\lambda,\kappa\in\mathbb{R}\}$ and $\{A_V^\pm(\tau,\kappa)|\tau,\kappa\in\mathbb{R}\}$, respectively. They are characterized by the positivity of the generators $\pm T_U^\pm$ and $\pm T_V^\pm$ of the unitary groups $\{U^\pm(\lambda)|\lambda\in\mathbb{R}^\pm\}$ and $\{V^\pm(\tau)|\tau\in\mathbb{R}^\pm\}$,

[181] The action α of a group is called ergodic if the only projections $P\in\mathcal{M}$ for which $\alpha(P)=P$ are $\mathbb{O}$ and $\mathbb{1}$. If a group acts ergodically on an algebra, this algebra is necessarily a *factor* (Kadison and Ringrose 1986, pp. 546f). Note that ergodic actions correspond to irreducible projective representations in a Hilbert-space realization.

$$U^{\pm}(\lambda) = e^{2\pi i \lambda T_U^{\pm}} \quad \text{with } \pm T_U^{\pm} > 0, \tag{9.94a}$$

$$V^{\pm}(\tau) = e^{-2\pi i \tau \Lambda_V^{\pm}} \quad \text{with } \pm \Lambda_V^{\pm} > 0. \tag{9.94b}$$

9.7.3 *Operator Realization of the Two Affine Subgroups*

9.7.3.1 Four Scale-Invariant Subalgebras

The transformation relations (9.81a), (9.81b) imply

$$e^{-2\pi i \kappa S} f(T, \Lambda) e^{2\pi i \kappa S} = f\big(e^{-\kappa} T, e^{\kappa} \Lambda\big), \tag{9.95}$$

so that with

$$T^- := \int_{-\infty}^{0} t F(dt), \qquad T^+ := \int_{0}^{\infty} t F(dt), \tag{9.96a}$$

$$\Lambda^- := \int_{-\infty}^{0} \ell E(d\ell), \qquad \Lambda^+ := \int_{0}^{\infty} \ell E(d\ell), \tag{9.96b}$$

we obtain

$$e^{-2\pi i \kappa S} T^{\pm} e^{2\pi i \kappa S} = T^{\pm}, \quad \kappa \in \mathbb{R}, \tag{9.97a}$$

$$e^{-2\pi i \kappa S} \Lambda^{\pm} e^{2\pi i \kappa S} = \Lambda^{\pm}, \quad \kappa \in \mathbb{R}. \tag{9.97b}$$

These relations imply that the projections $P_U^{\pm} = \vartheta(\pm T)$ and $P_V^{\pm} = \vartheta(\pm \Lambda)$ commute with the scale operator S,

$$\big[P_U^{\pm}, S\big]_- = \mathbf{0}, \qquad \big[P_V^{\pm}, S\big]_- = \mathbf{0}. \tag{9.98}$$

Both the two W*-subalgebras generated by the operators $P_U^{\pm}$ and S and the two W*-subalgebras generated by the operators $P_V^{\pm}$ and S,

$$\mathcal{A}_U^{\pm} := \big\{P_U^{\pm}, S\big\}'' \subset \mathcal{T}, \qquad \mathcal{A}_V^{\pm} := \big\{P_V^{\pm}, S\big\}'' \subset \mathcal{T}, \tag{9.99}$$

are invariant under the action of the dilation group $\mathfrak{D}$ and lead to *four different descriptions of the temporal domain.*

9.7.3.2 Past and Future

If we describe the temporal domain in terms of the W*-algebra

$$\mathcal{A}_U = \mathcal{A}_U^- \oplus \mathcal{A}_U^+, \tag{9.100}$$

the projectors P_U^- and P_U^+ are *classical observables*. Therefore the time operator T and the scale operator S can be decomposed as

$$T = T_U^- \oplus T_U^+, \qquad S = S_U^- \oplus S_U^+, \tag{9.101}$$

with the restrictions

$$T_U^\pm = T|_{\mathcal{A}_U^\pm}, \qquad S_U^\pm = S|_{\mathcal{A}_U^\pm}. \tag{9.102}$$

The spectra of both self-adjoint operators S_U^+ and S_U^- cover the entire real line. While the operator T_U^+ is positive, T_U^- is negative, so that the time concept related to the affine group $\mathcal{A}_U$ allows a precise characterization of "past" and "future":

$$\text{the time operator } T_U^- < 0 \text{ characterizes the } \textit{past}, \tag{9.103a}$$

$$\text{the time operator } T_U^+ > 0 \text{ characterizes the } \textit{future}. \tag{9.103b}$$

Since the classical observables P_U^- and P_U^+ generate the center of $\mathcal{A}_U$, the factors $\mathcal{A}_U^-$ and $\mathcal{A}_U^+$ are separated by a superselection rule that prohibits coherent superpositions between past and future. That is, if $\mathcal{A}_U^-$ or $\mathcal{A}_U^+$ is the relevant kinematical group, time-translation symmetry is broken. This situation corresponds to the traditional distinction between past and future, where holistic correlations between past and future are impossible.

In communication theory, a system is called *causal* if it is *non-anticipative*. That is, an output response signal $t \mapsto r(t)$ of a causal system at an arbitrary time t depends exclusively on past stimulus signals $t \mapsto s(t)$ and not on future stimuli.

9.7.3.3 Negative and Positive Frequencies

If we describe temporal phenomena in terms of the W*-algebra

$$\mathcal{A}_V = \mathcal{A}_V^- \oplus \mathcal{A}_V^+, \tag{9.104}$$

then the projectors P_V^- and P_V^+ are *classical observables*. Therefore the frequency operator Λ and the scale operator S can be decomposed as

$$\Lambda = \Lambda_V^- \oplus \Lambda_V^+, \qquad S = S_V^- \oplus S_V^+, \tag{9.105}$$

with the restrictions

$$\Lambda_V^\pm = \Lambda|_{\mathcal{A}_V^\pm}, \qquad S_V^\pm = S|_{\mathcal{A}_V^\pm}. \tag{9.106}$$

The spectra of the operators S_V^+ and S_V^- cover the entire real line. While the operator Λ_V^+ is positive, Λ_V^- is negative, so that the projectors P_V^+ and P_V^- generate the following *Boolean classification*:

$$\text{the projector } P_V^+ > 0 \text{ characterizes } \textit{spectral positivity}, \tag{9.107a}$$

$$\text{the projector } P_V^- < 0 \text{ characterizes } \textit{spectral negativity}. \tag{9.107b}$$

In this description, frequency-translation symmetry is broken. The factors $\mathcal{A}_V^-$ and $\mathcal{A}_V^+$ are separated by a superselection rule that prohibits coherent superpositions between positive and negative frequencies.

9.7.4 Logarithmic Canonical Operators

9.7.4.1 The Operators ln |T| and ln |Λ|

The restricted observables $T^\pm$ and $\Lambda^\pm$ (see Sect. 9.7.3.1) do not fulfill the Heisenberg commutation relation $[X, Y]_- = \text{const}\,\mathbb{1}$, but the affine commutation relation $[X, Y]_- = X$. In this case the appropriate kinematical group is no longer the Weyl-Heisenberg group but the affine group with the Lie-algebra relation $[X, Y]_- = X$. However, *logarithmic* time and frequency operators allow a Weyl-Heisenberg-type description in terms of functions on the half-plane $\mathbb{R}^\pm$.

The commutation relation (9.95) implies

$$e^{-2\pi i\kappa S} f(T) e^{2\pi i\kappa S} = f\big(e^{-\kappa} T\big), \tag{9.108a}$$

$$e^{-2\pi i\kappa S} g(\Lambda) e^{2\pi i\kappa S} = g\big(e^{\kappa} \Lambda\big), \tag{9.108b}$$

hence on the Schwartz space $\mathcal{S}(\mathbb{R})$:

$$\big[f(T), S\big]_- = (i/2\pi) f'(T) T, \tag{9.109a}$$

$$\big[g(\Lambda), S\big]_- = (1/2\pi i) g'(\Lambda) \Lambda. \tag{9.109b}$$

With $f(x) = \ln|x|$ and the relations (Kanwal 1983, p. 88)

$$\frac{d \ln|x|}{dx} = \mathcal{P}\left(\frac{1}{x}\right), \quad \mathcal{P}\left(\frac{1}{x}\right) x = 1, \tag{9.110}$$

where $\mathcal{P}$ denotes the Cauchy principal value, we obtain with $f(T) = \ln|T|$ and $g(\Lambda) = \ln|\Lambda|$:

$$[\ln|T|, S]_- = (i/2\pi)\mathbb{1}, \tag{9.111a}$$

$$[\ln|\Lambda|, S]_- = (1/2\pi i)\mathbb{1}. \tag{9.111b}$$

9.7.4.2 The Operators ln |T^±| and ln |Λ^±|

Since the scale operator S of Eq. (9.85) commutes with the projections $P_U^\pm$ and $P_V^\pm$ (see Eq. (9.98)), the operators $T^\pm$ and $S^\pm$,

$$T^\pm := P_U^\pm T P_U^\pm, \qquad S_U^\pm := P_U^\pm S P_U^\pm, \tag{9.112a}$$

$$\Lambda^{\pm} := P_V^{\pm} \Lambda P_V^{\pm}, \qquad S_V^{\pm} := P_V^{\pm} S P_V^{\pm}, \tag{9.112b}$$

still fulfill the affine commutation relation:

$$\left[S_U^{\pm}, T^{\pm}\right]_{-} = (1/2\pi i) T^{\pm}, \tag{9.113a}$$

$$\left[S_V^{\pm}, \Lambda^{\pm}\right]_{-} = (i/2\pi) \Lambda^{\pm}, \tag{9.113b}$$

and therefore on the Schwartz space $\mathcal{S}(\mathbb{R})$:

$$\left[f(T^{\pm}), S_U^{\pm}\right]_{-} = (i/2\pi) f'(T^{\pm}) T^{\pm}, \tag{9.114a}$$

$$\left[g(\Lambda^{\pm}), S_V^{\pm}\right]_{-} = (1/2\pi i) g'(\Lambda^{\pm}) \Lambda^{\pm}. \tag{9.114b}$$

The locally integrable functions $x \mapsto \vartheta(\pm x)\ln(\pm x)$ define the generalized functions

$$\left(\ln|x|\right)_{\pm} := \vartheta(\pm x)\ln|x| \tag{9.115}$$

with the distributional derivatives $d(\ln x)^{\pm}/dx = \pm\vartheta(\pm x)|x|^{-1}$ (see Kanwal 1983, pp. 86–88), so that

$$\frac{d(\ln x)^{\pm}}{dx} x^{\pm} = 1. \tag{9.116}$$

On the invariant subalgebras $\mathcal{A}_U^{\pm} = \{P_U^{\pm}, S\}'' \subset \mathcal{T}$, the operators $\ln|T^{\pm}|$and $S_U^{\pm}$ fulfill the canonical commutation relations

$$\left[\ln|T^{\pm}|, S_U^{\pm}\right]_{-} = (i/2\pi)\mathbb{1}, \tag{9.117}$$

while on the invariant subalgebras $\mathcal{A}_V^{\pm} = \{P_V^{\pm}, S\} \subset \mathcal{T}$, the operators $\ln|\Lambda^{\pm}|$ and $S_V^{\pm}$ fulfill the canonical commutation relations

$$\left[\ln|\Lambda^{\pm}|, S_V^{\pm}\right]_{-} = (1/2\pi i)\mathbb{1}. \tag{9.118}$$

The self-adjoint operators $S_U^{\pm}$ and $S_V^{\pm}$ have the simple Lebesgue spectrum $\mathbb{R}$.

9.7.4.3 Weyl's Commutation Relations for the Logarithmic Operators

The relations (9.95) and (9.112a), (9.112a) imply

$$e^{-2\pi i\kappa S_U^{\pm}} f(T^{\pm}) e^{2\pi i\kappa S_U^{\pm}} = f(e^{-\kappa} T^{\pm}), \tag{9.119a}$$

$$e^{-2\pi i\kappa S_V^{\pm}} g(\Lambda^{\pm})) e^{2\pi i\kappa S_V^{\pm}} = g(e^{\kappa} \Lambda^{\pm}), \tag{9.119b}$$

so that with

$$f(T) := e^{2\pi i\mu \ln T} = M(\mu), \qquad g(\Lambda) := e^{-2\pi i\nu \ln \Lambda} = N(\nu), \tag{9.120}$$

we get Weyl's formulation of the canonical commutation relations

$$M(\mu)D(\kappa) = e^{-2\pi i\mu\kappa} D(\kappa)M(\mu), \quad \mu, \kappa \in \mathbb{R}, \tag{9.121a}$$

$$N(\nu)D(\kappa) = e^{2\pi i\nu\kappa} D(\kappa)N(\nu), \quad \nu, \kappa \in \mathbb{R}. \tag{9.121b}$$

While the canonical pair $\{T, \Lambda\}$ generates the full W*-algebra $\mathcal{T}$ of the temporal domain, the canonical pairs $\{\ln|T^{\pm}|, S^{\pm}\}$ and $\{\ln|\Lambda^{\pm}|, S_V^{\pm}\}$ generate the W*-subalgebras $\mathcal{A}_U^{\pm}$ and $\mathcal{A}_V^{\pm}$ of two complementary time descriptions. Since every pair of canonical operators $\{T, \Lambda\}$, $\{\ln|T^{+}|, S_U^{+}\}$, $\{\ln|T^{-}|, S_U^{-}\}$, $\{\ln|\Lambda^{+}|, S_V^{+}\}$ and $\{\ln|\Lambda^{-}|, S_V^{-}\}$ generates ergodic Weyl systems, the Stone-von Neumann uniqueness theorem (Stone 1930; von Neumann 1931) implies that all these pairs are unitarily equivalent.

9.7.4.4 Unitary Equivalence

We sketch the proof of the unitary equivalence of the canonical pairs $\{\ln|T^{+}|, S_U^{+}\}$ and (T, Λ); the proofs for the other pairs are analogous. First we consider the coordinate transformation

$$\mathbb{R} \ni t \to e^{t} > 0, \tag{9.122}$$

which induces a transformation $R : \mathbf{L}^2(\mathbb{R}^+) \to \mathbf{L}^2(\mathbb{R})$ that takes functions in $\mathbf{L}^2(\mathbb{R}^+)$ and stretches them into functions in $\mathbf{L}^2(\mathbb{R})$:

$$\breve{f}(t) := \{Rf\}(t) = e^{t/2} f\big(e^{t}\big), \quad f \in \mathbf{L}^2\big(\mathbb{R}^+, d\theta\big),\ t \in \mathbb{R}, \tag{9.123a}$$

$$f(\theta) = \big\{R^{-1}\breve{f}\big\}(\theta) = \frac{\breve{f}(\ln\theta)}{\sqrt{\theta}}, \quad \breve{f} \in \mathbf{L}^2(\mathbb{R}, dt),\ \theta > 0. \tag{9.123b}$$

For every $f, g \in \mathbf{L}^2(\mathbb{R}^+)$, we obtain

$$\int_0^{\infty} f^*(\theta)g(\theta)d\theta = \int_{-\infty}^{\infty} f^*\big(e^{t}\big)g\big(e^{t}\big)e^{t}dt = \int_{-\infty}^{\infty} \breve{f}^*(t)\breve{g}(t)dt, \tag{9.124}$$

so that R is a unitary transformation.

On the Lebesgue-Hilbert space $\mathbf{L}^2(\mathbb{R}^+)$, the positive time operator T_+ acts as a multiplication operator:

$$\{T_+ g\}(\theta) = \theta g(\theta), \quad g \in \mathbf{L}^2\big(\mathbb{R}^+, d\theta\big), \tag{9.125}$$

so that

$$\begin{aligned}\int_0^{\infty} f^*(\theta)\{T_+ g\}(\theta)d\theta &= \int_0^{\infty} f^*(\theta)\theta g(\theta)d\theta \\ &= \int_{-\infty}^{\infty} f^*\big(e^{t}\big)e^{t} g\big(e^{t}\big)e^{t}dt\end{aligned}$$

$$= \int_{-\infty}^{\infty} \breve{f}^*(t) t \breve{g}(t) dt$$

$$= \int_{-\infty}^{\infty} \breve{f}^*(t) \{T\breve{g}\} dt, \tag{9.126}$$

where the time operator T acts as a multiplication operator in $\mathbf{L}^2(\mathbb{R}, dt)$,

$$\{T\breve{g}\}(t) = t\breve{g}(t), \quad \breve{g} \in \mathbf{L}^2(\mathbb{R}, dt). \tag{9.127}$$

In the multiplicative representation of T_+, the scale operator S_+ is given by

$$\{S_+ f\}(\theta) = \frac{1}{4\pi i}\left\{\frac{d}{d\theta}\theta + \theta\frac{d}{d\theta}\right\} f(\theta), \quad f \in \mathbf{L}^2(\mathbb{R}^+, d\theta), \tag{9.128}$$

so that

$$\begin{aligned}
\int_0^{\infty} f^*(\theta)\{S_+ g\}(\theta) d\theta &= \frac{1}{4\pi i}\int_0^{\infty} f^*(\theta)\left\{\frac{d}{d\theta}\theta + \theta\frac{d}{d\theta}\right\} g(\theta) d\theta \\
&= \frac{1}{4\pi i}\int_0^{\infty} f^*(\theta)\left\{g(\theta) + 2\theta g'(\theta)\right\} d\theta \\
&= \frac{1}{4\pi i}\int_{-\infty}^{\infty} f^*\left(e^t\right)\left\{g\left(e^t\right) + 2e^t g'\left(e^t\right)\right\} e^t dt \\
&= \frac{1}{2\pi i}\int_{-\infty}^{\infty} \breve{f}^*(t)\breve{g}'(t) dt \\
&= \int_{-\infty}^{\infty} \breve{f}^*(t)\{\Lambda\breve{g}\}(t) dt,
\end{aligned} \tag{9.129}$$

where Λ is the differential operator in $\mathbf{L}^2(\mathbb{R}, dt)$,

$$\{\Lambda\breve{g}\}(t) = \frac{1}{2\pi i}\frac{d\breve{g}(t)}{dt}, \quad \breve{g} \in \mathbf{L}^2(\mathbb{R}, dt). \tag{9.130}$$

Equations (9.126) and (9.129) imply that the logarithmic time operator $\ln|T_+|$ and the scale operator S_+ on $\mathbf{L}^2(\mathbb{R}^+)$ are unitarily equivalent to the time operator T and the frequency operator Λ on $\mathbf{L}^2(\mathbb{R})$:

$$R(\ln T_+)R^* = T, \qquad RS_+R^* = \Lambda. \tag{9.131}$$

Chapter 10
Representations of Temporal Groups and Semigroups

10.1 Hilbert-Space Representations

A continuous unitary representation of a locally compact Lie group $\mathfrak{G}$ is a mapping π of $\mathfrak{G}$ into the unitary operators on a Hilbert space $\mathcal{H}$ such that

- $\pi(\mathfrak{g} \circ \mathfrak{g}') = \pi(\mathfrak{g})\pi(\mathfrak{g}')$ for all $\mathfrak{g}, \mathfrak{g}' \in \mathfrak{G}$,
- $\pi(\mathfrak{e}) = \mathbb{1}$,
- the function $\mathfrak{g} \mapsto \langle \Phi | \pi(\mathfrak{g}) \Psi \rangle$ is continuous for every $\Phi, \Psi \in \mathcal{H}$.

There is a canonical unitary representation of $\mathfrak{G}$ in the Lebesgue-Hilbert space $\mathbf{L}^2(\mathfrak{X}, m)$ with a measure space $(\mathfrak{X}, m)$, given by

$$\pi(\mathfrak{g})\Phi(x) = \sqrt{\frac{dm(\mathfrak{g}^{-1}x)}{dm(x)}}\,\Phi\big(\mathfrak{g}^{-1}x\big), \quad \Phi \in \mathbf{L}^2(\mathfrak{X}, m), \tag{10.1}$$

In particular the so-called *left-regular representation* is defined by

$$\pi_{\mathrm{L}}(\mathfrak{g})\Phi(x) = \Phi\big(\mathfrak{g}^{-1}x\big), \quad \Phi \in \mathbf{L}^2(\mathfrak{G}, \mu_{\mathrm{L}}), \tag{10.2}$$

while the *right-regular representation* is given by

$$\pi_{\mathrm{R}}(\mathfrak{g})\Phi(x) = \Phi(\mathfrak{g}x), \quad \Phi \in \mathbf{L}^2(\mathfrak{G}, \mu_{\mathrm{R}}). \tag{10.3}$$

A unitary representation π of $\mathfrak{G}$ on $\mathcal{H}$ is said to be *irreducible* if there is no invariant proper subspace $\mathcal{K}$ of $\mathcal{H}$:

$$\pi \text{ is irreducible if } \pi(\mathcal{K}) = \mathcal{H} \text{ implies } \mathcal{K} = \mathcal{H} \text{ or } \mathcal{K} = \{0\}. \tag{10.4}$$

A strongly continuous irreducible unitary representation $\mathfrak{g} \to \pi(\mathfrak{g})$ of a locally compact separable Lie group $\mathfrak{G}$ on a separable Hilbert space $\mathcal{H}$ is said to be *square-integrable* if there exists a non-zero vector $\eta \in \mathcal{H}$ such that

$$\int_{\mathfrak{G}} \big|\langle \eta | \pi(\mathfrak{g})\eta \rangle\big|^2 d\mu(\mathfrak{g}) < \infty. \tag{10.5}$$

H. Primas, *Knowledge and Time*,
DOI 10.1007/978-3-319-47370-3_10

Here μ is either the left-invariant or the right-invariant Haar measure on $\mathfrak{G}$. Since it is marginal whether the left- or the right-invariant measure is used, we will use the left-invariant measure in the following. A vector $\eta \in \mathcal{H}$ satisfying Eq. (10.5) is called *admissible*. If there exists an admissible vector, then the set of all admissible vectors is dense in $\mathcal{H}$. The set of all admissible vectors equals $\mathcal{H}$ if and only if $\mathfrak{G}$ is unimodular.

For a square-integrable representation of $\mathfrak{G}$ there exists a unique unbounded self-adjoint positive operator C whose domain coincides with the dense set of admissible vectors and which fulfills, for any two admissible vectors η_1, η_2 and any two vectors $\Phi_1, \Phi_2 \in \mathcal{H}$, the relation[182]

$$\int_{\mathfrak{G}} \langle \pi(\mathfrak{g})\eta_1 | \Phi_1 \rangle \langle \Phi_2 | \pi(\mathfrak{g})\eta_2 \rangle d\mu(\mathfrak{g}) = \langle C^{-1/2}\eta_1 | C^{-1/2}\eta_2 \rangle \langle \Phi_2 | \Phi_1 \rangle. \tag{10.6}$$

If the group $\mathfrak{G}$ is unimodular, then C is a multiple of the identity.

The crucial importance of square-integrable representations lies in the generalization (10.6) of Schur's orthogonality relations for finite groups. Let $\eta \in \mathcal{H}$ be an admissible vector of a square-integrable representation π of the separable locally compact group $\mathfrak{G}$ with the left Haar measure μ_{L}. Then for all vectors $\Phi \in \mathcal{H}$ the following decomposition holds:

$$\Phi = \frac{1}{c_\eta} \int_{\mathfrak{G}} \langle \pi(\mathfrak{g})\eta | \Phi \rangle \pi(\mathfrak{g})\eta \, d\mu_{\mathrm{L}}(\mathfrak{g}), \tag{10.7a}$$

$$c_\eta := \langle \eta | C^{-1}\eta \rangle. \tag{10.7b}$$

Therefore, any function $\Phi \in \mathcal{H}$ can be decomposed as a linear combination of functions $|\mathfrak{g}\rangle := \pi(\mathfrak{g})\eta$ obtained by the action of the group $\mathfrak{G}$ on η.

The system of state vectors $\{\eta_{\mathfrak{g}} | \mathfrak{g} \in \mathfrak{G}\}$ with

$$\eta_{\mathfrak{g}} := \pi(\mathfrak{g})\eta, \tag{10.8}$$

is called the *coherent-state system* associated with the unitary representation $\{\pi(\mathfrak{g}) | \mathfrak{g} \in \mathfrak{G}\}$ of the group $\mathfrak{G}$. For a square-integrable representation the coherent vectors $\eta_{\mathfrak{g}}$ generate a *resolution of the identity*

$$\frac{1}{c_\eta} \int_{\mathfrak{G}} |\eta_{\mathfrak{g}}\rangle \langle \eta_{\mathfrak{g}}| d\mu_{\mathrm{L}}(\mathfrak{g}) = \mathbb{1}, \tag{10.9}$$

where $\mathbb{1}$ is the identity operator in $\mathcal{H}$, and the integral is taken in the weak operator sense.

---

[182]Compare Duflo and Moore (1976), or Grossmann et al. (1986).

10.2 Temporal Abelian Groups

10.2.1 Generalized Eigenfunction Expansion

There are three important unitarily equivalent Hilbert-space representations for the temporal domain $\mathcal{T}$. The time, frequency, and scale representation refer to the multiplication of the time operator T, the frequency operator Λ, and the scale operator S, respectively. The time and frequency representation refer to time translations and to frequency translations of a signal, respectively. The scale representation decouples the size and the shape of a signal. Each of these representations have their benefits and drawbacks, so that it is worthwhile to discuss all three of them.

We are interested in those generalized eigenfunctions of a self-adjoint operator A ($A = T, \Lambda, S$), which have eigenvalues in the spectrum $sp(A) \subseteq \mathbb{R}$. The corresponding eigenvalue problems can be formulated in a mathematically rigorous way in terms of the nuclear spectral theorem (cf. Gel'fand and Vilenkin 1964, Chap. I, Sect. 4.5):

$$A\Psi_a^A = a\Psi_a^A, \quad a \in sp(A) \subseteq \mathbb{R}. \tag{10.10}$$

In this interpretation the generalized eigenfunction Ψ_a^A is an element in the topological dual of an appropriate complete nuclear test-function space of a rigged Hilbert space (Gelfand triple). A rigged Hilbert space is a triplet of spaces $\mathcal{H}_+ \subset \mathcal{H} \subset \mathcal{H}_-$, where $\mathcal{H}$ is a Hilbert space, $\mathcal{H}_+$ is a dense subspace of the Hilbert space, and $\mathcal{H}_-$ is the topological dual of $\mathcal{H}_+$. That is, $\mathcal{H}_+$ is the set of anti-linear functionals over the space $\mathcal{H}_-$, where the canonical bilinear form is defined by

$$\langle\Psi|\Phi\rangle := \int_{sp(A)} \psi(a)^*\Psi(a)\mu(da) =: \langle\Phi|\Psi\rangle^*, \quad \Phi \in \mathcal{H}_+,\ \Psi \in \mathcal{H}_-. \tag{10.11}$$

The topology of the space $\mathcal{H}_+$ is finer than the topology inherited from $\mathcal{H}$. To discuss the eigenvalue problem of the self-adjoint operator A, we require that the domain of A is a subset of $\mathcal{H}_+$. By a generalized eigenvector of the operator A in the rigged Hilbert space $\mathcal{H}_+ \subset \mathcal{H} \subset \mathcal{H}_-$ we mean a functional $\Psi_a^A \in \mathcal{H}_-$ such that

$$\langle\Psi_a|A\Phi\rangle = a\langle\Psi_a^A|\Phi\rangle \quad \text{for every } \Phi \in \mathcal{H}_+. \tag{10.12}$$

The real number a is called a generalized eigenvalue corresponding to the generalized eigenvector Ψ_a^A. The generalized eigenvectors form a complete system, characterized by the Parseval-type identity

$$\langle\Phi_1|\Phi_2\rangle = \int_{sp(A)} \mu(da)\langle\Phi_1|\Psi_a\rangle\langle\Psi_a|\Phi_2\rangle, \quad \Phi_1, \Phi_2 \in \mathcal{H}_+. \tag{10.13}$$

The generalized eigenfunctions form an orthonormal basis,

$$\int_{sp(A)} \left(\Psi_a^A\right)^*(x)\Psi_{a'}(x)dx = \delta\left(a - a'\right), \tag{10.14a}$$

$$\int_{\mathbb{R}} (\Psi_a^A)^* (x') \Psi_a(x) da = \delta(x - x'), \tag{10.14b}$$

so that an arbitrary function Ξ can be expanded as

$$\Phi(x) = \int_{sp(A)} da \Psi_a(x) \int dy \Psi_a^*(y) \Phi(y) dy. \tag{10.15}$$

10.2.1.1 Eigenfunctions of the Operators T, Λ and S

On the Schwartz space $\check{\mathbb{S}}(\mathbb{R}) \subset \check{\mathbf{L}}^2(\mathbb{R}, dt)$ of all rapidly decreasing complex-valued infinitely differentiable functions on $\mathbb{R}$ the operators T, Λ and S are represented by

$$\{\check{T}\check{\Phi}\}(\tau) = t\check{\Phi}(\tau), \tag{10.16a}$$

$$\{\check{\Lambda}\check{\Phi}\}(\tau) = \frac{1}{2\pi i} \frac{\partial \check{\Phi}(\tau)}{\partial t}, \tag{10.16b}$$

$$\{\check{S}\check{\Phi}\}(x) = \frac{1}{4\pi i} \left\{ \frac{\partial}{\partial x} x + x \frac{\partial}{\partial x} \right\} \check{\Phi}(x), \tag{10.16c}$$

with $\check{\Phi} \in \check{\mathbb{S}}(\mathbb{R}) \subset \check{\mathbf{L}}^2(\mathbb{R}, dt)$.

The eigenfunctions of the generators T and Λ are the following tempered distributions on the Schwartz space $\check{\mathbb{S}}(\mathbb{R})$:

$$\check{\Psi}_t^T(x) = \delta(x - t), \qquad \left\{\check{T}\check{\Psi}_t^T\right\}(x) = t\check{\Psi}_t^T(x), \tag{10.17a}$$

$$\check{\Psi}_\ell^\Lambda(x) = e^{2\pi i x \ell}, \qquad \left\{\check{\Lambda}\check{\Psi}_\ell^\Lambda\right\}(x) = \ell\check{\Psi}_\ell^\Lambda(x). \tag{10.17b}$$

Using the representation (10.16c), the generalized eigenvalue problem for the scale operator S can be written as

$$\frac{t}{i} \frac{\partial \check{\Psi}_s^S(\tau)}{\partial t} = (\pi x_s + i/4)\check{\Psi}_s^S(\tau), \quad \check{\Phi}_s \in \mathbb{S}^*(\mathbb{R}) \subset \check{\mathbf{L}}^2(\mathbb{R}, dt), \ t \in \mathbb{R}. \tag{10.18}$$

The solutions are given by the Schwartz distributions

$$t_+^s := \begin{cases} t^s & \text{for } t > 0 \\ 0 & \text{for } t \leq 0 \end{cases}, \tag{10.19a}$$

$$t_-^s := \begin{cases} 0 & \text{for } t \geq 0 \\ |t|^s & \text{for } t < 0 \end{cases}, \tag{10.19b}$$

which are well defined for all complex values $s \neq -1, -2, -3, \ldots$.[183]

[183]That is, their functional values $s \to \langle t_\pm^s | \phi \rangle$, $\phi \in \mathfrak{X}(\mathbb{R}^\pm)$, are entire functions of the complex variable s. Compare Gel'fand and Shilov (1964, Sect. 3.2).

Using the relation (Gel'fand and Shilov 1964, Sect. 3.2)

$$\frac{dt_{\pm}^{s}}{dt} = \pm s t_{\pm}^{s-1}, \quad s \neq -1, -2, \ldots, \tag{10.20}$$

we obtain for the generalized eigenfunction ${}^{\pm}\check{\Psi}_{s}^{S} \in \check{\mathbb{S}}^{*}(\mathbb{R}^{\pm}) \subset \check{\mathbf{L}}^{2}(\mathbb{R}^{\pm}, dt)$ of the scale operator S:

$${}^{+}\check{\Psi}_{s}^{S}(\tau) = t_{+}^{2\pi i s - 1/2} = \frac{e^{2\pi i s \ln(\tau)}}{\sqrt{t}} \quad \text{for } t > 0,\ s \in \mathbb{R}, \tag{10.21a}$$

$${}^{-}\check{\Psi}_{s}^{S}(\tau) = t_{-}^{2\pi i s - 1/2} = \frac{e^{2\pi i s \ln(|t|)}}{\sqrt{|t|}} \quad \text{for } t < 0,\ s \in \mathbb{R}. \tag{10.21b}$$

That is, for each value $s \in \mathbb{R}$ there are two generalized eigenfunctions in the Schwartz space $\check{\mathbb{S}}^{*}(\mathbb{R})$ of tempered distributions. The generalized eigenfunction ${}^{+}\check{\Psi}_{s}^{S}$ has its support on $\mathbb{R}^{+}$, while the generalized eigenfunction ${}^{-}\check{\Psi}_{s}^{S}$ has its support on $\mathbb{R}^{-}$.

Using the relations

$$\int_{0}^{\infty} \{\check{\Phi}_{s}^{+}(\tau)\}^{*} \check{\Phi}_{s}^{\prime +}(\tau)^{*} dt = \int_{-\infty}^{\infty} e^{-2\pi i (s-s')x} dx = \delta(s - s'), \tag{10.22a}$$

$$\int_{-\infty}^{0} \{\check{\Phi}_{s}^{-}(\tau)\}^{*} \check{\Phi}_{s}^{-}(t')^{*} dt = \int_{-\infty}^{\infty} e^{-2\pi i (s-s')x} dx = \delta(s - s'), \tag{10.22b}$$

we find that the generalized eigenfunctions (10.21a), (10.21b) are orthogonal and complete. Note that the even and odd linear combinations

$$|t|^{2\pi i s - 1/2} = t_{+}^{2\pi i s - 1/2} + t_{-}^{2\pi i s - 1/2}, \tag{10.23a}$$

$$|t|^{2\pi i s - 1/2}\operatorname{sgn}(\tau) = t_{+}^{2\pi i s - 1/2} - t_{-}^{2\pi i s - 1/2}. \tag{10.23b}$$

are also generalized eigenfunctions of the scale operator S for the eigenvalue s.

10.2.1.2 Unitary Equivalence

The continuous orthonormal systems of generalized eigenfunctions of the time, frequency, and scale operators are unitarily equivalent. With respect to the basis of the time operator T we write

$$\check{\Psi}_{a}^{T} = R^{\Lambda} \hat{\Psi}_{a}^{\Lambda} = R^{S} \widetilde{\Psi}_{a}^{S}, \tag{10.24}$$

where R^{Λ} and R^{S} are the unitary operators which can be represented by the complex-valued kernels R^{Λ} and R^{S} via the integral transform

$$\check{\Psi}_{a}^{T}(\tau) = \int_{\mathbb{R}} R^{\Lambda}(t, \ell) \hat{\Psi}_{a}^{\Lambda}(\ell) d\ell = \int_{\mathbb{R}} R^{S}(t, s) \widetilde{\Psi}_{a}^{S}(s) ds. \tag{10.25}$$

With the generalized eigenfunctions (10.17a), (10.17b) we get

$$R^{\Lambda}(\ell, a) = e^{-2\pi i a \ell}, \tag{10.26}$$

so that

$$\hat{\Psi}_a^{\Lambda}(\ell) = \int_{\mathbb{R}} e^{-2\pi i a \ell} \check{\Psi}_a^{T}(\tau) dt \tag{10.27}$$

generates the Fourier transform F:

$$\mathsf{F}\{\check{\Phi}\}(\ell) = \hat{\Phi}(\ell) := \int_{\mathbb{R}} e^{-2\pi i t \ell} \check{\Phi}(\tau) dt, \quad \ell \in \mathbb{R},\ \check{\Phi} \in \check{\mathbf{L}}^2(\mathbb{R}, d\ell), \tag{10.28}$$

and its inverse

$$\mathsf{F}^{-1}\{\hat{\Phi}\}(\tau) = \check{\Phi}(\tau) := \int_{\mathbb{R}} e^{2\pi i \ell t} \hat{\Phi}(\ell) d\ell, \quad t \in \mathbb{R},\ \hat{\Phi} \in \hat{\mathbf{L}}^2(\mathbb{R}, d\ell). \tag{10.29}$$

With Eqs. (10.17a), (10.21a), (10.21b), the kernel of the unitary operator R^S is given by:

$$R^S(s, t) = \frac{e^{-2\pi i s \ln(|t|)}}{\sqrt{|t|}}, \quad t \in \mathbb{R},\ s > 0, \tag{10.30}$$

which generates a unitary map $\mathbf{L}^2(\mathbb{R}^+) \to \mathbf{L}^2(\mathbb{R})$, called the *scale transform* S:

$$\mathsf{S}\{\check{\Phi}\}(s) = \widetilde{\Phi}(s) = \int_0^{\infty} dt \check{\Phi}(\tau) t_+^{-2\pi i s - 1/2}. \tag{10.31}$$

The inverse scale transform is given by (cf. Sect. 10.2.4)

$$\mathsf{S}^{-1}\{\widetilde{\Phi}\}(\tau) = \check{\Phi}(\tau) = \int_{-\infty}^{\infty} ds \widetilde{\Phi}(s) t_+^{2\pi i s - 1/2}. \tag{10.32}$$

10.2.2 The Time-Domain Representations

The time-domain representation is defined by the complete set of generalized eigenfunctions Ψ_t^T of the time operator T,

$$T\Psi_t^T = t\Psi_t^T, \quad t \in \mathbb{R}. \tag{10.33}$$

On the Schwartz space $\check{\mathbb{S}}(\mathbb{R}) \subset \check{\mathbf{L}}^2(\mathbb{R}, dt)$ of all rapidly decreasing complex-valued infinitely differentiable functions on $\mathbb{R}$ the time operator T is represented as a multiplication operator $\check{T}$:

$$\{\check{T}\check{\Phi}\}(\tau) = t\check{\Phi}(\tau), \quad t \in \mathbb{R},\ \check{\Phi} \in \check{\mathbb{S}}(\mathbb{R}) \subset \check{\mathbf{L}}^2(\mathbb{R}, dt). \tag{10.34}$$

The eigenvalue problem for T is given by

$$\{\check{T}\check{\Psi}_t^T\}(t') = t\check{\Psi}_t^T(t'), \qquad \check{\Psi}_t^T \in \check{\mathbb{S}}(\mathbb{R}) \subset \check{\mathbf{L}}^2(\mathbb{R}, dt'), \tag{10.35a}$$

$$\check{\Psi}_t^T(t') = \delta(t - t'). \tag{10.35b}$$

In the *time-domain representation* on the Lebesgue-Hilbert space $\check{\mathbf{L}}^2(\mathbb{R}, dt)$ of square-integrable complex-valued functions $t \mapsto \check{\Phi}(\tau)$ on the real axis $\mathbb{R}$, the unitary operators $U(\lambda)$, $V(\tau)$ and $D^{\pm}(\kappa)$ are irreducibly represented by

$$\{\check{U}(\lambda)\check{\Phi}\}(\tau) = e^{2\pi i\lambda t}\check{\Phi}(\tau), \quad \check{\Phi} \in \check{\mathbf{L}}^2(\mathbb{R}, dt),\ \lambda \in \mathbb{R}, \tag{10.36a}$$

$$\{\check{V}(\tau)\check{\Phi}\}(\tau) = \check{\Phi}(t - \tau), \quad \check{\Phi} \in \check{\mathbf{L}}^2(\mathbb{R}, dt),\ \tau \in \mathbb{R}, \tag{10.36b}$$

$$\{\check{D}^+(\kappa)\check{\Phi}\}(\tau) = \sqrt{\kappa}\Phi(\kappa t), \quad \check{\Phi} \in \check{\mathbf{L}}^2(\mathbb{R}^+, dt),\ \kappa > 0, \tag{10.36c}$$

$$\{\check{D}^-(\kappa)\check{\Phi}\}(\tau) = \sqrt{\kappa}\Phi(\kappa t), \quad \check{\Phi} \in \check{\mathbf{L}}^2(\mathbb{R}^-, dt),\ \kappa > 0. \tag{10.36d}$$

On the Schwartz spaces $\check{\mathbb{S}}(\mathbb{R}) \subset \check{\mathbf{L}}^2(\mathbb{R}, dt)$ and $\check{\mathbb{S}}(\mathbb{R}^+) \subset \check{\mathbf{L}}^2(\mathbb{R}^+, dt)$ of all rapidly decreasing complex-valued infinitely differentiable functions on $\mathbb{R}$ the corresponding essentially self-adjoint generators, $\check{T}$, $\check{\Lambda}$, $\check{S}^{\pm}$, are represented by

$$\check{U}(\lambda) = e^{2\pi i\lambda\check{T}}, \qquad \{\check{T}\Phi\}(\tau) = t\check{\Phi}(\tau), \tag{10.37a}$$

$$\check{V}(\tau) = e^{-2\pi i\tau\check{\Lambda}}, \qquad \{\check{\Lambda}\check{\Phi}\}(\tau) = \frac{1}{2\pi i}\frac{\partial\check{\Phi}(\tau)}{\partial t}, \tag{10.37b}$$

$$\check{D}^{\pm}(\kappa) = e^{2\pi i\ln(\kappa)\check{S}}, \qquad \{\check{S}\check{\Phi}\}(\tau) = \frac{1}{4\pi i}\left\{\frac{\partial}{\partial t}t + t\frac{\partial}{\partial t}\right\}\check{\Phi}(\tau), \tag{10.37c}$$

for every $\check{\Phi} \in \check{\mathbb{S}}(\mathbb{R}) \subset \check{\mathbf{L}}^2(\mathbb{R}, dt)$. Since $\check{\mathbb{S}}(\mathbb{R})$ is dense in $\check{\mathbf{L}}^2(\mathbb{R}, dt)$, these operators can be extended to unbounded self-adjoint operators on the Hilbert space $\check{\mathbf{L}}^2(\mathbb{R}, dt)$. The operators $\check{S}$, $\check{T}$, and $\check{\Lambda}$ are unitarily equivalent and have the purely continuous spectrum $\mathbb{R}$.

10.2.3 The Frequency-Domain Representations

The frequency-domain representation is defined by the complete set of generalized eigenfunctions Ψ_ℓ^Λ of the frequency operator Λ,

$$\Lambda\Psi_\ell^\Lambda = \ell\Psi_\ell^\Lambda, \quad \ell \in \mathbb{R}. \tag{10.38}$$

On the Schwartz space $\hat{\mathbb{S}}(\mathbb{R}) \subset \hat{\mathbf{L}}^2(\mathbb{R}, d\ell)$ of all rapidly decreasing complex-valued infinitely differentiable functions on $\mathbb{R}$ the frequency operator Λ is represented as a multiplication operator $\hat{\Lambda}$:

$$\{\hat{\Lambda}\hat{\Phi}\}(\ell) = \ell\hat{\Phi}(\ell), \quad \ell \in \mathbb{R},\ \hat{\Phi} \in \hat{\mathbb{S}}(\mathbb{R}) \subset \hat{\mathbf{L}}^2(\mathbb{R}, d\ell). \tag{10.39}$$

The eigenvalue problem for Λ is given by

$$\{\hat{\Lambda}\Psi_\ell^\Lambda\}(\ell') = \ell\hat{\Psi}_\ell^\Lambda(\ell'), \quad \hat{\Psi}_\ell^\Lambda \in \hat{\mathbb{S}}(\mathbb{R}) \subset \hat{\mathbf{L}}^2(\mathbb{R}, d\ell'), \tag{10.40a}$$

$$\hat{\Psi}_\ell^\Lambda(\ell') = \delta(\ell - \ell'). \tag{10.40b}$$

The frequency-domain representation is the unitary Fourier transform of the time-domain representation. The Fourier transformation

$$\mathsf{F} : \check{\mathbf{L}}^2(\mathbb{R}, dt) \to \hat{\mathbf{L}}^2(\mathbb{R}, d\ell), \quad t, \ell \in \mathbb{R}, \tag{10.41}$$

is defined by $\check{\Phi} \mapsto \hat{\Phi} := \mathsf{F}\check{\Phi}$, where

$$\mathsf{F}\{\check{\Phi}\}(\ell) := \hat{\Phi}(\ell) := \int_{\mathbb{R}} e^{-2\pi i \ell t}\check{\Phi}(\tau)dt, \quad t \in \mathbb{R},\ \check{\Phi} \in \check{\mathbf{L}}^2(\mathbb{R}, dt), \tag{10.42a}$$

$$\mathsf{F}^{-1}\{\hat{\Phi}\}(\tau) := \check{\Phi}(\tau) := \int_{\mathbb{R}} e^{2\pi i \ell t}\hat{\Phi}(\ell)d\ell, \quad \ell \in \mathbb{R},\ \hat{\Phi} \in \hat{\mathbf{L}}^2(\mathbb{R}, d\ell). \tag{10.42b}$$

This results in the unitarily equivalent irreducible *frequency-domain representation* on the Lebesgue-Hilbert space $\hat{\mathbf{L}}^2(\mathbb{R}, d\ell)$. In this representation the unitary operators U, V and S are represented by

$$\{\hat{U}(\lambda)\hat{\Phi}\}(\ell) = \hat{\Phi}(\ell - \lambda), \quad \lambda \in \mathbb{R}, \tag{10.43a}$$

$$\{\hat{V}(\tau)\hat{\Phi}\}(\ell) = e^{-2\pi i \lambda \ell}\hat{\Phi}(\ell), \quad \tau \in \mathbb{R}, \tag{10.43b}$$

$$\{\hat{D}(\kappa)\hat{\Phi}\}(\ell) = \hat{\Phi}(\ell/\kappa)/\sqrt{\kappa}, \quad \kappa > 0, \tag{10.43c}$$

for every $\hat{\Phi} \in \hat{\mathbf{L}}^2(\mathbb{R}, d\ell)$.

On the Schwartz space $\hat{\mathbb{S}}(\mathbb{R}) \subset \hat{\mathbf{L}}^2(\mathbb{R}, d\ell)$ of all rapidly decreasing complex-valued infinitely differentiable functions on $\mathbb{R}$ the essentially self-adjoint generators $\hat{S}$, $\hat{T}$ and $\hat{\Lambda}$ are for every function $\ell \mapsto \hat{\Phi}(\ell) \in \hat{\mathbb{S}}(\mathbb{R}) \subset \hat{\mathbf{L}}^2(\mathbb{R}, d\ell)$ represented by

$$\{\hat{T}\hat{\Phi}\}(\ell) = \frac{i}{2\pi}\frac{d\hat{\Phi}(\ell)}{d\ell}, \quad t \in \mathbb{R}, \tag{10.44a}$$

$$\{\hat{\Lambda}\hat{\Phi}\}(\ell) = \ell\hat{\Phi}(\ell), \quad \ell \in \mathbb{R}, \tag{10.44b}$$

$$\{\hat{S}\hat{\Phi}\}(\ell) = \frac{i}{4\pi}\left\{\frac{d}{d\ell}\ell + \ell\frac{d}{d\ell}\right\}\hat{\Phi}(\ell), \quad \ell \in \mathbb{R}. \tag{10.44c}$$

10.2.4 The Scale-Domain Representation

10.2.4.1 The Scale Transform

The scale-domain representation is defined by the complete set of generalized eigenfunctions Ψ_s^S of the scale operator S,

$$S\Psi_s^S = s\Psi_s^S, \quad |s| \in \mathbb{R}. \tag{10.45}$$

On the Schwartz space $\widetilde{\mathbb{S}}(\mathbb{R}^+) \subset \widetilde{\mathbf{L}}^2(\mathbb{R}^+, ds')$ of all rapidly decreasing complex-valued infinitely differentiable functions on $\mathbb{R}^+$ the scale operator S is represented as a multiplication operator $\widetilde{S}$:

$$\{\widetilde{S}\Psi_s^S\}(s') = s\widetilde{\Psi}_s^S(s'), \qquad \widetilde{\Psi}_s^S \in \widetilde{\mathbb{S}}(\mathbb{R}) \subset \widetilde{\mathbf{L}}^2(\mathbb{R}, ds'), \tag{10.46a}$$

$$\widetilde{\Psi}_s^S(s') = \delta(s - s'). \tag{10.46b}$$

Every function $t \mapsto \check{\Phi}(\tau)$ in $\check{\mathbf{L}}^2(\mathbb{R}, dt)$ can be expanded in terms of scale eigenfunctions (10.21a), (10.21b),

$$\check{\Phi}(\tau) = \int_{-\infty}^{\infty} ds\, {}^+\check{\Psi}_s^S(\tau) \int_0^{\infty} dt' \check{\Phi}^+(t') ({}^+\check{\Psi}_s^S)^*(\tau) \check{\Phi}_{s'}^+(t')^*, \tag{10.47a}$$

$$\check{\Phi}(\tau) = \int_{-\infty}^{\infty} ds\, {}^-\check{\Psi}_s^S(\tau) \int_{-\infty}^{0} dt' \check{\Phi}^-(t') ({}^-\check{\Psi}_s^S)^*(\tau) \check{\Phi}_{s'}^-(t')^*. \tag{10.47b}$$

The function

$$s \mapsto \widetilde{\Phi}(s) := \int_0^{\infty} dt \check{\Phi}(\tau) ({}^-\check{\Psi}_s^S)^*(\tau) = \int_0^{\infty} dt \check{\Phi}(\tau) t_+^{-2\pi i s - 1/2} \tag{10.48}$$

is the *scale transform* (10.31) of $\check{\Phi}$ (Cohen 1993). The map $\check{\Phi} \to \widetilde{\Phi}$ is a unitary transformation $R^S : \mathbf{L}^2(\mathbb{R}^+) \to \mathbf{L}^2(\mathbb{R})$ with the kernel $R^S(x, y)$ according to:

$$R^S(x, y) = {}^+\check{\Psi}_y^S(x) = x_+^{2\pi i y - 1/2}. \tag{10.49}$$

The inverse transform is given by:

$$\check{\Phi}(\tau) = \int_{-\infty}^{\infty} ds \widetilde{\Phi}(s) \check{\Phi}_s^+(\tau) = \int_{-\infty}^{\infty} ds \widetilde{\Phi}(s) t_+^{2\pi i s - 1/2}. \tag{10.50}$$

The scale-domain representation is the *unitary scale transform* (10.31) of the time-domain representation, which provides the relation between the scale operator S and the time operator T.

The scale transform is a special version of the Mellin transform M,[184]

$$\widetilde{\Phi}(s) := \{\mathsf{M}\check{\Phi}\}\left(-2\pi i s + \frac{1}{2}\right), \tag{10.51}$$

---

[184]The Mellin transform of a complex-valued locally integrable function f on $[0, +\infty]$ is defined by the integral $\{\mathsf{M}f\}(z) = \int_0^{\infty} t^{z-1} f(\tau) dt$, $z \in \mathbb{C}$, on some strip $a < \mathrm{Re}(z) < b$, but the integral may admit an analytic continuation beyond the convergence domain. If $f(\tau)$ is of bounded variation in the neighborhood of t, then the inversion formula $f(t-0) + f(t+0) = 2\int_{c-i\infty}^{c+i\infty} t^z \{\mathsf{M}f\}(z) dz$ holds. For mathematical details compare Titchmarsh (1937) and Doetsch (1950). The Mellin transform of generalized functions is discussed by Brychkov and Prudnikov (1989, Sect. 3.6).

where

$$\{\mathsf{M}\check{\Phi}\}(z) := \int_0^\infty t^{z-1}\check{\Phi}(\tau)dy, \quad z \in \mathbb{C}. \tag{10.52}$$

10.2.4.2 Scale Invariance

When a signal $\tau \mapsto \check{\Phi}(\tau)$ is scaled in time, the scaled signal

$$\check{\Phi}_\kappa(\tau) := \{\check{D}(\kappa)\}\check{\Phi}(\tau) = \sqrt{\kappa}\check{\Phi}(\kappa t), \quad \Phi \in \check{\mathbf{L}}^2(\mathbb{R}, dx), \ \kappa > 0 \tag{10.53}$$

is only changed in size, while its shape is invariant. In particular, the scale transform $\widetilde{\Phi}_\kappa$ of $\check{\Phi}_\kappa$,

$$\widetilde{\Phi}_\kappa(s) := \int_0^\infty dt\check{\Phi}_\kappa(\tau)t_+^{-2\pi is-1/2} = \kappa^{2\pi is}\widetilde{\Phi}(s), \tag{10.54}$$

is only affected by a phase factor, so that the magnitude of $\widetilde{\Phi}(s)$ is *scale invariant*,

$$\left|\widetilde{\Phi}_\kappa(s)\right| = \left|\widetilde{\Phi}(s)\right| \quad \text{for every } \kappa > 0. \tag{10.55}$$

That is, the scale transformation segregates the *shape information* of the signal from the *size information*. This property makes the scale transform appropriate for the recognition of the shape of signals of arbitrary size (Cohen 1993, 1995).

10.2.4.3 Scaling Transforms of Temporal Observables

The self-adjoint generator S of the one-parameter dilation group fulfills the canonical commutation relations

$$[S, T]_- = (1/2\pi i)T, \tag{10.56a}$$

$$[S, \Lambda]_- = (i/2\pi)\Lambda, \tag{10.56b}$$

so that with $D(\kappa) = e^{2\pi i \ln(\kappa)S}$:

$$D(\kappa)f(T, \Lambda)D(\kappa)^* = f\left(\kappa T, \kappa^{-1}\Lambda\right). \tag{10.57}$$

In particular we obtain

$$D(\kappa)T_+^s D(\kappa) = \kappa^s T_+^s, \tag{10.58a}$$

$$D(\kappa)\Lambda_+^s D(\kappa) = \kappa^{-s}\Lambda_+^s, \tag{10.58b}$$

where the self-adjoint operator T_+^s is defined via the Hilbert-space representation

$$\left\{T_+^s\Phi\right\}(x) = x_+^s\Phi(x), \quad \Phi \in \mathbf{L}^2(\mathbb{R}, dx). \tag{10.59}$$

10.2.4.4 The Invariance of the Hartley Information of a Signal

The number of samples required to represent a signal with approximate bandwidth $\Delta\lambda$ and approximative time duration $\Delta\tau$ is approximately equal to $2(\Delta\lambda)(\Delta\tau)$ (compare Slepian and Pollak 1961; Landau and Pollak 1961, 1962; Slepian 1976, 1983). In 1928 Ralph Vinton Lyon Hartley showed that (Hartley 1928, p. 535)

> the amount of information which can be transmitted is proportional to the product of the width of the frequency range by the time it is available.

If we define a non-probabilistic Hartley information by $H = \log\{2(\Delta\lambda)(\Delta\tau)\}$, then this result implies that the amount of information of a signal is *invariant under scaling transformations*. That is, for every $a > 0$ the signals $t \mapsto s(\tau)$ and $t \mapsto \sqrt{a}s(at)$ have the same Hartley-information capacity: *Hartley information is the conserved quantity associated with scale invariance.*

10.2.5 Temporal Reversal Symmetries

10.2.5.1 Representation of the Time-Reversal Group

In an irreducible representation on a Hilbert space $\mathcal{H}$ the time-reversal map ι can be implemented on the algebra $\mathcal{B}(\mathcal{H})$ by an *anti-linear* and *anti-unitary* operator J (Wigner 1932b, 1959, Chap. 26)

$$\iota(A) = JAJ^{-1}, \quad A \in \mathcal{B}(\mathcal{H}). \tag{10.60}$$

The operator J fulfills

$$J(a\Psi + b\Phi) = a^* J\Psi + b^* J\Phi, \qquad \langle J\Phi | J\Psi\rangle = \langle\Phi|\Psi\rangle^* = \langle\Psi|\Phi\rangle, \tag{10.61}$$

for all $\Psi, \Phi \in \mathcal{H}$ and $a, b \in \mathbb{C}$. The square of J is a unitary operator with the property $J^2 = \pm\mathbb{1}$. The sign of J^2 is a property of the system considered (it is positive for spin-free systems). On the Hilbert space $\check{\mathbf{L}}^2(\mathbb{R}, dt)$ J is represented by the operator $\check{J}$ as

$$\{\check{J}\check{\Phi}\}(\tau) = \check{\Phi}(-t)^*, \quad \check{\Phi} \in \check{\mathbf{L}}^2(\mathbb{R}, dt), \tag{10.62a}$$

$$\check{J}(a\check{\Phi} + b\check{\Psi}) = a^*\check{J}\check{\Phi} + b^*\check{J}\check{\Psi}, \quad a, b \in \mathbb{C},\ \check{\Phi}, \check{\Psi} \in \check{\mathbf{L}}^2(\mathbb{R}, dt), \tag{10.62b}$$

$$\langle \check{J}\check{\Psi} | \check{J}\check{\Phi}\rangle = \langle\check{\Psi}|\check{\Phi}\rangle^* = \langle\check{\Phi}|\check{\Psi}\rangle, \qquad \check{\Phi}, \check{\Psi} \in \check{\mathbf{L}}^2(\mathbb{R}, dt). \tag{10.62c}$$

In an irreducible Hilbert-space representation on the temporal domain $\mathcal{T}$, time reversal can be implemented by the same *antilinear* and *antiunitary* operator J,

$$A \to JAJ^{-1} \quad \text{for all } A \in \mathcal{B}(\mathcal{T}), \tag{10.63}$$

which fulfills, as in (10.61)

$$J(a\Psi + b\Phi) = a^* J\Psi + b^* J\Phi, \qquad \langle J\Phi | J\Psi\rangle = \langle\Phi|\Psi\rangle^* = \langle\Psi|\Phi\rangle, \tag{10.64}$$

for all $\Psi, \Phi \in \mathcal{H}$ and $a, b \in \mathbb{C}$.

10.2.5.2 Representation of the Frequency-Reversal Group

In an irreducible representation on a Hilbert space $\mathcal{H}$ the time-reversal map ι can be implemented on the algebra $\mathcal{B}(\mathcal{H})$ by an *antilinear* and *antiunitary* operator K,

$$\kappa(A) = KAK^{-1}, \quad A \in \mathcal{B}(\mathcal{H}), \tag{10.65}$$

which fulfills

$$K(a\Psi + b\Phi) = a^* K\Psi + b^* K\Phi, \qquad \langle K\Phi | K\Psi\rangle = \langle\Phi|\Psi\rangle^* = \langle\Psi|\Phi\rangle, \tag{10.66}$$

for all $\Psi, \Phi \in \mathcal{H}$ and $a, b \in \mathbb{C}$. The square of K is a unitary operator with the property $K^2 = \pm\mathbb{1}$.

On the Hilbert space $\hat{\mathbf{L}}^2(\mathbb{R}, d\lambda)$ the operator K is defined by:

$$\{K\hat{\Phi}\}(\lambda) = \hat{\Phi}(-\lambda)^*, \quad \hat{\Phi} \in \hat{\mathbf{L}}^2(\mathbb{R}, d\lambda), \tag{10.67a}$$

$$K(a\hat{\Phi} + b\hat{\Psi}) = a^* K\hat{\Phi} + b^* K\hat{\Psi}, \quad a, b \in \mathbb{C}, \ \hat{\Phi}, \hat{\Psi} \in \hat{\mathbf{L}}^2(\mathbb{R}, d\lambda), \tag{10.67b}$$

$$\langle K\hat{\Psi} | K\hat{\Phi}\rangle = \langle\hat{\Psi}|\hat{\Phi}\rangle^* = \langle\hat{\Phi}|\hat{\Psi}\rangle, \quad \hat{\Phi}, \hat{\Psi} \in \hat{\mathbf{L}}^2(\mathbb{R}, d\lambda). \tag{10.67c}$$

The closed orthogonal subspaces $\hat{\mathbf{L}}^2(\mathbb{R}^+, d\lambda)$ and $\hat{\mathbf{L}}^2(\mathbb{R}^-, d\lambda)$ of $\hat{\mathbf{L}}^2(\mathbb{R}, d\lambda)$,

$$\hat{\mathbf{L}}^2\big(\mathbb{R}^+, d\lambda\big) \perp \hat{\mathbf{L}}^2\big(\mathbb{R}^-, d\lambda\big), \quad \hat{\mathbf{L}}^2(\mathbb{R}, d\lambda) = \hat{\mathbf{L}}^2\big(\mathbb{R}^+, d\lambda\big) \oplus \hat{\mathbf{L}}^2\big(\mathbb{R}^-, d\lambda\big). \tag{10.68}$$

are invariant under the action of the group $\{U(\lambda)|\lambda \in \mathbb{R}\}$ so that the frequency reversal swaps the subspaces $\hat{\mathbf{L}}^2(\mathbb{R}^+, d\lambda)$ and $\hat{\mathbf{L}}^2(\mathbb{R}^-, d\lambda)$,

$$K\hat{\mathbf{L}}^2\big(\mathbb{R}^+, d\lambda\big) = \hat{\mathbf{L}}^2\big(\mathbb{R}^-, d\lambda\big), \qquad K\hat{\mathbf{L}}^2\big(\mathbb{R}^-, d\lambda\big) = \hat{\mathbf{L}}^2\big(\mathbb{R}^+, d\lambda\big). \tag{10.69}$$

10.2.6 Stationary Processes on Commutative Groups

10.2.6.1 Stationary Processes on Locally Compact Groups

A commutative group is a group where any two elements commute. We define a stochastic process over a locally compact commutative group $\mathfrak{G}$ as a function over $\mathfrak{G}$ with values in the Hilbert space $\mathcal{H}$ of complex random variables with finite variance, where the inner product is given by their covariance.

A $\mathfrak{G}$-stochastic process $\{X(\mathfrak{g})|\mathfrak{g} \in \mathfrak{G}\}$ is called $\mathfrak{G}$-stationary if (Yaglom 1961; Yazici and Kashhyap 1997; Yazici 1997)

$$X(\mathfrak{g} \circ \mathfrak{h}) \stackrel{d}{=} X(\mathfrak{g}) \quad \text{for all } \mathfrak{g}, \mathfrak{h} \in \mathfrak{G}, \tag{10.70}$$

where $\stackrel{d}{=}$ denotes equality in terms of finite joint probability distributions. A weaker condition is the notion of second-order $\mathfrak{G}$-stationarity. A second-order process $\mathfrak{g} \mapsto X(\mathfrak{g})$ is said to be wide-sense time $\mathfrak{G}$-stationary if the expectation value

$$\mathcal{E}\{X(\mathfrak{g})X(\mathfrak{h})^*\} = C(\mathfrak{g} \circ \mathfrak{h}^{-1}), \quad \mathfrak{g}, \mathfrak{h} \in \mathfrak{G}, \tag{10.71}$$

where C is a positive-definite function on the group $\mathfrak{G}$. For a $\mathfrak{G}$-stationary process we find that $\mathcal{E}\{X(\mathfrak{g})\} = \text{constant}$, so that without loss of generality we can assume that $\mathcal{E}\{X(\mathfrak{g})\} = 0$.

10.2.6.2 Shift-Stationary Time Processes

Ordinary stationary stochastic processes $\tau \mapsto Y(\tau)$ are $\mathfrak{G}$-stationary with respect to the additive commutative time translation group $\mathfrak{V} = \{V(\tau)|\tau \in \mathbb{R}\}$,

$$\mathfrak{g}_1 \circ \mathfrak{g}_2 \Longleftrightarrow \tau_1 + \tau_2, \quad \mathfrak{g}_1, \mathfrak{g}_2 \in \mathfrak{G}, \tau_1, \tau_2 \in \mathbb{R}. \tag{10.72}$$

For $\mathcal{E}\{Y(\tau)\} = 0$, the positive-definite function C on the time translation group $\mathfrak{V}$ is given by the covariance function $(\tau, \tau') \mapsto C(\tau, \tau')$,

$$C(\tau, \tau') = \mathcal{E}\{Y(\tau)Y(\tau')^*\}, \quad \tau, \tau' \in \mathbb{R}. \tag{10.73}$$

10.2.6.3 Scale-Stationary Time Processes

Scale-stationary stochastic processes are stationary with respect to the multiplicative dilation group $\mathfrak{D} := \{D(\alpha)|\alpha > 0\}$,

$$\mathfrak{g}_1 \circ \mathfrak{g}_2 \Longleftrightarrow \alpha_1\alpha_2, \quad \mathfrak{g}_1, \mathfrak{g}_2 \in \mathfrak{G}, \; \alpha_1, \alpha_2 \in \mathbb{R}^+. \tag{10.74}$$

For $\mathcal{E}\{X(\alpha)\} = 0$, the positive-definite function C on the temporal dilation group $\mathfrak{D}$ is given by the scale autocorrelation function,

$$\mathcal{E}\{X(\tau_1)X(\tau_2)^*\} = C(\tau_1/\tau_2), \quad 0 < \tau_1, \tau_2 < \infty. \tag{10.75}$$

10.2.7 Self-Similar Processes

10.2.7.1 Basic Experimental Facts

When tiny particles of microscopic size are suspended in a fluid, they carry out a peculiar haphazard movement that never ceases. This phenomenon was already

Fig. 10.1 Sketch of a process undergoing Brownian motion

known to the inventor of the modern microscope, Antony van Leeuwenhoek (1632–1723). He and the early investigators seem to have assumed that these particles where alive.

Their first serious investigation was due to the Scottish botanist Robert Brown (1773–1858). While attempting to determine the behavior of cytoplasmic granules extracted from pollen dispersed in water on a microscopic slide, he discovered in 1827 what he called "active molecules". He found the grains to be in active chaotic motion, and at first associated the motion with the vitality of the pollen. But he soon found that all small particles in living as well as in non-living matter under the same conditions behaved the same way. He had no explanation for this but refuted experimentally facile explanations like convection currents or temperature effects (Brown 1828, 1829).

For a long time, the true nature and cause of this apparently spontaneous motion could not be clarified but after 1878 most scientists attributed it to thermal molecular motions.[185] Nevertheless, the majority of scientists still did not have a true understanding of Brownian motion. It was tempting to think that the curious zigzag trajectory of a Brownian particle is due to kicks from exceptionally energetic molecules. But this can be excluded since each collision with a molecule has an entirely negligible effect. The correct explanation was given by Carbonelle in 1874 and Delsaulx in 1877, who found that Brownian motion has to be considered as a fluctuation phenomenon, due to the fluctuations in the accumulation of an enormous number of collisions.

The motion of a Brownian particle as seen under the microscope consists apparently of steps in random directions (compare Fig. 10.1). Brown's observations were carried out with "a simple microscope, and indeed with one and the same lense, the focal length of which is about 1/32 of an inch" (Brown 1828). In 1912, the French physicist Jean Baptiste Perrin recorded the position of the particle every 30 seconds and plotted the corresponding random walk by joining the successive positions by straight segments (which have no physical reality whatsoever), but also noted that such diagrams give (Perrin 1913)

[185] For a historical account of the early history of Brownian motion compare Brush (1968).

> only a very meager idea of the extraordinary discontinuity of the actual trajectory. If indeed this particle's motion were marked down 100 times more frequently, each segment would be replaced by a polygon relatively just as complicated as the whole drawing, and so on. It is easy to see that in practice the notion of tangent is meaningless for such curves.

Increasing both the resolution of the microscope and the time resolution, the general structure of the original trajectory is reproduced. Since the intervals of time between successive collisions are far too short to be observed by microscopic methods, it becomes natural to idealize Brownian motion as self-similar. Moreover, Brownian motion is continuous since the particles never jump. The very irregular self-affine character of the Brownian trajectories suggests that they never have a well-defined direction. This led Perrin to think of continuous but nowhere differentiable curves. Foreshadowing fractal geometry, one of the major developments of 20th century mathematics, Perrin (1909) wrote:

> Now, however, an important school, developing with rigor the notion of continuity, has created a new mathematics, within which the old theory of functions is only the study (profound, to be sure) of a group of singular cases. It is curves with derivatives that are now the exceptions; or, if one prefers the geometrical language, curves with no tangents at any point become the rule, while the familiar regular curves become some kind of curiosities, doubtless interesting, but still very special.

The decisive breakthrough in the *physical* understanding of the phenomenon of Brownian motion was due to the theoretical work by Albert Einstein (1906) and, quite independently, Marian von Smoluchowsky (1906). Both Einstein and Smoluchowsky showed that in a first approximation the coordinate $x(\tau)$ of a Brownian particle at time τ defines a sample function of a stochastic process, and especially Smoluchowsky used a random walk picture and considered $\tau \mapsto x(\tau)$ as a Markov process.

10.2.7.2 On the Relevance of 1/f Noise

Many physical devices exhibit noise fluctuations with an approximative spectral density $|f|^{-\gamma}$ with $\gamma \approx 1$. Such fluctuations are usually referred to as $1/f$-noise, where f stands for the frequencies involved in the process considered (see Keshner 1982 for a review). The integral of the power spectral density for $1/f$ noise with the spectral density $|f|^{-\gamma}$ with $1 < \gamma < 2$ is infinite, so that it is a *non-stationary* stochastic process. If one introduces a low-frequency cutoff and a high-frequency cutoff, the modified $1/f$ noise behaves like a stationary stochastic process. Approximate $1/f$ noise was first observed as an excess low-frequency voltage fluctuation in electron tubes which is called *flicker noise*.

The $1/f$ power law often extends over decades of frequencies, down to the lowest frequencies accessible to experiment. No lower limit of the $1/f$ spectrum has ever been found, although measurements down to 10^{-7} Hz have been performed (Dutta and Horn 1981). This implies the existence of correlations over surprisingly long times.

$1/f$-noise is an ubiquitous phenomenon, yet the universal character of $1/f$ behavior is difficult to explain. It has been observed in all kinds of current-carrying matter (e.g. in metals, semiconductors, insulators, liquid electrolytes, biopolymers, etc.), in all solid-state devices, as voltage differences across cell membranes, as time difference between two atomic clocks, in the fluctuations of sunspot numbers, as fluctuations in traffic density on a highway, in economic time series, and for both level and pitch variations in all kinds of music, from Bach to Stockhausen and Ligeti.[186]

10.2.7.3 Long-Range Dependence

A second-order stationary stochastic process $\{X(\tau)|\tau \in \mathbb{R}\}$ of zero mean, covariance $\tau \mapsto \check{C}(\tau)$ and spectral density $\lambda \mapsto \hat{c}(\lambda)$,

$$\check{C}(\tau) = \mathcal{E}\{X(\tau)X(0)^*\} = \int_{-\infty}^{\infty} e^{2\pi i\lambda\tau}\hat{c}(\lambda)d\lambda, \tag{10.76}$$

is said to have *short-range dependence* if

$$\int_0^{\infty} \frac{\check{C}(\tau)}{\check{C}(0)} d\tau < \infty \quad \text{or equivalently if } \hat{c}(0) < \infty. \tag{10.77}$$

If the probability distribution of a stochastic process is heavy-tailed (that is, if its tails are not exponentially bounded), correlations are an improper measure of the length of memory. Nevertheless the formulation (10.77) can be generalized for non-stationary processes (Cox 1991; Flandrin et al. 2003): A second-order non-stationary stochastic process $\{X(\tau)|\tau \in \mathbb{R}\}$ with the covariance function $C(\tau, \tau') = \mathcal{E}\{x(\tau)x(\tau')^*\} - \mathcal{E}\{x(\tau)\}\mathcal{E}\{x(\tau')^*\}$ is said to have long-range-dependence if

$$\int_0^{\infty} \frac{\check{C}(\tau, \tau')}{\sqrt{\check{C}(\tau, \tau)\check{C}(\tau', \tau')}} d\tau' = \infty \quad \text{for all fixed } \tau \in \mathbb{R}. \tag{10.78}$$

Particular examples of stochastic processes with long-range dependence are processes with an autocovariance that decays slower than exponentially, for example if the autocovariance function is power-like so that the spectral density is $|f|^{-\gamma}$ with $0 < \gamma < 1$ for $f \to \infty$.

10.2.7.4 Self-Similarity and Fractals in Music

According to Manfred Schroeder the unifying concept underlying fractals, chaos and power laws is self-similarity (Schroeder 1991, p. xiii):

[186] György Ligeti wrote his cycle of 18 études for solo piano after studying fractal geometry. Referring to his piano concerto he explained: "The whole structure of the piece is self-similar and the impression we get is that of a giant interconnected web" (Ligeti 2011, p. 103).

> Self-similarity, or invariance against changes in scale or size, is an attribute of many laws of nature and innumerable phenomena in the world around us. Self-similarity is, in fact, one of the decisive symmetries that shapes our universe and our efforts to comprehend it.

Self-similarity is present in a wide variety of musical aspects which are cognitively relevant (Voss and Clarke 1978a, 1978b; Madden 2007; Pareyon 2011). Voss and Clarke analyzed selected musical compositions and found that their spectral density behaves like $1/f$ noise in a bandpass between 0.1 and 10 kHz. They also produced sounds using white noise, $1/f$ noise, and $1/f^2$ noise and played them to several hundreds of listeners. In comparison to music, white noise was classified as "too random" and $1/f^2$ noise as "too correlated", while noise with an approximate $1/f$ spectral density was judged as closest to what listeners expected of music.

Self-similarity and fractal structures of music can be found in both the frequency and the time ("rhythm") domain. Algorithmic systems have been developed to analyze music and to compose music using fractal methods (Mandelbrot 1982, 1999; Mazzola 2002; Madden 2007).

Fluctuations with $\gamma = 0$ characterize white noise with $\mathcal{E}\{n(\tau)\} \sim \delta(\tau)$, corresponding to processes with independently distributed events. Power spectra with values of γ in the range $-1 < \gamma < 0$ can arise from stationary Gaussian processes. For $\gamma \leq -1$ the spectrum is not integrable so that it cannot arise from a second-order stationary stochastic process. Yet such low-frequency behavior has been observed in many experimental studies. The best known example is *Brownian motion*, a Gaussian process with independently distributed increments.

10.2.7.5 The Hurst Parameter

An object is said to be self-similar if it resembles a part of itself. More precisely, self-similarity means to be *similar* at different scales, to be scale-invariant means to have the same properties at different scales. Self-similar processes are stochastic processes whose distributions are invariant under a suitable scaling of time.[187]

Self-similarity for a stochastic process means *scale invariance of the finite-dimensional distributions*, but does not imply scale invariance for the trajectories. That is, in contrast to deterministic self-similarity, the self-similarity of *stochastic* processes does not imply that the same trajectories repeat themselves by scaling. Furthermore, no characteristic time scale can be identified for a self-similar process.

More precisely, a stochastic process $\{X(\tau)|\tau \in \mathbf{T}\}$ (where $\mathbf{T}$ is either $\mathbb{R}$, or $\{\tau|\tau \geq 0\}$ or $\{\tau|\tau < 0\}$) is said to be *self-similar* (in the statistical sense) if for any $\alpha > 0$ there exists a $\beta > 0$ such that

$$\{X(\alpha\tau)|\tau \geq 0\} \stackrel{d}{=} \{\beta X(\tau)|\tau \in \mathbf{T}\}, \tag{10.79}$$

[187] The first rigorous probabilistic treatment of self-similar processes is due to Lamperti (1962) who used the term "semi-stable" instead of "self-similar". The term "self-similar" is due to Mandelbrot (1982).

where $\stackrel{d}{=}$ denotes equality for all finite-dimensional distributions. For any non-trivial self-similar process $\{X(\tau)|\tau \geq 0\}$, which is stochastically continuous at $\tau = 0$ (i.e. if $\lim_{h\to 0} = \mu\{|X(h) - X(0)| > \varepsilon\} = 0$ for every $\varepsilon > 0$), there exists a unique exponent $H \geq 0$ such that[188]

$$\{X(\alpha\tau)\big|\tau \geq 0\} \stackrel{d}{=} \{\alpha^H X(\tau)\big|\tau \in \mathbf{T}\} \quad \text{for all } \alpha > 0. \tag{10.80}$$

That is, in this case *self-similarity implies fractality*. The fractal scaling index H is called the *Hurst parameter* (Hurst 1951). The value of H provides information about correlations at different time scales.

Whenever $0 < H < 1/2$, increases in the values of the process are likely to be followed by decreases, and decreases are likely to be followed by increases. Such processes are called *anti-persistent*. For $H = 1/2$ the process is a *Wiener process* (a mathematical model for Brownian motion). It has no memory, changes in the values are uncorrelated with each other and all its correlation coefficients are equal to zero. For $H > 1/2$, a stronger trend exists in time series and long-range dependencies occur, which means that the correlation coefficients tend to infinity. For $1/2 < H < 1$, increases in the values of the process are likely to be followed by increases, and decreases are likely to be followed by decreases. Such processes have *long-term correlations* and are called *persistent*.

All non-trivial self-similar processes with stationary increments have $H > 0$. Moreover, $H > 0$ if and only if $X(0) = 0$ almost surely. If $H = 0$ and X is measurable, then $X(\tau) = X(0)$ for each $\tau \in \mathbb{R}$ with probability 1. Self-similarity in the sense of $\{X(a\tau)|\tau \in \mathbf{T}\} \stackrel{d}{=} \{a^H X(\tau)|\tau \in \mathbf{T}\}$ is a possibility which does not arise for processes with stationary increments. If $H < 0$, then $X(\tau) = 0$ with probability 1 for each real τ (compare Vervaat 1985, p. 4).

10.2.7.6 Relation Between Self-Similar and Stationary Processes

Non-degenerate self-similar processes cannot be stationary but there is a logarithmic correspondence between the additive time-translation group and the multiplicative time-scaling group. This implies that there is a strict correspondence between shift-stationary and scale-stationary stochastic processes. The bijective invertible map between self-similar and stationary processes can be realized by a nonlinear time change, called the Lamperti transformation (Lamperti 1962). This transformation is an important tool which allows us to transfer well-known results about stationary processes to corresponding theorems about self-similar processes.

Lamperti Transformation
If $\{X(\tau)|\tau \geq 0\}$ is a self-similar process with $H > 0$, then

$$Y(\tau) := e^{-\tau H} X(e^{\tau}), \quad \tau \in \mathbb{R}$$

is a strictly stationary process.

---

[188] See Lamperti (1962). The concept has been generalized by Maejima and Sato (1999). They called processes with the property $\{X(\alpha t)|t \geq 0\} \stackrel{d}{=} \{\alpha^H X(\tau) + ct|t \geq 0\}$ *semi-self-similar*.

Conversely, if $\{Y(\tau)|\tau \in \mathbb{R}\}$ is a strictly stationary process, then

$$X(\tau) := \tau^H Y(\ln \tau), \quad \tau > 0,\ X(0) = 0,\ H > 0$$

is a self-similar process with $H > 0$.

The Lamperti transformation establishes a relation between the statistical properties of stationary and self-similar processes. If $\{Y(\tau)|\tau \in \mathbb{R}\}$ is a stationary process with finite second moments and the covariance function

$$C_Y(\tau, \tau') := \mathcal{E}\{Y(\tau)Y(\tau')^*\} - \mathcal{E}\{Y(\tau)\}\mathcal{E}\{Y(\tau')^*\}, \tag{10.81}$$

then the covariance function

$$C_X(\tau, \tau') := \mathcal{E}\{X(\tau)X(\tau')^*\} - \mathcal{E}\{X(\tau)\}\mathcal{E}\{X(\tau')^*\} \tag{10.82}$$

of the self-similar process $X(\tau) = t^H Y(\ln t)$, $H > 0$, has the form

$$C_X(\tau, \tau') = (\tau\tau')^H f_H(\tau/\tau'), \quad \tau, \tau' > 0, \tag{10.83}$$

where f_H is a non-negative definite function. Or explicitly,

$$C_X(\tau, \tau') = e^{H[\ln \tau + \ln(\tau']} C_Y\{\ln(\tau) - \ln(\tau')\}. \tag{10.84}$$

In the following, we indicate three examples.

Ordinary Ornstein-Uhlenbeck Process
Let $\{X(\tau)|\tau \geq 0\}$ be the standard Wiener process. It is a non-stationary zero-mean self-similar Gaussian process with $H = \frac{1}{2}$. Then $\{Y(\tau)|\tau \in \mathbb{R}\}$ with $Y(\tau) := e^{-\tau/2}X(e^\tau)$ and $\tau \in \mathbb{R}$ is a stationary Gaussian Ornstein-Uhlenbeck process with covariance $\mathcal{E}\{X(\tau)X(s)\} = e^{|\tau - s|/2}$ (cf. Uhlenbeck and Ornstein 1930).

Stable Ornstein-Uhlenbeck Process
The Lamperti transformation of a stable Lévy process $\{X_\alpha(\tau)|\tau \geq 0\}$ leads to a generalization of the Ornstein-Uhlenbeck process, called the stable stationary Ornstein-Uhlenbeck process $Y(\tau) := e^{-\tau/\alpha}X_\alpha(e^\tau)$.

Fractional Brownian Motion
The inverse of the Lamperti transformation of fractional Brownian motion with the covariance

$$\mathcal{E}\{B_H(\tau)B_H(\tau')\} = \tfrac{1}{2}\sigma^2\{|\tau|^{2H} + |\tau'|^{2H} - |\tau - \tau'|^{2H}\} \tag{10.85}$$

results in a stationary process $\tau \mapsto Y(\tau)$ with the stationary covariance (Borgnat et al. 2002)

$$\mathcal{E}\{Y(\tau)Y(\tau + s)\} = \sigma^2\{\cosh(H|s|) - 2^{2H-1}\sinh(|s|/2)\}. \tag{10.86}$$

10.2.7.7 Fractional Brownian Motion

The generic example for self-similar processes is the Wiener process $\tau \mapsto w(\tau)$, often also called Brownian motion. The Wiener motion $\tau \mapsto w(\tau)$ is a process with independent and stationary Gaussian increments. Setting $w(0) = 0$, the Wiener processes $t \mapsto w(\tau)$ and $\sqrt{\kappa}\tau \mapsto w(\kappa\tau)$ are identical in distribution for every $\kappa > 0$,

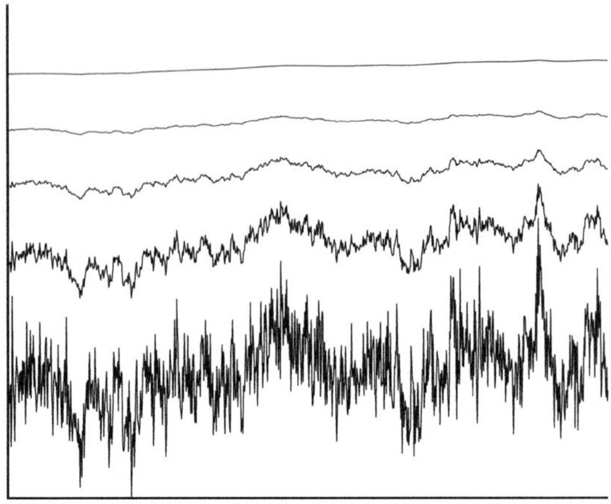

Fig. 10.2 Examples of fractional Brownian motion with Hurst parameters $H = 0.9, 0.7, 0.5, 0.3, 0.1$ from *top* to *bottom* (adapted from Sellerio et al. 2012)

so that $\tau \mapsto w(\tau)$ and $\sqrt{\kappa\tau} \mapsto w(\kappa\tau)$ are statistically self-affine (Mandelbrot 1967, p. 350).

The more general fractional Brownian motions $\tau \mapsto B_H t$ are Gaussian self-similar processes with stationary increments and belong to the class of *long-memory stochastic processes*.[189] Mandelbrot (1967) and Mandelbrot and van Ness (1968) introduced the concept of fractional Brownian motion $t \mapsto B_H(\tau)$ with Hurst parameter $H = (1 + \gamma)/2$ as the Weyl fractional integro-differential of the ordinary Brownian motion process $t \mapsto w(\tau)$ of Wiener (see Fig. 10.2)

For values $H \in]0, 1/2[$ the time series exhibit anti-persistent behavior, determined by *negatively correlated increments*. For $H = 1/2$, the correlation coefficients are equal to zero and the process is homogeneous and strictly scale invariant. For Hurst exponents larger than $1/2$, time series have a strong trend and long-range dependencies, so that the correlation coefficients tend to infinity.

Fractional Brownian motion with Hurst parameter $H \in (0, 1]$ is a non-stationary, almost surely continuous real-valued Gaussian process $\{B_H(\tau) | \tau \in \mathbb{R}\}$ with $\mathcal{E}\{B_H(\tau)\} = 0$, $\mathcal{E}\{B_H(1)^2\} < \infty$, and covariance

$$\mathcal{E}\big\{B_H(\tau)B_H(\tau')\big\} = \frac{1}{2}\sigma^2\big\{|\tau|^{2H} + |\tau'|^{2H} - |\tau - \tau'|^{2H}\big\}, \tag{10.87a}$$

$$\text{where } \sigma^2 = \mathcal{E}\big\{B_H(1)^2\big\}. \tag{10.87b}$$

With a proper regularization at $\tau = -\infty$, the fractional Brownian motion $\{B_H(\tau) | \tau \geq 0\}$ has a representation as a derivative of fractional order $(H - 1/2)$ of the

[189] They were first discussed by Kolmogorov (1940a, 1940b), who called these processes "Wiener spiral" and used similarity hypotheses for modeling the scaling properties of fully developed turbulence (Kolmogorov 1941c). The current term "fractional Brownian motion" was introduced by Mandelbrot and van Ness (1968) who gave a representation of fractional Brownian motion as an integral over standard Brownian motion.

Brownian motion $\tau \mapsto w(\tau)$,

$$B_H(\tau) = \mathcal{E}\{B_H(1)^2\} \int_{-\infty}^{0} \{(\tau - \tau')^{H-\frac{1}{2}} - (-\tau')^{H-\frac{1}{2}}\} dw(\tau')$$
$$+ \mathcal{E}\{B_H(1)^2\} \int_{0}^{\tau} (\tau - \tau')^{H-\frac{1}{2}} dw(\tau'). \tag{10.88}$$

Since the kernel of this integral is of power-law form, it has no intrinsic time scale, so that we find the scaling property

$$B_H(\kappa\tau) - B_H(0) = \kappa^H \{B_H(\kappa\tau) - B_H(0)\} \tag{10.89}$$

for any scale factor $\kappa > 0$.

Since fractional Brownian motion is a non-stationary stochastic process, one cannot define its power spectrum. But fractional Brownian motion has stationary increments. The variance of the increments of the fractional Brownian motion $t \mapsto B_H(\tau)$ is given by the power law

$$\mathcal{E}\{[B_H(\tau + \tau') - B_H(\tau)]^2\} = \sigma^2 |\tau|^{2H}. \tag{10.90}$$

Both ordinary and fractional Brownian motion have infinitely long correlations and variances that diverge with time. *Fractional Gaussian noise* is defined as the generalized derivative of fractional Brownian motion.

10.3 Representations of the Weyl Group

10.3.1 The Irreducible Representations

In the irreducible *time-domain representation* of the Weyl-Heisenberg algebra on the Lebesgue-Hilbert space $\check{\mathbf{L}}^2(\mathbb{R}, dt)$ of square-integrable complex-valued functions $x \mapsto \check{\Phi}(x)$ on the real axis $\mathbb{R}$, the unitary operators $U(\lambda)$ and $V(\tau)$ are represented by the multiplication operator $\check{U}(\lambda)$ and the shift operator $\check{V}(\tau)$, so that $(t, \tau, \lambda \in \mathbb{R})$

$$\{\check{U}(\lambda)\check{\Phi}\}(\tau) = e^{2\pi i\lambda t}\check{\Phi}(\tau), \quad \check{\Phi} \in \check{\mathbf{L}}^2(\mathbb{R}, dt), \tag{10.91a}$$

$$\{\check{V}(\tau)\check{\Phi}\}(\tau) = \check{\Phi}(t - \tau), \quad \check{\Phi} \in \check{\mathbf{L}}^2(\mathbb{R}, dt), \tag{10.91b}$$

$$\{\check{W}(\tau, \lambda)\check{\Phi}\}(\tau) = e^{-\pi i\lambda\tau} e^{+2\pi i\lambda t}\check{\Phi}(t - \tau), \quad \check{\Phi} \in \check{\mathbf{L}}^2(\mathbb{R}, dt). \tag{10.91c}$$

On the Schwartz space $\check{\mathcal{S}}(\mathbb{R}) \subset \check{\mathbf{L}}^2(\mathbb{R}, dt)$ of all rapidly decreasing complex-valued infinitely differentiable functions on $\mathbb{R}$ their essentially self-adjoint generators $\check{S}$, $\check{T}$ and $\check{\Lambda}$ are represented by

$$\check{U}(\lambda) = e^{2\pi i\lambda\check{T}}, \qquad \{\check{T}\check{\Phi}\}(\tau) = t\check{\Phi}(\tau), \tag{10.92a}$$

$$\check{V}(\tau) = e^{-2\pi i \tau \check{\Lambda}}, \qquad \{\check{\Lambda}\check{\Phi}\}(\tau) = \frac{1}{2\pi i}\frac{\partial \check{\Phi}(\tau)}{\partial t}, \tag{10.92b}$$

for $t \in \mathbb{R}$ and every $\check{\Phi} \in \check{\mathcal{S}}(\mathbb{R}) \subset \check{\mathbf{L}}^2(\mathbb{R}, dt)$. Since $\check{\mathcal{S}}(\mathbb{R})$ is dense in $\check{\mathbf{L}}^2(\mathbb{R}, dt)$, these operators can be extended to unbounded self-adjoint operators on the Hilbert space $\check{\mathbf{L}}^2(\mathbb{R}, dt)$. Both operators $\check{T}$ and $\check{\Lambda}$ are unitarily equivalent and have the purely continuous spectrum $\mathbb{R}$.

10.3.2 The Reducible Phase-Space Representation

The phase-space representation of the temporal Weyl commutation relations acts on the Hilbert space $\mathbf{L}^2(\mathbb{R} \times \mathbb{R})$ of Lebesgue square-integrable functions on the phase space $\mathbb{R} \times \mathbb{R}$. The phase-space representations $\overline{U}$, $\overline{V}$, $\overline{W}$ of the Weyl operators U, V, W are for every $\Phi \in \mathbf{L}^2(\mathbb{R} \times \mathbb{R})$ defined by

$$\big\{\overline{W}(\tau, \lambda)\Phi\big\}(t, \ell) = e^{i(\lambda|t)/\hbar - (\tau|\ell)/\hbar}\Phi\left(t - \frac{1}{2}\tau, \ell - \frac{1}{2}\lambda\right), \tag{10.93a}$$

$$\big\{\overline{U}(\lambda)\Phi\big\}(t, \ell) = e^{i(\lambda|t)/\hbar}\Phi\left(t, \ell - \frac{1}{2}\lambda\right), \tag{10.93b}$$

$$\big\{\overline{V}(\tau)\Phi\big\}(t, \ell) = e^{i(\tau|\ell)/\hbar}\Phi\left(t - \frac{1}{2}\tau, \ell\right). \tag{10.93c}$$

The unbounded self-adjoint generators $\overline{T}$ and $\overline{\Lambda}$ of the unitary groups $\lambda \mapsto \overline{U}(\lambda) = e^{2\pi i \lambda \overline{T}}$ and $\tau \mapsto \overline{V}(\tau) = e^{-2\pi i \tau \overline{\Lambda}}$ can be represented on the Schwartz space $\mathcal{S}(\mathbb{R}^{2N})$ of all rapidly decreasing complex-valued infinitely differentiable functions on the phase space $\mathbb{R}^2$ as

$$\{\overline{T}\Phi\}(t, \ell) = (t + \pi i \partial/\partial\ell)\Phi(t, \ell),$$

$$\{\overline{\Lambda}\Phi\}(t, \ell) = (\ell - \pi i \partial/\partial t)\Phi(t, \ell),$$

for every $\Phi \in \mathcal{S}(\mathbb{R}) \subset \mathbf{L}^2(\mathbb{R} \times \mathbb{R})$. The W*-algebra $\overline{\mathcal{T}}$ generated by the Weyl operators $\{\overline{U}(\lambda) | \lambda \in \mathbb{R}\}$ and $\{\overline{V}(\tau) | \tau \in \mathbb{R}^N\}$ is a factor of type I_∞. It is a very small subalgebra of the W*-algebra $\mathcal{B}(\mathbf{L}^2(\mathbb{R} \times \mathbb{R}))$.

On the Hilbert space $\mathbf{L}^2(\mathbb{R} \times \mathbb{R})$ we can define another system of Weyl operators $\overline{\overline{W}}(\tau, \lambda)$, $(\tau, \lambda) \in \mathbb{R} \times \mathbb{R}$, called the *opposite* Weyl system. Its infinitesimal generators $\overline{\overline{\Lambda}}$ and $\overline{\overline{T}}$ are represented as

$$\{\overline{\overline{T}}\Phi\}(t, \ell) = (t - \pi i \partial/\partial\ell)\Phi(t, \ell),$$

$$\{\overline{\overline{\Lambda}}\Phi\}(t, \ell) = (\ell + \pi i \partial/\partial t)\Phi(t, \ell),$$

for every $\Phi \in \mathcal{S}(\mathbb{R}^2)$. The W*-algebra $\overline{\overline{\mathcal{T}}}$ generated by the Weyl operators $\overline{\overline{U}}(\lambda) = e^{2\pi i\lambda\overline{\overline{T}}}$ and $\overline{\overline{V}}(\tau) = e^{-2\pi i\tau\overline{\overline{\Lambda}}}$ is again a factor of type I_∞ in the W*-algebra $\mathcal{B}(\mathbf{L}^2(\mathbb{R}\times\mathbb{R}))$.

On the Schwartz space $\mathcal{S}(\mathbb{R}^2)$ the self-adjoint generators fulfill the following commutation relations:

$$[T,\Lambda]_- = 2\pi i\mathbb{1}, \qquad [\overline{T},\overline{\Lambda}]_- = -2\pi i\mathbb{1},$$

$$[T,\overline{T}]_- = [T,\overline{\Lambda}]_- = [\Lambda,\overline{\Lambda}]_- = [\Lambda,\overline{T}]_- = \mathbb{O}.$$

These relations imply that the W*-algebra $\overline{\overline{\mathcal{T}}} \subset \mathcal{B}(\mathbf{L}^2(\mathbb{R}\times\mathbb{R}))$ is the commutant of the W*-algebra $\overline{\mathcal{T}} \subset \mathcal{B}(\mathbf{L}^2(\mathbb{R}\times\mathbb{R}))$,

$$\overline{\overline{\mathcal{T}}} = \left\{\overline{\overline{M}} \in \mathcal{B}\big(\mathbf{L}^2(\mathbb{R}\times\mathbb{R})\big) \,\big|\, \overline{\overline{M}}\,\overline{M} = \overline{M}\,\overline{\overline{M}},\ \overline{M} \in \overline{\mathcal{T}}\right\},$$

$$\overline{\mathcal{T}} = \left\{\overline{M} \in \mathcal{B}\big(\mathbf{L}^2(\mathbb{R}\times\mathbb{R})\big) \,\big|\, \overline{M}\,\overline{\overline{M}} = \overline{\overline{M}}\,\overline{M},\ \overline{\overline{M}} \in \overline{\overline{\mathcal{T}}}\right\}.$$

Moreover, the following relations hold:

$$\overline{\mathcal{T}} \vee \overline{\overline{\mathcal{T}}} = \mathcal{B}\big(\mathbf{L}^2(\mathbb{R}\times\mathbb{R})\big), \qquad \overline{\mathcal{T}} \wedge \overline{\overline{\mathcal{T}}} = \mathbb{1}\mathbb{C},$$

where $\overline{\mathcal{T}} \vee \overline{\overline{\mathcal{T}}}$ is the smallest W*-subalgebra of $B(\mathbf{L}^2(\mathbb{R}\times\mathbb{R}))$ which contains both $\overline{\mathcal{T}}$ and $\overline{\overline{\mathcal{T}}}$, while $\overline{\mathcal{T}} \wedge \overline{\overline{\mathcal{T}}}$ is the largest W*-subalgebra of $B(\mathbf{L}^2(\mathbb{R}\times\mathbb{R}))$ which is contained in all $\overline{\mathcal{T}}$ and $\overline{\overline{\mathcal{T}}}$. These relations imply that the phase-space representation is highly reducible.

10.3.3 Canonical Phase-Space Representation

In the phase-space representation a linear isometric involution $J : \mathbf{L}^2(\mathbb{R}\times\mathbb{R}) \to \mathbf{L}^2(\mathbb{R}\times\mathbb{R})$ can be defined by the complex conjugation,

$$\{J\Phi\}(t,\ell) = \Phi(t,\ell)^*, \quad \Phi \in \mathbf{L}^2(\mathbb{R}\times\mathbb{R}). \tag{10.94}$$

The following relations hold:

$$J\overline{W}(\tau,\lambda)J = \overline{\overline{W}}(\tau,\lambda)(\lambda,q), \qquad J\overline{\overline{W}}(\tau,\lambda)(\tau,\lambda)J = \overline{W}(\tau,\lambda), \tag{10.95a}$$

$$J\overline{\mathcal{T}}J = \overline{\mathcal{T}}^{\mathrm{P}}, \qquad J\overline{\mathcal{T}}^{\mathrm{P}}J = \overline{\mathcal{T}}. \tag{10.95b}$$

Every pure or non-pure state functional $\rho \in \mathcal{T}_*$ on the abstract W*-algebra $\mathcal{T}$ can be represented by a unique *vector* Ξ_ρ with $J\Xi_\rho = \Xi_\rho$ in the cone $\mathcal{C}$ of the canonical phase-space representation $(\mathcal{T}, \mathbf{L}^2(\mathbb{R}\times\mathbb{R}), \mathcal{C}, J)$,

$$\rho(M) = \langle \Xi_\rho | M\Xi_\rho \rangle, \quad M \in \mathcal{T},\ \Xi_\rho \in \mathcal{C} \subset \mathbf{L}^2(\mathbb{R}\times\mathbb{R}). \tag{10.96}$$

The self-dual cone $\mathcal{C}$ of the canonical phase-space representation can be obtained conveniently from the self-dual cone of the unitarily equivalent Hilbert-Schmidt representation via a Weyl transform. If in the irreducible representation the state functional $\rho \in \mathcal{T}_*$ is realized by a density operator D_ρ^{irr}, then the vector $\Xi_\rho \in \mathcal{C}$ is given by the following Fourier-Weyl transformation:

$$\Xi_\rho(t,\ell) = \int_{\mathbb{R}^2} d\ell dt e^{is(\ell,t|\ell',t')} \text{tr}\Big\{ W^{\text{irr}}(\ell',t')^* \sqrt{D_\rho^{\text{irr}}} \Big\}. \tag{10.97}$$

Here $W^{\text{irr}}(\ell,t)$ is the irreducible representation of the abstract Weyl operator $W(\ell,t)$. If ρ is a *pure* state functional, then D_ρ^{irr} is a projection, $D_\rho^{\text{irr}} = (D_\rho^{\text{irr}})^2$, so that $\text{tr}\{W^{\text{irr}}(\tau,\lambda)^*(D_\rho^{\text{irr}})^{1/2}\} = \text{tr}\{W^{\text{irr}}(\ell,t)^*(D_\rho^{\text{irr}})\}$ is simply the state-generating function $\hat{w}_\rho(\ell,t) := \rho\{W(\tau,\lambda)\}$. The symplectic Fourier transform of a state-generating function $\hat{w}_\rho$ is the Wigner function w_ρ of the state functional ρ,

$$\check{w}_\rho(\ell,t) := \int_{\mathbb{R}^2} d\ell dt e^{2\pi i s(\ell,t|\ell',t')} \hat{w}\big(\ell',t'\big). \tag{10.98}$$

Accordingly, a state functional $\rho \in \mathcal{T}_*$ is pure if and only if its Wigner function $w_\rho : \mathbb{R}^2 \to \mathbb{R}$ equals the implementing vector $\Xi_\rho \in \mathcal{C}$ of the canonical phase-space representation:

$$\rho \text{ is pure if and only if } \Xi_\rho = w_\rho. \tag{10.99}$$

10.4 Representations of Temporal States

10.4.1 Characteristic Functions

We consider a Boolean temporal system (9.70a), (9.70b) characterized by the (classical) temporal Weyl operators $\{W^{\text{cl}}(\lambda,\tau)|\lambda,\tau \in \mathbb{R}\}$,

$$W^{\text{cl}}(\lambda,\tau) := U^{\text{cl}}(\lambda)V^{\text{cl}}(\tau) = e^{2\pi i\lambda T^{\text{cl}} - 2\pi i\tau \Lambda^{\text{cl}}}, \tag{10.100}$$

over the phase space $\mathbb{R}^2$ with the symplectic form (9.25)

$$\sigma\big(\lambda,\tau\big|\lambda',\tau'\big) = \lambda\tau' - \tau\lambda', \quad \lambda,\lambda',\tau,\tau' \in \mathbb{R}. \tag{10.101}$$

For Boolean time concepts the temporal W*-algebra $\mathcal{T}^{\text{cl}}$ can be taken as the Lebesgue space $\mathbf{L}^\infty(\mathbb{R}\times\mathbb{R})$ of all complex-valued essentially bounded Lebesgue-measurable functions on $\mathbb{R}\times\mathbb{R}$,

$$\mathcal{T}^{\text{cl}} = \mathbf{L}^\infty(\mathfrak{V}\times\mathfrak{U}) \simeq \mathbf{L}^\infty\big(\mathbb{R}^2\big). \tag{10.102}$$

The epistemic description of a Boolean time is given by a *probability measure* μ on the phase space $\mathbb{R}\times\mathbb{R}$. An equivalent, but more convenient description is given by the *characteristic function* $\hat{f}$,

$$\hat{f}(\lambda,\tau) := \mathcal{E}\{W^{\text{cl}}(\lambda,\tau)\} = \int_{\mathbb{R}^2} W^{\text{cl}}(\lambda,\tau)\mu(d\lambda d\tau). \tag{10.103}$$

The symplectic Fourier-Stieltjes transform of the characteristic function is the probability density $\check{f}$ corresponding to the absolutely continuous probability measure μ,

$$\check{f}(\lambda,\tau) = \mu(d\lambda d\tau)/d\lambda d\tau, \tag{10.104}$$

$$\check{f}(\lambda,\tau) := \int_{\mathbb{R}^2} e^{2\pi i s(\lambda,\tau|\lambda',\tau')}\hat{f}(\lambda',\tau')d\lambda' d\tau'. \tag{10.105}$$

Since there are no normal pure state functionals on the non-atomic W*-algebra $\mathbf{L}^\infty(\mathbb{R}^2)$, the variances

$$\sigma_\Lambda^2 := \mathcal{E}\{(\Lambda^{\text{cl}})^2\} - \mathcal{E}\{(\Lambda^{\text{cl}})\}^2 \quad \text{and} \tag{10.106a}$$

$$\sigma_T^2 := \mathcal{E}\{(T^{\text{cl}})^2\} - \mathcal{E}\{(T^{\text{cl}})\}^2 \tag{10.106b}$$

cannot vanish, $\sigma_\Lambda^2 > 0$, $\sigma_T^2 > 0$, but the product $\sigma_\Lambda \sigma_T$ can be arbitrarily small,

$$0 < \sigma_\Lambda \sigma_T < \infty. \tag{10.107}$$

10.4.2 State-Generating Functions

An elementary non-Boolean temporal system is characterized by a factor $\mathcal{T}$ of type I_∞ and a non-commutative temporal Weyl system over the phase space $\mathbb{R}^2$ with the symplectic form (9.25),

$$\sigma(\lambda,\tau|\lambda',\tau') = \lambda\tau' - \tau\lambda', \quad \lambda,\lambda',\tau,\tau' \in \mathbb{R}. \tag{10.108}$$

The non-Boolean analog to the characteristic function (10.103) is called the *state-generating function* $\hat{w}$,

$$\hat{w}(\lambda,\tau) := \mathcal{E}\{e^{2\pi i\lambda T - 2\pi i\tau\Lambda}\} = \rho\{W(\lambda,\tau)\}, \tag{10.109}$$

where now $\mathcal{E}$ represents the quantum-mechanical expectation value with respect to the normal state functional $\rho \in \mathcal{T}_*$, and Λ and T are the infinitesimal generators of the temporal Weyl system $\{W(\lambda,\tau)|(\lambda,\tau) \in \mathbb{R}^2\}$.

Every state-generating function $\hat{w}$ on $\mathbb{R}^2$ is a complex-valued function with the following properties:

- $|\hat{w}(\lambda,\tau)| \le \hat{w}(0,0) = 1$ for all $\lambda,\tau \in \mathbb{R}^2$,
- $\hat{w}(-\lambda,-\tau) = \hat{w}(\lambda,\tau)^*$ for all $\lambda,\tau \in \mathbb{R}^2$,

- $\hat{w}$ is continuous on $\mathbb{R}^2$,
- $\hat{w}$ is σ-non-negative definite, that is for any $N = 1, 2, \ldots$, for any $\lambda_1, \lambda_2, \ldots, \lambda_N \in \mathbb{R}^n$, for any $\tau_1, \ldots, \tau_N \in \mathbb{R}^n$, and for any complex numbers $c_1, c_2, \ldots, c_N$ we have:[190]

$$\sum_{j=1}^{N}\sum_{k=1}^{N} c_j c_k^* \hat{w}(\lambda_j - \lambda_k, \tau_j - \tau_k) e^{is(\lambda_j,\tau_j|\lambda_k,\tau_k)/(2\hbar)} \geq 0.$$

For a normal state functional $\rho \in \mathcal{M}_*$ with vanishing variance, $\sigma = 0$, this characterization reduces to Bochner's characterization of classical characteristic functions. A generalization of Bochner's theorem characterizes the state-generating functions: A continuous complex-valued s-non-negative definite function $\hat{w} : \mathbb{R}^2 \to \mathbb{C}$ is a temporal state-generating function of a normal state functional $\rho \in \mathcal{M}_*$ if and only if it fulfills the normalization condition $\hat{w}(0, 0) = 1$, and if it is s-non-negative definite (Araki 1960, Theorem 4.3, Segal 1961, Theorem 1). Moreover, the state-generating function $\hat{w}_\rho : \mathbb{R}^\mathbf{2} \to \mathbb{C}$ *uniquely* determines a normal state functional $\rho \in \mathcal{M}_*$.[191]

There are important relations between state-generating and characteristic functions:

- $\lambda \mapsto \hat{w}(\lambda, 0)$ and $\tau \mapsto \hat{w}(0, \tau)$ are characteristic functions on $\mathbb{R}^n$,[192]
- $(\lambda, q) \mapsto \hat{w}(\lambda, q)\hat{w}'(\lambda, q)$ is a characteristic function on $\mathbb{R}^{2n}$ (Cushen and Hudson 1971, p. 464),
- $(\lambda, \tau) \mapsto \hat{w}(\lambda, \tau)\hat{w}'(\lambda, \tau)$ is not necessarily a state-generating function (Jagannathan et al. 1987),
- $(\lambda, \tau) \mapsto \hat{w}(\lambda, \tau)\hat{f}(\lambda, \tau)$ is a state-generating function on $\mathbb{R}^{2n}$ (Segal 1961).

10.4.3 The Fourier-Weyl Calculus

The Weyl calculus is a generalization of the classical Fourier calculus. Formally, the Weyl calculus is based on the heuristic relation $\mathrm{tr}\{W(\lambda, \tau)W(\lambda', \tau')^*\} = \delta(\lambda - \lambda')\delta(\tau - \tau')$. Just as the Fourier transform, the Weyl transform requires a detailed study of its domain and range.

In the first place, the Fourier-Weyl transform provides a unitary isomorphism from the Hilbert space $\mathbf{L}^2(\mathbb{R}^2 \times \mathbb{R}^2)$ onto the Hilbert space $\mathcal{B}^2(\mathcal{H})$ of Hilbert-Schmidt operators acting on a Hilbert space $\mathcal{H}$. For an arbitrary trace-class operator $A \in \mathcal{B}^1(\mathcal{H})$, the Weyl transform is defined by

$$\hat{a}(\lambda, \tau) := \mathrm{tr}\big\{AW(\lambda, \tau)^*\big\}. \tag{10.110}$$

---

[190] This relation follows from Weyl's commutation relations and the positivity of the operator $\{\sum_j c_j W(\lambda_j, \tau_j)\}\{\sum_k c_k W(\lambda_k, \tau_k)\}^*$.

[191] For a proof in a more general context see Segal (1961, Sect. 3).

[192] This follows from the fact that $\{W(\lambda, 0)|\lambda \in \mathbb{R}^n\}$and $\{W(0, \tau)|\tau \in \mathbb{R}^n\}$ are unitary Abelian groups.

For every $\hat{a} \in \mathbf{L}^2(\mathbb{R}^2) \cap \mathbf{L}^1(\mathbb{R}^2)$, the inverse transformation

$$A = \int_{\mathbb{R}^2} \hat{a}(\lambda, \tau) W(\lambda, \tau) d\lambda d\tau \tag{10.111}$$

exists, where the operator-valued integral is to be understood as a weak integral (Segal 1963; Pool 1966). This map extends uniquely to a unitary map from $\mathbf{L}^2(\mathbb{R}^2)$ to the Hilbert space $\mathcal{B}^2(\mathcal{H})$ of Hilbert-Schmidt operators on Hilbert space $\mathcal{H}$.[193] The transformation $\hat{a} \leftrightarrow A$ is called the *Fourier-Weyl transform*. If a Weyl representation $\hat{a}$ of an operator A exists, then its expectation value with respect to a state functional ρ is given by

$$\rho(A) = \int_{\mathbb{R}^2} \hat{a}(\lambda, \tau) \hat{w}(\lambda, \tau) d\lambda d\tau. \tag{10.112}$$

Since the square of a state-generating function is a characteristic function, every state-generating function is Lebesgue square-integrable, so that the Fourier-Weyl transform applies. If the state functional ρ is represented by a density operator $D \in \mathcal{B}(\mathcal{H})$, we can recover the density operator from the state-generating function by

$$\hat{w}(\lambda, \tau) = \mathrm{tr}\big\{D W(\lambda, \tau)\big\} \quad \Longleftrightarrow \quad D = \int_{\mathbb{R}^2} \hat{w}(\lambda, \tau) W(\lambda, \tau)^* d\lambda d\tau. \tag{10.113}$$

If $\hat{w}_\rho$ and $\hat{w}_\varphi$ are the state-generating functions associated with the density operators D_ρ and D_φ, this representation theorem implies

$$D_\rho D_\varphi = \int_{\mathbb{R}^2} d\lambda' d\tau' \int_{\mathbb{R}^2} d\lambda'' d\tau'' \hat{w}_\rho(\lambda', \tau') \hat{w}_\varphi(\lambda'', \tau'') W(\lambda', \tau')^* W(\lambda'', \tau'')^*, \tag{10.114}$$

and $\langle \cdot | \cdot \rangle_{\mathrm{P}}$ is the inner product of the phase-space representation (10.93a)–(10.93c),

$$\begin{aligned} \langle w_\rho | \overline{W}(\lambda, \tau) w_\varphi \rangle_{\mathrm{P}} &= \int_{\mathbb{R}^2} d\lambda' d\tau' w_\rho(\lambda', \tau') \big\{\overline{W}(\lambda, \tau) w_\varphi\big\}(\lambda', \tau') \\ &= \int_{\mathbb{R}^2} d\lambda' d\tau' e^{2\pi i(\lambda, \tau') - 2\pi i(\tau, \lambda')} w_\rho(\lambda', \tau') \\ &\quad \times w_\varphi(\lambda' - \lambda/2, \tau' - \tau/2). \end{aligned} \tag{10.115}$$

With $\mathrm{tr}\{W(\lambda, \tau)\} = \delta(\lambda, \tau)$ we get the following expression for the generalized transition probabilities:

$$\mathrm{tr}(D_\rho D_\varphi) = \int_{\mathbb{R}^2} \hat{w}_\rho(\lambda, \tau) \hat{w}_\varphi(-\lambda, -\tau) d\lambda d\tau. \tag{10.116}$$

[193] As in classical Fourier analysis, this map can be extended to an arbitrary tempered distribution $a \in \mathcal{S}^*(\mathbb{R}^{2n})$. Compare for example Peck (1979).

A state ρ is pure if and only if its density operator D_ρ is idempotent, which is equivalent to the condition $\mathrm{tr}\{D_\rho^2\} = 1$. Therefore we get the following pure-state condition:

$$\rho \text{ is pure if and only if } \int_{\mathbb{R}^2} \left|\hat{w}_\rho(\lambda, \tau)\right|^2 d\lambda d\tau = 1. \tag{10.117}$$

10.4.4 The Weyl Correspondence and Wigner Functions

With the symplectic Fourier transform

$$\check{a}(\lambda, \tau) = \int_{\mathbb{R}^2} \hat{a}(\lambda', \tau') e^{2\pi i s(\lambda,\tau|\lambda',\tau')} d\lambda' d\tau', \tag{10.118a}$$

$$\hat{a}(\lambda, \tau) = \int_{\mathbb{R}^2} \check{a}(\lambda', \tau') e^{2\pi i s(\lambda,\tau|\lambda',\tau')} d\lambda' d\tau', \tag{10.118b}$$

we get the so-called *Weyl correspondence* $\check{a} \leftrightarrow \hat{a}$.

The symplectic Fourier transform of a state-generating function $\hat{w}_\rho$ is a temporal *Wigner function* $\check{w}_\rho$,[194]

$$\check{w}_\rho(\lambda, \tau) = \int_{\mathbb{R}^2} \hat{w}_\rho(\lambda', \tau') e^{2\pi i s(\lambda,\tau|\lambda',\tau')} d\lambda' d\tau', \tag{10.119a}$$

$$\hat{w}_\rho(\lambda, \tau) = \int_{\mathbb{R}^2} \check{w}_\rho(\lambda', \tau') e^{2\pi i s(\lambda,\tau|\lambda',\tau')} d\lambda' d\tau'. \tag{10.119b}$$

The Wigner function is a bounded real-valued function in $\mathbf{L}^2(\mathbb{R}^2)$. It gives the correct marginal distribution densities for Λ and T,

$$\int_{\mathbb{R}} w_\rho(\lambda, \tau) d\lambda = \int_{\mathbb{R}^2} e^{-2\pi i \lambda' \tau} \rho\{e^{2\pi i \lambda' T}\} d\lambda' = \rho\{\delta(T - \tau)\}, \tag{10.120a}$$

$$\int_{\mathbb{R}} w_\rho(\lambda, \tau) d\tau = \int_{\mathbb{R}^2} e^{-2\pi i \tau' \lambda} \rho\{e^{-2\pi i \tau' \Lambda}\} d\tau' = \rho\{\delta(\Lambda - \lambda)\}, \tag{10.120b}$$

but in contrast to joint probability densities, the Wigner function is in general not non-negative, and in general not an element of $\mathbf{L}^1(\mathbb{R}^2)$.

---

[194] See Wigner (1932a), see also Moyal (1949). In spite of the fact that this phase-space function is usually called "Wigner function", it was first introduced by Dirac (1930b). The Wigner-Moyal representation is an example of the so-called standard representation of traditional quantum mechanics (see Appendix A.6).

10.4.5 The Wigner Function of a Pure State

If ρ is a pure state functional, the corresponding density operator D_ρ is idempotent, so that with Eqs. (10.114) and (10.93a)–(10.93c) we obtain

$$D_\rho = D_\rho^2 = \int_{\mathbb{R}^2} d\lambda' d\tau' \langle w_\rho | \overline{W}(\lambda, \tau) w_\rho \rangle_{\mathrm{P}} W(\lambda', \tau')^*. \tag{10.121}$$

Using this expression, the state-generating-function $\hat{w}_\rho(\lambda, \tau) = \mathrm{tr}\{D_\rho W(\lambda.\tau)\}$ can be written as

$$\begin{aligned} \hat{w}_\rho(\lambda, \tau) &= \int_{\mathbb{R}^2} d\lambda' d\tau' \langle w_\rho | \overline{W}(\lambda', \tau') w_\rho \rangle_{\mathrm{P}} \mathrm{tr}\{W(\lambda, \tau) W(\lambda', \tau')^*\} \\ &= \langle w_\rho | \overline{W}(\lambda, \tau) w_\rho \rangle_{\mathrm{P}}. \end{aligned} \tag{10.122}$$

This result says that the Wigner function $(\lambda, \tau) \mapsto \check{w}_\rho(\lambda, \tau)$ of a *pure* state equals the normalized state vector $(\lambda, \tau) \mapsto \overline{\Psi}(\lambda, \tau) \in \mathbf{L}^2(\mathbb{R}^2)$ in the reducible phase-space representation,

$$\overline{\Psi}_\rho(\lambda, \tau) = \check{w}_\rho(\lambda, \tau), \quad (\lambda, \tau) \in \mathbb{R}^2. \tag{10.123}$$

10.4.6 The Wigner Function of a Pure State in an Irreducible Representation

The expression of pure-state Wigner functions in terms of state vectors of irreducible representations are slightly more complicated. In the irreducible time-domain representation a normal pure-state functional ρ can be represented by a state vector $\check{\Psi}_\rho$, so that the state-generating function $\hat{w}_\rho$ is given by

$$\hat{w}_\rho(\lambda, \tau) = \langle \check{\Psi}_\rho | \check{W}(\lambda, \tau) \check{\Psi}_\rho \rangle. \tag{10.124}$$

In the irreducible frequency-domain representation, the state function $\hat{\Psi}_\rho$ is given by

$$\hat{\Psi}_\rho(\lambda) := \int_{\mathbb{R}} e^{-2\pi i \lambda \tau} \check{\Psi}_\rho(\tau) d\tau, \quad \tau \in \mathbb{R}. \tag{10.125}$$

Hence the state-generating function can be evaluated as

$$\hat{w}_\rho(\lambda, \tau) = \int_{\mathbb{R}} \check{\Psi}(\tau')^* e^{-\pi i \lambda \tau} e^{+2\pi i \lambda \tau'} \check{\Psi}(\tau' - \tau) d\tau' \tag{10.126a}$$

$$= \int_{\mathbb{R}} \hat{\Psi}(\lambda')^* e^{+\pi i \lambda \tau} e^{-2\pi i \tau \lambda'} \hat{\Psi}(\lambda' - \lambda) d\lambda'. \tag{10.126b}$$

With the Fourier transforms (10.119a), (10.119b) we obtain for the Wigner function,

$$\check{w}_\rho(\lambda,\tau) = \int_{\mathbb{R}^n} e^{2\pi i\lambda\tau'}\check{\Psi}\big(\tau+\tau'/2\big)^*\check{\Psi}\big(\tau-\tau'/2\big)d\tau' \tag{10.127a}$$

$$= \int_{\mathbb{R}^n} e^{-2\pi i\tau\lambda'}\hat{\Psi}\big(\lambda+\lambda'/2\big)^*\hat{\Psi}\big(\lambda-\lambda'/2\big)d\lambda'. \tag{10.127b}$$

10.4.7 The Parity Operator and the Weyl Calculus

Define an operator Π by

$$\Pi := \frac{1}{2}\int_{\mathbb{R}^2} W(\lambda,\tau)d\lambda d\tau, \tag{10.128}$$

where the integral is defined in the weak operator topology (Grossmann 1976). The operator Π is a self-adjoint and unitary involution,

$$\Pi = \Pi^* = \Pi^{-1}. \tag{10.129}$$

Using Weyl's commutation relations (9.47a), (9.47b) we get

$$\Pi W(\lambda,\tau)\Pi = \frac{1}{4}\int_{\mathbb{R}^2} d\tau' d\lambda' \int_{\mathbb{R}^2} d\tau'' d\lambda'' W\big(\lambda',\tau'\big)W(\lambda,\tau)W\big(\lambda'',\tau''\big), \tag{10.130}$$

and with the substitutions

$$x = \tau'+\tau'', \qquad x' = \tau'-\tau'', \qquad y = \lambda'+\lambda'', \qquad y' = \lambda'-\lambda'',$$

$$d\tau' d\tau'' d\lambda' d\lambda'' = \frac{1}{4}dxdx'dydy',$$

we obtain

$$\begin{aligned}
&\Pi W(\lambda,\tau)\Pi \\
&= \frac{1}{16}\int_{\mathbb{R}^2} dxdy \int_{\mathbb{R}^2} dx'dy' e^{2\pi i[y'(\tau/2-x/4)-x'(\lambda/2+y/4)]}\, W\{y+\lambda, x+\tau\} \\
&= \int_{\mathbb{R}^2} dxdy \int_{\mathbb{R}^2} \delta(2\tau+x)\delta(2\lambda+y)W\{y+\lambda,x+\tau\} = W(-\lambda,-\tau).
\end{aligned} \tag{10.131}$$

This relation implies

$$\Pi e^{+2\pi i\lambda T - 2\pi i\tau\Lambda}\,\Pi = e^{-2\pi i\lambda T + 2\pi i\tau\Lambda}, \tag{10.132}$$

so that the operator Π is the *temporal parity operator* which implements the transformation $\Lambda \to -\Lambda$, $T \to -T$,

$$\Pi\Lambda\Pi = -\Lambda, \qquad \Pi T\Pi = -T. \tag{10.133}$$

The displayed temporal parity operator $\Pi(\lambda, \tau)$,

$$\Pi(\lambda, \tau) := W(\lambda, \tau)^* \Pi W(\lambda, \tau), \quad \lambda, \tau \in \mathbb{R}, \tag{10.134}$$

is also a self-adjoint and unitary involution,

$$\Pi(\lambda, \tau) = \Pi(\lambda, \tau)^* = \Pi(\lambda, \tau)^{-1}, \tag{10.135}$$

and reflects with respect to the phase space point $(\lambda, \tau) \in \mathbb{R}^2$,

$$\Pi(\lambda, \tau)(\Lambda - \lambda)\Pi(\lambda, \tau) = -(\Lambda - \lambda), \tag{10.136a}$$

$$\Pi(\lambda, \tau)(T - \tau)\Pi(\lambda, \tau) = -(T - \tau). \tag{10.136b}$$

With the definition (10.128) and the commutation relations (9.47a), (9.47b) we find that the displayed parity operator equals the symplectic Fourier transform of the Weyl operator,

$$\Pi(\lambda, \tau) = \int_{\mathbb{R}^2} e^{2\pi i \lambda \tau' - 2\pi i \tau \lambda'} W(\lambda', \tau') d\lambda' d\tau' =: \hat{W}(\lambda', \tau'). \tag{10.137}$$

With the definitions (10.109) and (10.119a), (10.119b) we can express the Wigner function $(\lambda, \tau) \mapsto \check{w}_\rho(\lambda, \tau)$ as expectation value of the displayed temporal parity operator,

$$\check{w}_\rho(\lambda, \tau) = \rho\{\Pi(\lambda, \tau)\}. \tag{10.138}$$

10.5 Affine Temporal Weyl Systems

10.5.1 *Representations of the Affine Weyl Group*

The affine Weyl-Heisenberg group $\mathfrak{W}_{\text{aff}} = \{(\tau, \lambda, \kappa, \varphi) | \tau, \lambda, \varphi \in \mathbb{R}, \kappa > 0\}$ can be represented by a suitable combination of the unitary dilation operator $D(\kappa)$, the unitary frequency-shift operator $U(\lambda)$, and the unitary time-translation operator $V(\tau)$. The commutation relations (9.80a)–(9.80f) imply that the operators $V(\tau)$, $U(\lambda)$, and $D(\kappa)$ in (9.78a), (9.78b) fulfill the group law (9.74) of the affine Weyl-Heisenberg group so that (9.78a), (9.78b) is a unitary realization of the affine Weyl-Heisenberg group.

With (10.36a)–(10.36d) we obtain the unitary representation of the affine Weyl-Heisenberg group on the Lebesgue-Hilbert space $\check{\mathbf{L}}^2(\mathbb{R}, dt)$ for $\check{\Phi} \in \check{\mathbf{L}}^2(\mathbb{R}, dt)$ as

$$\check{W}_{\text{aff}}(\tau, \lambda, \kappa, \varphi)\check{\Phi}(\tau) = e^{2\pi i(\varphi + \lambda t - \lambda \tau)} \sqrt{\kappa}\, \check{\Phi}(\kappa t - \kappa \tau), \tag{10.139}$$

and the unitarily equivalent Fourier-transformed representation for $\hat{\Phi} \in \hat{\mathbf{L}}^2(\mathbb{R}, d\ell)$ is given by

$$\hat{W}_{\text{aff}}(\tau, \lambda, \kappa, \varphi)\hat{\Phi}(\ell) = e^{-2\pi i(\ell\tau - \varphi)} e^{2\pi i(\varphi - \lambda \tau)} \hat{\Phi}\left(\frac{\ell - \lambda}{\kappa}\right). \tag{10.140}$$

These representations are *irreducible*, but *not square-integrable* (Torrésani 1991; Kalisa and Torrésani 1991). This means that neither the decomposition theorem (10.7a), (10.7b) nor the resolution of the identity (10.9) are applicable.

The time-domain representation of the affine groups $\mathfrak{A}_{\mathfrak{U}} = \{A_U(\tau,\kappa)|\lambda \in \mathbb{R}\}$ and $\mathfrak{A}_{\mathfrak{V}} = \{A_V(\lambda,\kappa)|\lambda \in \mathbb{R}\}$ on the Lebesgue-Hilbert space $\check{\mathbf{L}}^2(\mathbb{R}, dt)$ is given by the unitary operators $\check{A}_U(\tau,\kappa)$ and $\check{A}_V(\lambda,\kappa)$,

$$\check{A}_U(\tau,\kappa)\check{\Phi}(\tau) := \check{W}_{\mathrm{aff}}(\tau,0,\kappa,0)\check{\Phi}(\tau) = \sqrt{\kappa}\check{\Phi}(\kappa t - \kappa\tau) \tag{10.141a}$$

$$\check{A}_V(\lambda,\kappa)\check{\Phi}(\tau) := \check{W}_{\mathrm{aff}}(0,\lambda,\kappa,0)\check{\Phi}(\tau) = e^{2\pi i\lambda t}\sqrt{\kappa}\check{\Phi}(\kappa t) \tag{10.141b}$$

for every $\check{\Phi} \in \check{\mathbf{L}}^2(\mathbb{R}, dt)$. The frequency-domain representation of the affine groups $\mathfrak{A}_{\mathfrak{U}} = \{A_U(\tau,\kappa)|\lambda \in \mathbb{R}\}$ and $\mathfrak{A}_{\mathfrak{V}} = \{A_V(\lambda,\kappa)|\lambda \in \mathbb{R}\}$ on the Lebesgue-Hilbert space $\hat{\mathbf{L}}^2(\mathbb{R}, d\ell)$ is given by the unitary operators $\hat{A}_U(\tau,\kappa)$ and $\hat{A}_V(\lambda,\kappa)$,

$$\hat{A}_U(\tau,\kappa)\hat{\Phi}(\ell) := \hat{W}_{\mathrm{aff}}(\tau,0,\kappa,0)\hat{\Phi}(\ell) = e^{-2\pi i(\ell\tau-\varphi)}\hat{\Phi}\left(\frac{\ell}{\kappa}\right), \tag{10.141c}$$

$$\hat{A}_V(\lambda,\kappa)\hat{\Phi}(\ell) := \hat{W}_{\mathrm{aff}}(0,\lambda,\kappa,0)\hat{\Phi}(\ell) = \frac{1}{\sqrt{\kappa}}\hat{\Phi}\left(\frac{\ell-\lambda}{\kappa}\right), \tag{10.141d}$$

for every $\hat{\Phi} \in \hat{\mathbf{L}}^2(\mathbb{R}, d\ell)$.

The representations (10.141a), (10.141b) are reducible. Up to unitary equivalence they have two non-trivial faithful inequivalent irreducible unitary representations (Gel'fand and Neumark 1947). The corresponding invariant subspaces $\Phi \in \mathbf{L}^2(\mathbb{R}^\pm, dt)$ of $\Phi \in \mathbf{L}^2(\mathbb{R}, dt)$ contain functions with support on $\mathbb{R}^\pm$.

10.5.2 Temporal Semigroups

10.5.2.1 Weak Weyl Representations

Let $\{A(x)|x \in \mathbb{R}\}$ be a strongly continuous one-parameter group on a complex Hilbert space $\mathcal{H}$. Let G_A be the self-adjoint generator of $\{A(x)|x \in \mathbb{R}\}$,

$$A(x) = e^{-2\pi i x G_A}, \tag{10.142}$$

and G_B a symmetric operator on $\mathcal{H}$ with the domain $\mathcal{D}(G_B)$. We say that the pair (G_A, G_B) obeys the *weak Weyl relation* if

$$e^{-2\pi i x G_A}\mathcal{D}(G_B) \subset \mathcal{D}(G_B), \tag{10.143a}$$

$$G_B e^{-2\pi i x G_A}\Phi = e^{-2\pi i x G_A}(G_B + x), \quad \text{for all } \Phi \in \mathcal{D}(G_B), \tag{10.143b}$$

holds for all $x \in \mathbb{R}$. Every pair (G_A, G_B) with $\Phi, \Psi \in \mathcal{D}(G_A) \cap \mathcal{D}(G_B)$ satisfying the weak Weyl relation satisfies the canonical commutation relation (Schmüdgen

1983; Arai 2005, 2007; Arai and Matsuzawa 2007):

$$\langle G_A \Phi | G_B \Psi \rangle - \langle G_B \Phi | G_A \Psi \rangle = \frac{1}{2\pi i} \langle \Phi | \Psi \rangle. \tag{10.144}$$

Under these conditions it follows that G_B has an absolutely continuous spectrum (Schmüdgen 1983).

If (G_A, G_B) is a weak Weyl representation on the Hilbert space $\mathcal{H}$, where the symmetric operator G_B is semi-bounded (that is, bounded below or bounded above), the operator G_B cannot be self-adjoint. In this case there are three possibilities, which are distinguished by the spectrum $sp(G_B)$ of G_B (Arai and Matsuzawa 2007):

(i) if G_B is bounded below,

$$\text{then either } sp(G_B) = \mathbb{C} \text{ or } sp(G_B) = \overline{\Pi}_+, \tag{10.145a}$$

(ii) if G_B is bounded above,

$$\text{then either } sp(G_B) = \mathbb{C} \text{ or } sp(G_B) = \overline{\Pi}_-, \tag{10.145b}$$

$$\text{(iii) if } G_B \text{ is bounded, then } sp(G_B) = \mathbb{C}. \tag{10.145c}$$

Here $\mathbb{C}$ is the set of complex numbers, $\overline{\Pi}_+ := \{z \in \mathbb{C} | \mathrm{Im}(z) \geq 0\}$ is the closed upper half-plane of $\mathbb{C}$, and $\overline{\Pi}_- := \{z \in \mathbb{C} | \mathrm{Im}(z) \leq 0\}$ is the closed lower half-plane of $\mathbb{C}$. If $\overline{G_B}$ denotes the closure of G_B, then $sp(G_B, \overline{G_B})$ is also a weak Weyl representation. Since symmetric operators are closable, we can without loss of generality assume that G_B is closed.

The affine commutation relations (9.86) for self-adjoint operators can be generalized in terms of the operators G_A and G_B (Arai and Matsuzawa 2007). First we replace the self-adjoint scale operator S by a symmetric operator $\breve{S}$:

$$\breve{S} := \frac{1}{2}(G_B G_A + \overline{G_A G_B}). \tag{10.146}$$

Then the pair $(\breve{S}, \ln|G_A|)$ is a weak Weyl representation,

$$\breve{S} e^{-2\pi i x \ln|G_A|} = e^{-2\pi i x \ln|G_A|} (\breve{S} + x), \tag{10.147}$$

for all $x \in \mathbb{R}$ and $e^{-2\pi i x \ln|G_A|} \mathcal{D}(\breve{S}) = \mathcal{D}(\breve{S})$.

The spectrum relations (10.145a)–(10.145c) imply:

(i) if $\ln|G_A|$ is bounded below,

$$\text{then either } sp(\breve{S}) = \mathbb{C} \text{ or } sp(\breve{S}) = \overline{\Pi}_+, \tag{10.148a}$$

(ii) if $\ln|G_A|$ is bounded above,

$$\text{then either } sp(\breve{S}) = \mathbb{C} \text{ or } sp(\breve{S}) = \overline{\Pi}_-, \tag{10.148b}$$

$$\text{(iii) if } \ln|G_A| \text{ is bounded, then } sp(\breve{S}) = \mathbb{C}. \tag{10.148c}$$

If $\breve{S}$ is essentially self-adjoint, then

$$sp(|G_A|) = [0, \infty), \qquad sp(\overline{\breve{S}}) = \mathbb{R}, \qquad sp_{\mathrm{p}}(\overline{\breve{S}}) = \emptyset. \tag{10.149}$$

10.5.2.2 Semigroup Weyl Systems

An important special case of a weak Weyl representation (10.143a), (10.143b) is given by *semigroup Weyl systems*. Let $\{\widetilde{A}(x)|x \in \mathbb{R}^+\}$ and $\{\widetilde{B}(y)|y \in \mathbb{R}^+\}$ be two strongly continuous one-parameter contraction semigroups on a Hilbert space $\widetilde{\mathcal{H}}$ and suppose that the two *non-unitary* operator-valued functions $x \to A(x)$ and $y \to B(y)$ defined by

$$\widetilde{A}(x) := \begin{cases} A(x) & \text{for } y \geq 0, \\ A^*(-x) & \text{for } y < 0, \end{cases} \tag{10.150a}$$

$$\widetilde{B}(y) := \begin{cases} B(y) & \text{for } y \geq 0, \\ B^*(-y) & \text{for } y < 0, \end{cases} \tag{10.150b}$$

satisfy the Weyl commutation relation

$$\widetilde{B}(y)\widetilde{A}(x)e^{\pi ixy} = \widetilde{A}(x)\widetilde{B}(y)e^{-\pi ixy}, \quad x, y \in \mathbb{R}. \tag{10.151}$$

Then the system $\{A(x), B(y)|x, y \in \mathbb{R}^+\}$ is called a *semigroup Weyl system* (Foiaş et al. 1960; Foiaş and Gehér 1963; Putnam 1967, §4.8). There is an extension Hilbert space $\mathcal{H}$ of $\widetilde{\mathcal{H}}$ in which the two strongly continuous *unitary groups*$\{A(x), B(y)|x, y \in \mathbb{R}^+\}$ act as *a unitary group Weyl system* and satisfy the unitary group Weyl commutation relation,

$$B(y)A(x)e^{\pi ixy} = A(x)B(y)e^{-\pi ixy}, \quad x, y \in \mathbb{R}. \tag{10.152}$$

The Hilbert space $\mathcal{H}$ can be chosen such that it is generated by the elements $A(x)B(y)\Phi$ with $\Phi \in \mathcal{H}$ and $x, y \in \mathbb{R}$. In this case the Weyl systems $\{A(x), B(y)| x, y \in \mathbb{R}^+\}$ and $\{A(x), B(y)|x, y \in \mathbb{R}\}$ are mutually determined up to isomorphisms.

The one-parameter contraction semigroups $x \mapsto A(x)$ and $y \mapsto B(y)$ have the representations

$$A(x) = e^{-2\pi ixG_A}, \quad 0 \leq x < \infty, \tag{10.153a}$$

$$B(y) = e^{2\pi iyG_B}, \quad 0 \leq y < \infty, \tag{10.153b}$$

where the infinitesimal generators G_A and G_B are *maximal dissipative operators*. The semigroups $\{A(x)|x \in \mathbb{R}^+\}$ and $\{B(y)|y \in \mathbb{R}^+\}$ are unitary if and only if the respective generator is self-adjoint.

10.5.3 The Semigroups $\mathfrak{S}_{\mathfrak{U}}$ and $\mathfrak{S}_{\mathfrak{V}}$

A special case of a semigroup Weyl system $\{\widetilde{A}(x)\widetilde{B}(y)|x, y \in \mathbb{R}^+\}$ is the *restricted half-line Weyl system* introduced by Jørgensen and Muhly (1980, Sect. 8), where $\{A(x)|x \in \mathbb{R}^+\}$ is a strongly continuous one-parameter group of unitary operators on a Hilbert space $\mathcal{H}$, and $\{B(y)|y \geq 0\}$ is a strongly continuous one-parameter semigroup of contractions. In this case the generator G_A is self-adjoint and the generator G_B is maximal dissipative.

The generators of a restricted half-line Weyl system can be realized by choosing G_A as multiplication operator $\check{G}_A$, and G_B as differential operator $\check{G}_B$ on the half-axis $\mathbb{R}^+ = (0, \infty)$,

$$\{\check{G}_A\Phi\}(x) := x\Phi(x), \qquad \{\check{G}_B\Phi\}(x) := \frac{1}{2\pi i}\frac{\partial\Phi(x)}{\partial x}, \qquad \Phi \in \mathcal{C}_0^\infty\big(\mathbb{R}^+\big). \tag{10.154}$$

The operator $\check{G}_A$ is self-adjoint and non-negative and has the spectrum $\check{G}_A = [0, \infty)$, while the generator $\check{G}_B$ of the semigroup $\{e^{ix\check{G}_B}|x \in \mathbb{R}^+\}$ is maximal dissipative.

With respect to the full temporal Weyl system, there are four canonical half-line restrictions of the Weyl operators,

$$U_U^\pm(\lambda) = P_U^\pm U(\lambda)P_U^\pm, \qquad V_U^\pm(\tau) = P_U^\pm V(\tau)P_U^\pm, \tag{10.155a}$$

$$U_V^\pm(\lambda) = P_V^\pm U(\lambda)P_V^\pm, \qquad V_V^\pm(\tau) = P_V^\pm V(\tau)P_V^\pm. \tag{10.155b}$$

Since the projector $P_U^\pm$ commutes with all $U(\lambda)$ $(0 \leq \lambda < \infty)$ and the projector $P_V^\pm$ commutes with all $V(\tau)$ $(0 \leq \tau < \infty)$, the generator $T_{U^\pm}$ of the unitary semigroup $\{U_U^\pm(\lambda)|\lambda \in \mathbb{R}^\pm\}$ and the generator $\Lambda_V^\pm$ of the unitary semigroup $\{V_V^\pm(\tau)|\tau \in \mathbb{R}^\pm\}$ are both semi-bounded *self-adjoint* operators,

$$U_U^\pm(\lambda) = \exp\big(2\pi i\lambda T_U^\pm\big), \qquad V_V^\pm(\tau) = \exp\big(-2\pi i\tau\Lambda_V^\pm\big). \tag{10.156a}$$

The generators $T_U^\pm$ and $\Lambda_V^\pm$ of the remaining completely non-unitary contractions $\tau \mapsto V_U(\tau)$ and $\lambda \mapsto U_V(\lambda)$,

$$V_U^\pm(\tau) = \exp\big(-2\pi i\tau\Lambda_U^\pm\big), \qquad U_V^\pm(\lambda) = \exp\big(2\pi i\lambda T_V^\pm\big), \tag{10.156b}$$

are non-unitary maximal dissipative operators with dense domains. The operator pairs $(U_U^\pm, V_U^\pm)$ and $(U_V^\pm, V_V^\pm)$ form two dual *restricted semigroup Weyl systems*:

$$V_U^\pm(\tau)U_U^\pm(\lambda)e^{\pi i\lambda\tau} = U_U^\pm(\lambda)V_U^\pm(\tau)e^{-\pi i\lambda\tau}, \qquad \lambda, \tau \in \mathbb{R}^\pm, \tag{10.157a}$$

$$V_V^\pm(\tau)U_V^\pm(\lambda)e^{\pi i\lambda\tau} = U_V^\pm(\lambda)V_V^\pm(\tau)e^{-\pi i\lambda\tau}, \qquad \lambda, \tau \in \mathbb{R}^\pm. \tag{10.157b}$$

Chapter 11
Complementary Temporal Descriptions

Bisection and symmetry reduction,
that is the gist of the matter.

Wolfgang Pauli[195]

11.1 Complementary Projections

In general, two orthogonal proper non-trivial projections P_1, P_2 acting on a complex Hilbert space $\mathcal{H}$ with $\dim \mathcal{H} \geq 2$ induce a *Halmos decomposition* (Halmos 1969) of $\mathcal{H}$ into an orthogonal sum

$$\mathcal{H} = \mathcal{H}_{00} \oplus \mathcal{H}_{01} \oplus \mathcal{H}_{10} \oplus \mathcal{H}_{11} \oplus \mathcal{R}, \tag{11.1}$$

where $\mathcal{H}_{nm}$ $(n, m = 0, 1)$ is spanned by the common eigenvectors Φ of P_1 and P_2,

$$P_1 \Phi = n\Phi, \qquad P_2 \Phi = m\Phi, \quad \Phi \in \mathcal{H}, \tag{11.2}$$

and $\mathcal{R}$ is the orthogonal complement of the sum of the first four subspaces. The subspace $\mathcal{H}_{00} \oplus \mathcal{H}_{01} \oplus \mathcal{H}_{10} \oplus \mathcal{H}_{11}$ is the maximal subspace in which P_1 and P_2 commute, while in the subspace $\mathcal{R}$ there are no common eigenvectors Φ of P_1 and P_2. One says that the restrictions $P_1|_{\mathcal{H}}$ and $P_2|_{\mathcal{H}}$ are in *generic position*.[196]

Without restrictions on generality we may assume that the four subspaces $\mathcal{H}_{nm}$ are trivial, so that we can discuss the case of two orthogonal projections P_1, P_2 in generic position acting on a Hilbert space $\mathcal{R}$ with $\dim \mathcal{R} \geq 2$. The prerequisite that the four subspaces $\mathcal{H}_{nm}$ are trivial implies that $P_1 \wedge P_2 = P_1 \wedge P_2^{\perp} = P_1^{\perp} \wedge P_2 =$

[195]Quoted from a letter to Heisenberg, see Heisenberg (1959, p. 663). German original, alluding to a passage from Goethe's *Faust*: "Zweiteilung und Symmetrieverminderung, das ist des Pudels Kern".

[196]Dixmier (1948) uses the French expression "in position p". Compare also Halmos (1969) and Lenard (1972).

H. Primas, *Knowledge and Time*,
DOI 10.1007/978-3-319-47370-3_11

$P_1^{\perp} \wedge P_2^{\perp} = \mathbb{O}$.[197] These relations lead to a precise definition of incompatibility and complementarity (Piron 1964, p. 447; Emch and Jauch 1965, pp. 269f):

General Mathematical Definition of Incompatibility
For two projections P_1 and P_2 in any W*-algebra, the projection

$$C(P_1, P_2) := (P_1 \wedge P_2) \vee \left(P_1 \wedge P_2^{\perp}\right) \vee \left(P_1^{\perp} \wedge P_2\right) \vee \left(P_1^{\perp} \wedge P_2^{\perp}\right) \tag{11.3}$$

measures the degree of incompatibility of P_1 and P_2. The following relations hold:

$$\mathbb{O} \leq C(P_1, P_2) \leq \mathbb{1}, \tag{11.4a}$$

$$C(P_1, P_2)P_1 = P_1 C(P_1, P_2), \tag{11.4b}$$

$$C(P_1, P_2)P_2 = P_2 C(P_1, P_2), \tag{11.4c}$$

$$C(P_1, P_2) = \mathbb{1} \quad \text{if and only if} \quad P_1 P_2 = P_2 P_1, \tag{11.4d}$$

$$C(P_1, P_2) = \mathbb{O} \quad \text{implies}$$

$$P_1 \wedge P_2 = P_1 \wedge P_2^{\perp} = P_1^{\perp} \wedge P_2 = P_1^{\perp} \wedge P_2^{\perp} = \mathbb{O}. \tag{11.4e}$$

If $C(P_1, P_2) = \mathbb{1}$, the projections P_1 and P_2 are called *compatible*, otherwise *incompatible*. If $C(P_1, P_2) = \mathbb{O}$, there exists no normal state ρ such that both P_1 and P_2 are truth-definite, that is, such that $\rho(P_1)$ and $\rho(P_2)$ are both 0 or 1 (Raggio and Rieckers 1983, Proposition 2.2). Such maximally incompatible projections are called *complementary*.

11.2 Complementary Temporal Projections

The two possible descriptions of the temporal domain in terms of the W*-algebras $\mathcal{A}_U$ and $\mathcal{A}_V$ are not only mutually exclusive in the sense that they cannot be captured in a single description. They are even *complementary* in the sense that $C(P_1, P_2)$ in Eq. (11.3) vanishes.

This implies that the descriptions of temporal phenomena in terms of the algebras $\mathcal{A}_U^{\pm}$ and $\mathcal{A}_V^{\pm}$ are *mutually complementary*. As Pauli (1954, p. 17) stated,

> [the contrast of pairs of opposites] can only be overcome in a symmetrical way. This means that one member of the pair is never eliminated in favor of the other, but both are taken over into a new kind of physical law which expresses properly the complementary character of the contrast.

The center of $\mathcal{A}_U$ is generated by the two orthogonal projectors P_U^+ and P_U^-, and the center of $\mathcal{A}_V$ is generated by the two orthogonal projectors P_V^+ and P_V^-. They are defined via the projection-valued spectral measure F of the time operator T and the spectral measure E of the frequency operator Λ,

$$T = \int_{-\infty}^{\infty} t F(dt), \qquad \Lambda = \int_{-\infty}^{\infty} \ell E(d\ell), \tag{11.5a}$$

[197] Here $P_1 \wedge P_2$ is defined by s-$\lim_{n\to\infty}(P_1 P_2)^n =$ s-$\lim_{n\to\infty}(P_2 P_1)^n$.

$$P_U^- := \int_{-\infty}^{0} F(dt) = \vartheta(-T), \qquad P_U^+ := \int_{0}^{\infty} F(dt) = \vartheta(T), \tag{11.5b}$$

$$P_V^- := \int_{-\infty}^{0} E(d\ell) = \vartheta(-\Lambda), \qquad P_V^+ := \int_{0}^{\infty} E(d\lambda) = \vartheta(\Lambda), \tag{11.5c}$$

where $x \mapsto \vartheta(x)$ is the Heaviside step function

$$\vartheta(x) = \begin{cases} 1 & \text{for } x > 0, \\ 0 & \text{for } x < 0. \end{cases} \tag{11.6}$$

The definitions (11.5a)–(11.5c) imply that the projectors P_U^+, P_U^-, P_V^+, and P_V^- are *maximally incompatible* with one another (Busch and Lahti 1986, Eq. 12):

$$P_U^+ \wedge P_V^+ = P_U^+ \wedge P_V^- = P_U^- \wedge P_V^+ = P_U^- \wedge P_V^- = \mathbb{O}. \tag{11.7}$$

Since

$$\mathcal{A}_U^{\pm} = P_U^{\pm} \mathcal{A} P_U^{\pm}, \qquad \mathcal{A}_V^{\pm} = P_V^{\pm} \mathcal{A} P_V^{\pm}, \tag{11.8}$$

the projectors $P_U^{\pm}$ commute with all observables in $\mathcal{A}_U$, the projectors $P_V^{\pm}$ commute with all observables in $\mathcal{A}_V$, and they have dispersion-free values in each pure state of the W*-algebra $\mathcal{A}_U$ and $\mathcal{A}_V$, respectively.

On the Hilbert-Lebesgue space $\check{\mathbf{L}}^2(\mathbb{R}, dt)$ of square-integrable complex-valued functions $t \mapsto \check{\Phi}(t)$ on the real axis $\mathbb{R}$, the projections $P_V^{\pm}$ are represented by

$$\{\check{P}_V^{\pm}\check{\Phi}\}(t) = \vartheta(\pm t)\check{\Phi}(t), \quad \check{\Phi} \in \check{\mathbf{L}}^2(\mathbb{R}, dt). \tag{11.9}$$

Analogously the projections $P_U^{\pm}$ can be represented on the Hilbert-Lebesgue space $\hat{\mathbf{L}}^2(\mathbb{R}, d\ell)$ of square-integrable complex-valued functions $t \mapsto \hat{\Phi}(\lambda)$ as

$$\{\hat{P}_U^{\pm}\hat{\Phi}\}(\ell) = \vartheta(\pm\ell)\hat{\Phi}(\ell), \qquad \hat{\mathcal{H}}_U^{\pm} = \hat{\mathbf{L}}^2(\mathbb{R}^{\pm}, d\ell), \quad \ell \in \mathbb{R}. \tag{11.10}$$

In the time-domain representation, the projections $\check{P}_V^{\pm}$ are given by the Fourier transform F of $\hat{P}_V^{\pm}$,

$$\check{P}_V^{\pm} = \mathsf{F}^{-1}\hat{P}_V^{\pm}\mathsf{F}, \tag{11.11}$$

the inverse of which can be written as:

$$\begin{aligned} \mathsf{F}^{-1}\{\{\hat{P}_V^{\pm}\hat{\Phi}\}(\ell)\}(t) &= \int_0^{\infty} e^{2\pi i \ell t}\hat{\Phi}(\ell)d\ell = \int_{\mathbb{R}} dt'\check{\Phi}(t')\int_0^{\infty} e^{2\pi i\ell(t-t')}d\ell \\ &= \frac{1}{2}\check{\Phi}(t) - \frac{i}{2\pi}\int_{\mathbb{R}} \frac{\check{\Phi}(t')}{t'-t}dt'. \end{aligned} \tag{11.12}$$

With the Hilbert transform H we obtain

$$\check{P}_V^{\pm} := \mathsf{F}^{-1}\hat{P}_V^{\pm}\mathsf{F} = \frac{1}{2}\{\mathbb{1} - i\mathsf{H}\}. \tag{11.13}$$

Now consider an irreducible representation of the temporal Weyl system (9.41) on some Hilbert space $\mathcal{H}$. Then the ranges of the associated projections $P_U^{\pm}$ and $P_V^{\pm}$ are given by

$$\mathcal{H}_U^{\pm} := \{P_U^{\pm}\Phi|\Phi \in \mathcal{H}\}, \tag{11.14a}$$

$$\mathcal{H}_V^{\pm} := \{P_V^{\pm}\Phi|\Phi \in \mathcal{H}\}. \tag{11.14b}$$

If we choose as Hilbert space $\mathcal{H}$ the Lebesgue space $\check{\mathbf{L}}^2(\mathbb{R}, dt)$ of square-integrable complex-valued functions $t \mapsto \check{\Phi}(t)$ on the real axis $\mathbb{R}$, then $P_V^{\pm}$ is represented by $\check{P}_V^{\pm}$, and $\mathcal{H}_V^{\pm}$ by $\check{\mathcal{H}}_V^{\pm}$:

$$\{\check{P}_V^{\pm}\check{\Phi}\}(t) = \vartheta(\pm t)\check{\Phi}(t), \qquad \check{\mathcal{H}}_V^{\pm} = \check{\mathbf{L}}^2(\mathbb{R}^{\pm}, dt), \quad t \in \mathbb{R}. \tag{11.15}$$

Analogously the projection $P_U^{\pm}$ can be represented in the frequency domain as

$$\{\hat{P}_U^{\pm}\hat{\Phi}\}(\ell) = \vartheta(\pm \ell)\hat{\Phi}(\ell), \qquad \hat{\mathcal{H}}_U^{\pm} = \hat{\mathbf{L}}^2(\mathbb{R}^{\pm}, d\ell), \quad \ell \in \mathbb{R}. \tag{11.16}$$

The time-domain projection $\check{P}_V^{\pm}$, the frequency-domain projection $\hat{P}_U^{\pm}$ and their spaces $\check{\mathcal{H}}_V^{\pm}$ and $\hat{\mathcal{H}}_U^{\pm}$ are related by a Fourier transform F:

$$\check{P}_V^{\pm} := \mathsf{F}^{-1}\hat{P}_U^{\pm}\mathsf{F}, \qquad \check{\mathcal{H}}_V^{\pm} := \mathsf{F}^{-1}\hat{\mathcal{H}}_U^{\pm}\mathsf{F}. \tag{11.17}$$

11.3 Four Elementary Affine Weyl Semigroups

Since the W*-algebra $\mathcal{T}$ of the temporal domain is a factor, the automorphisms generated by the affine subgroups $\mathfrak{S}_{\mathfrak{U}}$ and $\mathfrak{S}_{\mathfrak{V}}$ are inner, so that they can be implemented by unitary operators $A_U(\lambda, \kappa)$ and $A_V(\tau, \kappa)$:

$$\mathfrak{S}_{\mathfrak{U}} = \{A_U(\lambda, \kappa) \,|\, \lambda \in \mathbb{R}, \kappa > 0\}, \tag{11.18a}$$

$$\mathfrak{S}_{\mathfrak{V}} = \{A_V(\tau, \kappa) \,|\, \tau \in \mathbb{R}, \kappa > 0\}. \tag{11.18b}$$

These unitary operators generate the W*-algebras $\mathcal{A}_U$ and $\mathcal{A}_V$, respectively:

$$\mathcal{A}_U := \{A_U(\lambda, \kappa) \in \mathcal{T} \,|\, \lambda \in \mathbb{R}, \kappa > 0\}'', \tag{11.19a}$$

$$\mathcal{A}_V := \{A_V(\tau, \kappa) \in \mathcal{T} \,|\, \tau \in \mathbb{R}, \kappa > 0\}'', \tag{11.19b}$$

which implies that the W*-algebras $\mathcal{A}_U$ and $\mathcal{A}_V$ are direct sums (9.91a), (9.91b) of two non-trivial factors

$$\mathcal{A}_U = \mathcal{A}_U^- \oplus \mathcal{A}_U^+, \qquad \mathcal{A}_V = \mathcal{A}_V^- \oplus \mathcal{A}_V^+. \tag{11.20}$$

On the algebras $\mathcal{A}_U^{\pm}$ the affine group $\mathfrak{A}_U$ acts ergodically, and on the algebras $\mathcal{A}_V^{\pm}$ the affine group $\mathfrak{A}_V$ acts ergodically.

There are four *elementary* affine Weyl-Heisenberg semigroups which are generated by the operators

$$U_U^{\pm}(\lambda) = P_U^{\pm} U(\lambda) P_U^{\pm} \in \mathcal{A}_U, \quad \lambda \in \mathbb{R}^{\pm}, \tag{11.21a}$$

$$V_U^{\pm}(\tau) = P_U^{\pm} V(\tau) P_U^{\pm} \in \mathcal{A}_U, \quad \tau \in \mathbb{R}^{\pm}, \tag{11.21b}$$

$$U_V^{\pm}(\lambda) = P_V^{\pm} U(\lambda) P_V^{\pm} \in \mathcal{A}_V, \quad \lambda \in \mathbb{R}^{\pm}, \tag{11.21c}$$

$$V_V^{\pm}(\tau) = P_V^{\pm} V(\tau) P_V^{\pm} \in \mathcal{A}_V, \quad \tau \in \mathbb{R}^{\pm}, \tag{11.21d}$$

where $P_U^{\pm}$ and $P_V^{\pm}$ are the orthogonal projections (11.5a)–(11.5c),

$$P_U^{\pm} = \vartheta(\pm T), \qquad P_V^{\pm} = \vartheta(\pm \Lambda). \tag{11.22}$$

The semigroups $\mathfrak{U}_U^{\pm} = \{U_U^{\pm}(\lambda) | \lambda \in \mathbb{R}^{\pm}\}$ and $\mathfrak{V}_V^{\pm} = \{V_V^{\pm}(\lambda) | \lambda \in \mathbb{R}^{\pm}\}$ are strongly continuous one-parameter unitary semigroups with self-adjoint generators $T_U^{\pm}$ and $\Lambda_V^{\pm}$,

$$U_U^{\pm}(\lambda) = e^{2\pi i \lambda T_U^{\pm}}, \qquad V_V^{\pm}(\tau) = e^{-2\pi i \tau \Lambda_V^{\pm}}. \tag{11.23}$$

The semigroups $\mathfrak{U}_V^{\pm} = \{U_V^{\pm}(\lambda) | \lambda \in \mathbb{R}^{\pm}\}$ and $\mathfrak{V}_U^{\pm} = \{V_U^{\pm}(\tau) | \tau \in \mathbb{R}^{\pm}\}$ with

$$U_V^{\pm}(\lambda) = e^{2\pi i \lambda T_V^{\pm}}, \qquad V_U^{\pm}(\tau) = e^{-2\pi i \tau \Lambda_U^{\pm}}, \tag{11.24}$$

are strongly continuous one-parameter semigroups of contractions whose generators $T_V^{\pm}$ and $\Lambda_U^{\pm}$, are not self-adjoint but *maximal dissipative*. As a consequence there are four different affine semigroups $\mathfrak{W}_U^{\pm}$ and $\mathfrak{W}_V^{\pm}$ of $\mathfrak{W}_{\text{aff}} = \{(\tau, \lambda, \kappa, \varphi) | \tau, \lambda, \varphi \in \mathbb{R}, \kappa > 0\}$ with the elements

$$W_U^{\pm}(\tau, \lambda, \kappa, \varphi) = e^{2\pi i \varphi} e^{\pi i \lambda \tau} V_U^{\pm}(\tau) U_U^{\pm}(\lambda) D_U^{\pm}(\kappa), \tag{11.25a}$$

$$W_V^{\pm}(\tau, \lambda, \kappa, \varphi) = e^{2\pi i \varphi} e^{\pi i \lambda \tau} V_V^{\pm}(\tau) U_V^{\pm}(\lambda) D_V^{\pm}(\kappa). \tag{11.25b}$$

This gives us four distinct time concepts:

- The self-adjoint time operator $T_U^{+} > 0$ characterizes the *future* and represents *advanced sequential time.*
- The self-adjoint time operator $T_U^{-} < 0$ characterizes the *past* and represents *retarded sequential time.*
- The maximal dissipative time operator T_V^{+} represents *non-sequential time* associated with a temporal system characterized by *spectral positivity* $\Lambda_U^{+} > 0$.

- The maximal dissipative time operator T_V^- represents *non-sequential time* associated with a temporal system characterized by *spectral negativity* $\Lambda_U^- < 0$.

11.4 Some Basic Time Concepts

11.4.1 Introductory Remarks

As laid out in Sects. 1.4.2 and 8.3.3–8.3.5, we adopt the view that time is not a self-subsistent entity but a Leibnizian order parameter $\tau \in \mathbb{R}$, so that neither the time operator T nor the canonically conjugate frequency operator Λ belong to the atemporal domains of material and mental phenomena described by the W*-algebras $\mathcal{A}_{\text{mat}}$ and $\mathcal{A}_{\text{men}}$ as in Eqs. (8.4), (8.8).

Generic concepts of time are based on the affine full Weyl-Heisenberg group with the time operator T, the frequency operator Λ and the scale operator S as self-adjoint generators (see Eqs. (9.79a)–(9.79c)). These operators generate the W*-algebra $\mathcal{T}$ of the temporal domain. The two complementary commutative algebras generated by T and Λ constitute a partial Boolean description.

In a description based on the affine full Weyl-Heisenberg group arbitrary temporal superpositions are possible. In a sequential-time description there are no superpositions between past and future states, while in the non-sequential description there are no superpositions between negative and positive frequency states. Both descriptions have an integrity of their own and are not translatable into each other. They are complementary in the sense of Sect. 11.2 and, therefore, mutually irreducible.

The two *restricted* Weyl systems $(U_U^\pm, V_U^\pm)$ and $(U_V^\pm, V_V^\pm)$ according to (10.157a), (10.157b) can be understood as a contextual reduction of the full symmetry of the affine Weyl-Heisenberg group. They generate two W*-algebras $\mathcal{A}_U^\pm$ and $\mathcal{A}_V^\pm$,

$$\mathcal{A}_U^\pm := \left(U_U^\pm, V_U^\pm\right)'', \qquad \mathcal{A}_V^\pm := \left(U_V^\pm, V_V^\pm\right)''. \tag{11.26}$$

These non-Boolean systems define two pairs of complementary non-classical time operators: the *sequential time operators* T_U^+ and T_U^- and the *non-sequential time operators* T_V^+ and T_V^-.

The associated *partial* descriptions are distinguished by mutually exclusive superselection rules. The sequential description is based on the W*-algebra $\mathcal{A}_U^+ \overline{\oplus} \mathcal{A}_U^-$ characterized by the Boolean distinction between past and future, while the non-sequential description is based on the W*-algebra $\mathcal{A}_V^+ \overline{\oplus} \mathcal{A}_V^-$ characterized by the Boolean distinction between negative and positive frequencies.

In the following we will discuss sequential time in terms of the affine semigroup $\mathfrak{S}_{\mathfrak{U}} = \{A_U(\lambda, \kappa) | \lambda \in \mathbb{R}, \kappa > 0\}$ (see Eq. (11.18a)), and non-sequential time in terms of the affine semigroup $\mathfrak{S}_{\mathfrak{V}} = \{A_V(\tau, \kappa) | \tau \in \mathbb{R}, \kappa > 0\}$ (see Eq. (11.18b)). It is important to acknowledge that there is no sharp demarcation between the

Table 11.1 Generic time and generic frequency

<table>
<tr><td colspan="2">two basic groups: $\mathfrak{V} \sim \mathbb{R}$ and $\mathfrak{U} \sim \mathbb{R}$
$\mathbf{U}(\lambda)\,\mathbf{V}(\tau) \;=\; \mathbf{e}^{\mathbf{2\pi i\lambda\tau}}\,\mathbf{V}(\tau)\,\mathbf{U}(\lambda)$
$\mathbf{U}(\lambda) \;=\; \mathbf{e}^{\mathbf{2pii\lambda T}}, \quad \mathbf{V}(\tau) \;=\; \mathbf{e}^{-\mathbf{2\pi i\tau\Lambda}}$
$\mathbf{T\Lambda} \,-\, \mathbf{\Lambda T} \;=\; (\mathbf{i}/\mathbf{2\pi})\,\mathbb{1}$</td></tr>
<tr><td>time domain</td><td>frequency domain</td></tr>
<tr><td>• The generating group is the unitary group $U\{(\lambda)|\lambda \in \mathbb{R}\}$.
• The time operator T is the self-adjoint generator of the group $\{U\{(\lambda) = e^{2\pi i\lambda T}|\lambda \in \mathbb{R}\}$, where T has the spectrum $\sigma(T) = \mathbb{R}$.
• If a system is invariant under time-translation, then the time evolution is reversible and not invariant under time reversal.</td><td>• The generating group is the unitary group $V\{(\tau)|\tau \in \mathbb{R}\}$.
• The frequency operator Λ is the self-adjoint generator of the group $\{V\{(\tau) = e^{-2\pi i\tau\Lambda}|\tau \in \mathbb{R}\}$, where Λ has the spectrum $\sigma(\Lambda) = \mathbb{R}$.
• If a system is invariant under frequency-translation, then it is not invariant under frequency reversal.</td></tr>
</table>

concepts of sequential and non-sequential time but rather a continuum. Sequential und non-sequential time concepts are complementary so that there is, according to Eqs. (9.60a), (9.60b), a continuous family of complementary time concepts between sequential and non-sequential time. This fact leads to many mutually complementary partial Boolean descriptions of temporal phenomena.

11.4.2 Generic Time

The time-translation group $\mathfrak{V} = \{V(\tau)|\tau \in \mathbb{R}\}$ and the frequency translation group $\mathfrak{U} = \{U(\lambda)|\lambda \in \mathbb{R}\}$ generate two complementary descriptions of time phenomena with the unbroken symmetries of the Weyl-Heisenberg group. They define the generic concepts of time and frequency.

The time operator T associated with the affine Weyl-Heisenberg group represents a time concept which respects all continuous temporal symmetries. Since in this description no symmetry is broken, generic time is featureless. In particular there is no "direction of time" so that the concepts of "past" and "future" make no sense. Likewise, the notion of the "present" ("nowness") has no meaning. Table 11.1 summarizes some essential properties of generic time and the complementary generic frequency.

11.4.3 Logarithmic Time

The dilation group $\mathfrak{D} = \{D(\kappa)|\kappa \in \mathbb{R}\}$ with $D(\kappa) = e^{2\pi i\kappa S}$ and the logtime translation group $\mathfrak{M} = \{M(\mu)|\mu > 0\}$ with $M(\mu) = e^{2\pi i \ln T}$ generate complementary

Table 11.2 Logarithmic time and scale

two basic groups: $\mathfrak{M} \sim \mathbb{R}$ and $\mathfrak{D} \sim \mathbb{R}$

$$\mathbf{M}(\mu)\,\mathbf{D}(\kappa) \;=\; \mathbf{e}^{-2\pi \mathbf{i}\mu\kappa}\mathbf{D}(\kappa)\,\mathbf{M}(\mu)$$

$$\mathbf{M}(\mu) \;=\; \mathbf{e}^{2\pi \mathbf{i}\mu \ln \mathbf{T}}, \quad \mathbf{D}(\kappa) \;=\; \mathbf{e}^{2\pi \mathbf{i}\kappa \mathbf{S}}$$

$$\ln(\mathbf{T})\,\mathbf{S} \;-\; \mathbf{S}\,\ln(\mathbf{T}) \;=\; (\mathbf{i}/2\pi)\,\mathbb{1}$$

logtime domain	**scale domain**
• The generating group is the unitary group $M\{(\mu)\vert\mu \in \mathbb{R}\}$. • The logtime operator $\ln T$ is the self-adjoint generator of the group $\{M\{(\mu) = e^{2\pi i\mu \ln T}\vert\mu \in \mathbb{R}\}$ where $\ln T$ has the spectrum $\sigma(\ln T) = \mathbb{R}$. • The time structure is invariant under scale transformations.	• The generating group is the unitary dilation group $D\{(\kappa)\vert\kappa \in \mathbb{R}\}$. • The scale operator S is the self-adjoint generator of the group $\{D\{(\kappa) = e^{2\pi i\kappa S}\vert\tau \in \mathbb{R}\}$ where S has the spectrum $\sigma(S) = \mathbb{R}$. • The system is invariant under scale transformations.

Table 11.3 Logarithmic frequency and scale

two basic groups: $\mathfrak{N} \sim \mathbb{R}$ and $\mathfrak{D} \sim \mathbb{R}$

$$\mathbf{N}(\nu)\,\mathbf{D}(\kappa) \;=\; \mathbf{e}^{2\pi \mathbf{i}\nu\kappa}\mathbf{D}(\kappa)\,\mathbf{N}(\mu)$$

$$\mathbf{N}(\nu) \;=\; \mathbf{e}^{-2\pi \mathbf{i}\nu \ln \boldsymbol{\Lambda}}, \quad \mathbf{D}(\kappa) \;=\; \mathbf{e}^{2\pi \mathbf{i}\kappa \mathbf{S}}$$

$$\ln(\boldsymbol{\Lambda})\,\mathbf{S} \;-\; \mathbf{S}\,\ln(\boldsymbol{\Lambda}) \;=\; (\mathbf{1}/2\pi\mathbf{i})\,\mathbb{1}$$

logfrequency domain	**scale domain**
• The generating group is the unitary group $\{N\{(\nu)\vert\nu \in \mathbb{R}\}$. • The logfrequency operator $\ln \Lambda$ is the self-adjoint generator of the group $\{N\{(\nu) = e^{-2\pi i\nu \ln \Lambda}\vert\nu \in \mathbb{R}\}$, where the spectrum $\sigma(\ln \Lambda) = \mathbb{R}$. • The frequency structure is invariant under scale transformations.	• The generating group is the unitary dilation group $D\{(\kappa)\vert\kappa \in \mathbb{R}\}$. • The scale operator S is the self-adjoint generator of the group $\{D\{(\kappa) = e^{2\pi i\kappa S}\vert\tau \in \mathbb{R}\}$, where S has the spectrum $\sigma(S) = \mathbb{R}$. • The system is invariant under scale transformations.

descriptions of temporal scaling phenomena. Table 11.2 summarizes some essential properties of logarithmic time and the complementary scale domain.

11.4.4 Logarithmic Frequency

The dilation group $\mathfrak{D} = \{D(\kappa)|\kappa \in \mathbb{R}\}$ with $D(\kappa) = e^{2\pi i S}$ and the logfrequency group $\mathfrak{N} = \{N(\nu)|\nu \in \mathbb{R}\}$ with $N(\nu) = e^{2\pi i \nu \ln \Lambda}$ generate complementary descriptions of the scaling behavior of frequency phenomena. Table 11.3 shows an overview of some essential properties of logarithmic frequency and the complementary scale domain.

Table 11.4 Sequential semigroup $\mathfrak{S}_{\mathfrak{U}}$-time systems

$$\mathbf{U}_{\mathbf{U}}^{\pm}(\lambda)\,\mathbf{V}_{\mathbf{U}}^{\pm}(\tau) = \mathbf{V}_{\mathbf{U}}^{\pm}(\tau)\,\mathbf{U}_{\mathbf{U}}^{\pm}(\lambda)\,\mathbf{e}^{\mathbf{2\pi i\lambda\tau}}$$

Kinematics is given by the affine Weyl-Heisenberg semigroup $\mathfrak{S}_{\mathfrak{U}}$.
For most state functionals ρ all moments $\rho\{(T_U^{\pm})^n\}$ $(n = 1, 2, \cdots)$ exist.
Time-reversal symmetry is broken.
Past and future are separated by a superselecetion rule.
Present has no extension.
Causal interpretations and communication are possible.
It makes sense to speak about facts.

retarded sequential time

time domain	frequency domain
• The time operator $T_U^- < 0$ is the self-adjoint generator of the unitary semigroup $\{U_U^-(\lambda) \vert \lambda \in \mathbb{R}^-\}$ with $U_U^-(\lambda) = e^{2\pi i\lambda T_U^-}$ and the spectrum $\sigma(T_U^-) = \mathbb{R}^-$. • It represents sequential time with retarded (forward) causality.	• The frequency operator Λ_U^- generates the non-unitary semigroup $\{V_U^-(\tau) \vert \tau \in \mathbb{R}^-\}$. It is not self-adjoint, but maximal dissipative with the spectrum $\sigma(\Lambda_U^-) \subset \mathbb{C}^-$. • It corresponds to retarded sequential time.

advanced sequential time

time domain	frequency domain
• The time operator $T_U^+ > 0$ is the self-adjoint generator of the unitary semigroup $\{U_U^+(\lambda) \vert \lambda \in \mathbb{R}^+\}$ with $U_U^+(\lambda) = e^{2\pi i\lambda T_U^+}$ and the spectrum $\sigma(T_U^+) = \mathbb{R}^+$. • It represents sequential time with advanced (backward) causality.	• The frequency operator Λ_U^+ generates the non-unitary semigroup $\{V_U^+(\tau) \vert \tau \in \mathbb{R}^+\}$. It is not self-adjoint, but maximal dissipative with the spectrum $\sigma(\Lambda_U^+) \subset \mathbb{C}^+$. • It corresponds to advanced sequential time.

11.4.5 Sequential Time

The temporal system generated by the full affine Weyl-Heisenberg group is not elementary. If one divides this system into ergodically acting elementary time systems then certain symmetries of the affine Weyl-Heisenberg group will be broken so that we get four different pairs of mutually complementary time descriptions. In particular there are two complementary elementary systems, which describe sequential and non-sequential time.

Sequential time is based on the affine semigroup $\mathfrak{S}_{\mathfrak{U}}$. While in descriptions based on the affine full Weyl-Heisenberg group arbitrary temporal superpositions are possible, the sequential-time description by the *restricted* Weyl system $(U_U^{\pm}, V_U^{\pm})$ according to Eqs. (10.157a), (10.157b) does not permit superpositions of past and future.

Table 11.5 Non-sequential semigroup $\mathfrak{S}_{\mathfrak{V}}$-time systems

$$\mathbf{U}_{\mathbf{V}}^{\pm}(\lambda)\,\mathbf{V}_{\mathbf{V}}^{\pm}(\tau) = \mathbf{V}_{\mathbf{V}}^{\pm}(\tau)\,\mathbf{U}_{\mathbf{V}}^{\pm}(\lambda)\,\mathbf{e}^{\mathbf{2\pi i}\lambda\tau}$$

Kinematics is given by the affine Weyl-Heisenberg semigroup $\mathfrak{S}_{\mathfrak{V}}$.
Frequency-reversal symmetry is broken.
Positive and negative frequencies are separated by a superselecetion rule.
Time-reversal and time-translation symmetry are intact.
There is no past and no future.
Extended present, unfolding duration.
There is no causality, memory or anticipation.

positive-frequency non-sequential time

time domain	**frequency domain**
• The generating semigroup is the completely non-unitary semigroup $\{U_V^+(\lambda) \vert \lambda \in \mathbb{R}^+\}$. • The time operator T_V^+ is not self-adjoint but maximal dissipative with the spectrum $\sigma(T_V^+) \subset \mathbb{C}^+$.	• The generating semigroup is the unitary semigroup $\{V_V^+(\tau) \vert \tau \in \mathbb{R}^+\}$. • The frequency operator Λ_V^+ is self-adjoint with the spectrum $\sigma(\Lambda_V^+) = \mathbb{R}^+$.

negative-frequency non-sequential time

time domain	**frequency domain**
• The generating semigroup is the completely non-unitary semigroup $\{U_V^-(\lambda) \vert \lambda \in \mathbb{R}^-\}$. • The time operator T_V^- is not self-adjoint but maximal dissipative with the spectrum $\sigma(T_V^-) \subset \mathbb{C}^-$.	• The generating semigroup is the unitary semigroup $\{V_V^-(\tau) \vert \tau \in \mathbb{R}^-\}$. • The frequency operator Λ_V^- is self-adjoint with the spectrum $\sigma(\Lambda_V^-) = \mathbb{R}^-$.

Table 11.4 displays the two pairs of time and frequency domains for retarded and advanced solutions, typically interpreted as forward and backward time evolution. It also shows some basic properties of sequential time applying to all four domains. Detailed explanations of the synoptic representation in Table 11.4 will be given in Sect. 11.5.

11.4.6 Non-Sequential Time

Non-sequential time is based on the affine semigroup $\mathfrak{S}_{\mathfrak{V}}$. While in descriptions based on the affine full Weyl-Heisenberg group arbitrary frequency superpositions are possible, in the non-sequential description by the *restricted* Weyl system $(U_V^{\pm}, V_V^{\pm})$ according to Eqs. (10.157a), (10.157b) there are no superpositions of positive and negative frequencies. Table 11.5 shows a systematic summary of the two pairs of time and frequency domains for non-sequential time. Detailed explanations of the synoptic representation in Table 11.5 will be given in Sect. 11.6.

11.5 Sequential Time and the Affine Semigroup $\mathfrak{S}_{\mathfrak{U}}$

11.5.1 Representations of $\mathfrak{S}_{\mathfrak{U}}$ on $\check{\mathbf{L}}^2(\mathbb{R}, dt)$

Realizing the W*-algebra $\mathcal{T}$ of the temporal domain as the algebra of all bounded operators on the Lebesgue-Hilbert space $\check{\mathbf{L}}^2(\mathbb{R}, dt)$, we get the unitary *time-domain representation* of the affine group $\mathfrak{A}_{\mathfrak{U}}$ as

$$\{\check{A}_U(\lambda,\kappa)\check{\Phi}\}(t) = e^{2\pi i\lambda t}\sqrt{\kappa}\check{\Phi}(\kappa t), \quad t\in\mathbb{R},\ \check{\Phi}\in\check{\mathbf{L}}^2(\mathbb{R}, dt). \tag{11.27}$$

This representation of the affine group $\mathfrak{A}_{\mathfrak{U}}$ is *reducible*. The two closed subspaces $\check{\mathbf{L}}^2(\mathbb{R}^+, dt)$ and $\check{\mathbf{L}}^2(\mathbb{R}^-, dt)$,

$$\check{\mathbf{L}}^2\big(\mathbb{R}^\pm, dt\big) = \big\{\check{\Phi}^\pm\in\check{\mathbf{L}}^2(\mathbb{R}, dt)|\check{\Phi}^\pm(t) = 0 \text{ for } \pm t<0\big\}, \tag{11.28a}$$

$$\check{\mathbf{L}}^2(\mathbb{R}, dt) = \check{\mathbf{L}}^2\big(\mathbb{R}^-, dt\big)\oplus\check{\mathbf{L}}^2\big(\mathbb{R}^+, dt\big), \tag{11.28b}$$

are invariant under the action of the affine group $\mathfrak{A}_{\mathfrak{U}}$, so that its two faithful representations $\check{A}_U^+(\lambda,\kappa)$ and $\check{A}_U^-(\lambda,\kappa)$, given by

$$\{\check{A}_U^\pm(\lambda,\kappa)\check{\Phi}^\pm\}(t) = e^{2\pi i\lambda t}\sqrt{\kappa}\check{\Phi}^\pm(\kappa t), \quad \pm t>0,\ \check{\Phi}\in\check{\mathbf{L}}^2\big(\mathbb{R}^\pm, dt\big), \tag{11.29}$$

are irreducible. The unitary operators $\check{A}_U^\pm(\lambda,\kappa)$ can be decomposed as

$$\check{A}_U^\pm(\lambda,\kappa) = \check{U}_U^\pm(\lambda)\check{X}_U^\pm(\kappa) = e^{2\pi i\lambda\check{T}_U^\pm}e^{2\pi i\ln(\kappa)\hat{S}_U^\pm}, \quad \lambda\in\mathbb{R}, \kappa>0, \tag{11.30}$$

where the restrictions $\check{T}_\pm$ of the time operator $\check{T}$, and the restrictions $\check{S}_\pm$ of the scale operator $\check{S}$,

$$\check{T}_U^\pm := \check{T}|_{\check{\mathbf{L}}^2(\mathbb{R}^\pm, dt)}, \qquad \check{S}_U^\pm := \check{S}|_{\check{\mathbf{L}}^2(\mathbb{R}^\pm, dt)}, \tag{11.31}$$

are well-defined essentially self-adjoint operators on the Hilbert space $\check{\mathbf{L}}^2(\mathbb{R}^\pm, dt)$ with the inner product

$$\langle\check{\Phi}^\pm|\check{\Psi}^\pm\rangle = \int_{\mathbb{R}^\pm}\check{\Phi}^\pm(t)^*\check{\Psi}^\pm(t)dt, \quad \check{\Phi}^\pm,\check{\Psi}^\pm\in\check{\mathbf{L}}^2\big(\mathbb{R}^\pm, dt\big). \tag{11.32}$$

The two irreducible representations of the affine group $\mathfrak{A}_{\mathfrak{U}}$ are characterized by the spectrum $sp(\check{T}_U^\pm)$ of the time operator $\check{T}_U^\pm$,

$$sp\big(\check{T}_U^+\big)>0, \qquad sp\big(\check{T}_U^-\big)<0. \tag{11.33}$$

On the Schwartz space $\check{\mathcal{S}}(\mathbb{R}^\pm)\subset\check{\mathbf{L}}^2(\mathbb{R}^\pm, dt)$ of all rapidly decreasing complex-valued infinitely differentiable functions on $\mathbb{R}^\pm$ the essentially self-adjoint operators $\check{T}_U^\pm$ and $\check{S}_U^\pm$ are represented as

$$\{\check{T}_U^\pm\check{\Phi}^\pm\}(t) = t\check{\Phi}^\pm(t), \tag{11.34a}$$

$$\left\{\check{S}_{\pm}\check{\Phi}^{\pm}\right\}(t) = \frac{1}{4\pi i}\left\{\frac{\partial}{\partial t}t + t\frac{\partial}{\partial t}\right\}\check{\Phi}^{\pm}(t), \tag{11.34b}$$

for $\pm t > 0$ and $\Phi^{\pm} \in \check{\mathcal{S}}(\mathbb{R}^{\pm})$.

The direct sum decomposition (9.91a) or (11.28b)

$$\check{\mathcal{A}}_U = \check{\mathcal{A}}_U^- \oplus \check{\mathcal{A}}_U^+, \qquad \check{\mathcal{A}}_U^{\pm} = \mathfrak{B}\left\{\check{\mathbf{L}}^2\left(\mathbb{R}^{\pm}, dt\right)\right\}, \tag{11.35}$$

implies a *superselection rule between past and future*. The corresponding classical observables are given by the projectors

$$P_U^+ = \vartheta(\Lambda), \qquad P_U^- = \vartheta(\Lambda). \tag{11.36}$$

11.5.2 Sequential Time Breaks Time-Reversal Symmetry

While the *reducible* representation (11.27) of the affine group $\mathfrak{A}_{\mathfrak{U}}$ allows a faithful representation of the time-translation group and of the time-reversal group, this is no longer possible for the irreducible representations. In particular, the time-translation symmetry cannot be implemented in the irreducible representation of the affine group $\mathfrak{A}_{\mathfrak{U}}$ since the operator $(1/2\pi i)\partial/\partial t$ has no self-adjoint extension on a half-axis (cf., e.g., Kato 1966, Example 3.14, p. 270, or Smirnow 1967, p. 487).

But we can restrict the unitary time-translation group $\{V(\tau)|\tau \in \mathbb{R}\}$ either to the half-axis $\mathbb{R}^+$ or to the half-axis $\mathbb{R}^-$. Let $P_{\pm}$ be the orthogonal projection from $\mathbf{L}^2(\mathbb{R}, dt)$ onto the closed subspace $\mathbf{L}^2(\mathbb{R}^{\pm}, dt)$,

$$\{P_{\pm}\Phi\}(t) = \theta(\pm t)\Phi(t), \quad t \in \mathbb{R}, \ \Phi \in \mathbf{L}^2(\mathbb{R}, dt), \tag{11.37}$$

where $\vartheta : \mathbb{R} \to \mathbb{R}$ is Heaviside's unit step function with $\vartheta(t) = 1$ for $t > 0$ and $\vartheta(t) = 0$ for $t < 0$. The isometries $P_+V(\tau)P_+$ and $P_-V(-\tau)P_-$ generate two strongly continuous isometric semigroups

$$V_+(\tau) \oplus \mathbf{1} := P_+V(\tau)P_+, \qquad \mathbf{1} \oplus V_-(\tau) := P_-V(-\tau)P_-. \tag{11.38}$$

With (10.65) and $JP_{\pm}J = P_{\mp}$ we obtain

$$J\left\{V_+(\tau) \oplus \mathbf{1}\right\}J = \mathbf{1} \oplus V_-(-\tau) \quad , \tag{11.39}$$

so that the two isometric semigroups are related by time inversion.

If the affine group $\mathfrak{A}_{\mathfrak{U}}$ is extended by the time-reversal operation, then the representation (11.27) in the Hilbert space $\check{\mathbf{L}}^2(\mathbb{R}, dt) = \check{\mathbf{L}}^2(\mathbb{R}^+, dt) \oplus \check{\mathbf{L}}^2(\mathbb{R}^-, dt)$ is irreducible because the time reversal swaps the subspaces $\check{\mathbf{L}}^2(\mathbb{R}^+, dt)$ and $\check{\mathbf{L}}^2(\mathbb{R}^-, dt)$,

$$\check{J}\check{\mathbf{L}}^2\left(\mathbb{R}^+, dt\right) = \check{\mathbf{L}}^2\left(\mathbb{R}^-, dt\right), \qquad \check{J}\check{\mathbf{L}}^2\left(\mathbb{R}^-, dt\right) = \check{\mathbf{L}}^2\left(\mathbb{R}^+, dt\right), \tag{11.40a}$$

$$\hat{J}\hat{\mathbf{H}}^2\left(\mathbb{C}^+\right) = \hat{\mathbf{H}}^2\left(\mathbb{C}^-\right), \qquad \hat{J}\hat{\mathbf{H}}^2\left(\mathbb{C}^-\right) = \hat{\mathbf{H}}^2\left(\mathbb{C}^+\right). \tag{11.40b}$$

That is, the use of the two faithful irreducible representations of the affine group $\mathfrak{A}_{\mathfrak{U}}$ implies the breakdown of time-reversal symmetry.

11.5.3 Sequential Time Operators

The unitary groups $\{U_U^+(\lambda)|\lambda \in \mathbb{R}\}$ and $\{U_U^-(\lambda)|\lambda \in \mathbb{R}\}$ of the restricted Weyl system (10.157a), (10.157b) define two self-adjoint time operators T_U^+ and T_U^-,

$$U_U^+(\lambda) = e^{2\pi i\lambda T_U^+} \quad \text{with } T_U^+ > 0, \qquad U_U^-(\lambda) = e^{2\pi i\lambda T_U^-} \quad \text{with } T_U^- < 0, \quad (11.41)$$

so that the time operator T_U can be decomposed as

$$T_U = T_U^- \oplus T_U^+. \quad (11.42)$$

The operator T_U^+ is positive, while T_U^- is negative, so that the time concept related to the affine group $\mathfrak{U}$ allows a precise description of the concepts "past" and "future" of sequential time:

$$\text{the time operator } T_U^- < 0 \text{ characterizes the } \textit{past}, \quad (11.43a)$$

$$\text{the time operator } T_U^+ > 0 \text{ characterizes the } \textit{future}. \quad (11.43b)$$

The restriction $V_U^+(\tau) = P_U^+ V(\tau) P_U^+$ of the unitary operator $V(\tau)$ is a completely non-unitary[198] and strongly continuous semigroup of isometries, so that

$$V_U^+(\tau)V_U^+(\tau') = V_U^+(\tau + \tau') \quad \text{for } \tau, \tau' \geq 0, \quad (11.44a)$$

$$\|V_\pm(\tau)\| \leq 1 \quad \text{for all } \tau \geq 0. \quad (11.44b)$$

In $\check{\mathbf{L}}^2(\mathbb{R}, dt)$ the unitary group $\{V(\tau)|\tau \in \mathbb{R}\}$ has the translation representation

$$\{\check{V}(\tau)\check{\Phi}\}(t) = \check{\Phi}(t - \tau), \quad \tau, t \in \mathbb{R}, \ \check{\Phi} \in \check{\mathbf{L}}^2(\mathbb{R}, dt). \quad (11.45)$$

The so-called *outgoing subspace* $\check{\mathbf{L}}^2(\mathbb{R}^+, dt) \subset \check{\mathbf{L}}^2(\mathbb{R}, dt)$ is invariant under the *forward translation semigroup* $\{\check{V}_U^+(\tau)|\tau > 0\}$, so that we obtain the representation

$$\{\check{V}_U^+(\tau), \check{\Phi}_U^+\}(t) = \begin{cases} \check{\Phi}_U^+(t-\tau) & \text{for } t \geq \tau, \\ 0 & \text{for } t < \tau, \end{cases} \quad \check{\Phi}_U^+ \in \check{\mathbf{L}}^2(\mathbb{R}^+, dt). \quad (11.46)$$

The adjoint semigroup $\{\check{V}_U^+(\tau)^*|\tau > 0\}$ is the *backward translation semigroup*, given by

$$\{\check{V}_U^+(\tau)^*\check{\Phi}_U^+\}(t) = \check{\Phi}_U^+(t + \tau), \quad \tau, t \in \mathbb{R}^+, \ \check{\Phi}_U^+ \in \check{\mathbf{L}}^2(\mathbb{R}^+, dt). \quad (11.47)$$

It tends strongly to zero,

$$\|\check{V}_U^+(\tau)^*\Phi_U^+\|^2 = \int_\tau^\infty |\check{V}_U^+(\tau)^*\Phi_U^+|^2(\tau')d\tau' \to 0 \quad \text{as } \tau \to 0. \quad (11.48)$$

---

[198]The contraction semigroup $\{V_U^+(\tau)\}$ is said to be completely non-unitary if there exists no non-trivial subspace reducing all $V_U^+(\tau)$ in which $V_U^+(\tau)$ acts unitarily.

The elements of the forward translation semigroup $\{\check{V}_U^+(\tau)|\tau \geq 0\}$ are isometries $\check{V}_U^+(\tau)$,

$$\check{V}_U^+(\tau)^* \check{V}_U^+(\tau) = \mathbb{1} \quad \text{for all } \tau \geq 0, \tag{11.49}$$

while the elements of the backward translation semigroup $\{\check{V}_U^+(\tau)^*|\tau \geq 0\}$ are partially isometric,

$$\check{V}_U^+(\tau)\check{V}_U^+(\tau)^* = P \quad \text{with } P = P^* = P^2. \tag{11.50}$$

11.5.4 The Probability Density of Sequential Time and Its Conjugate

The probability density $t \mapsto \check{f}_U^\pm(t)$ of sequential time exists for every normalized state vector $\check{\Psi} \in \check{\mathbf{L}}^2(\mathbb{R}, dt)$. It is given by

$$\check{f}_U^\pm(t) = \frac{|\check{\Psi}_U^\pm(t)|^2}{\|\check{\Psi}_U^\pm\|^2} \quad \text{with } \check{\Psi}_U^\pm(t) := \Psi(t)\vartheta(\pm t). \tag{11.51}$$

Since the sequential time operator $\hat{T}_U$ is self-adjoint, the valuations of the positive time operator $\hat{T}_U^+$ and the negative time operator $\hat{T}_U^-$ are straightforward.

The conjugate of sequential time is given by the Fourier transform of the state vector $\check{\Psi}_U^\pm(t) := \Psi(t)\vartheta(\pm t)$,

$$\hat{\Psi}_U^\pm(\ell) := \int_{\mathbb{R}} e^{-2\pi i t \ell} \check{\Psi}_U^\pm(t) dt = \int_{\mathbb{R}^\pm} e^{-2\pi i t \ell} \check{\Psi}(t) dt. \tag{11.52}$$

This inverse Fourier transform can be written as

$$\hat{\Psi}_U^\pm(\ell) = \frac{1}{2}\check{\Psi}_U^\pm(\ell) \mp \frac{i}{2\pi} \int_{\mathbb{R}} \frac{\check{\Psi}_U^\pm(\ell')}{\ell' - \ell} d\ell', \tag{11.53}$$

so that the probability density $|\hat{\Psi}^\pm(\ell)|^2$ of the conjugate of sequential time is given by

$$\left|\hat{\Psi}_U^\pm(\ell)\right|^2 = \left|\frac{1}{2}\check{\Psi}(\ell) \mp \frac{i}{2\pi} \int_{\mathbb{R}} \frac{\check{\Psi}(\ell')}{\ell' - \ell} d\ell'\right|^2. \tag{11.54}$$

If $\ell \mapsto \hat{\Psi}(\ell)$ is real-valued, we obtain

$$\left|\hat{\Psi}_U^\pm(\ell)\right|^2 = \frac{1}{4}\hat{\Psi}^2(\ell) + \frac{1}{4}\left\{\mathsf{H}[\hat{\Psi}](\ell)\right\}^2. \tag{11.55}$$

11.6 Non-Sequential Time and the Affine Semigroup $\mathfrak{S}_{\mathfrak{V}}$

11.6.1 Representations of $\mathfrak{S}_{\mathfrak{V}}$ on $\hat{\mathbf{L}}^2(\mathbb{R}, d\ell)$

Realizing the W*-algebra $\mathcal{T}$ of the temporal domain as the algebra of all bounded operators on the Lebesgue-Hilbert space $\hat{\mathbf{L}}^2(\mathbb{R}, d\ell)$ we get the unitary *frequency-domain representation* of the affine group $\mathfrak{A}_{\mathfrak{V}}$ as

$$\{\hat{A}_V(\tau,\kappa)\hat{\Phi}\}(\ell) = e^{-2\pi i\tau\ell}\frac{1}{\sqrt{\kappa}}\hat{\Phi}(\ell/\kappa), \quad \ell \in \mathbb{R}, \ \hat{\Phi} \in \hat{\mathbf{L}}^2(\mathbb{R}, d\ell). \tag{11.56}$$

In this reducible representation of the affine group $\mathfrak{A}_{\mathfrak{V}}$, the closed subspaces $\hat{\mathbf{L}}^2(\mathbb{R}^\pm, d\ell)$,

$$\hat{\mathbf{L}}^2(\mathbb{R}^\pm, d\ell) := \{\hat{\Phi}^\pm \in \hat{\mathbf{L}}^2(\mathbb{R}, d\ell) | \hat{\Phi}^\pm(\ell) = 0 \text{ for } \pm\ell < 0\}, \tag{11.57}$$

$$\hat{\mathbf{L}}^2(\mathbb{R}, d\ell) = \hat{\mathbf{L}}^2(\mathbb{R}^-, d\ell) \oplus \hat{\mathbf{L}}^2(\mathbb{R}^+, d\ell), \tag{11.58}$$

are invariant subspaces supporting the irreducible representation $\hat{A}_V^\pm(\tau,\kappa)$,

$$\{\hat{A}_V^\pm(\tau,\kappa)\hat{\Phi}^\pm\}(\ell) = e^{-2\pi i\tau\ell}\kappa^{-1/2}\hat{\Phi}^\pm(\ell/\kappa), \quad \pm\ell > 0, \tag{11.59}$$

for every $\hat{\Phi}^\pm \in \mathbf{L}^2(\mathbb{R}^\pm, d\ell)$. The unitary operators $\hat{A}_V^\pm(\tau,\kappa)$ can be decomposed as

$$\hat{A}_V^\pm(\tau,\kappa) = \hat{V}_V^\pm(\tau)\hat{X}_V^\pm(\kappa) = e^{-2\pi i\tau\hat{\Lambda}_V^\pm}e^{2\pi i\ln(\kappa)\hat{S}_V^\pm}, \quad \tau \in \mathbb{R}, \ \kappa > 0, \tag{11.60}$$

where the restrictions $\hat{\Lambda}_\pm$ of the frequency operator $\hat{\Lambda}$, and the restrictions $\hat{S}_\pm$ of the scale operator $\hat{S}$,

$$\hat{\Lambda}_V^\pm := \hat{\Lambda}|_{\hat{\mathbf{L}}^2(\mathbb{R}^\pm, d\ell)}, \qquad \hat{S}_V^\pm := \hat{S}|_{\hat{\mathbf{L}}^2(\mathbb{R}^\pm, d\ell)}, \tag{11.61}$$

are well-defined essentially self-adjoint operators on the Hilbert space $\hat{\mathbf{L}}^2(\mathbb{R}^\pm, d\ell)$ with the inner product

$$\langle\hat{\Phi}^\pm|\hat{\Psi}^\pm\rangle = \int_{\mathbb{R}^\pm}\hat{\Phi}^\pm(\ell)^*\hat{\Psi}^\pm(\ell)d\ell, \quad \hat{\Phi}^\pm, \hat{\Psi}^\pm \in \hat{\mathbf{L}}^2(\mathbb{R}^\pm, d\ell). \tag{11.62}$$

The two irreducible representations of the affine group $\mathfrak{A}_{\mathfrak{V}}$ are characterized by the spectrum $sp(\Lambda_V^\pm)$ of the frequency operator $\Lambda_V^\pm$,

$$sp(\hat{\Lambda}_V^+) > 0, \qquad sp(\hat{\Lambda}_V^-) < 0. \tag{11.63}$$

On the Schwartz space $\hat{\mathcal{S}}(\mathbb{R}^\pm) \subset \hat{\mathbf{L}}^2(\mathbb{R}^\pm, d\ell)$ of all rapidly decreasing complex-valued infinitely differentiable functions on $\mathbb{R}^\pm$ the essentially self-adjoint operators $\hat{\Lambda}_V^\pm$ and $\hat{S}_V^\pm$ are represented as

$$\{\hat{\Lambda}_V^\pm\hat{\Phi}_V^\pm\}(\ell) = \ell\hat{\Phi}_V^\pm(\ell), \tag{11.64a}$$

$$\{\hat{S}_V \hat{\Phi}_V^{\pm}\}(\ell) = \frac{i}{4\pi}\left\{\frac{\partial}{\partial \ell}\ell + \ell\frac{\partial}{\partial \ell}\right\}\hat{\Phi}_V^{\pm}(\ell), \tag{11.64b}$$

for $\pm\ell > 0$ and $\hat{\Phi}_V^{\pm} \in \hat{\mathcal{S}}(\mathbb{R}^{\pm})$.

The direct sum decomposition (9.91b) or (11.58)

$$\mathcal{A}_V = \mathcal{A}_V^{-} \oplus \mathcal{A}_V^{+}, \quad \hat{\mathcal{A}}_V^{\pm} = \mathfrak{B}\{\hat{\mathbf{L}}^2(\mathbb{R}^{\pm}, d\ell)\}, \tag{11.65}$$

implies that the projectors $P_V^{+} = \vartheta(\Lambda_U)$ and $P_V^{-} = \vartheta(\Lambda_U)$ are classical observables allowing the following *Boolean classification*:

$$\text{the projector} P_U^{+} \text{characterizes } \textit{spectral positivity}, \tag{11.66a}$$

$$\text{the projector} P_U^{-} \text{characterizes } \textit{spectral negativity}. \tag{11.66b}$$

11.6.2 Non-Sequential Time Breaks Frequency-Reversal Symmetry

The closed orthogonal subspaces $\hat{\mathbf{L}}^2(\mathbb{R}^+, d\ell)$ and $\hat{\mathbf{L}}^2(\mathbb{R}^-, d\ell)$,

$$\hat{\mathbf{L}}^2(\mathbb{R}^+, d\ell) \perp \hat{\mathbf{L}}^2(\mathbb{R}^-, d\ell), \quad \hat{\mathbf{L}}^2(\mathbb{R}, d\ell) = \hat{\mathbf{L}}^2(\mathbb{R}^+, d\ell) \oplus \hat{\mathbf{L}}^2(\mathbb{R}^-, d\ell), \tag{11.67}$$

are invariant under the action of the group $\{U(\lambda)|\lambda \in \mathbb{R}\}$,

$$\{U(\lambda)\hat{\Phi}^+\}(\ell) \in \hat{\mathbf{L}}^2(\mathbb{R}^+, d\ell) \quad \text{for every } \hat{\Phi}^+ \in \hat{\mathbf{L}}^2(\mathbb{R}^+, d\ell), \tag{11.68a}$$

$$\{V(\tau)\hat{\Phi}^-(\ell) \in \hat{\mathbf{L}}^2(\mathbb{R}^-, d\ell) \quad \text{for every } \hat{\Phi}^- \in \hat{\mathbf{L}}^2(\mathbb{R}^-, d\ell). \tag{11.68b}$$

The subspaces $\hat{\mathbf{L}}^2(\mathbb{R}^+, d\ell)$ and $\hat{\mathbf{L}}^2(\mathbb{R}^-, d\ell)$ are related by frequency reversal. The frequency reversal operation K can be implemented by an *anti-linear* and *anti-unitary* operator K.[199]

On the Hilbert space $\hat{\mathbf{L}}^2(\mathbb{R}, d\ell)$ the operator K is defined by:

$$\{K\hat{\Phi}\}(\ell) = \hat{\Phi}(-\ell)^*, \quad \hat{\Phi} \in \hat{\mathbf{L}}^2(\mathbb{R}, d\ell), \tag{11.69a}$$

$$K(a\hat{\Phi} + b\hat{\Psi}) = a^* K\hat{\Phi} + b^* K\hat{\Psi}, \quad a, b \in \mathbb{C}, \ \hat{\Phi}, \hat{\Psi} \in \hat{\mathbf{L}}^2(\mathbb{R}, d\ell), \tag{11.69b}$$

$$\langle K\hat{\Psi}|K\hat{\Phi}\rangle = \langle\hat{\Psi}|\hat{\Phi}\rangle^* = \langle\hat{\Phi}|\hat{\Psi}\rangle, \quad \hat{\Phi}, \hat{\Psi} \in \hat{\mathbf{L}}^2(\mathbb{R}, d\ell). \tag{11.69c}$$

The frequency reversal operation fulfills the relations

$$KTK^{-1} = T, \qquad KU(\lambda)K^{-1} = U(-\lambda), \tag{11.70}$$

[199] In order that the canonical commutation relations are invariant under the operation of frequency reversal, the frequency reversal operator K has to be anti-unitary.

$$K\Lambda K^{-1} = -\Lambda, \qquad KV(\tau)K^{-1} = V(\tau), \tag{11.71}$$

$$KSK^{-1} = -S, \qquad KX(\kappa)K^{-1} = X(\kappa). \tag{11.72}$$

If the non-sequential time-related affine group $\mathfrak{A}_{\mathfrak{V}}$ is extended by the frequency-reversal operation, then the representation (11.56) in the Hilbert space $\hat{\mathbf{L}}^2(\mathbb{R}, d\ell) = \hat{\mathbf{L}}^2(\mathbb{R}^+, d\ell) \oplus \hat{\mathbf{L}}^2(\mathbb{R}^-, d\ell)$ is irreducible because the frequency reversal swaps the subspaces $\hat{\mathbf{L}}^2(\mathbb{R}^+, d\ell)$ and $\hat{\mathbf{L}}^2(\mathbb{R}^-, d\ell)$,

$$\hat{K}\hat{\mathbf{L}}^2(\mathbb{R}^+, d\ell) = \hat{\mathbf{L}}^2(\mathbb{R}^-, d\ell), \qquad \hat{K}\hat{\mathbf{L}}^2(\mathbb{R}^-, d\ell) = \hat{\mathbf{L}}^2(\mathbb{R}^+, d\ell), \tag{11.73a}$$

$$\hat{K}\hat{\mathbf{H}}^2(\mathbb{C}^+) = \hat{\mathbf{H}}^2(\mathbb{C}^-), \qquad \hat{K}\hat{\mathbf{H}}^2(\mathbb{C}^-) = \hat{\mathbf{H}}^2(\mathbb{C}^+). \tag{11.73b}$$

That is, the use of the two faithful irreducible representations of the affine group $\mathfrak{A}_{\mathfrak{V}}$ implies the breakdown of frequency-reversal symmetry.

11.6.3 Non-Sequential Time Operators

In the frequency-domain representation on the Hilbert space $\hat{\mathbf{L}}^2(\mathbb{R}, d\ell)$, the generator $\hat{T}$ of the group $\{\hat{U}(\lambda)|\lambda \in \mathbb{R}\}$ is the regular differential operator

$$\{\hat{T}\hat{\Phi}\}(\ell) = \frac{i}{2\pi}\frac{\partial\hat{\Phi}(\ell)}{\partial\ell} \quad \text{for every } \hat{\Phi} \in \hat{\mathcal{S}}(\mathbb{R}) \subset \hat{\mathbf{L}}^2(\mathbb{R}, d\ell). \tag{11.74}$$

In contrast, the generator $\hat{T}_V^+$ of the non-unitary semigroup $\{U_V^{\pm}(\lambda)| \pm \lambda \geq 0\}$ is singular and has no self-adjoint extension (cf., e.g., Akhiezer and Glazman 1993, Appendix II, §4, pp. 170–174). On the dense domain $\mathcal{D}(\hat{T}_V^+) \subset \hat{\mathbf{L}}^2(\mathbb{R}, d\ell)$,

$$\mathcal{D}(\hat{T}_V^+) = \left\{\hat{\Phi} \in \hat{\mathbf{L}}^2(\mathbb{R}^+) \,|\, \Phi \text{ is absolutely continuous and } \int_{\mathbb{R}^+}\left|\frac{d}{d\lambda}\hat{\Phi}(\lambda)\right|^2 < \infty\right\}, \tag{11.75}$$

the operator $\hat{T}_V^+$ is the minimal closed extension of the differential operator $d/d\ell$. It has the deficiency index $(0, 1)$, but is not self-adjoint. If in addition we require $\hat{\Phi}(0) = 0$, then $\hat{T}_V^+$ is maximally symmetric but admits no self-adjoint extension over $\hat{L}^2(\mathbb{R}^+, d\ell)$.

Only for the rather exceptional case when the state vector $\hat{\Phi}$ is not only in the domain $\mathcal{D}(\hat{T}_V^+)$ of the non-sequential time operator $\hat{T}_V^+$, but also fulfills the boundary condition $\hat{\Phi} = 0$, the first moment $\langle\hat{\Phi}|\hat{T}_V^{\pm}\hat{\Phi}\rangle$ and the second moment $\langle\hat{\Phi}|(\hat{T}_V^{\pm})^2\hat{\Phi}\rangle$ are finite. However, for $n > 2$ the n-th order moment $\langle\hat{\Phi}|(\hat{T}_V^{\pm})^n\hat{\Phi}\rangle$ does not necessarily exist. That is, the domain of definition of the maximally symmetric time operator $\hat{T}_V^{\pm}$ is too restrictive for an adequate discussion of non-sequential time.

Nevertheless, a comprehensive probabilistic description of non-sequential time is possible in terms of an operator-valued measure. The restriction

$$E_\tau^{\pm} := P_U^{\pm} E_\tau P_U^{\pm}, \quad \tau \in \mathbb{R}, \tag{11.76}$$

of the orthogonal spectral family $\{\hat{E}_\tau | \tau \in \mathbb{R}\}$ of the spectral representation (9.54) of the unrestricted self-adjoint time operator T is well-defined for all elements of the Hilbert space $\hat{\mathbf{L}}^2(\mathbb{R}^\pm, d\ell)$.

The eigenfunctions (10.17a), (10.17b) of the unrestricted time operator $\hat{T}$ can be represented as tempered distributions in the Schwartz space $\hat{\mathcal{S}}'(\mathbb{R}) \supset \hat{\mathbf{L}}^2(\mathbb{R}, d\ell)$ as

$$\hat{T}\hat{\alpha}_\tau = \tau\hat{\alpha}_\tau, \quad \hat{\alpha}_\tau \in \hat{\mathcal{S}}'(\mathbb{R}), \quad \hat{\alpha}_\tau(\ell) = e^{-2\pi i \tau \ell}. \tag{11.77}$$

These generalized eigenfunctions are orthogonal and complete in the sense that

$$\langle \hat{\alpha}_\tau | \hat{\alpha}_{\tau'} \rangle = \delta(\tau - \tau'), \quad \int_{\mathbb{R}} |\hat{\alpha}_\tau\rangle\langle\hat{\alpha}_\tau| \, d\tau = \mathbb{1}. \tag{11.78}$$

The spectral resolution of the unrestricted time operator can be represented as

$$\hat{T} = \int_{\mathbb{R}} \tau \, d\hat{E}_\tau = \int_{\mathbb{R}} \tau |\hat{\alpha}_\tau\rangle\langle\hat{\alpha}_\tau| d\tau. \tag{11.79}$$

The restriction (11.76) of the orthogonal spectral family $\{\hat{E}_\tau | \tau \in \mathbb{R}\}$ can be represented as

$$\hat{E}^\pm_\tau = |\hat{\alpha}^\pm_\tau\rangle\langle\hat{\alpha}^\pm_\tau| \quad \text{with } \hat{\alpha}_\tau(\ell) := e^{-2\pi i \ell \tau}\theta(\pm\ell), \ \ell \in \mathbb{R}. \tag{11.80}$$

This family of generalized functions is complete,

$$\int_{\mathbb{R}} |\hat{\alpha}^\pm_\tau\rangle\langle\hat{\alpha}^\pm_\tau| d\tau = \mathbb{1}, \tag{11.81a}$$

but not orthogonal,

$$\langle \hat{\alpha}^\pm_\tau | \hat{\alpha}^\pm_{\tau'} \rangle = \int_{\mathbb{R}^\pm} e^{-2\pi i \ell(\tau - \tau')} d\ell = \tfrac{1}{2}\delta(\tau - \tau') \mp \frac{i}{2\pi}\frac{\mathcal{P}}{\tau - \tau'}, \tag{11.81b}$$

where $\mathcal{P}$ stands for the principal value. This non-orthogonal spectral family defines for all Borel subsets $\mathcal{B}$ of $\mathbb{R}$ a unique *positive operator-valued measure* for non-sequential time.

Pro Memoria: Operator-Valued Measures

A normalized positive operator-valued measure (or *POV-measure* for short) $E : \Sigma \to \mathcal{B}(\mathfrak{H})$ is defined by the following properties:

- E is *non-negative*, that is, $E(\mathcal{B}) \geq 0$ for each Borel set $\mathcal{B}$.
- E is *additive*: if $\mathcal{B}_1 \cap \mathcal{B}_2 = \emptyset$, then $E(\mathcal{B}_1 \cup \mathcal{B}_2) = E(\mathcal{B}_1) + E(\mathcal{B}_2)$.
- E is *continuous* in the sense that $\sup E(\mathcal{B}_n) = E(\mathcal{B})$ whenever $\{\mathcal{B}_n\}$ is an increasing sequence of Borel sets of $\mathbb{R}$ whose union $\mathcal{B}$ is also a Borel set of $\mathbb{R}$.
- E is *normalized* in the sense that $E(\mathbb{R}) = \mathbb{1}$.

The non-orthogonal spectral family (11.80) defines for all Borel subset $\mathcal{B}$ of $\mathbb{R}$ a unique *positive operator-valued measure* $\mathcal{B} \mapsto \hat{E}^\pm(\mathcal{B})$ associated with the non-sequential time operator,

$$\hat{E}^\pm(\mathcal{B}) = \int_{\mathcal{B}} |\hat{\alpha}^\pm_\tau\rangle\langle\alpha^\pm_\tau| d\tau. \tag{11.82}$$

The weak Weyl relation (10.143a), (10.143b) implies that the operator-valued measure $\hat{E}^{\pm}$ is covariant under time displacements,

$$e^{2\pi i\tau\Lambda^{\pm}}\hat{E}^{\pm}(\mathcal{B})e^{-2\pi i\tau\Lambda^{\pm}} = \hat{E}^{\pm}(\mathcal{B}+\tau) \quad \text{for all Borel subset } \mathcal{B} \text{ of } \mathbb{R}. \tag{11.83}$$

For every normalized vector Φ in the Hilbert space $\hat{\mathbf{L}}^2(\mathbb{R}^{\pm}, d\ell)$ the operator-valued measure $\mathcal{B} \mapsto \hat{E}^{\pm}(\mathcal{B})$ defines a unique probability measure $\mu_V^{\pm}$ for non-sequential time,

$$\mu_V^{\pm}(\mathcal{B}) := \langle\hat{\Phi}|\hat{E}^{\pm}(\mathcal{B})|\hat{\Phi}\rangle, \tag{11.84}$$

which provides a comprehensive probabilistic description of non-sequential temporal phenomena.

11.6.4 The Probability Density of Non-Sequential Time

With the representation (11.82), we obtain

$$\mu_V^{\pm}(\mathcal{B}) = \int_{\mathcal{B}} d\tau \left|\langle\hat{\Phi}|\hat{\alpha}_\tau^{\pm}\rangle\right|^2 = \int_{\mathcal{B}} d\tau f_V^{\pm}(\tau), \tag{11.85}$$

where the probability density $f_V^{\pm}$ is given by

$$2f_V^{\pm}(\tau) = \left|\int_{\mathbb{R}^{\pm}} (\hat{\Phi}^{\pm})^*(\ell)\hat{\alpha}_\tau^{+}(\ell)d\ell\right|^2 = \left|\int_{\mathbb{R}^{\pm}} (\hat{\Phi}^{\pm})^*(\ell)e^{-2\pi i\ell\tau}d\ell\right|^2, \tag{11.86a}$$

$$= \left|\int_{\mathbb{R}^{\pm}} \hat{\Phi}^{\pm}(\ell)\cos(2\pi\ell\tau)d\ell\right|^2 + \left|\int_{\mathbb{R}^{\pm}} \hat{\Phi}^{\pm}(\ell)\sin(2\pi\ell\tau)d\ell\right|^2. \tag{11.86b}$$

The asymptotic expansions

$$\int_0^{\infty} f(\ell)\cos(2\pi\ell\tau)d\ell \approx -f'(0)(2\pi\tau)^{-2} + f'''(0)(2\pi\tau)^{-4} - O(\tau^{-6}), \tag{11.87a}$$

$$\int_0^{\infty} f(\ell)\sin(2\pi\ell\tau)d\ell \approx +f(0)(2\pi\tau)^{-1} - f''(0)(2\pi\tau)^{-3} + O(\tau^{-5}), \tag{11.87b}$$

imply that for $\hat{\Phi}(0) \neq 0$ the measure $\mu_V^{\pm}$ defines a *probability distribution without finite moments*. In this case $\mu_V^{\pm}$ belongs to the class of *heavy-tailed* probability distributions[200] whose tails are not exponentially bounded, so that

$$f_V^{\pm}(\tau) \sim \tau^{-(1+\kappa)} \quad \text{as } \tau \to \infty,\ \kappa > 0. \tag{11.88}$$

---

[200]Compare for example Rachev (2003). Since the late 1800s, researchers have been aware that the probability of events that fall on the tail ends of a statistical distribution cannot always be described by a Gaussian distribution. Heavy-tailed distributions were first introduced in 1906 by the economist Vilfredo Pareto in the context of income distributions. In 1920 the French mathemati-

The properties of heavy-tailed distributions are qualitatively different from commonly used distributions (such as Gaussian or Poisson distributions). They exhibit *long-range dependencies*, which has important implications for the interpretation of non-sequential time. The corresponding stochastic processes are *long-term memory processes*.

Example: Pareto Distribution
A Pareto distribution is a family of continuous power-law probability distributions specified by two parameters, the location parameter $\beta > 0$ and the shape parameter $\alpha > 0$,

$$f(x) := \begin{cases} \alpha\beta^{\alpha}x^{-\alpha-1} & \text{if } \beta \leq x < \infty, \\ 0 & \text{if } \beta > x. \end{cases} \tag{11.89}$$

so that the n-th moments are given by

$$\mathcal{E}\big(x^n\big) = \beta^n \frac{\alpha}{\alpha - n} \quad \text{if } 0 < n < \alpha, \tag{11.90a}$$

$$\mathcal{E}\big(x^n\big) = \infty \quad \text{if } n \geq \alpha. \tag{11.90b}$$

However, the characteristic function exists for all parameters of the Pareto distribution. It is given by

$$\mathcal{E}\big\{e^{i\kappa x}\big\} = \alpha(-i\beta\kappa)^{\alpha}\Gamma(-\alpha, -i\beta\kappa), \quad \kappa \in \mathbb{R}, \tag{11.91}$$

where Γ is the incomplete gamma function.
Pareto distributions are related to Zipf distributions by an exchange of variables,

$$y(x) \overset{\text{Zipf}}{\sim} x^{\alpha} \quad \Longleftrightarrow \quad x(y) \overset{\text{Pareto}}{\sim} x^{-1/\alpha}. \tag{11.92}$$

11.7 Examples of Temporal Distributions

11.7.1 Featureless Gaussian Time Concepts

If the state functional of the temporal domain is a Gaussian given by $\rho_{\mathrm{G}} \in \mathcal{T}_*$, it can be characterized by the mean values $\bar{t}$, $\bar{\ell}$ and the covariances σ_T^2, σ_Λ^2:

$$\rho_{\mathrm{G}}(T) := \bar{t}, \tag{11.93a}$$

$$\rho_{\mathrm{G}}(\Lambda) := \bar{\ell}, \tag{11.93b}$$

$$\rho_{\mathrm{G}}(T - \bar{t})^2 := \sigma_T^2, \tag{11.93c}$$

---

cian Paul Lévy introduced a broad class of probability distributions with heavy tails, which nowadays are usually called Lévy-Pareto stable distributions. Benoit Mandelbrot (1963) recognized the importance of the heavy-tailed nature of many financial time series.

For a long time, heavy-tailed distributions were largely considered as pathological. However, the concept of infinite moments of a probability distribution is in no way reprehensible. Current research acknowledges heavy-tailed distributions as instrumental for modeling real-world phenomena like the fluctuations of stock-market prices, the word frequencies in natural languages (Zipf's law), weather forecasts, catastrophic risks in climate science, or earthquake prediction.

$$\rho_{\rm G}(\Lambda-\bar{\ell})^2 := \sigma_\Lambda^2. \tag{11.93d}$$

This Gaussian state functional represents a pure state if and only if

$$\sigma_T\sigma_\Lambda = 1/(4\pi). \tag{11.93e}$$

For simplicity and without loss of generality we choose $\bar{t}=0$ and $\bar{\ell}=0$, so that in the time domain $\rho_{\rm G}$ can be represented on the Lebesgue-Hilbert space $\check{\mathbf{L}}^2(\mathbb{R}, dt)$ by the normalized state vector $\check{\Psi}_{\rm G} \in \check{\mathbf{L}}^2(\mathbb{R}, dt)$,

$$t \mapsto \check{\Psi}_{\rm G}(t) = \frac{1}{\sqrt[4]{2\pi\sigma_T^2}} e^{-t^2/(4\sigma_T^2)}, \tag{11.94}$$

or, equivalently, by its Fourier transform $\ell \mapsto \hat{\Psi}_{\rm G}(\ell)$ of $t \mapsto \check{\Psi}_{\rm G}(t)$:

$$\ell \mapsto \hat{\Psi}_{\rm G}(\ell) := \int_{\mathbb{R}} e^{-2\pi i\ell t}\check{\Psi}_{\rm G}(t)dt = \frac{1}{\sqrt[4]{2\pi\sigma_\Lambda^2}} e^{-\ell^2/(4\sigma_\Lambda^2)}. \tag{11.95}$$

The corresponding quantum state functional $\rho_{\rm G}(T)$ is uniquely determined by the state generating function

$$(\ell, t) \mapsto w_G(t,\ell) = \rho_{\rm G}(T)\{W(\ell, t)\}, \quad t, \ell \in \mathbb{R}, \tag{11.96}$$

where the unitary Weyl operator (9.46) is given by

$$W(\ell, t) = e^{2\pi i\ell T - 2\pi i t\Lambda} = V(t)U(\ell)e^{\pi i\ell t}. \tag{11.97}$$

With the state vector (11.94) we obtain

$$w_G(t,\ell) = e^{-2\pi^2\sigma_T^2\ell^2 - 2\pi^2\sigma_\Lambda^2 t^2}, \quad \sigma_T\sigma_\Lambda = 1/(4\pi), \tag{11.98}$$

so that the temporal probability density $t \mapsto \check{f}(t)$ and the temporal spectral density $\ell \mapsto \hat{f}(\ell)$ are:

$$\check{f}(t) = w_G(t,0) = e^{-2\pi^2\sigma_\ell^2 t^2} = \frac{1}{\sqrt{2\pi\sigma_T^2}} e^{-t^2/(2\sigma_T^2)}, \tag{11.99a}$$

$$\hat{f}(\ell) = w_G(0,\ell) = e^{-2\pi^2\sigma_T^2\ell^2} = \frac{1}{\sqrt{2\pi\sigma_\Lambda^2}} e^{-\lambda^2/(2\sigma_\Lambda^2)}. \tag{11.99b}$$

In terms of the state vector $\check{\Psi}_{\rm G}$ and its Fourier transform $\hat{\Psi}_{\rm G}$ we can also write

$$\check{f}(t) = \left|\check{\Psi}_{\rm G}(t)\right|^2, \qquad \hat{f}(\ell) = \left|\hat{\Psi}_{\rm G}(\ell)\right|^2. \tag{11.100}$$

In this generic description of the temporal domain all conceivable temporal symmetries are present. However, as Pierre Curie argued, for any observable effect a particular type of *symmetry must be broken.*[201] This asymmetry arises from the (explicit or spontaneous) breakdown of the symmetries of the fundamental laws. A corresponding fundamental description is featureless with respect to all conserved symmetries.

In the following we will discuss the symmetry breakings which are induced by the affine subgroups $\mathfrak{A}_{\mathfrak{U}}$ and $\mathfrak{A}_{\mathfrak{V}}$ (see Sect. 9.5.2) of the full temporal symmetry group $\mathfrak{T}$. They lead to various time concepts associated with the algebras $\mathcal{A}_U^{\pm}$ and $\mathcal{A}_V^{\pm}$. The latter in particular, as addressed, e.g., in Sect. 11.7.3, will be relevant from a different point of view explicated in Chap. 13.

11.7.2 Time-Frequency Systems Based on the Semigroup $\mathfrak{S}_{\mathfrak{U}}$

The state vector $\check{\Psi}_{\mathrm{U}}^{\pm} \in \check{\mathbf{L}}^2(\mathbb{R}^{\pm}, dt)$, which realizes *sequential time* is given by the restriction of the Gaussian state vector (11.94) to a half-axis

$$t \mapsto \check{\Psi}_{\mathrm{U}}^{\pm}(t) = \frac{\sqrt{2}}{\sqrt[4]{2\pi\sigma_T^2}} e^{-t^2/(4\sigma_T^2)} \vartheta(\pm t). \tag{11.101}$$

The probability density $t \mapsto \check{f}_U^{\pm}(t)$ of sequential time is then given by

$$\check{f}_U^{\pm}(t) = \left|\check{\Psi}_{\mathrm{U}}^{\pm}(t)\right|^2 = \frac{2}{\sqrt{2\pi\sigma_T^2}} e^{-t^2/(2\sigma_T^2)} \vartheta(\pm t), \quad t \in \mathbb{R}^{\pm}. \tag{11.102}$$

Since the sequential time operators T_U^+ and T_U^- are self-adjoint, the valuations of both the positive time operator $\hat{T}_U^+$ and the negative time operator $\hat{T}_U^-$ are straightforward.

The probability density $|\hat{\Psi}^{\pm}(\ell)|^2$ of the *conjugate* of sequential time is given by Eq. (11.54)

$$\left|\hat{\Psi}_U^{\pm}(\ell)\right|^2 = \left|\frac{1}{2}\check{\Psi}(\ell) \mp \frac{i}{2\pi}\int_{\mathbb{R}} \frac{\check{\Psi}(\ell')}{\ell' - \ell} d\ell'\right|^2, \tag{11.103}$$

where $\ell \mapsto \hat{\Psi}_G^{\pm}(\ell)$ is the Fourier-transformed Gaussian state vector

$$\hat{\Psi}_{\mathrm{G}}(\ell) := \int_{\mathbb{R}} e^{-2\pi i \ell t} \check{\Psi}_{\mathrm{G}}(t) dt = \frac{1}{\sqrt[4]{2\pi\sigma_\Lambda^2}} e^{-\ell^2/(4\sigma_\Lambda^2)}. \tag{11.104}$$

Since $\hat{\Psi}_{\mathrm{G}}$ is real-valued, we obtain

$$\left|\hat{\Psi}_{\mathrm{U}}^{\pm}(\ell)\right|^2 = \frac{1}{4}\left\{\hat{\Psi}_{\mathrm{G}}(\ell)\right\}^2 + \frac{1}{4}\left\{\mathsf{H}[\hat{\Psi}_G](\ell)\right\}^2. \tag{11.105}$$

[201]"C'est la dissymétrie qui crée le phénomène" (Curie 1894a, p. 400).

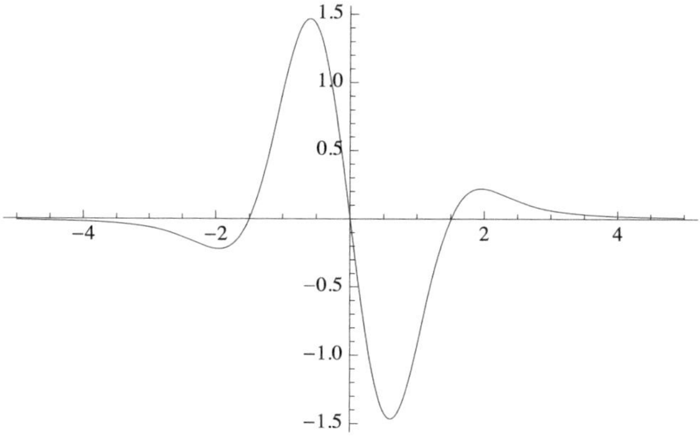

Fig. 11.1 Dawson's function $x \mapsto D(x)$ according to Eq. (11.106)

The Hilbert transform of the Gaussian function $x \mapsto f(-a^2x^2)$ is Dawson's function $x \mapsto D(x)$ (Weideman 1995, p. 257, Eq. 5.33),

$$\frac{1}{\pi} \fint_{-\infty}^{\infty} \frac{e^{-a^2\xi^2}}{\xi - x} d\xi = -\frac{2}{\sqrt{\pi}} D(ax). \tag{11.106}$$

Dawson's function (see Fig. 11.1) is an entire function of $z \in \mathbb{C}$ (see Olver et al. 2010, Chap. 7), which can be represented as

$$D(z) := e^{-z^2} \int_0^z e^{\xi^2} d\xi = \frac{i\sqrt{\pi}}{2} e^{-z^2} \{1 - \operatorname{erfc}(-iz)\}. \tag{11.107}$$

With Eq. (11.104) we obtain

$$\left|\hat{\Psi}_{\mathrm{U}}^{\pm}(\ell)\right|^2 = \frac{1}{\sqrt{2\pi\sigma_\Lambda^2}} \left\{ \frac{1}{4} e^{-\ell^2/(2\sigma_\Lambda^2)} + \frac{1}{4} \left| \frac{1}{\pi} \fint_{\mathbb{R}} \frac{e^{-\ell'^2/(4\sigma_\Lambda^2)}}{\ell' - \ell} d\ell' \right|^2 \right\}, \tag{11.108}$$

so that the probability density of the conjugate of sequential time is heavy-tailed and given by

$$\ell \mapsto \left|\hat{\Psi}_{\mathrm{U}}^{\pm}(\ell)\right|^2 = \left\{ \frac{1}{4} e^{-\ell^2/(4\sigma_\Lambda^2)} + \frac{1}{\pi} D^2(\ell/2\sigma_\Lambda) \right\}, \tag{11.109}$$

with

$$\int_{-\infty}^{\infty} \left|\hat{\Psi}_{\mathrm{U}}^{\pm}(\ell)\right|^2 d\ell = 1, \qquad \int_{-\infty}^{\infty} \ell^n \left|\hat{\Psi}_{\mathrm{U}}^{\pm}(\ell)\right|^2 d\ell = \infty, \quad n = 1, 2, 3, \ldots. \tag{11.110}$$

11.7.3 Time-Frequency Systems Based on the Semigroup $\mathfrak{S}_{\mathfrak{V}}$

Non-sequential Gaussian time can be described by the restriction of the Fourier transform $\Psi_{\mathrm{G}}(\ell)$ of the state vector $t \to \check{\Psi}_{\mathrm{G}}(t)$,

$$\check{\Psi}_{\mathrm{G}}(t) = \frac{1}{\sqrt[4]{2\pi\sigma_T^2}} e^{-t^2/(4\sigma_T^2)}, \quad t \in \mathbb{R}, \tag{11.111}$$

to the half-space $\ell \in \mathbb{R}^{\pm}$,

$$\ell \mapsto \hat{\Psi}_{\mathrm{V}}^{\pm}(\ell) := \frac{\sqrt{2}}{\sqrt[4]{2\pi\sigma_\Lambda^2}} e^{-\ell^2/(4\sigma_\Lambda^2)} \vartheta(\pm\ell). \tag{11.112}$$

Although this truncated Gaussian state vector $\hat{\Psi}_{\mathrm{V}}^{\pm}$ is not in the domain of the time operator $\hat{T}_V^{\pm}$, the spectral density $t \mapsto \check{f}_V^{\pm}(t)$ of non-sequential time is well-defined for all $\ell \in \mathbb{R}^{\pm}$:

$$\hat{f}_V^{\pm}(\ell) = \left|\hat{\Psi}_{\mathrm{V}}^{\pm}(\ell)\right|^2 = \frac{2}{\sqrt{2\pi\sigma_\Lambda^2}} e^{-\ell^2/(2\sigma_\Lambda^2)} \vartheta(\pm\ell), \quad \ell \in \mathbb{R}^{\pm}. \tag{11.113}$$

The probability density of non-sequential time is given by

$$t \mapsto \check{f}_V^{\pm}(t) = \left|\check{\Psi}_{\mathrm{V}}^{\pm}(t)\right|^2, \tag{11.114}$$

where $\check{\Psi}_{\mathrm{V}}^{\pm}$ is the Fourier transform of $\hat{\Psi}_{\mathrm{V}}^{\pm}$,

$$\check{\Psi}_{\mathrm{V}}^{\pm}(t) = \int_{\mathbb{R}} e^{2\pi i t \ell} \hat{\Psi}_{\mathrm{V}}^{\pm}(\ell) d\ell = \int_{\mathbb{R}^{\pm}} e^{2\pi i t \ell} \hat{\Psi}_{\mathrm{V}}^{\pm}(\ell) d\ell. \tag{11.115}$$

With the Hardy decomposition we obtain

$$\check{\Psi}_{\mathrm{V}}^{\pm}(t) = \frac{1}{2}\check{\Psi}_{\mathrm{V}}(t) \pm \frac{1}{2\pi i} \fint_{\mathbb{R}} \frac{\check{\Psi}_{\mathrm{V}}(t')}{t'-t} dt', \tag{11.116}$$

so that the probability density $|\check{\Psi}_{\mathrm{V}}^{\pm}(t)|^2$ of non-sequential time is given by

$$\left|\check{\Psi}_{\mathrm{V}}^{\pm}(t)\right|^2 = \left|\frac{1}{2}\check{\Psi}_{\mathrm{V}}(t) \mp \frac{i}{2\pi} \fint_{\mathbb{R}} \frac{\check{\Psi}_{\mathrm{V}}(t')}{t'-t} dt'\right|^2. \tag{11.117}$$

Since $t \mapsto \check{\Psi}_{\mathrm{V}}(t)$ is a real-valued function, we obtain the probability density $f_V^{\pm}(t) := \{\check{\Psi}_{\mathrm{G}}^{\pm}(t)\}^2$ of non-sequential Gaussian time with the Hilbert transform (11.106) and $\check{\Psi}_{\mathrm{G}}(t) = e^{-t^2/(4\sigma_T^2)} / \sqrt[4]{2\pi\sigma_T^2}$:

$$f_V^{\pm}(t) = \left|\check{\Psi}_{\mathrm{V}}^{\pm}(t)\right|^2 = \frac{1}{4}\left\{\check{\Psi}_{\mathrm{V}}(t)\right\}^2 + \frac{1}{4\pi^2}\left|\fint_{\mathbb{R}} \frac{\check{\Psi}_{\mathrm{V}}(t')}{t'-t} dt'\right|^2 \tag{11.118a}$$

$$= \left\{ \frac{1}{4} e^{-t^2/(4\sigma_T^2)} + \frac{1}{\pi} D^2(t/2\sigma_T) \right\}. \tag{11.118b}$$

The mean value and variance of $f_V^{\pm}$ are infinite,

$$\int_{-\infty}^{\infty} t f_V^{\pm}(t) dt = \infty, \qquad \int_{-\infty}^{\infty} t^2 f_V^{\pm}(t) dt = \infty, \tag{11.119}$$

so that the probability density of non-sequential Gaussian time is heavy-tailed. While a Markovian process with a Gaussian probability distribution $t \mapsto f(t)$ has short memory, a non-sequential Gaussian process whose variances diverges with time has a long memory and infinitely long-range correlations.

The top of Fig. 11.2 shows the Gaussian probability density $t \mapsto f(t)$ for the unrestricted Weyl system in a featureless Gaussian state (blue) and the probability density $t \mapsto f_V^+$ for Gaussian non-sequential time (red). The bottom figure illustrates how the densities behave for $f < 0.10$.

11.7.4 Gaussian Logarithmic Time

11.7.4.1 The Log-Normal Distribution

Consider a normally distributed random variable X with the probability density function $x \mapsto f_G(x)$

$$f_G(x) = \frac{1}{2\pi\sigma^2} e^{-(x-\mu)^2/(2\sigma^2)}, \qquad x \in \mathbb{R},\ \mu \in \mathbb{R},\ \sigma > 0. \tag{11.120}$$

Let $Y = e^X$ so that $X = \ln Y$. Then the probability density function of Y is a log-normal distribution (see Fig. 11.3):

$$f_{\mathrm{LN}}(y) dy = \begin{cases} \dfrac{1}{\sqrt{2\pi\sigma^2}} e^{-\{\ln(y/\mu)\}^2/(2\sigma^2)} \dfrac{dy}{y} & \text{for } y > 0, \\ 0 & \text{for } y \leq 0. \end{cases} \tag{11.121}$$

A random variable that is log-normally distributed takes only positive real values. With respect to a normal distribution with mean value μ and variance σ^2, the corresponding log-normal probability density function has

$$\text{mean value} \quad e^{\mu+\sigma^2/2}, \tag{11.122a}$$

$$\text{and variance} \quad e^{2\mu+\sigma^2}\left(e^{\sigma^2} - 1\right). \tag{11.122b}$$

Note that μ and σ^2 are *scale* and *shape parameters* of the log-normal distribution, and not its location and dispersion parameters.

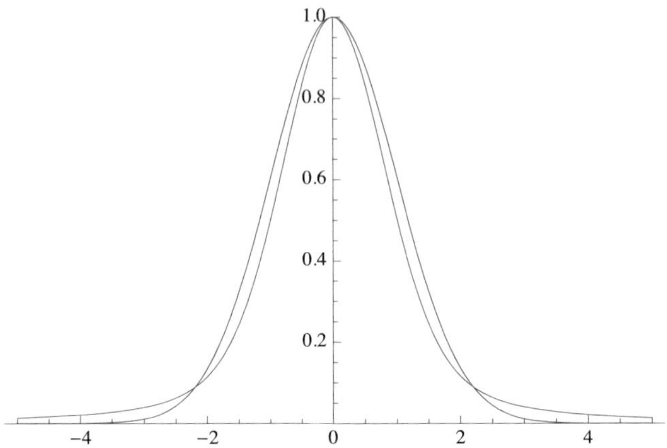

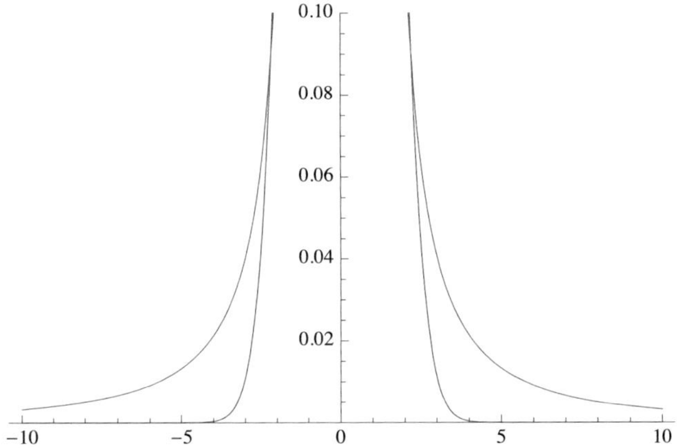

Fig. 11.2 The *blue curve* (*top figure*) shows the probability density $t \mapsto f(t)$ for the unrestricted Weyl system in a Gaussian state, normalized to $f(0) = 1$ (see Eq. (11.99a)). The *red curve* (*top figure*) shows the probability density $t \mapsto f_V^+$ for Gaussian non-sequential time normalized to $f_V^+(0) = 1$ (see Eq. (11.113)). The *bottom figure* magnifies how the densities behave for $f < 0.10$ and an increased time range. While the Gaussian probability density $t \mapsto f(t)$ of the unrestricted Weyl system is very small already for $t \approx 3$, the non-sequential time probability density $t \mapsto f_V^{\pm}(t)$ is still significant for $t > 3$

The log-normal distribution is a scale-invariant and positively-skewed distribution. Skewed distributions are particularly common for processes with conserved *positive* quantities with low mean and high variance.

While normal distributions arise when random effects are additive, log-normal distributions may develop when random effects are multiplicative. Francis Galton (1879) was the first to suggest that in certain cases the geometric mean has to be preferred over the arithmetic mean as a measure of location. Subsequently Donald McAllister (1879) gave a mathematical discussion of the log-normal distribution. For this reason, this distribution is sometimes referred to as the Galton-McAllister distribution.

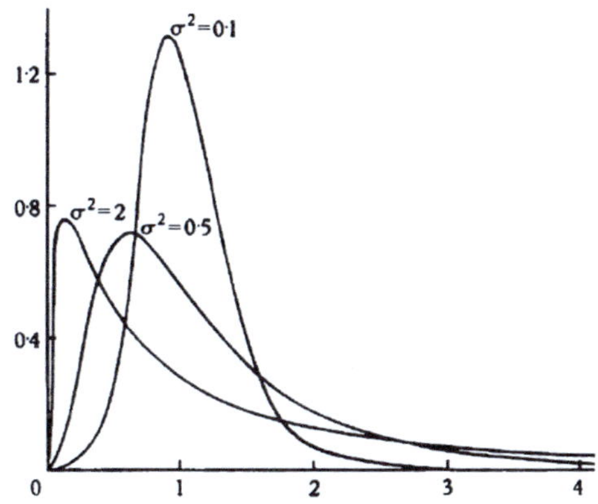

Fig. 11.3 Log-normal distribution f_{LN} according to Eq. (11.116), taken from Aitchison and Brown (1957, p. 10)

Log-normal statistical distributions are observed in a variety of scientific fields. e,g. in material sciences, medicine, ecology and economy. In particular a variety of biological mechanisms induce asymptotically log-normal distributions (see Koch 1966, 1969). For instance, the relationship between the physical strength of a stimulus and its strength as perceived by humans is not linear. The psychophysicists Ernst Heinrich Weber (1834) and Gustav Theodor Fechner (1860) proposed that the sensation (e.g., loudness) is approximately proportional to the *logarithm* of the external stimulus.[202] Another classic example is the Beer-Lambert law which relates the absorption of light to the properties of the material through which the light is traveling. It states that the light absorbance is proportional to the logarithm of the transmittance.

In addition to log-normal distributions, there is also log-log behavior, typical for instance in electric circuits. A so-called Bode log-log plot[203] is a standard format for plotting the frequency response of a linear input-output system, where the logarithm of the magnitude of the output is plotted against the logarithm of the frequency (see Fig. 11.4).

11.7.4.2 Log-Normal Stochastic Processes

Since a Gaussian stochastic process $t \mapsto g(t)$ can take on negative values, it is not appropriate for modeling the dynamics of positive random variables (like stock prices). Instead one can introduce a non-negative modification by $t \mapsto e^{g(t)}$. Such log-normal processes can be generated by multiplicative processes. In biology, multiplicative processes are used to describe the growth of organisms. As Mitzenmacher (2003) explained:

---

[202]Weber discovered that over a large dynamic range the threshold of discrimination between two stimuli increases linearly with stimulus intensity. Later Fechner proposed an integrated form of Weber's law by postulating that the external stimulus is scaled into a logarithmic internal representation of sensation.

[203]The term derives from Hendrik Wade Bode (1945), a pioneer of modern control theory.

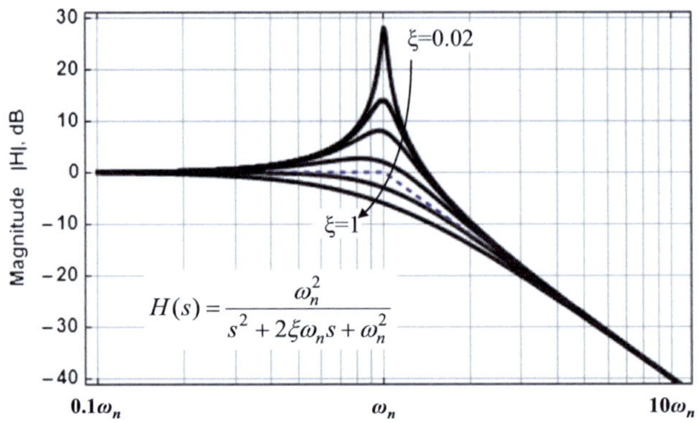

Fig. 11.4 Bode log-log plot for a second-order linear system with corner frequency ω_n, damping factor ξ and transfer function $s \mapsto H(s) = \int_0^\infty e^{-st} R(t)$, where $t \mapsto R(t)$ is the retarded linear response function (taken from p. 11 of https://web.njit.edu/~levkov/classes_files/ECE232/Handouts/Frequency%20Response.pdf)

Suppose we start with an organism of size X_0. At each step j, the organism may grow or shrink, according to a random variable F_j, so that $X_j = F_j X_{j-1}$. The idea is that the random growth of an organism is expressed as a percentage of its current weight, and is independent of its current actual size. If the F_k, $1 \le k \le j$, are all governed by independent lognormal distributions, then so is each F_j, inductively, since the product of lognormal distributions is again lognormal.
More generally, approximately lognormal distributions may be obtained even if the F_j are not themselves lognormal. Specifically, consider $\ln X_j = \ln X_0 + \sum_{k=1}^{j} \ln F_k$. Assuming the random variables $\ln F_k$ satisfy appropriate conditions, the central limit theorem says that $\sum_{k=1}^{j}$ converges to a normal distribution, and hence, for sufficiently large j, X_j is well approximated by a lognormal distribution. In particular, if the $\ln F_k$ are independent and identically distributed variables with finite mean and variance, then asymptotically X_j will approach a lognormal distribution.

A non-negative modification of a Wiener process is called a *geometric Brownian motion*. It is a multiplicative continuous-time stochastic process in which the logarithm of the randomly varying quantity follows a Wiener process with drift. More precisely, a stochastic process $\tau \mapsto Y_\tau$ is called a geometric Brownian motion if it satisfies the following stochastic differential equation:

$$dY_\tau = \mu Y_\tau d\tau + \sigma Y_\tau dW_\tau \quad \text{with initial value } Y_0 = y_0. \tag{11.123}$$

where $\tau \mapsto W_\tau$ is a Wiener process, $\mu \in \mathbb{R}$ is the drift and $\sigma > 0$ the volatility. Geometric Brownian motion is a Markov process, so that for a given present state the future state is independent of the past. It is often used to model stock price behavior (Black-Scholes model).[204]

[204] In this oversimplified model it is argued that the change dY_τ of the stock price Y_τ in a small time interval from τ to $\tau + d\tau$ is dY_τ / Y_τ. Black and Scholes (1973) decomposed dY_τ into a systematic part $\mu Y_\tau d\tau$ and a random part $\sigma dW(\tau)$, $dY_\tau = \mu Y_\tau d\tau + \sigma dW(\tau)$, where $dW(\tau)$ represents the noise term driving the stock price dynamics.

With Ito's lemma we get the unique solution

$$Y_\tau = e^{\ln y_0 + (\mu - \sigma^2/2)\tau + \sigma W_\tau}. \tag{11.124}$$

It follows that at each time τ the geometric Brownian motion has a log-normal distribution with

$$\text{mean value} \quad \mathcal{E}\{Y_\tau\} = y_0 e^{\mu\tau}, \tag{11.125a}$$

$$\text{variance} \quad \mathcal{E}\big\{Y_\tau^2 - \mathcal{E}\{Y_\tau\}^2\big\} = y_0^2 e^{2\mu\tau}\big\{e^{\tau\sigma^2} - 1\big\}. \tag{11.125b}$$

11.7.4.3 Gaussian Logarithmic Time

The logarithm $\mathfrak{L} := \ln T_U^+$ of the sequential time operator T_U^+ is well defined (see Sect. 9.7.4), so that for every normalized Gaussian state vector Ψ_U^+ (see Eq. (11.96)) the probability distribution is log-normal with the characteristic function $\hat{f}_{\rm LN}$,

$$\hat{f}_{\rm LN}(\ell) = \big\langle \Psi_U^+ \big| e^{2\pi i \ell \ln T_U^+} \Psi_U^+ \big\rangle \tag{11.126}$$

with the probability density $t \mapsto \check{f}_{\rm LN}(t)$,

$$\begin{aligned} \check{f}_{\rm LN}(t) &= \int_{-\infty}^{\infty} e^{2\pi i \ell t} \hat{f}_{\rm LN}^{+}(\ell) d\ell \\ &= \frac{1}{t\sqrt{2\pi\sigma_T^2}} e^{-\{\ln(t/\bar{t})\}^2/(2\sigma_T^2)}, \quad t > 0. \end{aligned} \tag{11.127}$$

It is appropriate to introduce a logarithmic time scale via the substitution

$$\mathfrak{l} := \ln t, \quad t = e^{\mathfrak{l}}, \tag{11.128}$$

where the parameter $\mathfrak{l}$ is the eigenvalue of the generalized eigenvalue problem

$$\mathfrak{L}\Phi = \mathfrak{l}\Phi, \quad \mathfrak{l} \in \mathbb{R}. \tag{11.129}$$

With $g_{\rm GLT}(\mathfrak{l}) := \check{f}_{\rm GLT}(t)$ we obtain (GLT stands for Gaussian logarithmic time):

$$g_{\rm GLT}(\mathfrak{l}) := \frac{e^{-\mathfrak{l}}}{\sqrt{2\pi\sigma_T^2}} e^{-(\mathfrak{l}-\bar{\mathfrak{l}})^2/(2\sigma_T^2)}, \quad \mathfrak{l} \in \mathbb{R}, \tag{11.130}$$

so that

$$g_{\rm GLT}(\mathfrak{l}) \sim e^{-\mathfrak{l}} \quad \text{or} \quad f_{\rm LN}(t) \sim t^{-1} \quad \text{for } \sigma_T^2 \text{ large}. \tag{11.131}$$

11.7.4.4 Gaussian Logarithmic Frequency

We consider a Gaussian pure state with the state functional $\rho_G(T) = 0$, $\rho_G(\Lambda) = \mu$ and $\rho_G(\Lambda^2) = \sigma_\Lambda^2$. The characteristic function of the probability distribution of Λ with respect to ρ_G is given by

$$t \mapsto \rho_G\big(e^{2\pi i \Lambda t}\big) = e^{2\pi i \mu t}\, e^{-\sigma_\Lambda^2 t^2/2}, \tag{11.132}$$

so that the corresponding probability density is given by

$$\hat{f}_G(\ell) = \frac{1}{\sqrt{2\pi\sigma_\Lambda^2}} e^{-(\ell-\mu)^2/(2\sigma_\Lambda^2)}, \quad \ell \in \mathbb{R}. \tag{11.133}$$

Since Λ is normally distributed with respect to ρ_G, $\ln\Lambda$ is log-normally distributed with $\bar{\ell} := \ln\mu$ and the Gaussian logarithmic frequency (GLF) is distributed as:

$$\hat{f}_{GLF}(\ell)d\ell = \begin{cases} \dfrac{1}{\sqrt{2\pi\sigma_\Lambda^2}} e^{-\{\ln(\ell/\bar{\ell}\}^2/(2\sigma_\Lambda^2)} \dfrac{d\ell}{\ell} & \text{for } \ell > 0, \\ 0 & \text{for } \ell \le 0, \end{cases} \tag{11.134}$$

with mean value $\bar{\ell} e^{\sigma_\Lambda^2/2}$ and variance $\bar{\ell}^2 e^{\sigma_\Lambda^2}(e^{\sigma_\Lambda^2} - 1)$. In the limit $\sigma_\Lambda \to 0$ the log-normal probability density approaches a Gaussian probability density with mean value $\bar{\ell}$ and variance $\bar{\ell}^2\sigma_\Lambda^2$. If we write Eq. (11.134) in the form

$$\begin{aligned} &\ln\{\hat{f}_{GLF}(\ell)\} \\ &= -\frac{\{\ln(\ell)\}^2}{2\sigma_\Lambda^2} + \ln(\ell)\left\{\frac{\ln(\bar{\ell})}{\sigma_\Lambda^2} - 1\right\} - \frac{\{\ln(\bar{\ell})\}^2}{2\sigma_\Lambda^2} - \ln\Big(\sqrt{2\pi\sigma_\Lambda^2}\Big), \end{aligned} \tag{11.135}$$

and if σ_Λ^2 is sufficiently large, then the first term will be small for a large range of ℓ, so that the logarithm of the density function $\ell \mapsto \hat{f}_{GLF}(\ell)$ is approximately linear for a large range of ℓ. Hence the log-normal spectral density looks, for not too small and not too large values of ℓ, like a $1/\ell$-distribution for several orders of magnitude:

$$\hat{f}_{GLF}(\ell) \sim \ell^{-1}. \tag{11.136}$$

Chapter 12
The Relevance of Sequential Time

The difference that exists between now and past and future points in time cannot, however, be reconstructed from laws of nature that are formally valid at every point in time; it is an assumption but not a consequence of the general laws of nature. Strangely enough, almost all physicists recoil emotionally from this conclusion.

Carl Friedrich von Weizsäcker (2006, p. 4)

12.1 Origin of Sequential Time

12.1.1 The Emergence of Facts[205]

A crucial precondition for the existence of facts is the possibility to discriminate between past and future. A fact refers to an event in the past which retains its facticity in the future. That is, factuality implies broken time-reversal symmetry. The registration of a fact requires the existence of non-anticipative measuring instruments in a *Boolean* domain of discourse, which can be represented by the center of the algebra of observables of the universe of discourse.

The objective past is characterized by facts. Without the possibility to discriminate between past and future it makes no sense to speak about facts. Carl Friedrich von Weizsäcker stressed the facticity of the past and the openness of the future in the sense that there are documents of the past but not of the future (von Weizsäcker 1958, p. 283, editor's translation):

> At every instant all past is an accomplished fact, which is to be regarded as known in principle; the future, however, is yet indeterminate.

[205]This subsection and some more material of Sect. 12.1 are based on a planned but unfinished chapter on *Disentangling and Entangling Systems* of Primas and Müller-Herold (1978). The draft version of that chapter remained unpublished since is was still in a not quite correct preliminary form.

H. Primas, *Knowledge and Time*,
DOI 10.1007/978-3-319-47370-3_12

Or (von Weizsäcker 2006, p. 646):

> Future events are possible, past events are facts. ... The future has not happened and cannot therefore be remembered.

The facticity of the past requires the existence of a *memory*, for example realized by material documents or fossils. While in traditional Hilbert-space quantum theory it is not evident how permanent documents can be defined, in algebraic quantum theory a representation of memories is possible in terms of classical observables. Any set of classical observables which are constant in the actual past defines a memory. Events can be defined as changes of the values of classical observables.

Despite the fact that the underlying basis of engineering science is given by the fundamental laws of physics, which have no preferred direction of time, *in every engineering application the time-reversal symmetry is broken*. The ordering of sequential engineering time is given by the one-parameter semigroup $\{V^+(\tau)|\tau \geq 0\}$. The relation between the fundamental time evolution group $\mathcal{V} = \{V(\tau)|\tau \in \mathbb{R}\}$ and the semigroup $\{V^+(\tau)|\tau \geq 0\}$ can be elucidated by a structure that was introduced into scattering theory by Lax and Phillips (1967).

A *Lax-Phillips structure* deals with a strongly continuous unitary one-parameter time evolution *group* $\mathcal{V} = \{V(\tau)|\tau \in \mathbb{R}\}$, acting on some Hilbert space $\mathcal{H}$ with two closed subspaces $\mathcal{H}_-$ and $\mathcal{H}_+$, distinguished by the following double K-structure:

$$V(\tau)\mathcal{H}_+ \subset \mathcal{H}_+ \quad \text{for } \tau \geq 0, \tag{12.1a}$$

$$V(\tau)\mathcal{H}_- \subset \mathcal{H}_- \quad \text{for } \tau \leq 0, \tag{12.1b}$$

$$\bigwedge_{\tau>0} V(\tau)\mathcal{H}_+ = \bigwedge_{\tau<0} V(\tau)\mathcal{H}_- = \{0\}, \tag{12.1c}$$

$$\bigvee_{\tau>0} V(\tau)\mathcal{H}_+ = \bigvee_{\tau<0} V(\tau)\mathcal{H}_- = \mathcal{H}. \tag{12.1d}$$

If the subspaces $\mathcal{H}_+$ and $\mathcal{H}_-$ are orthogonal, we can decompose $\mathcal{H}$ into three orthogonal components,

$$\mathcal{H} = \mathcal{H}_- \oplus \mathcal{H}_0 \oplus \mathcal{H}_+. \tag{12.2}$$

In the Lax-Phillips theory the subspaces $\mathcal{H}_-$, $\mathcal{H}_+$, and $\mathcal{H}_0$ are referred to as the *incoming subspace*, the *outgoing subspace*, and the *interaction subspace*, respectively. If $\mathcal{H}_0 = \{0\}$, then the Lax-Phillips system is called *trivial*.

Let $(\mathcal{H}, \{V(\tau)\}, \mathcal{H}_+, \mathcal{H}_-)$ be a non-trivial Lax-Phillips structure, and let $P_\pm$ be the orthogonal projection onto $\mathcal{H}_\pm^\perp$. Then (Lax and Phillips 1967, Sec. I.3):

- The one-parameter family of operators

$$S(\tau) := P_+ V(\tau) P_-, \quad \tau \geq 0, \tag{12.3}$$

  forms a strongly continuous semigroup of contractions on the subspace $\mathcal{K} = (\mathcal{H}_+ \oplus \mathcal{H}_-)^\perp$.
- Since P_- eliminates the distant past and P_+ eliminates the distant future, the semigroup $\{S(t)\}$ annihilates $\mathcal{H}_+$ and $\mathcal{H}_-$, and maps $\mathcal{K}$ onto itself.
- For each $\phi \in \mathcal{K}$ we have $\lim_{\tau\to\infty} S(\tau)\phi = 0$.

12.1.2 Automorphisms Cannot Generate New Facts

Classical observables and disjoint states of a quantum system exist only if it has infinitely many degrees of freedom. Since any realistic description of matter has to include the electromagnetic radiation field, our universe of discourse has infinitely many degrees of freedom and infinitely many inequivalent representations and classical observables.

A general result due to Klaus Hepp shows that automorphisms preserve the disjointness of states and therefore cannot change the values of classical observables (Hepp 1972, Lemma 2, p. 246):

> If φ' and φ'' are two disjoint state functionals on a C*-algebra and if α is an automorphism of this C*-algebra, then the transformed state functionals $\varphi' \circ \alpha$ and $\varphi'' \circ \alpha$ are disjoint. (12.4)

This theorem is not surprising since an automorphism is a symmetry which transforms a description into a fully equivalent description. However, the fact that an automorphic time evolution on any C*-algebra cannot generate new disjoint states is not a "no-go theorem" for the emergence of new facts. Postulating an automorphic dynamics may be justified for strictly closed systems with finitely many degrees of freedom, but not for infinite systems.[206]

No general theory is available for non-automorphic time evolutions. But Hepp proved that for appropriate quantum systems there exists a one-parameter automorphism group $\{\alpha_\tau | \tau \in \mathbb{R}\}$ such that for equivalent initial state functionals $\varphi_1, \varphi_2, \ldots$ the asymptotic limits $\varphi_1 \circ \alpha_\tau, \varphi_2 \circ \alpha_t, \ldots$ exist for $\tau \to \infty$ and are disjoint. Such state functionals are called *asymptotically disjoint.*

Since facts are permanent by definition and cannot be undone, a process which creates new facts has to be irreversible and must break the time-inversion symmetry. Lockhart and Misra (1986) discussed a dynamics which generates asymptotically disjoint final states and is, in addition, non-anticipative. It can be described as a K-flow generated by a weakly contractive semigroup. In this case the center of the algebra of observables grows irreversibly. We can argue that this growth of the center is a fundamental feature of the physical universe. (See also Primas 1997 as well as the discussion in Sect. 6.5.4.)

In classical physics there is no distinction between past and future, so that classical physics does not provide an acceptable characterization of the "flow of time".[207]

---

[206]The dynamics of group-theoretically defined elementary systems is usually automorphic, but this does not imply that the dynamics of interacting systems is automorphic. The dynamics of interacting systems must be *derived* in an unbiased manner from the known interactions without presupposing automorphic actions. Although this has not been achieved in general, many physically reasonable C*-algebraic systems without an automorphic dynamics are known.

[207]In classical thermostatics, the entropy of a system is well-defined only if the system is in thermal equilibrium. Therefore, entropy is not a generally appropriate tool to order past events in a temporal sequence: non-equilibrium entropy cannot be defined in a unique way, see Meixner (1968).

Since in quantum systems the center increases with every new event, a quantum universe of discourse allows us to *define* sequential time and the one-way "flow of time" by a parameter which measures the "size" of the center of the universe of discourse.[208]

12.1.3 Fact-Creating W*-Systems

12.1.3.1 Disjoint and Classically Equivalent States

The central part of a quantum system provides an important classification of states. The *support* S_ρ of a normal state functional ρ of a W*-algebra $\mathcal{M}$ is defined as the smallest projection operator $S \in \mathcal{M}$ such that $\rho(S) = 1$.

The *central support* C_ρ of a state functional ρ is defined as the smallest projection operator $C \in \mathcal{Z}(\mathcal{M})$ such that $\rho(C) = 1$. With this definition we get the following classification of state functionals on a W*-algebra $\mathcal{M}$ with the center $\mathcal{Z}(\mathcal{M})$:

- A state functional ρ on the W*-algebra $\mathcal{M}$ is *pure* if its support S_ρ is an *atom*.
- Two state functionals ρ' and ρ'' are called *orthogonal* if their supports $S_{\rho'}$ and $S_{\rho''}$ are orthogonal, $S_{\rho'} S_{\rho''} = 0$.
- Two state functionals ρ' and ρ'' are called *disjoint* if their *central* supports $C_{\rho'}$ and $C_{\rho''}$ are orthogonal, $C_{\rho'} C_{\rho''} = 0$.
- Two state functionals ρ' and ρ'' are called *classically equivalent* if their central supports $C_{\rho'}$ and $C_{\rho''}$ are equal, $C_{\rho'} = C_{\rho''}$.
- A state functional ρ which is dispersion-free with respect to every classical observable, $\varphi(Z^2) = \{\varphi(Z)\}^2$ for every $Z = Z^* \in \mathcal{Z}(\mathcal{M})$, is called a *factor state functional*.

Disjointness implies orthogonality, but only in commutative algebras orthogonality implies disjointness. Disjointness is a much stronger condition than orthogonality. Two pure states are disjoint if and only if there exists a classical observable such that the expectation values with respect to these states are different. Therefore, mutually disjoint states can be distinguished and classified in a classical manner.

12.1.3.2 Conditional State Changes

Conditional state changes can be described as mappings of states on states, which need not conserve the normalization condition. Completely positive normal linear maps, not necessarily conserving normalization, are called *operations*.[209] In the

---

[208] Editor's note: The remainder of Sect. 12.1 derives how this can be formulated in detail. It connects directly to Sect. 8.4.

[209] Operations in this sense were introduced by Haag and Kastler (1964). Compare also Davies and Lewis (1970) and Davies (1976).

framework considered here, an operation T_* is a positive, norm-non-increasing linear mapping from the predual $\mathcal{T}_*$ to itself, where $\mathcal{T}$ is the W*-algebra of the temporal domain,

$$\mathsf{T}_* : \mathcal{T}_* \to \mathcal{T}_*. \tag{12.5}$$

An operation is called *pure* if it transforms pure states into pure states. It is called *primary* if it transforms factor states into factor states. It is called *disjoint* if it transforms any pair of disjoint states into a pair of disjoint states (see Appendix A.4). In the W*-algebraic codification, states are represented by normal state functionals ρ in the predual $\mathcal{T}_*$ of the underlying W*-algebra $\mathcal{T}$, so that

$$\mathsf{T}_* \text{ is pure } \begin{cases} \text{if for every pure } \rho \in \mathcal{T}_* \text{ the transformed} \\ \text{state functional } \mathsf{T}_*(\rho) \in \mathcal{T}_* \text{ is pure,} \end{cases} \tag{12.6a}$$

$$\mathsf{T}_* \text{ is primary } \begin{cases} \text{if for every factor state } \rho \in \mathcal{T}_* \text{ the transformed} \\ \text{state functional } \mathsf{T}_*(\rho) \in \mathcal{T}_* \text{ is a factor state,} \end{cases} \tag{12.6b}$$

$$\mathsf{T}_* \text{ is disjoint } \begin{cases} \text{if for every pair of disjoint } \rho_1, \rho_2 \in \mathcal{T}_* \\ \text{the transformed state functionals} \\ \mathsf{T}_*(\rho_1), \mathsf{T}_*(\rho_2) \in \mathcal{T}_* \text{ are disjoint.} \end{cases} \tag{12.6c}$$

12.1.3.3 Hilbert-Space Representation of Operations

The predual $\mathcal{B}(\mathcal{H})_*$ of the W*-algebra $\mathcal{B}(\mathcal{H})$ of bounded operators on the Hilbert space $\mathcal{H}$ is isomorphic to the Banach space $\mathcal{B}^1(\mathcal{H})$ of trace-class operators,

$$\mathcal{B}(\mathcal{H})_* \sim \mathcal{B}^1(\mathcal{H}) := \left\{ B \mid B \in \mathcal{B}(\mathcal{H}), \operatorname{tr}\left|\sqrt{B^*B}\right| \leq \infty \right\}. \tag{12.7}$$

In the Hilbert-space formalism of traditional quantum mechanics every normal state functional ρ can be uniquely represented by a positive and normalized trace-class operator $D \in \mathcal{B}(\mathcal{H})_1$, called *density operator*,

$$\rho(B) = \operatorname{tr}\{DB\} \quad \text{for every } B \in \mathcal{B}(\mathcal{H}),\ D \geq \mathbb{O},\ \operatorname{tr}(D) = 1. \tag{12.8}$$

In this case, an operation T_* is a norm-non-increasing positive map $\mathcal{B}^1(\mathcal{H}) \to \mathcal{B}^1(\mathcal{H})$ with

$$\operatorname{tr}\left\{\mathsf{T}_*(D)\right\} \leq \operatorname{tr}\{D\} \quad \text{for all } D \geq \mathbb{O} \text{ with } D \in \mathcal{B}(\mathcal{H})_1. \tag{12.9}$$

The dual T of the operation T_* is a positive weak*-continuous linear map $\mathsf{T} : \mathcal{B}(\mathcal{H}) \to \mathcal{B}(\mathcal{H})$ satisfying $\mathbb{O} \leq \mathsf{T}(\mathbb{1}) \leq \mathbb{1}$. An operation T and its dual T^* can be represented (not uniquely) in a form known as the Kraus representation (Kraus 1983, p. 42),

$$\mathsf{T}_*(D) = \sum_k T_k D T_k^* \quad \text{for any } D \in \mathcal{B}^1(\mathcal{H}), \tag{12.10a}$$

$$\mathsf{T}(B) = \sum_k T_k^* B T_k \quad \text{for any } B \in \mathcal{B}(\mathcal{H}), \tag{12.10b}$$

with $T_k \in B(\mathcal{H})$ and $\sum_k T_k^* T_k \leq \mathbb{1}$. A *pure* operation (Hellwig and Kraus 1969) is representable with a single operator $T \in \mathcal{B}(\mathcal{H})$ with $T^*T \leq \mathbb{1}$,

$$\mathsf{T}_*(D) = T D T^* \quad \text{for any } D \in \mathcal{B}^1(\mathcal{H}), \tag{12.11a}$$

$$\mathsf{T}(B) = T^* B T \quad \text{for any } B \in \mathcal{B}(\mathcal{H}). \tag{12.11b}$$

12.1.3.4 Fact-Creating Operations

Every factor state functional ρ is dispersion-free with respect to an observable Z in the center $\mathcal{Z}$ if $\rho(Z^2) = \{\rho(Z)\}^2$ for every $Z \in \mathcal{Z}$. If we represent a normal state functional ρ by a (not necessarily unique) density operator $D \in \mathcal{T}$ as $\rho(X) = \operatorname{tr}\{DX\}$ for all $X \in \mathcal{T}$, we see that the unitarity of T implies that the functional $\rho^T := \mathsf{T}_*(\rho) = T\rho T^*$ is also dispersion-free on the center. That is, an operation T_* generated by a unitary operator T transforms factor states into factor states.

Let $\mathcal{T}$ be the W*-algebra of the temporal domain of discourse and $\mathcal{M} \subset \mathcal{T}$ a W*-subalgebra with the center $\mathcal{Z} = \mathcal{M}' \cap \mathcal{M}$,

$$\mathcal{Z}' = \mathcal{M}, \qquad \mathcal{M}' = \mathcal{Z}. \tag{12.12}$$

We consider an operation T generated by a unitary operator $V(\tau) \in \mathcal{M}$ by

$$X(\tau) := V^*(\tau) X V(\tau) \quad \text{for all } X \in \mathcal{T}, \tag{12.13}$$

and the W*-algebra $\mathcal{M}(\tau)$ generated by the operators $X(\tau)$,

$$\mathcal{M}(\tau) := \big\{X(\tau) \,\big|\, X \in \mathcal{M}\big\}. \tag{12.14}$$

If the W*-algebra $\mathcal{M}(\tau)$ decreases with increasing τ,

$$\mathcal{M}(\tau') \subset \mathcal{M}(\tau'') \quad \text{for } \tau' > \tau'', \tag{12.15}$$

then the center $\mathcal{Z}(\tau)$ increases with increasing τ,

$$\mathcal{Z}(\tau') \supset \mathcal{Z}(\tau'') \quad \text{for } \tau' > \tau''. \tag{12.16}$$

With $\mathcal{M}(0) = \mathcal{M}$ and $\mathcal{Z}(0) = \mathcal{Z}$, we obtain

$$\mathcal{M}(\tau) \subset \mathcal{M} \quad \text{for } \tau > 0, \tag{12.17a}$$

$$\mathcal{Z} \subset \mathcal{Z}(\tau) \quad \text{for } \tau > 0. \tag{12.17b}$$

Since facts are represented in terms of central observables, we call a system (12.17a), (12.17b) for which the center increases with increasing order parameter τ a *fact-creating W*-system*.

12.1.3.5 A Fact-Creating Weyl System

The temporal Weyl system $\{U(\lambda), V(\tau)|\lambda, \tau \in \mathbb{R}\}$ generates the W*-algebra $\mathcal{T}$ of the temporal domain. The spectral resolution of the unitary operator $U(\lambda)$,

$$U(\lambda) = \int_{-\infty}^{\infty} e^{2\pi i \lambda \mu} dF_{\mu}, \tag{12.18}$$

defines a projection-valued spectral family $\{F_{\mu}|\mu \in \mathbb{R}\}$, fulfilling

$$F_{\mu} \leq F_{\nu} \quad \text{if } \mu < \nu, \qquad F_{\mu}F_{\nu} = F_{\nu}F_{\mu} = F_{\mu\wedge\nu}, \tag{12.19a}$$

$$\operatorname*{s-lim}_{\mu\to-\infty} F_{\mu} = \mathbb{0}, \qquad \operatorname*{s-lim}_{\mu\to\infty} F_{\mu} = \mathbb{1}, \tag{12.19b}$$

where $\mu \wedge \nu$ denotes the minimum of μ, ν. The spectral family $\{F_{\mu}|\mu \in \mathbb{R}\}$ defines an increasing family of commutative W*-algebras $\{\mathcal{Z}(\tau)|\tau \in \mathbb{R}\}$,

$$\mathcal{Z}(\tau) := \{F_{\mu} \,|\, \infty < \mu < \tau\}'' \subset \mathcal{T}, \tag{12.20}$$

and its commutant $\mathcal{F}(\tau)$,

$$\mathcal{F}(\tau) := \mathcal{Z}'(\tau). \tag{12.21a}$$

Since $\mathcal{Z}$ is a W*-algebra, the double commutant theorem implies $\mathcal{Z} = \mathcal{Z}''$ and

$$\mathcal{F}'(\tau) := \mathcal{Z}(\tau). \tag{12.21b}$$

Now we define

$$\mathcal{M} := \mathcal{F}(0), \qquad \mathcal{Z} := \mathcal{Z}(0) = \mathcal{M}', \tag{12.22}$$

and, with the unitary group $\{V(\tau)|\tau \in \mathbb{R}\}$ of the Weyl system, the τ-evolved operator

$$M(\tau) := V^*(\tau)MV(\tau) \quad \text{for all } M \in \mathcal{M}, \tag{12.23}$$

for the W*-subalgebra $\mathcal{M}(\tau) \subset \mathcal{M}$,

$$\mathcal{M}(\tau) := \big(M(\tau) \in \mathcal{M}\big)''. \tag{12.24}$$

Since the subalgebras fulfill the relation

$$\mathcal{M}(\tau') \subset \mathcal{M}(\tau'') \quad \text{for } \tau' > \tau'', \tag{12.25}$$

this Weyl system is a fact-creating W*-system in the sense of Eqs. (12.17a), (12.17b).

The motion $\tau \to \mathcal{Z}(\tau)$ defines a one-parameter group if outer automorphism ϑ_{τ} of the commutative W*-algebra $\mathcal{Z} := \mathcal{Z}(\infty)$, so that $(\mathcal{Z}, \{\vartheta_{\tau}|\tau \in \mathbb{R}\})$ is a classical dynamical W*-system. With the projection operator P^+,

$$P^+ := \int_0^{\infty} dF_{\tau} \tag{12.26}$$

we define the so-called *past algebra* $\mathcal{Z}^+$,

$$\mathcal{Z}^+ := P^+ \mathcal{Z} P^+, \tag{12.27}$$

which generates a *commutative W*-algebraic K-system*[210] $\{\mathcal{Z}, \mathcal{Z}^+, \vartheta\}$ fulfilling

- $\vartheta_\tau\{\mathcal{Z}^+\} \subset \mathcal{Z}^+ \quad \text{for } \tau \geq 0,$ (12.28a)
- $\bigwedge_{\tau>0} \vartheta_\tau\{\mathcal{Z}^+\} = \mathbb{C}\mathbb{1},$ (12.28b)
- $\bigvee_{\tau>0} \vartheta_\tau\{\mathcal{Z}^+\} = \mathcal{Z}.$ (12.28c)

12.1.4 The Emergence of Sequential Time

The concept of sequential time is meaningless if no facts change. To describe this situation we introduce an increasing family of commutative W*-algebras $\mathcal{Z}(\tau) \in \mathcal{T}$, where the temporal domain is characterized by the temporal W*-algebra $\mathcal{T}$ and its commutant $\mathcal{F}(\tau) \in \mathcal{T}$,

$$\mathcal{F}(\tau) := \mathcal{Z}'(\tau), \quad \tau \in \mathbb{R}. \tag{12.29}$$

The W*-algebras $\{\mathcal{Z}(\tau)|\tau \in \mathbb{R}\}$ together with the past algebra $\mathcal{Z}^+$ defined by Eq. (12.27) generates the commutative W*-algebraic K-system $\{\mathcal{Z}, \mathcal{Z}^+, \theta\}$, where θ is the automorphism group generated by the unitary group

$$\mathfrak{V} = \{V(\tau)|\tau \in \mathbb{R}\}, \quad V(\tau) \in \mathcal{T}, \tag{12.30a}$$

$$\theta_\tau(X) = V^*(\tau) X V(\tau) \quad \text{for every } X \in \mathcal{T}. \tag{12.30b}$$

A description in terms of sequential time can then be generated by the fact-creating W*-system $\{\mathcal{T}, \mathcal{Z}^+, \mathfrak{V}\}$ and an increasing family (12.17a), (12.17b) of commutative W*-algebras $\{\mathcal{Z}^+(\tau)|\tau \geq 0\}$, where $\mathcal{Z}^+(\tau) \in \mathcal{T}(t)$.

The relations

$$\mathcal{Z}(-\tau) \subset \mathcal{Z}^+ \subset \mathcal{Z}(\tau) \quad \text{for } \tau > 0, \tag{12.31a}$$

$$\mathcal{F}(-\tau) \supset \mathcal{F}^+ \supset \mathcal{F}(\tau) \quad \text{for } \tau > 0, \tag{12.31b}$$

imply that all facts of the past remain facts in the future so that the center $\mathcal{Z}(\tau)$ represents a permanent *memory* for the order parameter τ. Moreover, the commutative

[210]Compare Cornfeld et al. (1982, Theorem 2, p. 457), Schröder (1984), and Narnhofer and Thirring (1990).

W*-algebra $\mathcal{Z}(\tau)$ *increases with increasing* τ, while the non-commutative factorial W*-algebra $\mathcal{F}(\tau)$ represents *non-factual but predictable* quantities. It *decreases with increasing* τ.

Since facts never disappear but can be generated, the center increases with every new event. If we assume that we know what we mean by the "past" and the "future" we can say that with every new event the center increases with time. But in the most fundamental physical theories there is no distinction between past and future. This situation suggests that we reverse the argument, so that we may argue that the objective past is characterized by facts and that the growth of the center of the universe of discourse defines the arrow of time by an increase of the center $\mathcal{Z}^{+}(\tau)$ with increasing τ. Sequential time and its one-way flow can then be *defined* by a parameter $t(\tau) := \rho\{\mathbb{1} \otimes T_\tau\}$, which measures the *size of the set of facts* and the *extension of the center of the universe of discourse*.[211]

We conclude that the parameter τ of the fact-generating K-flow (12.28a)–(12.28c) defines, via the relations discussed in Sects. 8.3.3 and 8.3.4, the empirically observable time t. This characterization is reminiscent of Leibniz's theory of time as the "order of successive phenomena"[212] such that "time, considered without the things, being an impossible fiction".[213]

12.2 Causation and Experiments

12.2.1 On the Connection of Cause and Effect

12.2.1.1 Causal Nexus

A process within which one event is a necessary condition for another event is described by a *causal relation*. The producing event is known as the *cause* and the event produced as its *effect*.

An event $\mathcal{E}_1$ is the cause of an event $\mathcal{E}_2$ if the occurrence of $\mathcal{E}_2$ depends on the occurrence $\mathcal{E}_1$. A causal relationship is a binary relation between two events which is *irreflexive* ("no event can be the cause of itself"), *antisymmetric* ("if $\mathcal{E}_1$ is the cause of $\mathcal{E}_2$, then $\mathcal{E}_2$ cannot be the cause of $\mathcal{E}_1$"), and *transitive* "if $\mathcal{E}_1$ is the cause of $\mathcal{E}_2$, and $\mathcal{E}_2$ the cause of $\mathcal{E}_3$, then $\mathcal{E}_1$ is the cause of $\mathcal{E}_3$"). The antisymmetry requirement implies that causation has a *direction*. That is, a causal nexus requires some universal order.

---

[211] Editor's note: A recent study of real-world systems along these lines is due to Antoniou et al. (2016). Much in the spirit of the author, Antoniou and colleagues utilize the theory of so-called *innovation eigenspaces of time operators* to define an internal time scale ("age") for the growth of facts in the evolution of dynamical systems.

[212] See Leibniz's third letter to Clarke, quoted from Alexander (1956, pp. 25–26).

[213] See Leibniz's fifth letter to Clarke, quoted from Alexander (1956, p. 77).

A fundamental issue in this respect is the relation of causality to time: Are the direction of time and the direction of causation logically independent or not? Usually it is said that effects never occur before their causes. Yet, these concepts are far from unproblematic.

12.2.1.2 Philosophical Views

For David Hume (1748, Sect. VII, part 2) causality is a central element in scientific understanding:

> For surely, if there be any relation among objects which it imports to us to know perfectly, it is that of cause and effect. On this are founded all our reasonings concerning matters of fact or existence.

An extensive philosophical debate started with Hume's belief that causality is not a matter of fact, but a "mental habit of association" (Hume 1739, book 1, part III, 1748, Sec. VII, part I). According to Hume the phrase "is a cause of" is simply shorthand for something like "is earlier than and causally related to". That is, Hume took causal direction and temporal direction to be related *by definition*. Hume argued that all content of our ideas has to come from experience, but that there is no experience of causation. What passes as causation is just the constant conjunction of cause and effect.[214] Later, Kant (1783) concluded in his *Prolegomena* that Hume's empiricist claim is false and that causation is an *a priori* concept *not* drawn from experience.

In his *On the Notion of Cause*, Bertrand Russell (1913, p. 1) wanted to dispose of Kantian *a priori* concepts and maintained

> that the word "cause" is so inextricably bound up with misleading associations as to make its complete extrusion from the philosophical vocabulary desirable. ... All philosophers, of every school, imagine that causation is one of the fundamental axioms or postulates of science, yet, oddly enough, in advanced sciences such as gravitational astronomy, the word "cause" never occurs. ... To me it seems that philosophy ought not to assume such legislative functions, and that the reason why physics has ceased to look for causes is that, in fact, there are no such things. The law of causality, I believe, like much that passes muster among philosophers, is a relic of a bygone age, surviving, like the monarchy, only because it is erroneously supposed to do no harm.

What Russell had in mind is that the concept of causation does not turn up in *first-principle physics*. However, the notion of cause is indispensable for any *experimental* science. Causation is a mode of *event propagation*. Although there is no justification by the first principles of physics, engineers have good reasons to take causation as intentional and temporally forward directed.

---

[214]Compare also Mach (1902, p. 555): "The concept of cause is replaced ... by the concept of function; the determining of the dependence of phenomena on one another", and (p. 483): "There is no cause nor effect in nature; nature has but an individual existence; nature simply is."

12.2.1.3 Relations Between Causation and Time

Most ideas about causation are—in an overt or hidden way—of temporal character.[215] Yet, the legitimacy of a causal theory depends on the possibility of avoiding circular explanations that already presuppose an understanding of temporal order. It seems that we cannot independently establish causal priorities without already knowing temporal priorities. If the state of the universe of discourse would be perfectly temporally symmetric, any distinction between past and future would disappear and no temporal asymmetry of the cause-effect relation would be possible.

A causal nexus requires some universal order. Leibniz (1989), in his *Metaphysical Foundations of Mathematics*, proposed a relation of co-existence that he called "contemporaneity":[216]

Time is the order of non-contemporaneous things.

In his causal theory of time he tried to define temporal relations in terms of sequential causal relations. By ranking events from earlier to later, Leibniz adopted a *backward causal* theory of time ("the temporal order is determined by the causal order, and not the other way round"):

> When one of two non-contemporaneous elements contains the ground for the other, the former is regarded as the antecedent, and the latter as the consequent.

Leibniz's attempt to define the earlier-later relation by means of causal connections was the forerunner to modern versions of a causal theory of time.[217]

12.2.1.4 Predictions and Retrodictions

An irreflexive, antisymmetric and transitive relation on the set of all events occurring in a universe of discourse is said to be

- backward causal if it maps present events onto past events,
- forward causal if it maps past events onto future events,
- deterministic if it is a bijective map from the set of past events onto the set of future events.

If the causal nexus is not deterministic, causal regularities can still be used for probabilistic inferences, either for situations in the past or in the future.[218] Prediction is the inference of future observational data from present ones, while retrodiction is the inference of past observational data from present ones.

---

[215]It is even difficult to express purely logical implications in a non-temporal manner. For example, the statement "a implies b" is usually explained by the sentence "when a occurs, then b follows", or by the sentence "if I can prove a, then I can show that b is valid".

[216]Leibniz's definition of co-existence is: "Given the existence of a multiplicity of concrete circumstances which are not mutually exclusive, we designate them as *contemporaneous* or *co-existing*." Quoted from Wiener (1951, p. 202).

[217]Compare for example Reichenbach (1956), Grünbaum (1973), van Fraassen (1970).

[218]For a careful analysis of the problems of prediction and retrodiction cf. Watanabe (1969).

12.2.1.5 Antecedence Postulate

The *antecedence postulate* is a time honored principle which postulates that causes must be prior to effects. It is substantiated by the success of traditional phenomenological physical and engineering descriptions. Also, conscious perception and cognition presuppose the usual forward direction of time, implying a memory of the past, but no anticipation of the future. Of course, if the correlations between two events $\mathcal{E}_1$ and $\mathcal{E}_2$ are temporally forward directed, this does not imply that $\mathcal{E}_1$ is the cause of $\mathcal{E}_2$: the two events can be effects of a third common cause preceding both of them. John Stuart Mill (1882, p. 277) proposed the following empirical criterion:

> Suppose that, by a comparison of cases of the effect, we have found an antecedent which appears to be, and perhaps is, invariably connected with it: we have not yet proved that antecedent to be the cause until we have reversed the process, and produced the effect by means of that antecedent. If we can produce the antecedent artificially and if, when we do so, the effect follows, the induction is complete; that antecedent is the cause of that consequent. ... Observation, in short, without experiment (supposing no aid from deduction) can ascertain sequences and co-existences, but can not prove causation.

That is, only an experiment can demonstrate that a particular response is the consequence of some specified cause. In spite of this empirical criterion, backward causation is not ruled out by any fundamental physical law. To avoid any prejudices we have to consider the following two possibilities:[219]

- *Forward determinism*: the thesis that, given the past of a physical system, there is a *unique future*,
- *Backward determinism*: the thesis that, given the future of a physical system, there is a *unique past*.

12.2.1.6 Causation Is an Intentional Action

Experimental science investigates which particular effect is the consequence of some specified cause. Stimulus-response experiments are essential parts of physics, chemistry and medicine. Thereby it is taken for granted that the actions of experimenters are *intentional* rather than something that merely happens to them. That is, the framework of experimental science requires as a *constitutive presupposition* that the experimenter has some *freedom of action* which is not governed by deterministic first principles of physics. Moreover, every experiment requires a carefully controlled *intervention* based on an experimental protocol that *precedes* the experimenter's actions.

Many philosophers accept a metaphysical doctrine known as the *causal closure of the physical*, according to which all physical effects can be accounted for by

[219]Editor's note: See Primas (1997) for more details about forward and backward concepts of causation and determinism. It should be noted, though, that there are subtle differences between causation and determinism that the author does not address here. See for instance Atmanspacher and Filk (2012).

physical causes (cf. Sect. 7.1.3). However, they do not find it necessary to give a tenable definition of the concept of "the physical". For instance, Jaegwon Kim (1993, p. 280) states the following doctrine:

> Any physical event that has a cause at time *t* has a physical cause at time *t*. This is the assumption that if we trace the causal ancestry of a physical event, we need never go outside the physical domain. To deny this assumption is to accept the Cartesian idea that some physical events need nonphysical causes, and if this is true there can in principle be no complete and self-sufficient physical theory of the physical domain. If the causal closure failed, our physics would need to refer in an essential way to nonphysical causal agents, perhaps Cartesian souls and their psychic properties, if it is to give a complete account of the physical world. I think most physicalists would find that picture unacceptable.

However, no experimental physicist is a metaphysical physicalist—he knows that all *causal relations* in experimental physics are established by observing the response to *deliberately* chosen stimuli. Since *present-day fundamental physics* has no room for intentionality, many philosophers think that "intentional" causation is nothing more than anthropocentric reasoning. For instance, James Woodward (2003, p. 150) defends an "agency" theory of causation without a constitutive role of human agency:

> Unfortunately, however, standard philosophical statements of the manipulability theory lead to accounts of causation that are unacceptably anthropocentric and subjectivist.

Yet, every meaningful experiment is intentional so that any requirement that experimental interventions should make no reference to human agency makes no sense. This does not entail any inconsistency with the first principles of physics because the fundamental laws of physics refer to strictly closed dynamical systems whose dynamics can be taken as bi-deterministic.[220] Every experiment depends on intentional external influences and cannot be discussed in terms of a universal bi-deterministic theory.

12.2.2 *Manipulability Accounts of Causation*

12.2.2.1 Experiments

The *manipulability theory* or *agency theory* maintains that the notions of cause and effect are related to our ability to intervene as agents, so that causal relations are po-

[220] A *dynamical system* is a system whose *state* changes in time. A dynamical system is bi-deterministic if fixing its state at some time *t* implies that every state before or after *t* is uniquely determined. A system is called *closed* if all variables that can influence it have been taken into account in the specification of its initial state at time *t* (compare Havas 1965, 1968). Any other system is called *open*.

tentially usable to *control* the effect by changing the cause in the right way (cf. Gasking 1955; Price 1992; Menzies and Price 1993). The possibility of intervention and manipulation is deeply embedded in experimental science.[221]

Many philosophers feel uneasy with manipulation theories of causation because they lead to an anthropocentric conception of causation. Yet, the possibility of intervention and manipulation is essential for experimental science. Also, as Satosi Watanabe (1986, p. 276) emphasized,

> we should eliminate the misunderstanding that causality denies the human freedom of action. It is my theory that causality presupposes the freedom of action. The reason is very simple. The causality gives us an advice of the form: if you do A, you will get B. This implies that you can choose an A. More precisely, this means that if you want to have B, you should prepare A. Seen from the side of causality, A is the cause and B is the effect but seen from the teleological point of view, B is the purpose, and A is the means. The freedom of choosing a B is the freedom of will, and the freedom of choosing an A is the freedom of action. If there were no such thing as freedom, every causal law would become an empty statement.

12.2.2.2 Response Follows Stimulus

Time-reversal symmetry is represented by a group of order two. If time-reversal symmetry is broken one gets *two* representations, one satisfying the generally accepted rules of retarded causation and the other one the strange rules of advanced causation. Since the first principles of physics are time-reversal symmetric, they do not provide a decision for which of the two possibilities is appropriate.

Nevertheless, the anisotropy of time is a *precondition* for the possibility of engineering science and any kind of communication. Communications are (Rosenblueth et al. 1943, p. 18)

> directed to the attainment of a goal—i.e. to a final condition in which the behaving object reaches a definite correlation in time or in space with respect to another object or event.

Communication requires conscious perception and cognition, hence a temporal structure which distinguishes past, present and future and which provides systems with a memory of the past, but no anticipation of the future. Moreover, as observed by Norbert Wiener (1948, p. 45),

> it is a very interesting intellectual experiment to make the fantasy of an intelligent being whose time should run the other way to our own. To such a being, all communication with

---

[221] According to Collingwood (1998, p. 296), the term "cause" in experimental science "expresses an idea relative to human conduct, because that which causes is something under human control and this control serves as means whereby human beings can control that which is caused. In this sense, the cause of an event in nature is the handle, so to speak, by which human beings can manipulate it."

> us would be impossible. Any signal he might send would reach us with a logical stream of consequents from his point of view, antecedents from ours. ... *Within any world with which we can communicate, the direction of time is uniform.*

In engineering science we consider the past as factual so that it produces documents, while the future is open. Yet we can make future-directed probability assignments. Every scientific experiment can be described phenomenologically by a map which relates the response obtained from a *classical* output system with the external stimulus generated by a *classical* input system. This also holds for descriptions of experiments with quantum objects. Experiments are based upon the following three fundamental preconceptions:

(i) The execution of an experiment requires an *intentional action.*
(ii) In contrast to a mere observation, the initial conditions for the experiment have to be *created.*
(iii) Every experiment requires *non-anticipative measuring instruments.*

A physical input-output system is called *non-anticipative* if the past of its output does not depend on the future of its input, and if the future of its input is not restricted by the past of its output. A system with outputs and internal states that depend solely on future input values is called an *anticipative system.* In physics anticipative systems are usually rejected because it is assumed that effects have to precede their causes. Engineers call the presupposition that "response follows stimulus" the *principle of retarded causation.*

12.2.3 Forward and Backward Causation

According to Sect. 12.2.1.5, *forward determinism* is the thesis that, given the past of a physical system, there is a *unique future*, and *backward determinism* is the thesis that, given the future of a physical system, there is a *unique past.* For a forward deterministic system the possibility of prediction is a criterion for *forward causation.* In the words of Moritz Schlick (1961):

> The true criterion of law, the essential sign of causality, is the success of prediction.

A non-anticipative input-output channel is called *forward causal*, while an anticipative input-output channel is called *backward causal* or *anti-causal.* In contrast to retarded causation, advanced causation with anti-causal behavior is not forbidden by the first principles of physics: forward causal and backward causal descriptions have the same logical status a priori.

Example: Simulation of an Anti-Causal Filter without Pre-Echoes
Using a delay line one can record a signal so that its values are available both forward and backward in time with respect to external clock time. In this sense one can mimic an anticipative filter. A widely known example is an oversampling CD player. Fig. 12.1 shows the causal response of CD players without pre-echoes and their anti-causal response with pre-echoes.

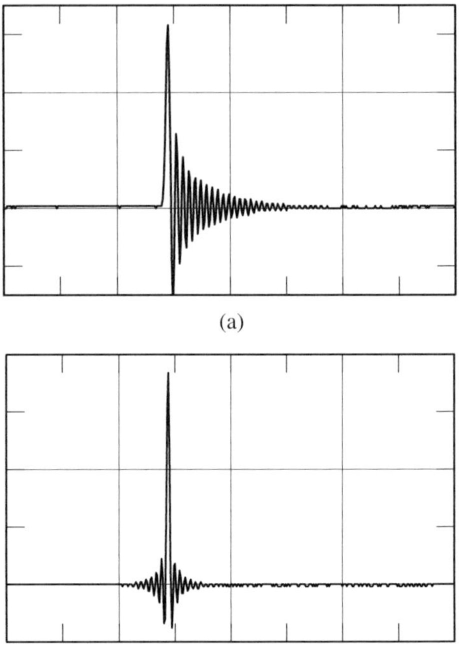

Fig. 12.1 (**a**) Causal impulse response of an analog low-pass filter in real time (as realized in CD players with analog filters without pre-echoes). (**b**) Seemingly anti-causal impulse response of a low-pass filter with an additional delay line for the input pulse (as realized in oversampling CD-players with pre-echoes)

12.2.4 Causal Structures and Time-Reversal

Time-reversal symmetry is a fundamental symmetry in all first-principle theories. More precisely, time-reversal symmetry holds if it is combined with parity-reversal and the interchange of particles and anti-particles.[222] All experiments so far have confirmed that all *fundamental* physical laws are invariant under the combined time reversal, charge conjugation, and space reflection symmetry. It is known that the laws governing gravity, electromagnetism, and the strong interaction are even individually invariant under these symmetry operations.

On the other hand, every experiment requires non-anticipative measuring instruments, hence a distinction between past and future. In a time-reversal symmetric world there is no memory of the past and no anticipation of the future. In engineering physics the direction of causation is always assumed from past to future. That is, the anisotropy of time is a precondition for experimental science. To derive experimental physics from first principles, the time-reversal symmetry of the fundamental laws of physics has to be broken. Even if we can explain the breaking of time-reversal symmetry, we still have to decide which of the two one-sided realizations to choose.

[222] A remarkable consequence of local quantum field theory is the PCT theorem, implying that invariance under proper Lorentz transformations implies invariance under the product of time reversal (T), charge conjugation (C), and space reflection (P). For a review, compare Grawert et al. (1959). If PCT-symmetry is broken, then Lorentz symmetry is also broken. See also Sect. 7.3.3.3.

The key idea of establishing cause and effect is based on the *antecedence principle* which postulates that causes must be prior to their effects. In realizable engineering systems with memory the present state may depend on past states but never on future states. This principle is often incorrectly elevated to the status of a natural law.

A system that has its output as a controlling factor for its input is called a *closed-loop system*. A closed-loop system has a feedback loop from its output to its input and can possibly anticipate future inputs. There are anticipative systems in which forward causation is not violated. According to Robert Rosen (1985, p. 339), a *causal anticipative system* is

> a system containing a predictive model of itself and/or of its environment, which allows it to change its state at an instant in accord with the model's predictions pertaining to a latter instant.

Causal systems that can encode and build models of themselves and their environment and predict future events are well known in engineering and provide models for learning and evolutionary processes. Another example is a human being who can prepare for his future behavior.

In contrast, a *non-anticipative open-loop system* is an input-output system that does not have its own output as a controlling factor for its performance. In this case a "future input" cannot influence the "present output". That is, all causes are external, with no backreaction.

It is well known that time-reversal symmetric equations of motion can exhibit solutions with broken time-reversal symmetry. If the time-reversal symmetry is not broken, then there is no place for causal asymmetries. If a solution is not invariant under the action of the time-reversal group, then there exist necessarily *two* asymmetric solutions which are related through the action of the symmetry transformations:

(i) a non-anticipative solution (any future state of the system depends only on the present and past states),
(ii) an anticipative solution (any future state determines changes of the present and past states).

We may select one of these solutions and reject the other if the time-reversal symmetry of the basic theory is broken. It is a remarkable empirical fact that "*all* experienced events can be accommodated within a *single* temporal order" (Denbigh 1981, p. 21). For quantum systems this fact can be explained by the completely positive character of the time-reversal operation (see Sect. 9.2.2). In *macroscopic* quantum systems, time-inversion symmetry is usually broken. A simple example for a spontaneously broken time-reversal symmetry is superconductivity.[223]

---

[223] A direct consequence of time-reversal symmetry is the following famous theorem by Felix Bloch (around 1933, unpublished; compare Brillouin 1933, p. 360f): *Every system in a state of lowest free energy and not subject to external fields has vanishing current density everywhere.* Or, in a provocative statement ascribed to Wolfgang Pauli: *All theories of superconductivity are wrong.*

12.3 Determinism and Predictability

In modern mathematical physics determinism is always understood in the sense of Hadamard. A motion is said to be *deterministic* if the state transition map φ is governed by the two-parameter semigroup law $\varphi_{r,s} \circ \varphi_{s,t} = \varphi_{r,t}$. That is, in a deterministic dynamical system every initial state determines all future states of a given system. In some simple systems the deterministic laws of motion may allow us to forecast its observable behavior in the near future with great accuracy. However, determinism implies neither predictability nor retrodictability in general.

Pierre Simon Laplace discussed the doctrine of determinism in the introduction to his famous *Essai Philosophique sur les Probabilités*, in which he imagined a superhuman intelligence capable of grasping the initial conditions at any fixed time of all bodies and atoms of the universe, and all forces acting upon them. For such a superhuman intelligence "nothing would be uncertain and the future, as the past, would be present to its eyes" (de Laplace 1814, translation from the Dover edition, p. 4).

Laplace's reference to the future and the past indicates that he refers to a fundamental theory with an unbroken time-reversal symmetry. His reference to a "superhuman intelligence" suggests that he is not referring to our possible knowledge of the world, but to things "as they really are". Laplace knew quite well that a perfect measurement of initial conditions is impossible, and he never asserted that deterministic systems are empirically predictable.

Many positivists tried to define determinism by predictability.[224] Such attempts are based on the notorious category mistake of conflating ontic and epistemic statements.[225] *Determinism does not deal with predictions.* Determinism refers to a hypothetical *ontic* description while predictability is an *epistemic* concept. Epistemic statements are often confused with ontic assertions. For a long time, there have been physicists who considered the distinction between determinism and predictability as hairsplitting.[226]

A related mistake is the claim is that "an underlying deterministic mechanism would refute a probabilistic theory by contradicting the randomness which ... is demanded by such a theory" (Gillies 1973, p. 135). The classic counterexample

---

[224]For example Herbert Feigl (1953, p. 408) claimed: "The clarified (purified) concept of causation is defined in terms of predictability according to a law (or, more adequately, according to a set of laws)."

[225]According to Fetzer and Almeder (1993, pp. 100f.), "ontic questions concern what is the case" while epistemic questions refer to our beliefs. "An ontic answer to an epistemic question (or vice versa) normally commits a *category mistake*." In the context discussed here, see also Atmanspacher (2002).

[226]For example, Max Born (1955a, 1955b) claimed that classical point mechanics is not deterministic since there are unstable mechanical systems which are epistemically unpredictable. For a critique of Born's view compare von Laue (1955). Similarly, Leon Brillouin claimed that "we should never discuss what happens while we are not making any observation", and that "initial conditions are not 'given'; they must be measured and observed" (Brillouin 1960, p. 92, 1962, p. 314).

is the fact that "white noise"—the most random of all possible processes—can be generated by a strictly deterministic dynamical system. It is well known that there is a close relationship between completely unpredictable stochastic processes and irreversible input-output systems.

In order to understand the chaotic behavior of some deterministic systems it is of crucial importance to distinguish between individual and statistical descriptions (cf. Atmanspacher 2002). The theory of individual chaotic dynamical systems does neither require statistical hypotheses nor measure-theoretical mathematical tools. The long-term unpredictability of the behavior of chaotic individual trajectories is a *consequence* of the sensitive dependence of their time evolution on small changes in the initial conditions.

The notion of "deterministic chaos" refers to categorically different ontic and epistemic descriptions. In detail, it expresses the behavior of a dynamical system for which every epistemic description lacks long-term predictability in spite of the fact that the mathematical description of an individual trajectory is strictly deterministic. *Determinism is a concept pertaining exclusively to the hypothetical individual states of an idealized mathematical description, while predictability pertains to epistemic states of a statistical description.*

The time evolution of an individual state of a smooth classical dynamical system can be represented by the motion of a point in a phase space. Yet, the information available to an observer is never given by a point of the phase space, but by a Borel set of non-vanishing Lebesgue measure. For every Borel set representing some knowledge there exists a smaller Borel set representing more detailed knowledge, but there is no smallest Borel set of non-vanishing Lebesgue measure representing all possible knowledge. Since a point (representing an individual state) is a Borel set of Lebessgue measure zero, it does not represent an epistemic state. That is, individual states are epistemically hidden. In the best case, an epistemic state specifies a measure μ which gives the probability $\mu(\mathcal{B}_t)$ that an individual state lies in the Borel set $\mathcal{B}_t$.

If the dynamics of a system is given by autonomous first-order differential equations, the individual state ω_t is uniquely determined by an individual initial state, say at $t = 0$, so that the dynamics is strictly deterministic in an individual description. In a chaotic system, neighboring initial points generically lead to *qualitatively* different behavior. In order to discuss this instability, the trajectories of the individual description have to be replaced by bundles of trajectories, corresponding to arbitrary small Borel sets of the phase space as epistemic initial states.

Mathematically this means that we replace the differentiable dynamical system by a measure-theoretic dynamical system while conceptually it means to replace an individual by a statistical description. To say that a dynamical system $(\Gamma, \Sigma, \mu, \{\tau_t\})$ is in the epistemic state $\mathcal{B}_t \in \Sigma$ is just shorthand for the assertion that we know (or believe to know) that at time t the individual state $\omega_t := \tau_t(\omega)$ lies in the Borel set $\mathcal{B}_t$ with probability $\mu(\mathcal{B}_t)$.

12.4 Predictions and Retrodictions

12.4.1 Statistical and Individual Descriptions

The problem of predicting weakly stationary stochastic processes was solved in the early 1940s by Andrei Nikolaevich Kolmogorov (compare Sect. 12.4.2). Kolmogorov's theory of prediction is formulated in terms of the conditional expectation $\mathcal{E}(\cdot|\mathcal{B}_0)$ relative to the past Borel subalgebra $\mathcal{B}_0$ generated by the family of equivalence classes $[x(t)]$ with $t \leq 0$.

An optimal predictor in the sense of Kolmogorov is given by the conditional expectation $\mathcal{E}([x(t)]|\mathcal{B}_0)$. Since this is an equivalence class of functions, any two of which may differ on a variable set of zero measure, the symbol $\mathcal{E}\{[x(t)]|\mathcal{B}_0\}(\omega)$ has in general no clear-cut meaning. Therefore the application of Kolmogorov's prediction theory to individual trajectories does in general not allow us to predict the future values of a single trajectory of a stochastic process. In a review of Fürstenberg's (1960) *Stationary Processes and Prediction Theory*, Masani (1963) stated:

> It is impossible to derive directly from the prediction theory of stochastic processes a prediction theory for general individual time sequences.

From a different point of view the prediction problem was also solved by Norbert Wiener. During World War II, Wiener worked on the design of fire control radar systems for anti-aircraft guns and noise filtration in radar (compare Sect. 12.4.3). Because Wieners's interpolation and extrapolation theory of *individual* functions is conceptually very different from Kolmogorov's statistical approach, we will discuss them separately.

For simplicity, we will focus on the important special case of the optimal *linear* least-squares prediction of univariate *weakly stationary* second-order processes. *Linear* prediction theory uses only the mean and the covariance of the stochastic process. Yet, an optimal prediction of a stochastic process requires complete information about the process, and this leads in general to a *nonlinear* prediction theory.

12.4.1.1 Birkhoff's Individual Ergodic Theorem

A crucial link between Wiener's individual and Kolmogorov's statistical approach is provided by the fundamental individual ergodic theorem by George David Birkhoff. Ergodic theorems provide conditions for the equality of time averages and ensemble averages. Johann von Neumann (1932c) was the first to establish the *mean ergodic theorem* which refers to $\mathbf{L}^2$-convergence. Stimulated by these ideas, one month later George David Birkhoff (1931) obtained the even more fundamental *individual (or pointwise) ergodic theorem* which refers to pointwise convergence.[227] The discrete

[227]The publication dates (von Neumann: communicated December 10, 1931, published 1932; Birkhoff: communicated December 1, 1931, published 1931) do not give the proper chronolog-

version of the pointwise ergodic theorem is a generalization of the strong law of large numbers.

There are a few attempts to deal with the general problem of predicting the future of a stochastic process given the past, but no systematic approach for nonlinear prediction has been established so far. Nonlinear predictions will not be discussed here.

12.4.1.2 Non-Robust Extrapolations

In classical mechanics there are many models which exhibit an extreme instability of motion so that even extremely weak external disturbances can cause tremendous effects. Such dynamical systems have unique solutions if the initial data is absolutely exact, but they give extremely poor predictions if errors or uncertainties are present. In 1914, Émile Borel argued that the extreme sensitivity of some classical Hamiltonian systems is the proper justification for the use of probabilistic methods in statistical mechanics and explains the irreversible behavior of large mechanical systems.[228]

The autocorrelation function

$$t \mapsto A(t) = \lim_{T\to\infty} \frac{1}{2T} \int_{-T}^{T} x(s)^* x(s+t)\,ds, \quad t \in \mathbb{R}, \tag{12.32}$$

provides a measure of the memory: if the individual function x has a particular value at one moment, its autocorrelation tells us the extent to which we can guess that it will have about the same value some time later. It is tempting but erroneous to conjecture that a dynamical system with asymptotically vanishing autocorrelations loses the memory of the initial conditions so that the future of such a process is not determined by its past. The following example shows that this conjecture is not true.

Example: Perfect Extrapolations Can Be Non-Robust

The function $\ell \mapsto e^{-\tau^2\ell^2/4}$ $(0 < \tau < \infty)$ qualifies as a spectral density so that $t \mapsto A(t) = 1/(2\pi)\int_{-\infty}^{\infty} e^{-\tau^2\ell^2/4} e^{-i\ell t}\,d\ell = 1/(\tau\sqrt{\pi})e^{-t^2/\tau^2}$ is the autocorrelation function of an individual function $t \mapsto x(t)$. Since $t \mapsto A(t)$ is an entire function, it can be represented by an infinite power series $A(t) = A(0) + A''(0)t^2/2! + A''''(0)t^4/4! + \cdots$, converging for every

---

ical order. As Birkhoff and Koopman (1932) explain, von Neumann communicated his results to them on October 22, 1931, and "raised at once the important question as to whether or not ordinary time means exist along the individual path-curves excepting for a possible set of Lebesgue measure zero". Shortly thereafter Birkhoff proved his individual ergodic theorem.

[228] Borel (1914, p. 98) discussed a purely classical mechanical model of an ideal gas and assumed that a relative change of the gravitation potential of 10^{-100} is not under our control. Such a change corresponds to an external perturbation due to a translation by 1 cm of a particle of mass 1 gram located on Sirius, 8.3×10^{16} meters distant from earth. Then the prediction from classical mechanics for the positions of the molecules in a macroscopic sample becomes completely wrong after 10^{-6} seconds. This instability derives from the fact that a slight change in the direction of motion of a particle is amplified at each collision.

finite value of t. This fact implies that the process $t \mapsto x(t)$ has derivatives of all orders, and that for all $\tau \in \mathbb{R}$ the individual prediction error $e_{\text{ind}}^{(n)}(\tau)$,

$$e_{\text{ind}}^{(n)}(\tau) := \lim_{T\to\infty} \frac{1}{T} \int_0^T \left| x(t+\tau) - \sum_{k=0}^{n} x^{(n)}(t) \frac{\tau^n}{n!} \right|^2,$$

vanishes in the limit $n \to \infty$, $\lim_{n\to\infty} e_{\text{ind}}^{(n)}(\tau) = 0$. Therefore a perfect extrapolation indefinitely ahead can be achieved by a Taylor expansion. However, this extrapolation cannot be used for a *robust* prediction since a small change in the spectral density changes drastically the extrapolation.

12.4.2 Measure-Theoretic Prediction Theory

12.4.2.1 Kolmogorov's Extrapolation Theory

The problem of predicting weakly stationary stochastic processes was solved in the early 1940s by Andrei Nikolaevich Kolmogorov (1939, 1941a, 1941b). His mathematically rigorous theory dealt with curves in Hilbert space in terms of *equivalence classes* of trajectories. Since an optimal predictor in the sense of Kolmogorov is an equivalence class, the corresponding optimal *statistical ensemble prediction* does in general not allow the prediction of future values of a single trajectory of a stochastic process. Individual trajectories play no role at all in Kolmogorov's prediction theory.

A complex-valued stochastic process $\{x(t|\omega)|t \in \mathbb{R}, \omega \in \Omega\}$ in the sense of Kolmogorov is a family of complex-valued random variables $\omega \mapsto x(\cdot|\omega)$ on the usual probability space (Ω, Σ, μ). For a fixed $\omega \in \Omega$, the function $t \mapsto x(t|\omega)$ is called a *trajectory* (or a *realization*) of the stochastic process. That is, a stochastic process is an *equivalence class* $[x(t)]$ of individual trajectories $t \mapsto x(t|\omega)$ with the same family of joint probability densities. In particular, the mean value $m(t)$ and the covariance $C(t,t')$ of the Kolmogorov stochastic process $\{x(t|\omega)|t \in \mathbb{R}, \omega \in \Omega\}$ are defined by

$$m(t) := \mathcal{E}\{x(t|\cdot)\} = \int_\Omega x(t|\omega)\mu(d\omega), \tag{12.33a}$$

$$\begin{aligned} C(t,t') &:= \mathcal{E}\{x(t|\cdot)^* x(t'|\cdot)\} - \mathcal{E}\{x(t|\cdot)\}^* \mathcal{E}\{x(t'|\cdot)\} \\ &= \int_\Omega x(t|\omega)^* x(t|\omega)\mu(d\omega) - m(t')^* m(t). \end{aligned} \tag{12.33b}$$

12.4.2.2 Past and Future of a Stochastic Process

Let $\{[x(t)]|t \in \mathbb{R}\}$ be an arbitrary stochastic process on a probability space (Ω, Σ, μ), which is not necessarily stationary or of second order. Let $\Sigma(a,b) \subset \Sigma$ be the Borel field generated by the stochastic process $t \mapsto [x(t)]$ in the interval $a \le t \le b$. That is, $\Sigma(a,b)$ is the smallest Borel field with respect to which every member of the set $\{[x(t)]|a \leq t \leq b\}$ is measurable, so that the Borel field $\Sigma(a, b)$ represents the information provided by the stochastic process in the time interval $[a, b]$. Since $\{\Sigma(-\infty, t)|t \in \mathbb{R}\}$ is a monotonically increasing, and $\{\Sigma(t, +\infty)|t \in \mathbb{R}\}$ a monotonically decreasing family of Borel fields,

$$\Sigma(-\infty, t) \subseteq \Sigma\big(-\infty, t'\big) \quad \text{if } t < t', \tag{12.34}$$

$$\Sigma(t, +\infty) \subseteq \Sigma\big(t', +\infty\big) \quad \text{if } t > t', \tag{12.35}$$

we can define the *remote past* $\Sigma(-\infty)$ and the *remote future* $\Sigma(+\infty)$ by

$$\Sigma(-\infty) := \bigcap_{t \leq 0} \Sigma(-\infty, t), \qquad \Sigma(+\infty) := \bigcap_{t \geq 0} \Sigma(t, +\infty). \tag{12.36}$$

These concepts are important for the prediction and retrodiction of stochastic processes.

12.4.2.3 Prediction and Retrodiction

Distinguishing between the past and the future of a stochastic process, we can pose the following two extrapolation problems:

- Given the past of a stochastic process, can we *predict* its future behavior?
- Given the future of a stochastic process, can we *retrodict* its past behavior?

If the underlying dynamics is invariant under time reversal, these two problems are related by the time reversal $t \leftrightarrow -t$, so that it is sufficient to consider the problem of *prediction*.

Let $\mathbf{L}^2\{\Omega, \Sigma(-\infty, t), \mu\}$ be the Hilbert space consisting of all $\Sigma(-\infty, t)$-measurable functions that are square-integrable with respect to the probability measure μ. Knowing the equivalence class $[x_{t'}|t' \leq t]$ of the past trajectories in $\mathbf{L}^2\{\Omega, \Sigma(-\infty, t), \mu\}$, one wants to forecast the equivalence class $[x_{t+\Delta}|\Delta > 0]$ of the future trajectories, using only elements $t \mapsto \widetilde{x}(t|\Delta)$ in $\mathbf{L}^2\{\Omega, \Sigma(-\infty, t), \mu\}$. If one adopts the least-squares criterion, the statistical prediction error $e_{\text{stat}}(\Delta)$ is given by

$$e_{\text{stat}}(t, \Delta) := \mathcal{E}\big|\big[t + \Delta) - \widetilde{x}(t|\Delta)\big]\big|^2. \tag{12.37}$$

If the process is stationary, the error for the optimum (not necessarily linear) predictor does not depend on t, $e_{\text{stat}}(t, \Delta) = e_{\text{stat}}(\Delta)$. The optimal mean-squared predictor is the *conditional expectation* (Masani and Wiener 1959a; Rosenblatt 1971, p. 164)

$$\widetilde{x}(t|\Delta) = \mathcal{E}\big\{x_{t+\Delta}\big|\Sigma(-\infty, t)\big\}. \tag{12.38}$$

A stochastic process $\{x(t)|t \in \mathbb{R}\}$ is called *forward predictable* if the optimal (in general nonlinear) predictor in terms of the past $\Sigma(-\infty, 0)$ allows an error-free prediction. It follows that in this case the process is in fact already determined by the remote past $\Sigma(-\infty)$. If an error-free prediction is not possible, the process is called *forward unpredictable*. A process is called *forward completely unpredictable* if the unconditional expectation is the best forecast.[229] In this case the remote past $\Sigma(-\infty)$ is the trivial Borel field $\{\Omega, \emptyset\}$ consisting only of the certain event Ω and the impossible event $\emptyset$. The present state of a forward predictable process determines all its future states, while a forward completely unpredictable process contains no components that can be predicted exactly from an arbitrarily long past record.

In this framework we arrive at the following classification (Krengel 1971, 1973): A stochastic process $\{[x(t)] | t \in \mathbb{R}\}$ is called

- forward predictable if and only if $\Sigma(-\infty) = \Sigma(-\infty, +\infty)$,
- forward completely unpredictable if and only if $\Sigma(-\infty) = \{\Omega, \emptyset\}$,
- backward predictable if and only if $\Sigma(+\infty) = \Sigma(-\infty, +\infty)$,
- backward completely unpredictable if and only if $\Sigma(+\infty) = \{\Omega, \emptyset\}$,
- bidirectionally predictable if and only if $\Sigma(-\infty) = \Sigma(+\infty)$.

The classification of stochastic processes in terms of the Borel field $\Sigma(a,b)$ generated by the process $t \mapsto [x(t)]$ in the time interval $[a,b]$ is appropriate for most general nonlinear prediction. Since up to now there is no generally applicable algorithm for nonlinear prediction (cf. Kallianpur 1961), we consider next the case of linear prediction, which can be worked out in much detail.

12.4.2.4 Linear Prediction Theory

Linear prediction theory uses only the information which is codified in the mean and the covariance of a stochastic process. Therefore, *Gaussian* stochastic processes play a distinguished role: for stationary *Gaussian* processes the optimal linear predictor coincides with the optimal nonlinear predictor. For the sake of simplicity, we discuss here only the important special case of the optimal *linear* least-squares prediction of univariate *weakly stationary* second-order processes. Extensions of these ideas to multivariate processes are possible.[230]

A *linear* predictor uses only *linear combinations*, including the limits in quadratic mean, of the past of the stochastic process. Let $\{x(t) | t \in \mathbb{R}\}$ be a real- or complex-valued weakly stationary second-oder process with mean value zero, and let $\mathcal{H}(a,b)$ be the Hilbert space generated by $t \mapsto x_t$ in the time interval $[a,b]$,

$$\mathcal{H}(a,b) = \text{closed span}\,\{x_t | a \leq t \leq b\}. \tag{12.39}$$

[229]Hermann Wold (1938) called a stationary forward predictable process *singular* if knowledge of its past allows an error-free prediction. It is called *regular* if it is not singular and if the conditional expectation is the best forecast. Unfortunately, Doob (1944) renamed singular processes "deterministic" and regular processes "nondeterministic". Since determinism has nothing to do with predictions, this now popular terminology is based on a notorious category mistake (see Sect. 12.3), which we will avoid in what follows.

[230]Compare Wiener's papers and the editorial comments by Masani (1981, pp. 164–370).

Note that the Hilbert space $\mathbf{L}^2\{\Omega, \Sigma(a,b), \mu\}$ generated nonlinearly by a stochastic process in the interval $[a,b]$ is in general much larger than the Hilbert space $\mathcal{H}(a,b)$ generated linearly,

$$\mathcal{H}(a,b) \subseteq \mathbf{L}^2\{\Omega, \Sigma(a,b), \mu\}, \tag{12.40}$$

so that the optimal prediction is in general nonlinear (see Masani and Wiener 1959a; Kallianpur 1961; Rosenblatt 1971).

For the important special case of stationary *Gaussian* processes, the Hilbert space $\mathcal{H}(a,b)$ coincides with the Hilbert space $\mathbf{L}^2\{\Omega, \Sigma(a,b), \mu\}$,

$$\mathcal{H}(a,b) = \mathbf{L}^2\{\Omega, \Sigma(a,b), \mu\}, \tag{12.41}$$

so that in this case the optimal predictor is linear. As long as linearity is retained, prediction theory is fairly complete and uses only Hilbert-space concepts. It applies to any weakly stationary process characterized by its first and second moment.[231]

First we note that $\{\mathcal{H}(-\infty, t) | t \in \mathbb{R}\}$ is a monotonically increasing family of Hilbert spaces, and that $\{\mathcal{H}(t, +\infty) | t \in \mathbb{R}\}$ is a monotonically decreasing family of Hilbert spaces. The *remote past* $\mathcal{H}(-\infty)$ in the linear sense and the *remote future* $\mathcal{H}(+\infty)$ in the linear sense are defined by

$$\mathcal{H}(-\infty) := \bigcap_{t \le 0} \mathcal{H}(-\infty, t), \qquad \mathcal{H}(+\infty) := \bigcap_{t \ge 0} \mathcal{H}(t, +\infty). \tag{12.42}$$

A process is called *forward singular in the linear sense* if the optimal linear predictor in terms of the past allows an error-free prediction. If an error-free prediction is not possible, the process is called *forward regular in the linear sense*. Corresponding definitions hold for retrodictions in the linear sense. Accordingly, a stochastic process $\{x(t) | t \in \mathbb{R}\}$ is

- forward singular in the linear sense if and only if $\mathcal{H}(-\infty) = \mathcal{H}(-\infty, \infty)$, hence forward predictable in the linear sense;
- forward regular in the linear sense if and only if $\mathcal{H}(-\infty) = \mathbb{C}$, hence forward completely unpredictable in the linear sense;
- backward singular in the linear sense if and only if $\mathcal{H}(\infty) = \mathcal{H}(-\infty, \infty)$, hence backward predictable in the linear sense;
- backward regular in the linear sense if and only if $\mathcal{H}(\infty) = \mathbb{C}$, hence backward completely unpredictable in the linear sense;
- bidirectionally singular in the linear sense if and only if $\mathcal{H}(-\infty) = \mathcal{H}(\infty)$, hence bidirectionally predictable in the linear sense.

Here $\mathbb{C}$ is the trivial Hilbert space, consisting only of constant functions. Since $\Sigma(-\infty) = \{\Omega, \emptyset\}$ implies $\mathcal{H}(\infty) = \mathbb{C}$, every completely unpredictable process in the nonlinear sense is also completely unpredictable in the linear sense. However,

---

[231] For an elementary introduction, compare for example Cramér and Leadbetter (1967, Sects. 5.7 and 7.9).

there are completely unpredictable processes in the linear sense which are predictable in the nonlinear sense.[232]

12.4.2.5 Spectral Representation of Stationary Processes

Let $\{[\check{x}(t)] | t \in \mathbb{R}\}$ be a weakly stationary second-order process with mean $t \mapsto m$ and covariance function $t \mapsto \check{C}(t)$,

$$m := \int_{\Omega} \check{x}(t|\omega)\mu(d\omega), \tag{12.43a}$$

$$\check{C}(t) := \int_{\Omega} \{\check{x}(t|\omega) - m\}\{\check{x}(0|\omega) - m\}^{*}\mu(d\omega). \tag{12.43b}$$

If the covariance function $t \mapsto \check{C}(t)$ is continuous at the point $t = 0$ (i.e. if $\lim_{t\to 0} \check{C}(t) = \check{C}(0)$), then the corresponding stochastic process is *mean-square continuous*, that is

$$\mathcal{E}\{(\check{x}(t) - \check{x}(t'))^2\} \to 0 \quad \text{for } t - t' \to 0. \tag{12.44}$$

In this case $t \mapsto \check{C}(t)$ is the characteristic function with the representation[233]

$$\check{C}(t) = \int_{\mathbb{R}} e^{2\pi i t \ell} \mu(d\ell), \tag{12.45}$$

where μ is a unique non-negative bounded Borel measure, called the *spectral measure* of the process. If the spectral measure μ is absolutely continuous, the *spectral density* is given by

$$\ell \mapsto \hat{C}(\ell) = \mu(d\ell)/d\ell \geq 0, \quad \hat{C}(\ell) = \int_{\mathbb{R}} e^{-2\pi i t \ell} \check{C}(t) dt. \tag{12.46}$$

Every wide-sense stationary mean-square continuous and zero-mean process $\{\check{x}(t) | t \in \mathbb{R}\}$ has a *spectral representation* of the form[234]

$$\check{x}(t) = \int_{\mathbb{R}} e^{2\pi i t \ell} d\hat{x}(\ell), \quad t \in \mathbb{R}, \tag{12.47}$$

where the orthogonal increments $d\hat{x}(\ell)$ of the spectral process $\{\hat{x}(\ell) | \ell \in \mathbb{R}\}$ are uncorrelated random variables. The process $\ell \mapsto \hat{x}(\ell)$ is fixed up to an additive random variable. If we choose $\hat{x}(-\infty) = 0$, we have

---

[232] An example can be found in Ash and Gardner (1975, problem 3 on p. 68).

[233] Compare for example Cramér and Leadbetter (1967, Sect. 7.4).

[234] For details, compare the example discussed by Doob (1953, Chap. XI, §4); or Cramér and Leadbetter (1967, Chap. 7.5).

$$\mathcal{E}\{\hat{x}(\ell)\} = 0, \qquad \mathcal{E}\{|d\hat{x}(\ell)|^2\} = \hat{C}(\ell)d\ell, \tag{12.48a}$$

$$\mathcal{E}\{d\hat{x}(\ell)d\hat{x}(\ell')\} = \delta(\ell - \ell')\hat{C}(\ell)d\ell. \tag{12.48b}$$

12.4.2.6 Unitary Groups Associated with Regular Processes

For a regular stochastic processes $\{[\check{x}(t)]|t \in \mathbb{R}\}$ there exists a time domain representation, which is generated by the time operator T characterized by

$$\{T\check{x}\}(t) = t\check{x}(t), \quad t \in \mathbb{R}. \tag{12.49}$$

The precise definition goes as follows:[235] Let $\{[\check{x}(t)]|t \in \mathbb{R}\}$ be a continuous, weakly stationary, regular complex-valued stochastic process with zero mean. Let $\mathcal{H} := \bigcup_{t\in\mathbb{R}} \mathcal{H}(-\infty, t)$ be the Hilbert space spanned by this process. The projections E_t from $\mathcal{H}$ onto $\mathcal{H}(-\infty, t+0)$,

$$E_t\mathcal{H} = \mathcal{H}(-\infty, t+0), \quad t \in \mathbb{R}, \tag{12.50}$$

generate a spectral family of projections with $E_s \leq E_t$ for $s < t$, and with $E_t(-\infty) = 0$, $E_t(\infty) = \mathbb{1}$. The associated self-adjoint operator

$$T := \int_{-\infty}^{\infty} t dE_t, \tag{12.51}$$

is the time operator associated with the process $\{[\check{x}(t)]|t \in \mathbb{R}\}$. It generates a strongly continuous one-parameter group $\{U(\lambda)|\lambda \in \mathbb{R}\}$ of unitary operators

$$U(\lambda) := \int_{-\infty}^{\infty} e^{2\pi i\lambda t} dE(t) = e^{2\pi i\lambda T}. \tag{12.52}$$

In addition we can define a family of unitary shift operators $V(t)$,

$$V(t)\check{x}(s) := \check{x}(t-s), \quad t, s \in \mathbb{R}, \tag{12.53}$$

which generate a strongly continuous one-parameter group $\{V(t)|t \in \mathbb{R}\}$. The unitary operators $V(t)$ and $U(\lambda)$ are related by Weyl's canonical commutation relation

$$U(\lambda)V(t) = e^{2\pi i\lambda t}V(t)U(\lambda), \quad \lambda, t \in \mathbb{R}. \tag{12.54}$$

---

[235] The crucial imprimitivity relation has been recognized by Hanson (1958, Eq. 2.3 on p. 163) and by Kallianpur and Mandrekar (1965, p. 560). The time operator itself has been introduced by Tjøstheim (1976), who also discussed the canonical commutation relation between the time and frequency operator. Independently, these relations have also been found by Gustafson and Misra (1976).

12.4.2.7 Wold's Decomposition and Szegö's Theorem

The linear prediction problem is greatly simplified by a decomposition theorem due to Hermann Wold (1938), which implies that every weakly stationary stochastic process is the sum of a singular and a regular process, both understood in the linear sense (where, of course, one component may be absent).

Wold Decomposition of Stationary Stochastic Processes[236]
For every weakly stationary second-order process $\{[\check{x}(t)] | t \in \mathbb{R}\}$ there exists a unique orthogonal decomposition into a so-called *singular process* $\{[\check{x}(t)]^{\mathrm{s}} | t \in \mathbb{R}\}$ that allows a perfect prediction, and a so-called *regular process* $\{[\check{x}(t)]^{\mathrm{r}} | t \in \mathbb{R}\}$ that does not contain any component allowing an error-free prediction,

$$[\check{x}(t)] = [\check{x}(t)]^{\mathrm{s}} + [\check{x}(t)]^{\mathrm{r}}, \quad t \in \mathbb{R}. \tag{12.55}$$

The Hilbert space $\mathcal{H}(t)$ spanned by $[\hat{x}(t')]$ with $t' \leq t$ decomposes into a direct sum of two mutually orthogonal Hilbert spaces $\mathcal{H}^{\mathrm{s}}(t)$ and $\mathcal{H}^{\mathrm{r}}(t)$,

$$\mathcal{H}(t) = \mathcal{H}^{\mathrm{s}}(t) \oplus \mathcal{H}^{\mathrm{r}}(t), \quad \mathcal{H}^{\mathrm{s}}(t) \perp \mathcal{H}^{\mathrm{r}}(t), \tag{12.56}$$

where $\mathcal{H}^{\mathrm{s}}(t)$ is the Hilbert space spanned by $[\check{x}(t')]^{\mathrm{s}}$ with $t' \leq t$, and $\mathcal{H}^{\mathrm{r}}(t)$ is the Hilbert space spanned by $[\check{x}(t')]^{\mathrm{r}}$ $(t' \leq t)$.

An important theorem by Szegö (1915, 1920) allows us to express the dichotomy between singular and regular stochastic processes analytically in terms of a Paley-Wiener criterion (Paley and Wiener 1934, pp. 16f). A Paley-Wiener criterion relates decay properties of a function or distribution at infinity to the analyticity of its Fourier transform (cf. Sect. 3.3.3.1).

Wiener-Kreĭn Criterion for Regular Processes[237]
Let $\{\check{x}(t) | t \in \mathbb{R}\}$ be a weakly stationary second-order process with the spectral distribution function $\hat{C}$. Since the function $\ell \mapsto \hat{C}(\ell)$ is non-decreasing and bounded, its derivative $\ell \mapsto d\hat{C}(\ell)/d\ell$ exists except for a set of Lebesgue measure zero.

- The process is (forward or backward) *singular* if and only if either $d\hat{C}(\ell)/d\ell$ is zero on a set of nonzero Lebesgue measure, or if $d\hat{C}(\ell)/d\ell$ is not zero almost everywhere but

$$\int_{-\infty}^{\infty} \frac{\ln\{d\hat{C}(\ell)/d\ell\}}{1+\ell^2} d\ell = -\infty. \tag{12.57a}$$

- The process is (forward or backward) *regular* if and only if its spectral distribution function is absolutely continuous and if

$$\int_{-\infty}^{\infty} \frac{\ln\{d\hat{C}(\ell)/d\ell\}}{1+\ell^2} d\ell > -\infty. \tag{12.57b}$$

---

[236] Wold's decomposition theorem refers to the special case of discrete-time stationary processes, and was extended by Hanner (1950) to the case of continuous-time processes. The general decomposition theorem is due to Cramér (1961).

[237] See Kreĭn (1945a, 1945b) and Wiener (1949). Wiener's original work appeared as a classified report in February 1942. Editor's note: The notion of a Wiener-Kreĭn criterion, unusual in the standard literature, was adopted by the author to appreciate essential ideas by Kreĭn in the development of Paley-Wiener criteria.

If the process $\{[\check{x}(t)] | t \in \mathbb{R}\}$ is *singular*, all $\mathcal{H}(-\infty, t)$ are independent of t, $\mathcal{H}(-\infty) = \mathcal{H}(-\infty, t) = \mathcal{H}(+\infty)$, so that the remote past contains all information about the present and the future. In this case the process $\{\check{x}(t) | t \in \mathbb{R}\}$ is *predictable* in the linear sense. The optimal linear predictor $\tilde{x}(t|\Delta)$ agrees almost everywhere with $\check{x}(t + \Delta)$, so that the mean-squared error of the optimal predictor is zero for all Δ.[238] If the process $\{[\check{x}(t)] | t \in \mathbb{R}\}$ is *regular*, then it is *linearly completely unpredictable*, that is, it contains no component that can be predicted exactly from an arbitrarily long past record.

Note that for a regular process the spectral measure μ is necessarily absolutely continuous so that the covariance function $t \mapsto \check{C}(t)$ vanishes for $t \to \infty$. However, there are regular processes with an asymptotically vanishing covariance function, so that *an asymptotically vanishing covariance function is not sufficient for regular behavior.*

Examples

Consider three second-order stationary stochastic processes with the covariance functions $t \mapsto C_1(t) = e^{-|t|}$, $t \mapsto C_2(t) = e^{-t^2/2}$ and $t \mapsto C_3(t) = \sin(t)/t$. They have absolutely continuous distribution functions with the following spectral densities:

$$\hat{C}_1(\ell) = \int_{\mathbb{R}} e^{-2\pi i t \ell} e^{-|t|} dt = \frac{2}{1 + (2\pi\ell)^2}, \tag{12.58a}$$

$$\hat{C}_2(\ell) = \int_{\mathbb{R}} e^{-2\pi i t \ell} e^{-t^2/2} dt = \sqrt{2\pi} e^{-2\pi^2 \ell^2}, \tag{12.58b}$$

$$\hat{C}_3(\ell) = \int_{\mathbb{R}} e^{-2\pi i t \ell} \frac{\sin(t)}{t} dt = \begin{cases} \pi/2 & \text{for } 2\pi\ell < 1, \\ 0 & \text{for } 2\pi\ell > 1. \end{cases} \tag{12.58c}$$

The spectral density (12.58a) fulfills the Wiener-Kreĭn regularity criterion (12.57b) so that the corresponding stochastic process is regular. Even though the covariance functions $t \mapsto C_2(t)$ and $t \mapsto C_3(t)$ tend to 0 for $t \to \infty$, the spectral densities (12.58b) and (12.58c) fulfill Eq. (12.57a), so that they characterize singular processes. The reasons for the divergence of the integral (12.57a), however, are different. The covariance $t \mapsto C_2(t) = e^{-t^2/2}$ falls off too fast, while the covariance $t \mapsto C_3(t) = \sin(t)/t$ falls off too slow.

12.4.2.8 Outline of the Kolmogorov-Kreĭn Linear Prediction Theory

We consider the problem of linear extrapolation of a complex-valued weakly stationary second-order stochastic process $\{[\check{x}(t)] | t \in \mathbb{R}\}$ on a probability space (Ω, Σ, μ) which is strongly continuous in t.[239] This process has a spectral representation of the form

$$\check{x}(t) = \int_{\mathbb{R}} e^{2\pi i t \ell} d\hat{x}(\ell), \quad t \in \mathbb{R}, \tag{12.59}$$

---

[238] An algorithm which calculates the future values of a trajectory $t \mapsto x(t|\omega)$, ω fixed, of a singular stochastic process for almost all $t > 0$ provided the past $\{x(t|\omega) | t \le 0\}$ is given has been discussed by Scarpellini (1978a, 1978b, 1978c, 1978d).

[239] For details compare for example Doob (1953, Chap. XII), Yaglom (1962), Rozanov (1967), Cramér (1961, Sect. 7.9).

where $\{[\hat{x}(\ell)] | \ell \in \mathbb{R}\}$ is the associated spectral process on $\mathbb{R}$ with orthogonal increments.

For the purpose of *linear* prediction the history of a stochastic process is given by the family of Hilbert spaces $\mathcal{H}(-\infty, t)$ generated by $[\check{x}(t')]$ with $t' \leq t$. The linear prediction problem is concerned with predicting the future $\{[\check{x}(t)] | t > 0\}$ by a linear functional of the past $\{[\check{x}(t)] | t \leq 0\}$ of the process. The minimal statistical prediction error $e_{\text{stat,min}}(\Delta)$ is given by

$$e^2_{\text{stat,min}}(\Delta) = \inf_{[\tilde{x}] \in \mathcal{H}_t} \left\| \left[\check{x}(t+\Delta) - \tilde{x} \right] \right\|^2,$$

so that the optimal linear predictor is the orthogonal projection of $[\check{x}(t)]$ on the subspace $\mathcal{H}(-\infty, 0)$.

In contrast to the singular case, for a regular stochastic process an analytic expression for the optimal predictor can be found. The optimal least-squares predictor $[\tilde{x}(t, \Delta)]$ is given by

$$\tilde{x}(t, \Delta) = -2\pi \int_{-\infty}^{\infty} e^{+2\pi i \ell t} \hat{R}_{\Delta}(\ell) dz(\ell) \tag{12.60a}$$

with

$$\hat{R}_{\Delta}(\ell) := \frac{1}{2\pi H(\ell)} \int_0^{\infty} dt e^{-2\pi i \ell t} \int_{-\infty}^{\infty} d\ell' e^{2\pi i \ell' (t+\Delta)} H(\ell'), \tag{12.60b}$$

where $\ell \mapsto H(\ell)$ is the boundary value of an analytic function $H^{\pm}(\zeta)$ which is holomorphic and has no zeros in the upper/lower half-plane of the complex variable ζ,

$$H(\ell) := \lim_{\varepsilon \downarrow 0} H^{\pm}(\ell \mp i\varepsilon), \tag{12.61a}$$

$$H^{\pm}(\zeta) := \exp\left\{ \frac{1}{2\pi^2 i} \int_{-\infty}^{\infty} \frac{1 - (2\pi)^2 \ell \zeta}{\zeta - \ell} \frac{\ln\{d\hat{C}(\ell)/d\ell\}}{1 + (2\pi\ell)^2} d\ell \right\}. \tag{12.61b}$$

With the relation $|H(\ell)|^2 = (-1/2\pi) d\hat{C}(\ell)/d\ell$ one finds for the minimal statistical mean-squared prediction error:

$$e^2_{\text{stat,min}}(\Delta) = \int_{-\infty}^{\infty} \left| e^{2\pi i \ell \Delta} - \hat{R}_{\Delta}(\ell) \right|^2 d\hat{C}(\ell) \tag{12.62a}$$

$$= \mathcal{E}\left\{ \left| \left[\check{x}(t) \right] \right|^2 \right\} - \int_{-\infty}^{\infty} \left| \hat{R}_{\Delta}(\ell) \right|^2 d\hat{C}(\ell). \tag{12.62b}$$

12.4.2.9 Moving Average of Stochastic Processes

The basic stochastic process is white noise. It is defined as the generalized derivative of the one-dimensional Wiener process $t \mapsto \check{w}(t)$ with incremental covariance

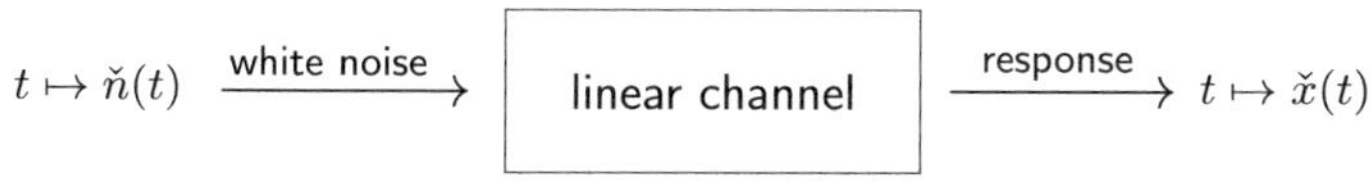

Fig. 12.2 Sketch of a linear input-output channel transforming white noise $\hat{n}(t)$ into a response $\hat{x}(t)$

$\mathcal{E}\{|d\check{w}(t)|^2\} = dt$. White noise $t \mapsto \check{n}(t) := dw(t)/dt$ is a generalized weakly stationary real-valued stochastic process with zero mean whose power spectral density is constant at all frequencies,

$$\mathcal{E}\{\check{n}(t)\} = 0, \qquad \mathcal{E}\{\check{n}(t)\check{n}(s)\} = \delta(t-s). \tag{12.63}$$

For a white-noise input process $t \mapsto \check{n}(t)$, the output process has an absolutely continuous spectral distribution function with the density

$$\hat{C}(\ell) = \frac{1}{2\pi}\left|H(\ell)\right|^2, \tag{12.64}$$

which satisfies the Wiener-Kreĭn criterion. For a given covariance function $t \mapsto \check{C}(t)$ with the spectral density

$$\ell \mapsto \hat{C}(\ell) = \int_{-\infty}^{\infty} e^{-2\pi i \ell t}\check{C}(t)dt, \tag{12.65}$$

one can always find a time-translation invariant linear input-output system such that the response to a white-noise input is a real-valued weakly stationary process $t \mapsto \check{x}(t)$ (see Fig. 12.2),

$$\check{x}(t) = \check{x}(0) + \int_{-\infty}^{\infty} dt' \check{R}(t-t')\check{n}(t'), \tag{12.66}$$

with the covariance

$$\mathcal{E}\{d\check{x}(t)d\check{x}(s)\} = \check{C}(t-s). \tag{12.67}$$

This representation is far from unique. If the Paley-Wiener criterion

$$\int_{-\infty}^{\infty} \frac{\ln|\hat{R}(\ell)|^2}{1+\ell^2} d\ell > -\infty \tag{12.68}$$

is met, the function $\ell \mapsto |\hat{R}(\ell)|^2$ can be factorized,

$$\left|\hat{R}(\ell)\right|^2 = \hat{R}^+(\ell)\hat{R}^-(\ell), \quad \hat{R}^+(\ell) = \{\hat{R}^-(\ell)\}^*, \ \ell \in \mathbb{R}. \tag{12.69}$$

The function $\ell \mapsto \hat{R}^+(\ell)$ is the boundary value of an analytic function $z \mapsto \hat{R}^+(z)$, holomorphic in the open upper half-plane $\mathbb{C}^+ := \{z \in \mathbb{C} | \mathrm{Im}(z) > 0\}$, while the function $\ell \mapsto \hat{R}^-(\ell)$ is the boundary value of another analytic function $z \mapsto \hat{R}^-(z)$, holomorphic in the open lower half-plane $\mathbb{C}^- := \{z \in \mathbb{C} | \text{Im}(z) < 0\}$. The Fourier transforms of $\hat{R}^{\pm}$ are given by

$$\check{R}^+(t) = \int_{\mathbb{R}} e^{2\pi i \ell t} \hat{R}^+(\ell) d\ell, \quad \check{R}^+(t) = 0 \text{ for } t > 0, \tag{12.70a}$$

$$\check{R}^-(t) = \int_{\mathbb{R}} e^{2\pi i \ell t} \hat{R}^-(\ell) d\ell, \quad \check{R}^-(t) = 0 \text{ for } t < 0, \tag{12.70b}$$

$$\check{R}^-(t) = \left\{\check{R}^+(-t)\right\}^*. \tag{12.70c}$$

The time-reversal invariant two-sided moving average representation (12.65) involves the values of the input Wiener process for all $t \in \mathbb{R}$. For *regular* processes there are two distinguished inequivalent *one-sided* moving-average representations, The retarded solution $t \mapsto \check{x}^-(t)$ does not depend on the future and is appropriate for prediction. The advanced solution $t \mapsto \check{x}^+(t)$ does not depend on the past and is appropriate for retrodiction. Both versions break the time-inversion symmetry and have the same logical status. The decision which of the two possibilities is appropriate cannot come from the first principles of physics but depends on the context.

The canonical representation (12.61a), (12.61b) characterizes a linear filter of minimum phase and defines *two* different one-sided moving-average realizations of the stochastic process $\{\check{x}(t) | t \in \mathbb{R}\}$ with the same covariance $\check{C}$: the *forward moving-average representation* $\check{x}^-$ and the *backward moving-average representation* $\check{x}^+$:

$$\check{x}^-(t) = \int_{-\infty}^{t} \check{R}^-(t-s) d\check{w}(s) = \int_{0}^{\infty} \check{R}^-(s) d\check{w}(t-s), \tag{12.71a}$$

$$\check{x}^+(t) = \int_{t}^{\infty} \check{R}^-(t-s) d\check{w}(s) = \int_{-\infty}^{0} \check{R}^+(s) d\check{w}(t-s), \tag{12.71b}$$

with

$$\mathcal{E}\left\{\check{x}^+(t)\check{x}^+(s)\right\} = \mathcal{E}\left\{\check{x}^-(t)\check{x}^-(s)\right\} = \check{C}(t-s). \tag{12.71c}$$

The stochastic process $\{\check{x}^-(t) | t \in \mathbb{R}\}$ is *canonical* in the sense that the closed linear manifold generated by $\{\check{x}^-(s) | s \leq t\}$ equals for every $t \in \mathbb{R}$ the closed linear manifold generated by a standard Wiener process $\{\check{w}(s) | s \leq t\}$ (cf. Hida 1970). The process $\{\check{x}^-(t) | t \in \mathbb{R}\}$ is forward regular in the linear sense. Quite generally, a weakly stationary process is forward regular in the linear sense if and only if it has a backward moving-average representation.

A non-anticipative input-output system produces the trajectories of a novelty-acquiring process in response to the application of a particular white noise trajectory at the input.[240] In contrast to engineering systems, the "input" cannot be assigned

[240] In the context of statistical prediction theory, Masani and Wiener (1959b) called a non-anticipative forward moving-average representation an *innovation representation*. For details compare the monograph by Rozanov (1967) and the review by Kailath (1974).

arbitrarily from the outside. The fact that every novelty-acquiring non-material process has a representation in terms of white noise suggests that the *memoryless* K-system generated by white noise can be considered as an *intrinsic structure element of the non-material (mental) domain.*

12.4.3 Wiener's Prediction Theory

From a point of view different from Kolmogorov's, the prediction problem was also solved by Norbert Wiener (1942, 1949), see also the monograph by Gardner (1988). Kolmogorov's theory of stochastic process refers to *equivalence classes* of trajectories, so in Kolmogorov's prediction theory the optimal predictor is the conditional expectation (12.38), i.e. an equivalence class of trajectories. Any two trajectories $t \mapsto x(t|\omega_1)$ and $t \mapsto x(t|\omega_2)$ in the same equivalence class may differ on a set of μ-measure zero so that except in special cases $\mathcal{E}\{x_{t+t'}|\Sigma(-\infty, t)\}$ cannot be interpreted as an individual trajectory. Therefore it is in general not possible to use Kolmogorov's prediction theory for predictions of individual trajectories.[241]

By contrast, Wiener developed a mathematical theory of great generality for predicting the future of individual trajectories on the basis of incomplete information about the past. In contradistinction to engineering practice, Wiener considered communication signals quite generally as *stochastic processes* and so initiated statistical communication theory.[242] In particular the following two problems can be discussed in terms of Wiener's theory:

- Given the trajectory of a stochastic process on the half-axis $\mathbb{R}^-$, can we *predict* the trajectory of the process on $\mathbb{R}^+$?
- Given the trajectory of a stochastic process on the half-axis $\mathbb{R}^+$, can we *retrodict* the trajectory of the process on $\mathbb{R}^-$

The great importance of Wiener's work lies in his insights about how estimation filters are practically realizable.

12.4.3.1 Wiener's Theory of Individual Trajectories

Wiener started his work on prediction in the winter of 1940 with the engineering problem of forecasting the future trajectories of fast flying airplanes on the basis

---

[241] Compare the review of Furstenberg (1960) by Masani (1963, p. 196).

[242] Wiener's early work was classified, his book *Extrapolation, Interpolation and Smoothing of Stationary Time Series* appeared in 1942 as *Classified Report to Section D2, National Defense Research Committee* of the USA. This report had an extraordinary influence in engineering circles and was released for general use in 1949.

of radar tracks of their past movements. His goal was to achieve optimal control of anti-aircraft fire and to maximize the chances of hitting the target.[243]

Wiener's work initiated the mathematical theory of stochastic processes and functional integration. In his celebrated memoir on generalized harmonic analysis, Wiener (1930) extended harmonic analysis of almost periodic functions to the study of individual processes with *continuous spectra* which allowed the definition of *deterministic irregular individual functions*. He based his theory not on equivalence classes of Lebesgue square-integrable functions but on *individual* measurable functions $t \mapsto x(t)$ for which the *individual autocorrelation function*

$$t \mapsto \breve{A}(t) = \lim_{T\to\infty} \frac{1}{2T} \int_{-T}^{T} x(s)^* x(s+t)ds, \quad t \in \mathbb{R}, \tag{12.72}$$

exists.[244] It follows that $\breve{A}(0)$ is real and that $|\breve{A}(t)| \leq \breve{A}(0)$ for all $t \in \mathbb{R}$.

The relations between Kolmogorov's and Wiener's theory can be explained as follows. If for a fixed $\omega_0 \in \Omega$ the individual function $t \mapsto x(t|\omega_0)$ is a trajectory of a zero-mean stationary ergodic stochastic process $t \mapsto [x(t)(\cdot)]$ on a Kolmogorov probability space (Ω, Σ, μ), then

$$\int_{\Omega} x(t|\omega)x(s+t|\omega)^* \mu(d\omega) = \lim_{T\to\infty} \frac{1}{2T} \int_{-T}^{T} x(s|\omega_0)x(s+t|\omega_0)^* ds, \tag{12.73}$$

so that the ensemble covariance function $t \mapsto \breve{C}(t)$ equals the individual autocorrelation function $t \mapsto \breve{A}(t)$,

$$\breve{C}(t) = \breve{A}(t), \quad t \in \mathbb{R}. \tag{12.74}$$

Wiener established the existence of a bounded non-decreasing function $\ell \mapsto \hat{A}(\ell)$,

$$\hat{A}(\ell) = \int_{-\infty}^{\infty} e^{-2\pi i \ell t} d\breve{A}(t), \quad \ell \in \mathbb{R}, \tag{12.75}$$

called the *spectral distribution function*.

In contrast to Kolmogorov's prediction theory of stochastic processes, Wiener's theory is non-probabilistic and reflects the point of view of communication engineering. In his monograph *Extrapolation, Interpolation, and Smoothing of Stationary Time Serie*, Wiener (1949) used individual functions $t \mapsto x(t)$ rather than equivalence classes. Based on his earlier work on ergodic theorems, integral equations and Fourier analysis, he addressed the detection of individual signals in noise and the design of optimal filters. In his own words (Wiener 1949, p. 47):

---

[243] For a summary of Wiener's work on anti-aircraft fire control, compare his autobiography (Wiener 1956) and the biography by Masani (1990, Chap. 14).

[244] Such averages were first used by Harald August Bohr (1932), Abram Samoilovitch Besicovitch (1932) and Vyacheslav Vassilievich Stepanov (1925) in their investigations of almost periodic functions and their discrete spectra. See also Maak (1950).

> This allows us to identify averages made on the observable past of a time series with averages to be subsequently obtained from the now unattainable future. It is at precisely this point that the ensemble of time series, as contrasted with the individual time series, becomes important. This step (legitimate for the ensemble) is not legitimate for the individual series. It is this step which constitutes the logical process of induction.

Wiener's approach refers to the interpolation and extrapolation of a single well-defined but highly irregular *individual* function. In Kolmogorov's terminology, Wiener's theory deals with the prediction of a *single trajectory* $t \mapsto x(t|\omega)$, ω fixed, of a stochastic process $t \mapsto [x(t, \cdot)]$. This approach has been criticized by Kakutani (1950) as unnecessarily arduous since it was not based on Kolmogorov's more effortless definition of measure-theoretic stochastic processes. Yet, it has to be stressed that for a serviceable prediction Wiener's individual approach cannot be replaced by the measure-theoretic Kolmogorov method. In practical applications there is only a *single* individual trajectory available, from whose past behavior one would like to predict its future behavior.

12.4.3.2 Reliable Filters and Past-Indeterminate Processes

We consider a physically realizable linear input-output system, called *filter*, with the response function $t \mapsto \check{R}(t)$, $\check{R}(t) = 0$ for $t < 0$, and the associated frequency-response function $\hat{R} : \mathbb{R} \to \mathbb{C}$,

$$\hat{R}(\ell) := \int_{-\infty}^{\infty} e^{-2\pi i \ell t} \check{R}(t)dt = \int_{0}^{\infty} e^{-2\pi i \ell t} \check{R}(t)dt, \quad \ell \in \mathbb{R}. \tag{12.76}$$

We assume that both the input function $t \mapsto \check{u}(t)$ and the output function $t \mapsto \check{y}(t)$ are real-valued irregular functions in the sense of Wiener. They are related by

$$\check{y}(t) = \int_{0}^{\infty} \check{R}(s)\check{u}(t-s)ds, \qquad \hat{y}(\ell) = \hat{R}(\ell)\hat{u}(\ell). \tag{12.77}$$

Suppose that the individual irregular input function $t \mapsto \check{u}(t)$ is past-indeterminate, so that the Szegö-Paley-Wiener criterion of Sect. 12.4.2.7 is fulfilled,

$$\int_{-\infty}^{\infty} \frac{|\ln\{\hat{A}_{u,u}(\ell)\}|}{1+\ell^2} d\ell < \infty.$$

Then we may ask whether the output $t \mapsto \check{y}(t)$ is also past-indeterminate. Using the relation $\hat{A}_{y,y}(\ell) = |\hat{R}(\ell)|^2 \hat{A}_{u,u}(\ell)$, the Szegö-Paley-Wiener criterion for $\check{y}$ can be written as:

$$\int_{-\infty}^{\infty} \frac{|\ln\{\hat{A}_{y,y}(\ell)\}|}{1+\ell^2} d\ell = \int_{-\infty}^{\infty} \frac{|\ln\{|\hat{R}(\ell)|^2(\ell)\}|}{1+\ell^2} d\ell + \int_{-\infty}^{\infty} \frac{|\ln\{\hat{A}_{u,u}(\ell)\}|}{1+\ell^2} d\ell.$$

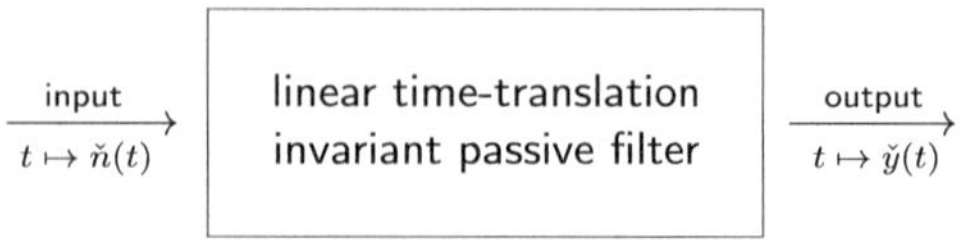

Fig. 12.3 Sketch of a linear time-translation invariant passive filter transforming white noise $\check{n}(t)$ into a past-indeterminate response $\check{y}(t)$

It follows that the output function is past-indeterminate if and only if the filter is non-anticipative,

$$\int_{-\infty}^{\infty} \frac{|\hat{R}(\ell)|^2}{1+\ell^2} d\ell < \infty. \tag{12.78}$$

Conclusion: The output of a linear filter is past-indeterminate if and only if the filter is physically realizable and if the input is past-indeterminate.

12.4.3.3 Filtered White Noise Is a Past-Indeterminate Process

White-noise processes are basic in the sense that they generate general past-indeterminate processes. Consider a linear, time-invariant non-anticipative filter with the response function $\check{R}$, driven by a real-valued stationary white noise process $t \mapsto \check{n}(t)$ as in Fig. 12.3. The output $t \mapsto \check{y}(t)$

$$\check{y}(t) = \int_0^{\infty} \check{R}(s)\check{n}(t-s)ds \tag{12.79}$$

of this filter is a past-indeterminate function with the spectral density

$$\hat{A}_{y,y}(\ell) = \left|\hat{R}(\ell)\right|^2 \hat{A}_{u,u}(\ell) = \left|\hat{R}(\ell)\right|^2. \tag{12.80}$$

Conclusion: Every past-indeterminate process can be realized as the output of a non-anticipative linear filter with an appropriate individual white-noise input function. The corresponding response function can be constructed by a Wiener factorization of the spectral density of the process.

Example: The Retarded Ornstein-Uhlenbeck Process
Consider a simple RC low-pass filter with resistance R and capacitance C. If the input of this filter is a white-noise Gaussian process $t \mapsto \check{w}(t)$, then its output is an Ornstein-Uhlenbeck process $t \mapsto \check{u}(t)$ (see Fig. 12.4). The output voltage $\check{e}_{\text{out}}$ is related to the input voltage $\check{e}_{\text{in}}$ via the differential equation $d\check{e}_{\text{out}}(t)/dt = \{\check{e}_{\text{out}}(t) + \check{e}_{\text{in}(t)}\}/RC$ with the general solution:

$$\check{e}_{\text{out}}(t) = e^{-(t-T)/(RC)}\check{e}_{\text{out}}(T) + \frac{1}{RC}\int_T^t e^{-(t-s)/(RC)}\check{e}_{\text{in}}(s)ds.$$

The retarded solution $t \mapsto \check{e}^{-}_{\text{out}}(t)$ is determined by $\check{e}_{\text{out}}(-\infty) = 0$, so that

$$\check{e}^{-}_{\text{out}}(t) = \frac{1}{RC}\int_0^{\infty} e^{-t/(RC)}\check{e}_{\text{in}}(s-t)ds,$$

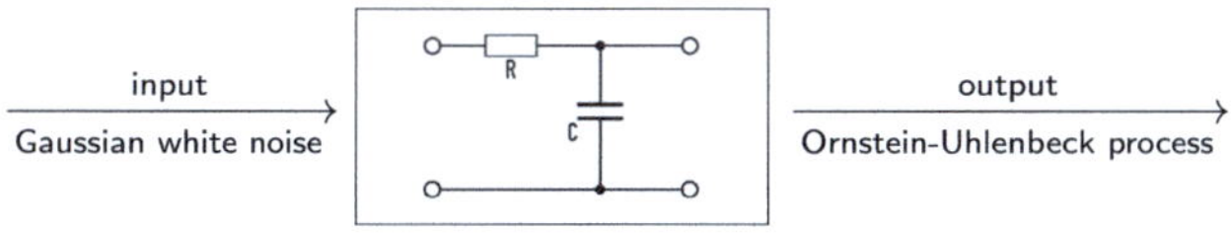

Fig. 12.4 Sketch of a RC low-pass filter transforming white noise into an Orstein-Uhlenbeck process. For more details on Ornstein-Uhlenbeck processes see Sect. 12.4.4

and the retarded frequency-response function $t \mapsto \check{R}^-(t)$ is given by

$$\check{R}^-(t) = \vartheta(t)\frac{1}{RC}e^{-t/(RC)}, \qquad \hat{R}^-(\omega) = \frac{1}{1 - i\omega RC}.$$

The advanced solution $t \mapsto \check{e}^+_{\text{out}}(t)$ is given by $\check{e}_{\text{out}}(+\infty) = 0$, so that

$$\check{e}^+_{\text{out}}(t) = -\frac{1}{RC}\int_{-\infty}^{0} e^{t/(RC)}\check{e}_{\text{in}}(s-t)ds,$$

and the advanced frequency-response function $\check{R}^+(t)$ is determined by

$$\check{R}^+(t) = -\vartheta(-t)\frac{1}{RC}e^{+t/(RC)}, \qquad \hat{R}^+(\omega) = \frac{-1}{1 + i\omega RC}.$$

The advanced solution $t \mapsto \check{e}^+_{\text{out}}(t)$ cannot be realized by a physical filter in *real time* since such a filter would have prognostic abilities. But using delay lines, the advanced solution can be approximately realized in shifted real time. In fact, such digital filters are used in many CD players to get a response behavior better adapted to human music perception.

12.4.3.4 Wiener Filter

Wiener's approach was originally restricted to linear filtering and linear prediction of time-translation invariant processes. The observed signal $t \mapsto s(t)$ (for example, the incoming radar signal from an airplane) is assumed to be the sum of an individual real-valued message $t \mapsto m(t)$ (for example, the plane's true position) and an individual real-valued noise process $t \mapsto n(t)$ (for example due to tracking error),[245]

$$s(t) = m(t) + n(t), \quad t \in \mathbb{R}. \tag{12.81}$$

The goal of Wiener's approach is to analyze the received signal $t \mapsto s(t)$ in such a way as to retrieve the message $t \mapsto m(t)$ as precisely as possible. The problem of predicting the future of the message $\{m(t)|t > 0\}$ from the past observed noisy signal $\{u(t)|t > 0\}$ has two parts:

(1) The *pure smoothing problem*: How can the message $\{m(t)|t < 0\}$ be optimally extracted from the observed noisy signal $\{s(t)|t < 0\}$?

[245] For convenience, we shall consider scalar continuous-time processes. However, vector-valued processes and discrete-time problems can be addressed in the same framework.

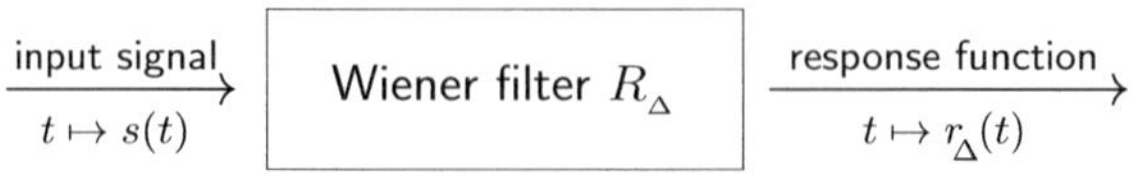

Fig. 12.5 Sketch of a Wiener filter transforming an input signal $s(t)$ into a filtered signal $r_\Delta(t)$ (see Eq. (12.82))

(2) The *pure prediction problem*: If there is no noise, $s(t) = m(t)$, how can the future message $\{m(t)|t > 0\}$ be optimally predicted from the past message $\{m(t)|t < 0\}$?

Wiener's technique has the merit of dealing with filtering and prediction at one stroke. The problem is to find a physically realizable on-line filter such that the output response function $t \mapsto r_\Delta(t)$ is the optimal prediction of the signal $s(t + \Delta)$ where Δ is the prediction time. "On-line" means that the estimate $r_\Delta(t)$ should be available at time t. In order to be physically realizable, the filter has to be non-anticipative. Since all involved processes are assumed to be time-translation invariant, the optimal filter is time-invariant, so that the linear input-output map can be described by (cf. Fig. 12.5):

$$r_\Delta(t) = \int_0^\infty R_\Delta(t')s(t - t')dt'. \tag{12.82}$$

As a criterion for measuring the difference between the single function $t \mapsto x(t + \Delta)$ and its estimator $t \mapsto x(t, t)$, Wiener's method uses the *individual* temporal mean-squared error $e_{\text{ind}}(\Delta)$,

$$e_{\text{ind}}(\Delta) := \frac{1}{2T} \lim_{T\to\infty} \int_{-T}^{T} \left|m(t + \Delta) - r_\Delta(t)\right|^2 dt. \tag{12.83}$$

For ergodic stationary stochastic processes Wiener's analytical representation theorem for a single function can be related to Khintchine's ensemble representation theorem for stochastic processes via Birkhoff's ergodic theorem.

Birkhoff's Individual Ergodic Theorem for Autocorrelations[246]
Let $\{x(t|\omega)|t \in \mathbb{R}\}$ be a complex-valued strictly stationary zero-mean stochastic process over the probability space (Ω, Σ, μ), and let $x(t|\cdot) \in \mathbf{L}^2(\Omega, \Sigma, \mu)$. Then for μ-almost all $\omega \in \Omega$, the trajectory $t \mapsto x(t|\omega)$ belongs to the Wiener class $\mathcal{W}_c$ of individual measurable functions, for which the individual autocorrelation function

$$A(t|\omega) := \lim_{T\to\infty} \frac{1}{2T} \int_{-T}^{T} x(t|\omega)^* x(t + s|\omega)ds, \quad t \in \mathbb{R}, \tag{12.84}$$

exists and is continuous on $\mathbb{R}$. For μ-almost all $\omega \in \Omega$ the autocorrelation function $t \mapsto A(t|\omega)$ satisfies the equality

$$A(t|\omega) = \mathcal{E}_\mathcal{J}\{[x(t|\cdot)x(0|\cdot)]\}, \tag{12.85}$$

[246]This formulation is taken from Masani (1990, p. 139f).

where $\mathcal{E}_{\mathfrak{J}}$ is the conditional expectation with respect to the σ-algebra $\mathfrak{J}$ of sets in Σ, invariant under the flow associated with the stochastic process $\{[x(t|\cdot)]|t \in \mathbb{R}\}$.

The stochastic process is *ergodic* if and only if $\mathfrak{J}$ is trivial. In this case $\mathcal{E}_{\mathfrak{J}}$ equals the unconditional expectation $\mathcal{E}$ so that for μ-almost all $\omega \in \Omega$ the autocorrelation function $t \mapsto A(t|\omega)$ of an individual function $t \mapsto x(t|\omega)$ equals the covariance function $t \mapsto C(t)$ of the stochastic process $\{[x(t|\cdot)]|t \in \mathbb{R}\}$,

$$A(t|\omega) = C(t) := \mathcal{E}\{[x(t|\cdot)x(0|\cdot)]\}. \tag{12.86}$$

The requirement of ergodicity is not a severe restriction. First of all, if a stochastic process is not ergodic, it can be decomposed under very general conditions into ergodic components (von Neumann 1932b, p. 617, see also von Plato 1982 for a historical account). Moreover, if the spectral distribution function of a strictly continuous stochastic process is continuous, then the process is ergodic.[247]

The individual ergodic theorem warrants that

$$\lim_{T\to\infty} \frac{1}{2T} \int_{-T}^{T} x(t|\omega)^* x(t+s|\omega)ds$$
$$= \lim_{T\to\infty} \frac{1}{T} \int_{-T}^{0} x(t|\omega)^* x(t+s|\omega)ds, \tag{12.87}$$

so that the autocorrelation function $t \mapsto A(t|\omega)$ can be evaluated from the *past* behavior of the function $t \mapsto x(t|\omega)$. This implies that for an ergodic stochastic process $\{[x(t|\cdot)]|t \in \mathbb{R}\}$ the covariance function $t \mapsto C(t)$ can in principle be determined from the past behavior of a *single* typical trajectory $t \mapsto x(t|\omega)$.

12.4.3.5 Wiener's Individual Processes as Trajectories of Stochastic Processes

Kolmogorov's statistical description and Wiener's individual description are not only conceptually disparate but also mathematically inequivalent. Already the statistical ensemble expectation value $\mathcal{E}(\cdot)$ and the temporal mean value $\mathcal{M}(\cdot)$ for individual functions, $\mathcal{M}(f) := \lim_{T\to\infty} \frac{1}{2T} \int_{-T}^{T} f(t)dt$ have different properties. For example, if for two zero-mean stationary stochastic processes $\{[f(t)]|t \in \mathbb{R}\}$ and $\{[g(t)]|t \in \mathbb{R}\}$ on the same probability space the covariance functions $\mathcal{E}\{[f(0)^* f(t)]\}$ and $\mathcal{E}\{[g(0)^* g(t)]\}$ exist, then Schwarz's inequality warrants the existence of the covariance function $\mathcal{E}\{[f(0)^* g(t)]\}$. If f and g are two locally integrable complex-valued individual functions, then the existence of the autocorrelation functions $\mathcal{M}\{f(0)^* f(t)\}$ and $\mathcal{M}\{g(0)^* g(t)\}$ does *not* imply the existence of the crosscorrelation function $\mathcal{M}\{f(0)^* g(t)\}$ (cf. Bass 1996).

Moreover, in a genuine statistical theory only the equivalence class $t \mapsto [f(t)]$ as a whole is of interest; an individual trajectory $t \mapsto f(t)$ plays no role in it. In spite of the fact that there are no general relations between Kolmogorov's statistical

[247] Wiener and Akutowicz (1957) proved that such a process is weakly mixing, which implies ergodicity.

theory and Wiener's altogether non-statistical theory, Birkhoff's theorem implies that μ-almost every trajectory of an ergodic stochastic process on a Kolmogorov probability space is an individual function in the Wiener class $\mathcal{W}_c$. Consequently, the individual sample function of second-order stationary stochastic processes provides a rich source of examples for irregular processes in the sense of Wiener. However, to say that (Mackey 1974, p. 217ff)

> from a practical point of view, Wiener's theory is a theory of sample functions ... [and] it is better to work with the whole ensemble of possible sample functions than to deal with any one of them

reflects the narrow viewpoint of many statisticians that misses the essential importance of Wiener's theory for a non-statistical theory of individual physical phenomena.[248]

Up to today there are only few genuine physical applications of Wiener's irregular individual functions to physical problems. Notable examples are attempts towards a deterministic theory of the turbulent velocity fluctuations arising in nonlinear partial differential equations of the Navier-Stokes type.[249]

12.4.4 Example: Ornstein-Uhlenbeck Processes

12.4.4.1 Three Inequivalent Ornstein-Uhlenbeck Processes

Any zero-mean real-valued stationary stochastic process $\mathbb{R} \ni t \mapsto \check{u}(t) \in \mathbb{R}$ with continuous paths and the *statistical covariance function*

$$\check{C}(t-t') = \mathcal{E}\{\check{u}(t)\check{u}(t')\} = \sigma^2 e^{-\gamma|t-t'|}, \quad \sigma > 0,\ \gamma > 0, \tag{12.88}$$

is called an *Ornstein-Uhlenbeck process.*[250]

The spectral density of a generic Ornstein-Uhlenbeck process is absolutely continuous and given by

$$\int_{\mathbb{R}} e^{-2\pi i \ell} \mathcal{E}\{\check{u}(t)\check{u}(0)\}dt = 2\sigma^2 \frac{\gamma}{\gamma^2 + (2\pi\ell)^2}. \tag{12.89}$$

---

[248] Of course, Wiener did realize the virtue of thinking of certain irregular individual functions as trajectories of stochastic processes. He also introduced Gibbs' statistical viewpoint into communication engineering.

[249] Compare Bass (1974) and Bass (1984, Chap. IX). In his valuable commentary, Masani (1979) stressed the importance of the role of generalized harmonic analysis for non-stochastic approaches to turbulence and for the quest for randomness.

[250] This stochastic process was introduced by Uhlenbeck and Ornstein (1930) as a physical model of Brownian motion.

Almost every trajectory of an Ornstein-Uhlenbeck process is an individual function in the Wiener class $\mathcal{W}_c$ with the *individual* autocorrelation function

$$t \mapsto \lim_{T\to\infty} \frac{1}{2T}\int_{-T}^{T} \breve{u}(s)\breve{u}(s+t)ds = \sigma^2 e^{-\gamma|t-t'|}, \quad t \in \mathbb{R}. \tag{12.90}$$

The function $\ell \mapsto 2\gamma\sigma^2(\gamma^2+4\pi^2\ell^2)^{-1}$ fulfills the Wiener-Kreĭn criterion (12.57b), so that all versions of an Ornstein-Uhlenbeck process are regular.

Depending on the factorization of the spectral density $\ell \mapsto \hat{C}(\ell)$ there are the following three inequivalent realizations of the moving-average representation of an Ornstein-Uhlenbeck process,

$$\breve{u}^{\mu}(t) = \int_{-\infty}^{\infty} \breve{R}^{\mu}(t-t')d\breve{w}(t'), \quad \left|\hat{R}^{\mu}(\ell)\right|^2 = \hat{C}(\ell), \; \mu = -1, 0, +1, \tag{12.91}$$

which are distinguished by their behavior under time-reversal.

(i) *The trivial factorization*

$$\left|\hat{R}(\ell)\right|^2 = \frac{2\sigma^2\gamma}{\gamma^2+4\pi^2\ell^2} = \sqrt{\frac{2\sigma^2\gamma}{\gamma^2+4\pi^2\ell^2}}\sqrt{\frac{2\sigma^2\gamma}{\gamma^2+4\pi^2\ell^2}}, \tag{12.92}$$

with

$$\breve{R}^{(0)}(t) := \int_{-\infty}^{\infty} e^{-2\pi i\ell t}\sqrt{\frac{2\sigma^2\gamma}{\gamma^2+4\pi^2\ell^2}}d\ell = \sqrt{2\gamma}(\sigma/\pi)K_0(\gamma t), \tag{12.93}$$

leads to the *time-reversal invariant* two-sided moving-average representation of the Ornstein-Uhlenbeck process,

$$\breve{u}^{(0)}(t) = \frac{1}{\pi}\sqrt{2\gamma\sigma^2}\int_{-\infty}^{\infty} K_0(2\pi\gamma|t-t'|)d\breve{w}(t'), \tag{12.94}$$

with the covariance

$$\begin{aligned}\mathcal{E}\{\breve{u}^0(t)\breve{u}^0(t')\} &= \frac{2\gamma\sigma^2}{\pi^2}\int_{-\infty}^{\infty} K_0(2\pi\gamma|t-t''|)K_0(2\pi\gamma|t'-t''|)dt'' \\ &= \sigma^2 e^{-\gamma|t-t'|},\end{aligned} \tag{12.95}$$

where K_0 is the modified Bessel function of the third kind. The response function $t \mapsto \breve{R}^0(t)$ of the two-sided process $t \mapsto \breve{u}^0$ involves the values of the input Wiener process for *all* $t \in \mathbb{R}$, hence it involves neither anticipation nor memory.

(ii) *The canonical Hardy factorization*

The function $\ell \mapsto (\gamma^2 + 4\pi^2\ell^2)^{-1}$ fulfills the Paley-Wiener criterion (12.68), so that the Ornstein-Uhlenbeck process is regular and allows a canonical Hardy factorization into a product of two functions with holomorphic extensions into complex half-planes. Then the function $\ell \mapsto |\hat{R}(\ell)|^2$ is given by

$$\left|\hat{R}(\ell)\right|^2 = \frac{2\sigma^2\gamma}{\gamma^2 + 4\pi^2\ell^2} = \hat{R}^+(\ell)\hat{R}(\ell)^-, \quad \hat{R}^\pm(\ell) := \frac{\sigma\sqrt{2\gamma}}{\gamma \pm 2\pi i\ell}, \tag{12.96}$$

and the response is:

$$\check{R}^\pm(t) = \int_{-\infty}^{\infty} e^{-2\pi i\ell t}\frac{\sigma\sqrt{2\gamma}}{\gamma \pm 2\pi i\ell}d\ell = \sigma\sqrt{2\gamma}e^{\pm\gamma t}\vartheta(\mp t). \tag{12.97}$$

The function $\ell \mapsto \hat{R}^-(\ell)$ is the boundary value of an analytic function $z \mapsto \hat{R}^-(z)$, holomorphic in the open lower half-plane $\mathbb{C}^-$, and $\ell \mapsto \hat{R}^+(\ell)$ is the boundary value of another analytic function $z \mapsto \hat{R}^+(z)$, holomorphic in the open upper half-plane $\mathbb{C}^+$.

The corresponding two time-reversal symmetry-breaking moving-average representations of the Ornstein-Uhlenbeck process are one-sided and given by

$$\check{u}^-(t) = \sigma\sqrt{2\gamma}\int_{-\infty}^{t} e^{-\gamma(t-t')}dw(t'), \tag{12.98}$$

$$\check{u}^+(t) = \sigma\sqrt{2\gamma}\int_{t}^{\infty} e^{\gamma(t-t')}dw(t'). \tag{12.99}$$

The retarded trajectory $t \mapsto \check{u}^-(t)$ is the *past-determined non-anticipative solution*, and the advanced trajectory $t \mapsto \check{u}^+(t)$ is the *future-determined anticipative solution* of the following linear stochastic differential equation

$$d\check{u}^\pm(t) = \mp\gamma\check{u}^\pm(t)dt + \sigma\sqrt{2\gamma}dw(t). \tag{12.100}$$

12.4.4.2 Prediction of an Ornstein-Uhlenbeck Process

The spectral distribution function $\hat{C}$ of an Ornstein-Uhlenbeck process is absolutely continuous with the spectral density $\ell \mapsto \hat{C}(\ell)$,

$$\hat{C}(\ell) = \int_{-\infty}^{\infty} e^{i\ell t}\check{C}(t)dt = \sigma^2\int_{-\infty}^{\infty} e^{-i\ell t}e^{-\gamma|t-s|}dt = 2\sigma^2\frac{\gamma}{\gamma^2+\ell^2}. \tag{12.101}$$

This spectral density fulfills the Paley-Wiener criterion (12.68), so that the Ornstein-Uhlenbeck process is regular. With the relation $|\varphi(\ell)|^2 = \hat{C}(\ell)$ and the unique factorization $\hat{C}(\ell) = 2\gamma\sigma^2(\gamma + i\ell)^{-1}(\gamma - i\ell)^{-1}$ we find:

$$\hat{R}(\ell) = \frac{\sqrt{2\gamma\sigma^2}}{\gamma + i\ell}, \qquad \hat{R}_t(\ell) = e^{-\gamma t}. \tag{12.102}$$

Therefore the best linear predictor of $[\check{x}(t+t)]$ based on $\{[\check{x}(s)]|s \leq t\}$ is given by

$$\check{x}(t,t) = \int_{-\infty}^{\infty} e^{i\ell t} e^{-\gamma t} Z(d\ell) = e^{-\gamma t} x(t), \tag{12.103}$$

and the minimal prediction is

$$e^2_{\text{stat,min}}(t) = \sigma^2 \{1 - e^{-2\gamma t}\}. \tag{12.104}$$

Chapter 13
The Relevance of Non-Sequential Time

The sequential notion of time
applies only to the factual aspect of reality.

Albrecht von Müller (2010, p. 61).

13.1 Complementarity of Sequential and Non-Sequential Time

The harmonic analysis of the non-commutative affine Weyl-Heisenberg group provides a natural setting and a unifying language for defining and analyzing complementary time concepts. The complementarity of the descriptions in terms of the affine subgroup $\mathfrak{A}_{\mathfrak{U}}$ and the affine subgroup $\mathfrak{A}_{\mathfrak{V}}$ implies that the concepts of a sequential $\mathfrak{A}_{\mathfrak{V}}$-description are not in the conceptual repertoire of a non-sequential $\mathfrak{A}_{\mathfrak{U}}$-description.

Any attempt to "clarify" the structure of non-sequential time by means of concepts of sequential thinking is inappropriate since logical arguments and intuitive insight are fundamentally different kinds of knowledge which cannot be translated into one another. Descriptions in terms of sequential time and descriptions in terms of non-sequential time are not contradictory but complementary. None of the two is more fundamental than the other, both are necessary for a complete picture.

Notions that can be explained in terms of a sequential language cannot be explained in terms of a non-sequential language. Difficulties and contradictions arise when we try to describe non-sequential temporal phenomena in terms of the familiar sequentially-ordered language. The significance of non-sequential thinking in science has not received full appreciation yet. The development of an appropriate language for non-sequential time for problems of natural science is an important and timely desideratum.

13.2 Sequential Time in Science

Since sequential thinking is deeply ingrained in our natural language we have hardly a proper language to speak about non-sequential time so that we may face severe dif-

H. Primas, *Knowledge and Time*,
DOI 10.1007/978-3-319-47370-3_13

ficulties to appreciate non-sequential insight. Science is far richer than is accessible by a purely analytic-logical approach. Noncausal narrative reasoning is also an important part of our scientific culture. It reflects cultural beliefs and meaning systems that cannot be simply rejected as "merely subjective". In the words of Kripal (2011, p. 210):

> Reality sometimes behaves as if it were a narrative or story. It organizes itself around metaphors and symbols and chains of meaning and not simply around mathematics and chains of causes.

Many scientists rely heavily on a one-sided approach of sequential thinking, preferring exclusively rational arguments. For example, Lord Kelvin (William Thomson 1891, pp. 80f) postulated in a lecture of May 3, 1883, that all scientific inquiry has to be *quantitative*:

> When you can measure what you are speaking about, and express it in numbers, you know something about it; but when you cannot express it in numbers, your knowledge is of a meagre and unsatisfactory kind; it may be the beginning of knowledge, but you have scarcely, in your thoughts, advanced to the stage of science, whatever the matter may be.

That sequential logical analysis comprises only a part of the machinery of our intelligence and that knowledge is not synonymous with sequential thinking is well known in every scientific discipline. In his *La science et l'hypothèse*, Poincaré asked:[251]

> If this science is only deductive in appearance, from whence is derived that perfect rigour which is challenged by none? If, on the contrary, all the propositions which it enunciates may be derived in order by the rules of formal logic, how is it that mathematics is not reduced to a gigantic tautology?

Non-sequential time is invariant under time-translation and time-reversal, and it has no direction. Therefore it is impossible to define the tensed concepts of past, present and future of sequential time in a non-sequential time description. As discussed in Sect. 11.6.2, the affine group $\mathfrak{A}_{\mathfrak{U}}$ breaks frequency-reversal symmetry. Non-sequential time is tenseless and represents an unfolding duration which cannot and must not be interpreted using the notions of sequential time. Non-sequential time is neither reducible to sequential time nor to causality.

In a description based on non-sequential time it is impossible to define a punctual present without extension. Non-sequential time is homogeneous with a parallel ordering, implying that all basic principles remain the same at all times—which is why it can be considered as invariant under time-translation and time-reversal. In this description there is no memory. Non-sequential time is related to *non-sequential knowledge*. Any awareness of past and future is essentially withdrawn. Both sequential and non-sequential time are mathematically well-defined, and they are complementary in the sense of Eq. (3.52).

---

[251] English translation quoted from Poincaré (1905, p. 1).

13.3 Philosophical Approaches to Non-Sequential Time[252]

In contrast to the sciences, non-sequential time has been discussed in a variety of contexts in philosophy. Virtually the first among them was that of the *specious present*, which William James characterized succinctly in his *Principles of Psychology*:[253] "the original paragon and prototype of all conceived times is the specious present, the short duration of which we are immediately and incessantly sensible" (James 1890, p. 631). In another passage, James estimates the duration of the specious present (p. 642):

> We are constantly aware of a certain duration—the specious present—varying from a few seconds to probably not more than a minute, and this duration (with its content perceived as having one part earlier and another part later) is the original intuition of time.

With this emphasis, James expresses the immediacy of the awareness of nowness, on which awareness of anything else is inevitably based. For Edmund Husserl, this became the core of his philosophical phenomenology. It starts with what he called phenomenological reduction, the process of "bracketing" any sensations so to achieve the experience of the present as pure and unperturbed as possible.

In the years from 1893 to 1917 Husserl worked out his ideas on the *Phenomenology of the Consciousness of Internal Time* (Husserl 1991), in which the central thesis is a threefold structure of temporal consciousness: *protention, retention, primal impression*. While the experience of the present *in toto* is a temporally extended whole, this whole can be abstractly decomposed into the "currently-given-as-new" (primal impression), the "experienced-as-just-past (retention) and the "shortly-to-be-experienced" (protention).

An illustrative example is the perception of a musical phrase, such as the initial motif in the first movement of Beethoven's fifth symphony, g g g e-flat (Fig. 13.1), forming a musical gestalt as a whole. The primal impression of the second g differs from the primal impression of the first g, since this is already in retention when the second is perceived, and the third g is then in protention. Without retention (and protention) every one of the g notes would be experienced the same. However, retention is not recollection, and protention is not anticipation. They are both contained in the present, and do neither belong to the past nor to the future.

Non-sequential time in the sense of the specious present entails a radical consequence: the experience of the present does ultimately not consist of single events that are juxtaposed along a time axis. Insofar as the present offers a temporal whole, everything experienced within it allows no earlier-later relation. This implies that efficient causation in the sense of causes temporally preceding their effects becomes inapplicable.

---

[252]This subsection has been added by the editor, in order to cover some material showing that non-sequential time has been a desideratum for philosophers for more than a century.

[253]As Andersen and Grush (2009) found, there are a number of precursors to James, most notably an anonymous source who published the book *The Alternative: A Study in Psychology* in 1882 under the pseudonym of Robert Kelly, to whom William James referred as E.R. Clay.

Fig. 13.1 The initial motif of the first movement of Beethoven's fifth symphony

This was one of the early insights of Henri Bergson in this doctoral thesis *Essai sur les données immédiates de la conscience* at Paris in 1889,[254] which laid the foundations of his theory of duration (French: *durée*). For Bergson, duration is a qualitative (rather than quantitative) multiplicity, a temporal heterogeneity, in which "conscious states are organized into a whole, permeate one another, [and] gradually gain a richer content" (Bergson 2001, p. 122). In his article *Introduction to Metaphysics* of 1903,[255] he discussed the analogy with the physical example of a color spectrum, which we already met in a remarkable quotation by Max Frisch in Sect. 4.3.

Bergson's philosophy was quite influential for a number of later French thinkers, such as Merleau-Ponty, Sartre, Levinas and, above all, Deleuze whose concept of "becomings" is modeled after Bergson's qualitative multiplicity in duration. However, there is one other philosopher (and mathematician) outside the French tradition, who focused on non-sequential time in his major opus *Process and Reality* of 1929: Alfred North Whitehead.[256]

The key primitive concept in Whitehead's metaphysics is that of *actual occasions*—events that are spatiotemporally extended and resemble the monads of Leibniz with the additional feature that they interact, while Leibnizian monads are interaction-free. Even more to the point, actual occasions *are* interactions (hence the term "process" in the title of his monograph) rather than substances or objects. Whitehead (1929, p. 27) describes them as "drops of experience, complex and interdependent", but experience in this sense must not be confused with consciousness. Whitehead's actual occasions are paradigmatic examples of psychophysically neutral entities from which mind and matter arise as abstractions.

13.4 Experienced Durational Time

A dreadful but nevertheless enlightening example of durational, non-sequential time comes from Langer's[257] sensitive reading of testimonies of survivors of the Holocaust and his distinction between chronological time, which is made up of past,

[254]The complete title of the English translation is *Time and Free Will: An Essay on the Immediate Data of Consciousness* (Bergson 2001).

[255]This article became the center piece of the monograph *The Creative Mind* (French: *La penssée et le mouvant*) of 1934.

[256]Although there are many thematic links between Whitehead, Bergson and James (see for instance Robinson 2009), Lowe (1949) doubts that Whitehead's work was substantially *influenced* by Bergson or James.

[257]Lawrence L. Langer is a foremost scholar of Holocaust literature and remembrance.

present, and future, and non-sequential *durational* time (Langer 1995, p. 20; cf. also Langer 1991, 1998):

> Durational time does not believe in or allow for any sort of foreclosure toward the past: its foreclosure expels the future.

Claude Lanzmann (1994) says about the Shoah:

> There is no greater mistake you can make about the Shoah than to regard it as history.

And Elie Wiesel (1978) ascertains:

> Auschwitz cannot be explained ... the Holocaust transcends history. ... The Holocaust can never be comprehended or transmitted.

According to Georg Steiner (1970):

> The world of Auschwitz lies outside speech as it lies outside reason.

Personally experienced durational time without past or future sets in when extreme crises become the norm of life. According to Langer (1998, p. 141), durational time

> does not enter into the chronology of events that have a past, a present, and a future, but possesses a non-sequential integrity that exists outside the flow of normal time.

Langer's notion of durational time refers to the singular sense of time as experienced by Holocaust survivors where events exist within a frame of temporal referentiality of its own—it does not exist in the past against which a sense of present or future can be postulated (Langer 1995, p. 15):

> In the realm of durational time, no one recovers because nothing is recovered, only uncovered and then recovered, buried again beneath the fruitless struggle to expose "the way it was".

Or (Langer 1998, p. 69):

> Anyone who hears [the survivors' testimonies] will understand that for the witnesses time is durational as well as chronological and that durational time is experienced continuously, not sequentially as a memory from which one can be liberated.

13.5 Temporal Nonlocality[258]

Although a non-sequential concept of time is unusual (to say the least) in scientific discussions of time, precursors to it arise from the discussion of nonlocality in quantum mechanics.[259] The basic test for this nonlocality is the violation of a Bell-type inequality (Bell 1966; see also d'Espagnat 1979), which provides an upper bound for

[258] This subsection has been added by the editor. It addresses some more recent advances concerning non-sequential time, which the author was not aware of at the time he finished the manuscript.

[259] Nonlocality of this type is also known as non-separability or entanglement and was introduced in Sect. 6.3.5. For a non-technical reference for the history of the concept up to current applications see Gilder (2008).

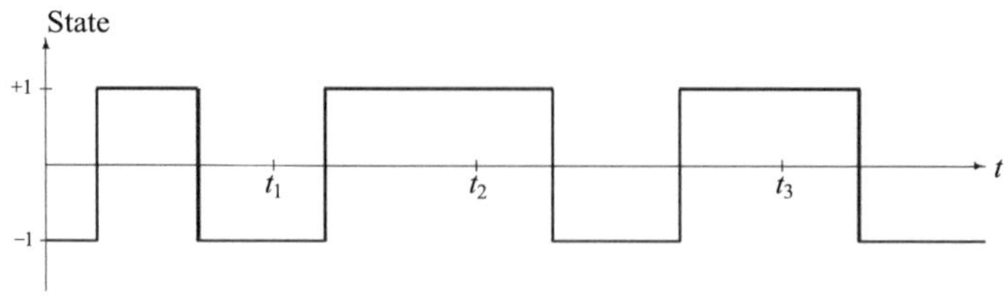

Fig. 13.2 A classical trajectory assumes a definite state (here one of two possible states, labeled $+1$ and -1) at each moment in time. With respect to three instances t_1, t_2, and t_3 the history of states falls into one of $2^3 = 8$ possible classes. For the shown history the class is $(-1, +1, +1)$, for which the first two states as well as the first and third state are anti-correlated

correlations between observables measured at spatially distant subsystems *if those correlations are purely classical*. Quantum correlations can exceed this upper bound and indicate that the measured correlations are stronger than a classical theory is capable of explaining.

Nonlocalities of this kind are the hallmark of quantum physics as compared to classical physics. For empirical tests of their underlying quantum correlations (Aspect et al. 1982), it is important to exclude physical interactions (signaling) between the subsystems. Such interactions would introduce additional (classical) correlations that can easily lead to violations of Bell inequalities, but have nothing to do with quantum correlations as indicators of nonlocality.

While most tests of nonlocality refer, implicitly or explicitly, to correlations over spatial distance, it is possible to set up a Bell-type inequality for temporal distance as well. This was first proposed by Leggett and Garg (1985),[260] who considered the temporal evolution of a two-state system at three instances t_1, t_2, t_3, for which the probability p to find anti-correlations between states at different times is given by

$$p(t_1, t_3) \leq p(t_1, t_2) + p(t_2, t_3). \tag{13.1}$$

For instance, the scheme shown in Fig. 13.2 exhibits anti-correlations between the states at t_1 and t_2 as well as t_1 and t_3, so (13.1) is clearly satisfied. Inequality (13.1) can be generalized to n-state systems.

In order to test such a temporal Bell inequality empirically, the crucial condition of no signaling means that one must be able to exclude interactions between measurements. In other words, it must be guaranteed that a measurement at any time does not influence the time evolution of the system such that classical correlations arise. This is known as the condition of non-invasive measurements.

Since it is hard to conceive of temporally successive measurements satisfying this condition, non-invasivity has been a major stumbling block for empirical tests of temporal Bell inequalities so far (for reviews see Emary et al. 2014; Mahler 2016). A possible way out of this dilemma was recently suggested by Kujala and Dzhafarov

[260] Their result was first proven, in a more general framework, by Suppes and Zanotti (1981).

(2016), who proposed a strategy to disentangle classical and non-classical correlations, so that one can evaluate the identified quantum correlations alone. See also the illustrative discussion of this proposal by Bacciagaluppi (2015).

Irrespective of the problem of non-invasiveness and its possible solution, we may ask the question what it actually would mean if a temporal Bell inequality were found to be violated by provably non-classical correlations. A main assumption for deriving (13.1) obviously is that the measured state can always be assigned to a single instant t in time. A violation of (13.1) implies that this assumption does not hold: States are "nonlocally" extended along the time axis as "temporal distributions", and cannot be properly characterized in terms of sharply defined classical observables. Section 11.7 presents how examples of such temporal distributions can be formally constructed.

Temporally nonlocal states are "stretched" over an extended time interval with a length that will typically depend on the specific system considered. Within this interval, relations such as "earlier" or "later" are illegitimate designators of the system state. This is just another way of saying that it is impossible to define causal relationships in the usual sense of efficient causation within such a time interval (Filk and von Müller 2009).

Temporally nonlocal states as a consequence of violated temporal Bell inequalities are evidently related to the idea of an extended present, or nowness, as expressed in Sect. 13.3. To make this relation more explicit, a well-defined option to apply temporal Bell inequalities to mental rather than physical systems would be desirable. This was first suggested by Atmanspacher and Filk (2010) for the scenario of bistable perception, a well-known phenomenon in cognitive science. The perception of ambiguous stimuli, which can be interpreted in two different ways, switches stochastically and spontaneously between the two interpretations on a time scale of typically a few seconds.

The formal model of this scenario describes bistable perception as a two-state system that is unstable under its dynamics, very much analogous to the quantum Zeno effect (Misra and Sudarshan 1977). An intriguing feature of this model is that it predicts violations of a temporal Bell inequality for certain parameters. This means that temporally nonlocal *mental* states have to be expected in bistable perception. As indicated above, empirical realizations of a corresponding test are difficult due to invasivity.[261]

Nevertheless, it is interesting to speculate how the concrete experience of temporally nonlocal mental states could be imagined. An illuminating remark by Sudarshan (1983) about a hypothetical direct perception of quantum states is illustrative in this respect. He indicates a mode of awareness in which

> sensations, feelings, and insights are not neatly categorized into chains of thoughts, nor is there a step-by-step development of a logical-legal argument-to-conclusion. Instead, patterns appear, interweave, coexist; and sequencing is made inoperative. Conclusion, premises, feelings, and insights coexist in a manner defying temporal order.

---

[261]For other possible applications of Bell-type inequalities in behavioral and social systems see Dzhafarov et al. (2015).

The exploration of non-sequential time in this sense offers huge potential for future work in physics, psychology, and philosophy. It has been one of the great challenges for human scholarship over centuries. The relationship between temporal and atemporal universes of discourse lies at its heart. Eventually it is not unlikely that it may be a pivotal key to unravel the mystery of the relationship between the mental and the physical.

Appendix
C*- and W*-Algebras

A.1 Algebras of Observables

A.1.1 Abstract *-Algebras

An algebra is called abstract if it is defined without reference to operators acting on some vector space. An abstract *-algebra $\mathcal{A}$ endowed with a Banach-space norm $\|\cdot\| : \mathcal{A} \to \mathbb{R}^+$, with the properties

$$\|A + B\| \le \|A\| + \|B\|, \qquad \|AB\| \le \|A\| \; \|B\|, \qquad \|A^*\| = \|A\|, \tag{A.1}$$

for every $A, B \in \mathcal{A}$, is called a *Banach *-algebra*. The crucial property that characterizes abstract *C*-algebras* among all Banach *-algebras is the so-called C*-identity:

$$\|A^* A\| = \|A\|^2 \quad \text{for all } A \in \mathcal{A}. \tag{A.2}$$

In contrast to a general Banach *-algebra, the norm of a C*-algebra is uniquely determined by the *purely algebraic* property

$$\|A\|^2 = \text{spectral radius of } A^* A = \sup\big\{|\lambda| \,|\, A^* A\lambda \text{ is not invertible in } A\big\}. \tag{A.3}$$

This result implies that an algebraic formulation of physical theories in terms of C*-algebras is *purely algebraic*.

A.1.2 W*-Algebras and Their Preduals

A C*-algebra $\mathcal{M}$ is called a *W*-algebra* (also called von Neumann algebra) if it is a dual space of a Banach space (Sakai 1971, p. 1). That is, for every W*-algebra there exists a unique Banach space $\mathcal{M}_*$ such that $\mathcal{M} = (\mathcal{M}_*)^*$, where $(\mathcal{M}_*)^*$ is the dual Banach space of $\mathcal{M}_*$. The Banach space $\mathcal{M}_*$ is called the *predual* of the

H. Primas, *Knowledge and Time*,
DOI 10.1007/978-3-319-47370-3

W*-algebra $\mathcal{M}$. Every W*-algebra $\mathcal{M}$ can be represented as a subalgebra of a W*-algebra $\mathcal{B}(\mathcal{H})$ of all bounded linear operators on some Hilbert space $\mathcal{H}$.

Contextual descriptions of any material universe of discourse can be formulated algebraically in terms of W*-algebras. In analogy to the concept of additivity of a measure in classical probability theory, a linear positive functional φ on a W*-algebra $\mathcal{M}$ is said to be completely additive if it satisfies $\varphi(\bigvee F_n) = \sum \varphi(F_n)$ for every set $\{F_n\}$ of pairwise orthogonal projections in $\mathcal{M}$, $F_n F_m = 0$ for $n \neq m$.

In analogy to Lebesgue's classical monotone convergence theorem, a linear positive functional φ is said to be *normal* if $\varphi(M_n) \uparrow \varphi(M)$ for every monotonically increasing net $\{M_n\}$ of operators M_n in $\mathcal{M}$ with upper bound M. Operationally accessible states are represented by *normal* state functionals; they are elements of the predual $\mathcal{M}_* \subset \mathcal{M}^*$ of the W*-algebra $\mathcal{M}$ of contextual observables.

A.1.3 W-Algebras and Commutants*

A.1.3.1 Commutants and Double Commutants

Let $\mathcal{H}$ be a separable Hilbert space and $\mathcal{B}(\mathcal{H})$ the algebra of all bounded operators acting on $\mathcal{H}$. The commutant $\mathcal{M}$ of any self-adjoint subalgebra $\mathcal{N}$ of $\mathcal{B}(\mathcal{H})$,

$$\mathcal{M} := \big\{M | M \in \mathcal{B}(\mathcal{H}), MN = NM \text{ for every } N \in \mathcal{N}\big\}, \tag{A.4}$$

is a W*-algebra, that is a self-adjoint and weakly closed subalgebra of $\mathcal{B}(\mathcal{H})$. Clearly, the commutant of $\mathcal{M} \subset \mathcal{B}(\mathcal{H})$, denoted by $\mathcal{M}'$, is also a W*-algebra,

$$\mathcal{M}' := \big\{B | B \in \mathcal{B}(\mathcal{H}), MB = BM \text{ for every } M \in \mathcal{M}\big\}. \tag{A.5}$$

Moreover, the double commutant theorem holds,

$$\mathcal{M}'' := \big(\mathcal{M}'\big)' = \mathcal{M}. \tag{A.6}$$

If $\mathcal{M} \subseteq \mathcal{B}(\mathcal{H})$ is a W*-algebra with a bi-cyclic vector, then every normal state functional $\rho \in \mathcal{M}_*$ can be represented by a unit vector $\Phi \in \mathcal{H}$ as $\rho(M) = \langle \Phi | M\Phi \rangle$ for all $M \in \mathcal{M}$. If ρ is faithful, Φ can be chosen to be cyclic for $\mathcal{M}$ (Sakai 1971, Theorem 2.7.9).

A.1.3.2 The Commutant of a Set of Operators

Let $\mathcal{O}$ be an arbitrary set of operators within the algebra $\mathcal{B}(\mathcal{H})$ of all bounded operators acting on a Hilbert space $\mathcal{H}$. Let $\mathcal{O}'$ be the set of all operators in $\mathcal{B}(\mathcal{H})$ commuting with every element in $\mathcal{O}$,

$$\mathcal{O}' := \big\{A \in \mathcal{B}(\mathcal{H}) \,|\, TA = AT \text{ for every } T \in \mathcal{O}\big\}. \tag{A.7}$$

The set $\mathcal{O}'$ is called the *commutant* of $\mathcal{O}$. If $\mathcal{O}'$ is a self-adjoint subset of $\mathcal{B}(\mathcal{H})$, then $\mathcal{O}'$ is a W*-algebra. The double commutant $\mathcal{O}''$ is defined as the commutant of $\mathcal{O}'$, $\mathcal{O}'' = (\mathcal{O}')'$. It follows that $\mathcal{O} \subseteq \mathcal{O}''$. If $\mathcal{O}'$ is a self-adjoint subset of $\mathcal{B}(\mathcal{H})$, then $\mathcal{O}''$ is the smallest W*-algebra containing $\mathcal{O}$, i.e. the W*-algebra generated by $\mathcal{O}$.

The center $\mathcal{Z}(\mathcal{M})$ of a W*-algebra $\mathcal{M}$ is a *commutative* W*-subalgebra,

$$\mathcal{Z}(\mathcal{M}) := \{Z \,|\, z \in \mathcal{M}, ZM = MZ \text{ for every } M \in \mathcal{M}\}. \tag{A.8}$$

The center is said to be *trivial* if it consists of the multiples of the identity element only. A W*-algebra with a trivial center is called a *factor*. Non-trivial self-adjoint elements of the center are called *classical observables*. They commute with all elements of the W*-algebra $\mathcal{M}$.

- If a W*-algebra $\mathcal{M}$ is contained in its commutant $\mathcal{M}'$, $\mathcal{M} \subseteq \mathcal{M}'$, then $\mathcal{M}$ is a *commutative* W*-algebra. If $\mathcal{M} = \mathcal{M}'$, then $\mathcal{M}$ is called *maximally commutative*.
- The center $\mathcal{Z}$ of a W*-algebra $\mathcal{M}$ is given by $\mathcal{Z} := \mathcal{M} \cap \mathcal{M}'$. If the center is trivial, $\mathcal{M} \cap \mathcal{M}' = 1\mathbb{C}$, the W*-algebra $\mathcal{M}$ and its commutant $\mathcal{M}'$ are factors.
- If the commutant $\mathcal{M}'$ of a W*-algebra $\mathcal{M}$ is trivial, $\mathcal{M}' = 1\mathbb{C}$, then $\mathcal{M}$ is called *irreducible*. For example, the algebra $\mathcal{B}(\mathcal{H})$ of all bounded operators in a Hilbert space is irreducible.
- *Double commutant theorem*: A set $\mathcal{O}$ of self-adjoint operators acting on a Hilbert space $\mathcal{H}$ containing the identity operator is a W*-algebra if and only if $\mathcal{O} = \mathcal{O}''$.

A.1.4 W-Algebras of Type I, II, and III*

Von Neumann (1949) showed that every W*-algebra on a separable Hilbert space is isomorphic to a direct integral of factors. This decomposition is essentially unique, so that the classification of W*-algebras can be reduced to the classification of factors. Murray and von Neumann (1936) showed that every factor is either of type I, type II or type III. They are defined as follows.

- A factor is of type I if it contains an atom. A factor is of type I_n with $n < \infty$ if it is *-isomorphic to the algebra of all complex $n \times n$ matrices. A factor is of type I_∞ if it is *-isomorphic to the algebra of all bounded operators on some infinite-dimensional Hilbert space.
  Every W*-algebra $\mathcal{M}$ of type I with a separable predual $\mathcal{M}_*$ can be represented either as an algebra of $n \times n$ matrices (so-called W*-algebras of type I_∞) or by the algebra $\mathcal{B}(\mathcal{H})$ of all bounded operators on some infinite-dimensional Hilbert space.
- A factor is of type II if it does not contain atoms but some non-zero finite projections.
- A factor is of type III if it does not contain any non-zero finite projection.

This classification can be extended to general W*-algebras: A W*-algebra is of type X if it can be decomposed as a direct integral of type-X factors. Every W*-algebra can be written uniquely as a direct sum of W*-algebras of types I, II, and III.

A.2 Reduced Descriptions

A.2.1 Completely Positive Maps

A.2.1.1 Abstract Definition

Let $\mathcal{A}$ and $\mathcal{B}$ be C*-algebras. Recall that a linear map T of $\mathcal{A}$ into $\mathcal{B}$ is *positive* if $\mathrm{T}(A) \geq 0$ for every positive $A \in \mathcal{A}$. A linear map $\mathrm{T}:\mathcal{A} \to \mathcal{B}$ is *n-positive* if the linear map

$$\mathrm{T} \otimes 1_n : \mathcal{A} \otimes \mathcal{M}_n \to \mathcal{B} \otimes \mathcal{M}_n \tag{A.9}$$

is positive. Here, $\mathcal{M}_n$ is the algebra of all complex $n \times n$ matrices and 1_n is the identity transformation of $\mathcal{M}_n$ onto itself. If T is n-positive for all n, then T is *completely positive*.[262] If $\mathcal{A}$ or $\mathcal{B}$ is commutative, every positive linear map is completely positive (cf. Kadison and Ringrose 1992, p. 698). However, if both $\mathcal{A}$ and $\mathcal{B}$ are non-commutative, not every positive map is completely positive. For instance, each *-anti-homomorphism of a non-commutative C*-algebra is a positive linear mapping which is not 2-positive (see, again, Kadison and Ringrose 1992, p. 698). The paradigmatic physical example is the anti-automorphic *time-reversal map* which is positive but not completely positive.

A.2.1.2 Representation of Completely Positive Maps

We consider completely positive linear maps between an arbitrary C*-algebra $\mathcal{A}$ and the W*-algebra $\mathcal{B}(\mathcal{H})$ of all bounded operators on some Hilbert space $\mathcal{H}$. If $(\pi, \mathcal{H})$ is a representation of the C*-algebra $\mathcal{A}$ on the Hilbert space $\mathcal{H}$, and if V is a bounded linear operator of $\mathcal{H}$ into another Hilbert space $\mathcal{K}$, then the map $\mathrm{S}:\mathcal{A} \to \mathcal{B}(\mathcal{H})$ defined by

$$\mathrm{S} : A \to V^* \pi(A)\, V, \quad A \in \mathcal{A}, \tag{A.10}$$

is completely positive (Takesaki 1979, p. 194ff). If π is normal, then S is normal.

The converse is the important generalization of Naimark's dilation theorem due to Stinespring (1955); see also Takesaki (1979, Sect. IV.3), Bratteli and Robinson (1981, p. 229):

Stinespring's Dilation Theorem
Let S be a completely positive map from a C*-algebra $\mathcal{A}$ with identity into $\mathcal{B}(\mathcal{H})$. Then every completely positive map of $\mathcal{A}$ into $\mathcal{B}(\mathcal{H})$ has the form

$$\mathrm{S}(A) = V^* \pi(A)\, V, \quad A \in \mathcal{A}, \tag{A.11}$$

where π is a representation of $\mathcal{A}$ on some Hilbert space $\mathcal{H}$, and V is a bounded operator from $\mathcal{H}$ into $\mathcal{K}$. If $\mathcal{A}$ and $\mathcal{H}$ are separable, then $\mathcal{K}$ can be taken to be separable.

---

[262]The notion of a completely positive map between two C*-algebras was introduced by Stinespring (1955). Compare also Arveson (1967).

If $S(1) = 1$, then V is the projection $P_{\mathcal{H}}$ of $\mathcal{K}$ onto $\mathcal{H}$ so that

$$S(A) = P_{\mathcal{H}}\pi(A)|_{\mathcal{H}}, \quad A \in \mathcal{A}. \tag{A.12}$$

If $\mathcal{A}$ is a W*-algebra, the map S is normal if and only if π is normal.

The special case of a normal map S of a W*-algebra $\mathcal{M}$ into itself is of particular interest for the statistical description of the dynamics of open systems. It follows from Stinespring's theorem that a normal map S: $\mathcal{M} \to \mathcal{M}$ is completely positive if and only if

$$S(M) = \sum_n V_n M V_n, \quad M \in \mathcal{A},\ V_n \in \mathcal{M}, \tag{A.13}$$

where the convergence is in the weak operator topology (Bratteli and Robinson 1981, p. 230). If the predual $\mathcal{M}_*$ is separable, the set $\{V_n\}$ can be taken to be countable. If $\mathcal{M}$ is isomorphic to a matrix algebra, the set $\{V_n\}$ is finite-dimensional.

A.2.1.3 Reduced Description and Conditional Expectation

Consistent reduced descriptions are generated by *conditional expectations*. Let $\mathcal{N}$ be a W*-subalgebra of a W*-algebra $\mathcal{M}$. A conditional expectation from $\mathcal{M}$ onto $\mathcal{N}$ is a positive linear mapping E: $\mathcal{M} \to \mathcal{N}$ such that (Kadison and Ringrose 1986, p. 834)

$$E(1) = 1, \tag{A.14a}$$

$$E(N_1 M N_2) = N_1 E(M) N_2, \tag{A.14b}$$

whenever $M \in \mathcal{M}$ and $N_1, N_2 \in \mathcal{N}$. It follows that $E(M)^*E(M) \le E(M^*M)$, and that a conditional expectation is a projection of norm 1,

$$E(M) = M \quad \text{for every } M \in \mathcal{M}, \qquad \|E(N)\| \le \|N\| \quad \text{for every } N \in \mathcal{N}. \tag{A.15}$$

Conversely, every projection of norm 1 from a W*-algebra $\mathcal{N}$ onto a W*-algebra $\mathcal{M}$ is a conditional expectation.

If one restricts the one-parameter group $\{\Theta_\tau | \tau \in \mathbb{R}\}$ of *-automorphisms $\Theta_\tau \in$ Aut($\mathcal{M}$) to a W*-subalgebra $\mathcal{N} \subset \mathcal{M}$, then the reduced action $\tau \to S_\tau$ on $\mathcal{N}$ is given by

$$S_\tau = E \circ \Theta_\tau|_{\mathcal{N}}. \tag{A.16}$$

Since the composition of completely positive maps is completely positive, the complete positivity of the automorphism $\Theta_\tau : \mathcal{M} \to \mathcal{M}$ and the complete positivity of the conditional expectation imply that the time evolution map $S_\tau : \mathcal{N} \to \mathcal{N}$ is completely positive (see Kadison and Ringrose 1992, p. 699; Tomiyama 1957; Evans and Lewis 1977, pp. 35f).

The reduced dynamics of the W*-algebra $\mathcal{N}$ can be described by the map $\tau \to \rho_\tau^{\mathcal{N}}$, where the reduced state functional $\rho_\tau^{\mathcal{N}} \in \mathcal{N}_*$ is defined by

$$\rho_\tau^{\mathcal{N}} := \rho_\tau(N) \quad \text{for every } N \in \mathcal{N} \subset \mathcal{M}. \tag{A.17}$$

A.3 The GNS-Construction

Let ρ be an arbitrary faithful state over an abstract C*-algebra $\mathcal{A}$. The so-called *GNS-construction*, named after Gel'fand and Neumark (1943) and Segal (1947), allows to construct a Hilbert space $\mathcal{H}_\rho$, a representation $\pi_\rho(\mathcal{A})$ of $\mathcal{A}$, and a bi-cyclic unit vector $\Omega_\rho \in \mathcal{H}_\rho$, such that $\rho(A) = \langle \Omega_\rho | A\Omega_\rho \rangle$ for all $A \in \mathcal{A}$. Recall that a vector $\Omega_\rho \in \mathcal{H}_\rho$ is said to be

- cyclic if $\pi_\rho(\mathcal{A})\, \Omega_\rho$ is dense in $\mathcal{H}_\rho$,
- separating if for all $A \in \pi_\rho(\mathcal{A})$ the relation $A\Omega_\rho = 0$ implies $A = 0$.
- bi-cyclic if it is cyclic and separating simultaneously (or equivalently, cyclic for the factor and its commutant).

This representation $(\pi_\rho, \mathcal{H}_\rho)$ is unique up to unitary equivalence. The concrete C*-algebra $\pi_\rho(\mathcal{A})$ is called the *GNS-representation* of the abstract C*-algebra $\mathcal{A}$ induced by the state functional ρ.

The GNS-representation $\pi_\rho(\mathcal{A})$ of a C*-algebra $\mathcal{A}$ is norm-closed. By closing the C*-algebra $\pi_\rho(\mathcal{A})$ in the *weak topology* of the algebra $\mathcal{B}(\mathcal{H}_\rho)$ of all bounded operators acting on the Hilbert space $\mathcal{H}_\rho$, one gets the ρ-dependent W*-algebra $\mathcal{M}_\rho$, $\pi_\rho(\mathcal{A}) \subset \mathcal{M}_\rho \subset \mathcal{B}(\mathcal{H}_\rho)$. The new topology on $\mathcal{A}$, which is induced by the reference state functional ρ, corresponds to the weak operator topology on $\mathcal{B}(\mathcal{H}_\rho)$. A state φ on the W*-algebra $\mathcal{M}_\rho$ that can be represented as

$$\varphi(M) = \mathrm{tr}(DM), \tag{A.18}$$

where $D \in \mathcal{B}(\mathcal{H}_\rho)$ is a density operator (i.e. a positive trace-class operator in $\mathcal{H}_\rho$), is called a *normal state*. In general, not all states of the C*-algebra $\pi_\rho(\mathcal{A})$ are normal.

A.4 Disjoint States

The *support* S_ρ of a normal state ρ is defined as the smallest projection operator S in the W*-algebra $\mathcal{M}$ such that $\rho(S) = 1$,

$$\mathrm{S}_\rho := \inf\big\{ S \in \mathcal{M} \,\big|\, \rho(S) = 1,\ S = S^* = S^2 \big\}. \tag{A.19}$$

The *central support* C_ρ of a normal state ρ on a W*-algebra $\mathcal{M}$ is defined via the center $\mathcal{Z}(\mathcal{M})$ of $\mathcal{M}$ as

$$\mathrm{C}_\rho := \inf\big\{ C \in \mathcal{Z}(\mathcal{M}) \,\big|\, \rho(C) = 1,\ C = C^* = C^2 \big\}. \tag{A.20}$$

Let ρ, ϕ be normal states on a W*-algebra $\mathcal{M}$. We adopt the following terminology:

- ρ is *pure* if its support S_ρ is an atom;
- ρ is *primary* if its central support $C(\rho)$ is an atom;
- ρ and ϕ are *orthogonal*, $\rho \perp \phi$, if $S_\rho S_\phi = 0$;

- ρ and ϕ are *disjoint*, $\rho \top \phi$, if $C_\rho C_\phi = 0$;
- ρ and ϕ are *equivalent*, $\rho \sim \phi$, if $C_\rho = C_\phi$.

Disjointness implies orthogonality, but only in commutative algebras orthogonality implies disjointness. In general, disjointness is a much stronger condition than orthogonality.

Criteria for Orthogonality (Pedersen 1979, pp. 44f)
Let ρ and ϕ be normal states on a W*-algebra $\mathcal{M}$ with the supports S_ρ and S_ϕ. The following statements are equivalent:

(I) ρ and ϕ are orthogonal, $S_\rho S_\phi = 0$;
(II) $\|\rho - \phi\| = \|\rho\| + \|\phi\|$;
(III) for every $\varepsilon > 0$ there exists a positive element $M \in \mathcal{M}$ such that $\rho(M) < \varepsilon$ and $\phi(1 - M) < \varepsilon$.

Criteria for Disjointness (Baker 1980, Theorem 3.4)
Let ρ and ϕ be normal states on a W*-algebra $\mathcal{M}$ with the supports S_ρ and S_ϕ. The following statements are equivalent:

(I) ρ and ϕ are disjoint, $\rho \top \phi$;
(II) for every unitary element $U \in \mathcal{M}$, the equality $\|\rho_U - \phi\| = 2$ holds, where $\rho_U(M) = \rho(U^* M U)$ for every $M \in \mathcal{M}$;
(III) for every unitary element $U \in \mathcal{M}$ we have $U^* S_\rho U S_\rho = 0$.

If ρ and ϕ are two state functionals with different values for some element $Z \in \mathcal{Z}(\mathcal{M})$, these state functionals are disjoint.

A.5 Modular Theory

A.5.1 Tomita-Takesaki Theory

The modular theory for non-commutative algebras is a generalization of the modular function in the theory of non-unimodular groups. Every locally compact topological group has (up to a positive multiplicative constant) a unique non-zero left invariant countably additive measure. If the group is Abelian or compact, then this measure is also right-invariant and known as the Haar measure (Haar 1933).

The modular function accounts for the difference between right and left Haar measure for non-unimodular groups. A group whose left Haar measure equals its right Haar measure is called unimodular. In this case, the modular function is identically 1. Minoru Tomita (1967) and later Masamichi Takesaki (1970) succeeded to convert these concepts into a powerful tool for the investigation of W*-algebras.

Given a W*-algebra $\mathcal{M}$ with a separable predual together with a faithful normal state functional ρ, the Tomita-Takesaki theory associates to the pair $(\mathcal{M}, \rho)$ a positive *modular operator* Δ_ρ and an anti-unitary *modular conjugation* J_ρ that maps the algebra $\mathcal{M}$ into its commutant $\mathcal{M}'$,

$$J_\rho \mathcal{M} J_\rho = \mathcal{M}', \qquad J_\rho \Delta_\rho J_\rho = \Delta_\rho^{-1}, \qquad J_\rho = J_\rho^{-1} = J_\rho^*. \tag{A.21}$$

The self-adjoint modular operator Δ_ρ generates the *modular group* $\{\Delta_\rho^{i\tau}|\tau\in\mathbb{R}\}$, which gives the *modular automorphism group* $\{\sigma_{\rho,\tau}|\tau\in\mathbb{R}\}$,

$$\sigma_{\rho,\tau}(M)=\Delta_\rho^{i\tau}M\Delta_\rho^{-i\tau},\quad \tau\in\mathbb{R},\ M\in\mathcal{M}. \tag{A.22}$$

The Tomita-Takesaki modular theory plays a crucial role in the structure theory of W*-algebras and in quantum physics (cf. Haag 1992; Baumgärtel and Wollenberg 1992; Connes 1994).

In the following we state without proof some of the basic facts about the Tomita-Takesaki modular theory. Let $\mathcal{M}$ be a W*-algebra with a separable predual and ρ a faithful normal state on $\mathcal{M}$. We assume that $\mathcal{M}$ is already in its GNS-representation on the separable Hilbert space $\mathcal{H}_\rho$, $\mathcal{M}\subset\mathcal{B}(\mathcal{H}_\rho)$, where ρ is represented by the bi-cyclic vector $\Omega_\rho\in\mathcal{H}_\rho$,

$$\rho(M)=\langle\Omega_\rho|M\Omega_\rho\rangle\quad\text{for every } M\in\mathcal{M}. \tag{A.23}$$

Since the vector Ω_ρ is cyclic both for $\mathcal{M}$ and $\mathcal{M}'$, the domains $\mathcal{D}_\rho:=\mathcal{M}\,\Omega_\rho\subset\mathcal{H}$ and $\mathcal{D}'_\rho:=\mathcal{M}'\Omega_\rho\subset\mathcal{H}$ are both dense in $\mathcal{H}$. The anti-linear operator S_ρ from $\mathcal{D}_\rho$ to $\mathcal{D}'_\rho$ and the anti-linear operator F_ρ from $\mathcal{D}'{}_\rho$ to $\mathcal{D}_\rho$, defined by

$$S_\rho M\Omega_\rho=M^*\Omega_\rho,\quad M\in\mathcal{M},\qquad F_\rho M'\Omega_\rho=\big(M'\big)^*\Omega_\rho,\quad M'\in\mathcal{M}', \tag{A.24}$$

are in general both unbounded but closable (cf., e.g., Kadison and Ringrose 1986, Lemma 9.2.1, p. 597). We denote their closures by the same symbols S_ρ and F_ρ. Both have unique polar decompositions,

$$S_\rho=J_\rho\,|S_\rho|=J_\rho\,\Delta_\rho^{1/2},\qquad F_\rho=J_\rho\,|F_\rho|=J_\rho\,\Delta_\rho^{-1/2}, \tag{A.25}$$

where the unique positive operator $\Delta_\rho=|S_\rho|^2=|F_\rho|^2$ is called the *modular operator*, and the unique anti-linear operator $J_\rho=S_\rho|S_\rho|^{-1}=F_\rho|F_\rho|^{-1}$, which fulfills $J_\rho^2=\mathbb{1}$, is called the *modular conjugation* associated with the pair $(\mathcal{M},\Omega_\rho)$. The following relations hold (cf. Takesaki 1970; Araki 1974; Kadison and Ringrose 1986; Sunder 1987; Takesaki 2003):

(i) Δ_ρ is a linear positive non-singular self-adjoint operator such that $\mathcal{D}_\rho(\Delta_\rho^{1/2})=\mathcal{D}'{}_\rho$ and $\Delta_\rho\Omega_\rho 0=\Omega_\rho$, $J_\rho\Omega_\rho=\Omega_\rho$, $\Delta_\rho=F_\rho S_\rho$, $\Delta_\rho^{-1}=S_\rho F_\rho$.

(ii) The operator J_ρ is an anti-unitary involution:

- $J_\rho(c_1\Phi+c_2\Psi)=c_1^*J_\rho\,\Phi+c_2^*J_\rho\,\Psi$ for every $\Phi,\Psi\in\mathcal{H}$ and $c_1,c_2\in\mathbb{C}$,
- $J_\rho=J_\rho^*=J_\rho^{-1},\quad J_\rho^2=\mathbb{1},\quad J_\rho\Delta_\rho\,J_\rho=J_\rho^{-1}$,
- $\langle\Omega_\rho\,|\,MJMJ\Omega_\rho\rangle\geq 0$ for every $M\in\mathcal{M}$, where the equality holds if and only if $M=0$.

(iii) The bi-cyclic vector Ω_ρ is invariant under the action of the operators S_ρ, F_ρ, Δ_ρ, J_ρ.

(iv) The Tomita-Takesaki theorem states that the strongly continuous unitary one-parameter group $\{\Delta_\rho^{i\tau}|\tau\in\mathbb{R}\}$ with $\Delta_\rho^{i\tau}\Omega_\rho=\Omega_\rho$ defines the modular group of automorphisms of $\mathcal{M}$ and $\mathcal{M}'$, given by $\Delta_\rho^{i\tau}\mathcal{M}\,\Delta_\rho^{-i\tau} = \mathcal{M}$ and $\Delta_\rho^{i\tau}\mathcal{M}'\Delta_\rho^{-i\tau} = \mathcal{M}'$ for $\tau \in \mathbb{R}$.

(v) The matrix elements of the modular group have the following analytic properties: If $A, B \in \mathcal{M}$ and $\sigma_{\rho,\tau}(A) = \Delta_\rho^{i\tau} A\,\Delta_\rho^{-i\tau}$, then the continuous function $\tau \mapsto \langle \Omega_\rho \,|\, B\sigma_{\rho,\tau}(A)\,\Omega_\rho\rangle$ has a bounded analytic continuation into the strip $\{z \in \mathbb{C} | -1 < \operatorname{Im} z < 0\}$, which fulfills $\langle \Omega_\rho \,|\, \sigma_{\rho,\tau}(A) B\,\Omega_\rho\rangle = \langle \Omega_\rho | B\,\sigma_{\rho,\tau-i}(A)\,\Omega_\rho\rangle$.

(vi) The relation $\langle \Omega_\rho \,|\, \sigma_{\rho,\tau}(A) B\,\Omega_\rho\rangle = \langle \Omega_\rho | B\,\sigma_{\rho,\tau-i}(A)\,\Omega_\rho\rangle$ implies that the state ρ satisfies the β-KMS-condition $\{\sigma_{\rho,t}(A)B\} = \rho\,\{B\sigma_{\rho,t+i\hbar\beta}(A)\}$ for the dynamical automorphism group $\alpha_{\rho,t}(M) = e^{itH_\rho/\hbar}Me^{-itH_\rho/\hbar}$. Here, $M \in \mathcal{M}$, $t = -\hbar\beta\tau$ and β is the inverse temperature $\beta := (kT)^{-1}$ with Boltzmann constant k and absolute temperature T. In this case, ρ is referred to as a β-KMS state. If $\mathcal{M}$ is a factor, then $\mathcal{M}$ possesses a unique β-KMS state.

A.5.2 *Outer Automorphisms*

An automorphism $\alpha \in \operatorname{Aut}(\mathcal{M})$ of the W*-algebra $\mathcal{M}$ is called *inner* if there is a unitary element U in $\mathcal{M}$ such that $\alpha(M) = U^*MU$:

$$\alpha \in \operatorname{Inn}(\mathcal{M}) \quad\Longleftrightarrow\quad \alpha(M) = U^*MU, \quad M \in \mathcal{M}. \tag{A.26}$$

Two automorphisms α' and α'' are *equivalent*, $\alpha' \sim \alpha''$, if they are related by an inner automorphism, that is, if there is a unitary element $U \in \mathcal{M}$ with $U^*\alpha'(M)U = \alpha''(M)$:

$$\alpha' \sim \alpha'' \quad\Longleftrightarrow\quad U^*\alpha'(M)U = \alpha''(M) \quad \text{for every } M \in \mathcal{M}, \tag{A.27}$$

The equivalence relation $\sim$ generates equivalence classes of automorphisms.

Not all automorphisms are inner. The quotient group $\operatorname{Aut}(\mathcal{M})/\operatorname{Inn}(\mathcal{M})$ of all automorphisms with respect to inner ones is called the group of *outer automorphisms* $\operatorname{Out}(\mathcal{M})$. The modular automorphisms of a W*-algebra $\mathcal{M}$ is inner if and only if $\mathcal{M}$ is semi-finite, i.e. if it contains only factors of type I or type II (Takesaki 1970, Sect. 14). For inner automorphisms $\operatorname{Out}(\mathcal{M})$ is trivial. Every type III factor with a separable predual has an outer automorphism (Takesaki 1970, Corollary 14.3, p. 100). Factors of type III_1 yield $\operatorname{Out}(\mathcal{M}) = \mathbb{R}$.

The modular automorphism group of a W*-algebra $\mathcal{M}$ depends on the state ρ, $\rho(M) := \langle \Omega_\rho | M\Omega_\rho\rangle$, $M \in \mathcal{M}$, but the modular groups of automorphisms are all the same modulo inner automorphisms. That is, changing the state does not change the image of the modular automorphism in the *outer* automorphism group of $\mathcal{M}$.

Two faithful state functionals ρ and $\tilde{\rho}$ of $\mathcal{M}$ with the associated modular automorphisms $\sigma_{\rho,\tau}$ and $\sigma_{\tilde{\rho},\tau}$ are mutually related by a non-commutative analog of the Radon-Nikodým theorem (Connes 1973):

Radon-Nikodým Cocycle Theorem
There exists a σ-strongly continuous map $\mathbb{R} \ni \tau \to U_\tau \in \mathcal{M}$ such that

- U_τ is unitary for all $\tau \in \mathbb{R}$,
- $U_{\tau+\tau'} = U_\tau\, \sigma_{\rho,\tau}(U_{\tau'})$ (cocycle identity),
- $\sigma_{\tilde{\rho},\tau}(M) = U_\tau\, \sigma_{\rho,\tau}\, U_\tau^*$.

The one-cocycle $\{U_\tau | \tau \in \mathbb{R}\}$ is called the *cocycle derivative* of $\tilde{\rho}$ with respect to ρ and one writes:

$$U_\tau = (D\tilde{\rho} : D\rho)_\tau = \tilde{\rho}^{i\tau} \rho^{-i\tau}. \tag{A.28}$$

Besides the cocycle identity, the cocycle derivative fulfills the chain rule

$$(D\tilde{\rho} : D\rho)_\tau = (D\tilde{\rho} : D\hat{\rho})_\tau (D\hat{\rho} : D\rho)_\tau \quad \text{for all } t \in \mathbb{R}, \tag{A.29}$$

where ρ, $\tilde{\rho}$, $\hat{\rho}$ are faithful states on $\mathcal{M}$.

A.5.3 The Modular Structure of W-Algebras*

A.5.3.1 The Algebraic Invariant of Connes

The modular structure of a W*-algebra encodes essential physical information. Let $\mathcal{M}$ be a factor with a separable predual $\mathcal{M}_*$, $\rho \in \mathcal{M}_*$ a faithful normal state on $\mathcal{M}$, and Δ_ρ the associated modular operator, and denote by $sp(\Delta_\rho)$ the spectrum of the self-adjoint operator Δ_ρ. The *proper modular spectrum* $\mathcal{S}_0(\mathcal{M})$ is defined as the intersection of the spectra of all possible Δ_ρ (Connes 1973, Sect. III; Connes 1994, Sect. V.5; Takesaki 2003, Chap. XII, §1),

$$\mathcal{S}_0(\mathcal{M}) := \bigcap \left\{ sp(\Delta_\rho) \,\middle|\, \rho \in \mathcal{M}_* \right\}, \tag{A.30}$$

where ρ runs over all faithful normal states on $\mathcal{M}$. The modular spectrum $\mathcal{S}_0(\mathcal{M})$ is an algebraic invariant of $\mathcal{M}$ with the following properties:

(i) If $\mathcal{M}$ is a factor of type I or II, then $\mathcal{S}_0(\mathcal{M}) = \{1\}$.
(ii) If $\mathcal{M}$ is of type III_0, then $\mathcal{S}_0(\mathcal{M}) = \{0, 1\}$.
(iii) $\mathcal{M}$ is of type III_λ, $0 < \lambda < 1$, if and only if $\mathcal{S}_0(\mathcal{M}) = \{0\} \cup \{\lambda^n | n \in \mathbb{Z}\}$.
(iv) $\mathcal{M}$ is of type III_1, if and only if $\mathcal{S}_0(\mathcal{M}) = [0, \infty)$.

A.5.3.2 Positive Energy

Let $\mathcal{M}$ be a W*-algebra with a bi-cyclic vector Ω_ρ, and let $\{U(t) | t \in \mathbb{R}\}$ be a continuous unitary group satisfying $U(t)\, \mathcal{M}\, U(-t) \subset \mathcal{M}$ for all $t \geq 0$. Then any two of the following conditions imply the third (Wiesbrock 1992):

(i) $U(t) = e^{itH}$ with $H \geq 0$.
(ii) $U(t)\Omega_\rho = \Omega_\rho$ for all $t \in \mathbb{R}$.
(iii) $\Delta^{it} U(s) \Delta^{-it} = U(e^{-2\pi t} s)$ and $JU(s)J = U(-s)$ for all $s, t \in \mathbb{R}$.

A.5.3.3 Time Reversal

The modular involution J implements the time-reversal operation.

A.6 Standard Representation of W*-Algebras

A.6.1 Introduction

Von Neumann's irreducible Hilbert-space representation of the quantum mechanics of the pioneers has often been misunderstood. The Stone-von Neumann uniqueness theorem says that for finitely many degrees of freedom the representation of the canonical commutation relations is unique up to unitary equivalence and multiplicity. Von Neumann eliminated the multiplicity by choosing an *irreducible* representation. Recall that a set $\mathcal{O}$ of operators acting on a Hilbert space $\mathcal{H}$ is called *irreducible* if and only if the only operators in $\mathcal{B}(\mathcal{H})$ that commute with all elements of $\mathcal{O}$ are the scalar operators $c\mathbb{1}$, $c \in \mathbb{C}$.

Clearly, the algebra $\mathcal{B}(\mathcal{H})$ of all bounded operators acting on a Hilbert space $\mathcal{H}$ forms an irreducible set. Accordingly, two irreducible representations of a quantum system on the Hilbert spaces $\mathcal{H}_1$ and $\mathcal{H}_2$ are physically equivalent if they are unitarily equivalent, that is, if there exists a unitary operator U of $\mathcal{H}_1$ onto $\mathcal{H}_2$.

For many discussions this setting is convenient, but it is important to know that *irreducibility has no physical meaning*. We can also realize traditional quantum mechanics reducibly in a physically fully equivalent way on any direct sum $\mathcal{H} \oplus \mathcal{H} \oplus \cdots \oplus \mathcal{H}$, where $\mathcal{H}$ is the Hilbert space used for the irreducible realization. The number of irreducible components of a reducible representation is called its *multiplicity*. The choice between the usual irreducible or a reducible representation is purely a matter of convenience with no interpretational consequences whatsoever.

To get a clear conceptual picture, it is best to formulate all physically relevant concepts in a representation-independent way. The proper mathematical objects for such an abstract formulation of physical theories are *abstract* C*- and W*-algebras. Since the multiplicity of the representation is physically immaterial, there is algebraic uniqueness. In a W*-algebraic formulation, the Stone-von Neumann theorem implies that the algebra of observables of a quantum system with finitely many degrees of freedom is a unique factor of type I.

There are many different possibilities for a faithful representation of abstract C*- or W*-algebras of observables as algebras of operators acting on some Hilbert space. All these representations are physically equivalent. However, physical equivalence does not imply that all representations are equally convenient for the discussion of a particular problem. For this purpose it is important to carefully select the most convenient representation.

Representations of a W*-algebra $\mathcal{M}$ on a Hilbert space $\mathcal{H}$ can be characterized in terms of the so-called commutant $\mathcal{M}'$ of $\mathcal{M}$, defined as the set of all operators in

$\mathcal{B}(\mathcal{H})$ commuting with every element of $\mathcal{M}$,

$$\mathcal{M}' := \{A \in \mathcal{B}(\mathcal{H}) \mid MA = AM \text{ for every } M \in \mathcal{M}\}. \tag{A.31}$$

The commutant $\mathcal{M}'$ of a W*-algebra $\mathcal{M}$ is also a W*-algebra over the same Hilbert space $\mathcal{H}$.

A W*-algebra $\mathcal{M}$ is *irreducibly* represented if its commutant $\mathcal{M}'$ is trivial, that is, if $\mathcal{M}' = \mathbb{1}\mathbb{C}$. Irreducible representations exist only for factors of type I. In particular, for thermodynamic systems (described by W*-algebras of type III) and for classical systems (described by commutative W*-algebras) there are no irreducible representations. In contrast, every W*-algebra has faithful *reducible* representations. Among all possible reducible representations, the so-called *standard representation* is distinguished by many exquisite properties. A W*-algebra $\mathcal{M}$ is said to be in *standard form* if its commutant $\mathcal{M}'$ is anti-isomorphic to $\mathcal{M}$.[263]

Every system describable W*-algebraically can be described by a standard form. Nothing substantial is lost or gained by switching between two isomorphic algebraic descriptions, but calculations may become simpler or insight might be facilitated. Standard forms provide a unifying framework for classical and quantum-theoretical descriptions in terms of the underlying phase space, including the quantum statistics of infinite systems with Kubo-Martin-Schwinger (KMS) states.[264]

A.6.2 Canonical Standard Form of a W*-Algebra

Every W*-algebra has a faithful reducible representation on a Hilbert space $\mathcal{K}$ in standard form. The standard form is characterized by a self-dual closed cone (called *natural positive cone*) $\mathfrak{C}$ in the Hilbert space $\mathcal{K}$. This positive cone is a generalization

---

[263] W*-algebras in standard form were first introduced by Jacques Dixmier (1969, Sect. I.5.5) in the context of his studies of Hilbert algebras. The Tomita-Takesaki theory of modular Hilbert algebras allows to drop the restriction to type I and type II algebras. Compare Takesaki (1970), Araki (1974), Connes (1974).

[264] Implicitly, without taking advantage of the highly developed mathematical tools of the standard representation, and not realizing the physical equivalence to the irreducible formulation, standard forms have been used in many classical and quantum mechanical descriptions. Examples are the Koopman representation of classical statistical mechanics (Koopman 1931; Reed and Simon 1975, pp. 313–318), the Wigner-Moyal representation of traditional quantum mechanics (Wigner 1932a; Moyal 1949), the description of thermofields and their dynamics by Umezawa et al. (1981).

Rather unconventionally, Umezawa and coworkers interpreted their reducible representation as a doubling of the numbers of degrees of freedom. Vitiello (1995) extended Umezawa's thermofield dynamics to dissipative dynamics and suggested that the doubling of collective modes represents forward and backward arrows of time. He even speculates that this doubling leads to a permanent dialog of the brain with its "double", implying that "consciousness might emerge from dissipation" (Vitiello 2001, p. 125; Freeman and Vitiello 2016).

Also, the "extension" of traditional quantum mechanics by the so-called Liouville superoperator formalism is nothing but a standard representation of ordinary quantum mechanics (George and Prigogine 1979; Prigogine 1980, Chap. VIII and Appendix D).

of positivity, it allows to construct a *unique* canonical standard form. A set $\mathcal{C}$ in a vector space $\mathcal{K}$ is called a *cone* if $\Phi \in \mathcal{C}$ implies $c\,\Phi \in \mathcal{C}$ for every $c > 0$. If $\mathcal{K}$ is a Hilbert space with the inner product $\langle\cdot|\cdot\rangle$, then a cone $\mathcal{C}^{\#}$ dual to a cone $\mathcal{C}$ is defined by $\mathcal{C}^{\#} := \{\Phi \in \mathcal{K} \mid \langle\Phi|\Psi\rangle \geq 0, \Psi \in \mathcal{C}\}$. A cone is called self-dual if $\mathcal{C} = \mathcal{C}^{\#}$.

Definition: Self-Dual Cone in a Hilbert Space
A self-dual cone in Hilbert space $\mathcal{K}$ is a subset $\mathcal{C} \subset \mathcal{K}$ such that

$$\mathcal{C} := \big\{\Phi \in \mathcal{K} \,\big|\, \langle\Phi|\Psi\rangle \geq 0, \Psi \in \mathcal{C}\big\}. \tag{A.32}$$

Self-dual cones have the following properties (Haagerup 1975; Bös 1976):

- There is a unique antiunitary involution $J : \mathcal{K} \to \mathcal{K}$ satisfying $J^2 = \mathbb{1}$ and $J\Phi = \Phi$ for every $\Phi \in \mathcal{C}$.
- $\mathcal{C} \cap \{-\mathcal{C}\} = \{0\}$.
- $\mathcal{K}$ is linearly spanned by $\mathcal{C}$.
- Every element $\Phi \in \mathcal{K}$ with $J\Phi = \Phi$ has the unique decomposition $\Phi = \Phi_1 - \Phi_2$, where $\Phi_1, \Phi_2 \in \mathcal{C}$ and $\langle\Phi_1|\Phi_2\rangle = 0$.

Two Familiar Examples of Self-Dual Cones:

- If (Ω, Σ, μ) is a probability space, then the set of positive functions in the Hilbert space $\mathbf{L}^2(\Omega, \Sigma, \mu)$ is a self-dual cone in $\mathbf{L}^2(\Omega, \Sigma, \mu)$.
- An operator $T \in \mathcal{B}(\mathcal{H})$ on a Hilbert space $\mathcal{H}$ belongs to the Hilbert-Schmidt class $\mathcal{B}^2(\mathcal{H})$ if and only if $\operatorname{tr}(T^*T) < \infty$. With the inner product $(A|B) := \operatorname{tr}(A^*B)$, the Hilbert-Schmidt class $\mathcal{B}^2(\mathcal{H})$ becomes a Hilbert space, and the set of positive operators in $\mathcal{B}^2(\mathcal{H})$ is a self-dual cone.

Let $\mathcal{M} \subseteq \mathcal{H}$ be a W*-algebra with a separable predual, let ρ be a faithful normal state functional associated with a bi-cyclic vector $\Omega_\rho \in \mathcal{H}$, and let J_ρ be the modular involution (A.21) of the Tomita-Takesaki theory. Define a cone $\mathcal{C}_\rho$ by

$$\mathcal{C}_\rho := \operatorname{clos}\{M J_\rho M J_\rho \Omega | M \in \mathcal{M}\}. \tag{A.33}$$

Then it follows that (Connes 1974):

- $\mathcal{C}_\rho$ is a self-dual cone, i.e. $\mathcal{C}_\rho = \{\Psi \in \mathcal{H} | \langle\Psi|\Phi\rangle \geq 0\}$ for all $\Phi \in \mathcal{C}_\rho$.
- $\mathcal{C}_\rho$ is a unitary invariant of $\mathcal{M}$, that is, if $\tilde{\rho}$ is another faithful normal state functional associated with a bi-cyclic vector $\tilde{\Omega}_\rho \in \mathcal{H}$, then there exists a unique unitary operator $V \in \mathcal{M}'$ such that $V\Omega_\rho = \tilde{\Omega}_\rho$. Furthermore, if $\tilde{\Omega}_\rho \in \mathcal{C}_\rho$, then $\mathcal{C}_{\tilde{\rho}} = \mathcal{C}_\rho$, and $J_{\tilde{\rho}} = J_\rho$.

Definition: Canonical Standard Form of an Arbitrary W*-Algebra[265]
A quadruple $(\mathcal{M}, \mathcal{K}, \mathcal{C}, J)$, consisting of a W*-algebra $\mathcal{M}$ on a complex Hilbert space $\mathcal{K}$, a linear isometric involution J of $\mathcal{K}$, and a self-dual cone $\mathcal{C}$ in $\mathcal{K}$, is said to be of canonical standard form, if

- $\mathcal{M}' = J\,\mathcal{M}\,J$, $\mathcal{M}' := \{B \in \mathcal{B}(\mathcal{K}) | BM = MB,\, M \in \mathcal{M}\}$,
- $JZJ = Z^*$ for every central element $Z \in \mathcal{Z} := \mathcal{M} \cap \mathcal{M}'$,

[265] Compare Haagerup (1975), Rousseau et al. (1977). This canonical standard form is called hyperstandard by Strătilă and Zsidó (1979, p. 312).

- $J\Phi = \Phi$ for every $\Phi \in \mathcal{C}$,
- $MJMJ\,\Phi \in \mathcal{C}$ for every $\Phi \in \mathcal{C}$ and every $M \in \mathcal{M}$.

The canonical standard form is *unique* up to unitary equivalence. That is, if $(\mathcal{M}, \mathcal{K}, \mathcal{C}, J)$ and $(\tilde{\mathcal{M}}, \tilde{\mathcal{K}}, \tilde{\mathcal{C}}, \tilde{J})$ are two standard representations of the same W*-algebra, then the *-isomorphism $\phi : \mathcal{M} \to \tilde{\mathcal{M}}$ is implemented by a *unique* unitary operator $U : \mathcal{K} \to \tilde{\mathcal{K}}$ satisfying $\phi(M) = UMU^*$ for every $M \in \mathcal{M}$, $\tilde{J} = UJU^*$, and $\tilde{\mathcal{C}} = U(\mathcal{C})$.

Every W*-algebra is *-isomorphic to a canonical W*-algebra in standard form. The Hilbert space $\mathcal{K}$ of a W*-algebra $\mathcal{M}$ in standard form is separable if and only if $\mathcal{M}$ has a separable predual $\mathcal{M}_*$ (Schmitt 1990, pp. 561f). Moreover, for every W*-algebra in canonical standard form on a separable Hilbert space there exists a bi-cyclic vector in $\mathcal{C}$ (Haagerup 1975, p. 277).

A.6.3 Implementation of State Functionals

State functionals and isomorphisms are special examples of completely positive maps. Their implementation follows from the implementation theorems referring to general completely positive maps (Demoen and Vanheuverzwijn 1980).

Canonical Vectorial Implementation of State Functionals (Araki 1974)
Let $(\mathcal{M}, \mathcal{K}, \mathcal{C}, J)$ be the canonical standard representation of an arbitrary W*-algebra and let $\rho \in \mathcal{M}_*$ be any normal state functional. Then there exists a *unique* vector $\Xi \in \mathcal{C}$ such that

$$\rho(M) = \langle \Xi \,|\, M\Xi \rangle \quad \text{for every } M \in \mathcal{M},\ \Xi \in \mathcal{C}. \tag{A.34}$$

Note that this implementation of state functionals by vectors is very different from that of the traditional irreducible representation. First of all, the implementation is by *vectors* and not by *rays*, which determine state vectors only up to a phase factor. In the standard representation, the prerequisite that the implementing vector has to be an element of the self-dual cone is a *positivity requirement* which eliminates phase factors.

Furthermore, in the standard representation every normal state, *pure or non-pure*, on any W*-algebra has a unique implementation by a vector. By contrast, in the irreducible representation (which only exists for type-I factors) an implementation by a ray is possible only for pure states. The possibility to represent non-pure states by vectors in the standard representation has many advantages, but it also makes it difficult to distinguish between pure and non-pure states. Also, the construction of coherent superpositions of pure states is mathematically rather awkward.

A.6.4 Implementation of Symmetry Groups

The *implementation of automorphisms* in the standard representation is also different from the irreducible representation. In the traditional irreducible representation of quantum theory, the implementation of a symmetry group of automorphisms is given by a group of *unitary operator rays*, that is by unitary *ray representations* (also called *projective unitary representations*).[266] In contrast, in the standard representation of any physical system, every symmetry group of automorphisms can be implemented by a unique group of unitary operators, that is, by *genuine vector representations*. This fact simplifies many mathematical discussions and permits deeper conceptual insight.

Canonical Implementation of Automorphisms (Haagerup 1975)
Let $(\mathcal{M}, \mathcal{K}, \mathcal{C}, J)$ be the canonical standard representation of an arbitrary W*-algebra. Every automorphism $\alpha \in \mathrm{Aut}(\mathcal{M})$ has a unique unitary canonical implementation $\alpha \to U_\alpha \in \mathcal{B}(\mathcal{K})$ such that

- $\alpha(M) = U_\alpha M U_\alpha^*$
- $U_\alpha J = J U_\alpha$
- $U_\alpha \mathcal{C} = \mathcal{C}$.

The canonical implementation gives rise to he following properties:

- If V is a unitary operator in $\mathcal{M}$, then the canonical implementation of the inner automorphism group $M \mapsto VMV^*$ with $M \in \mathcal{M}$ is given by

$$\alpha(M) = U_\alpha M U_\alpha^* \quad \text{with } U_\alpha = VJVJ. \tag{A.35}$$

- If $\alpha : \mathfrak{G} \to \mathrm{Aut}(\mathcal{M})$ is a σ-weakly continuous action of a locally compact group $\mathcal{G}$, then the canonical implementation $g \to U_g$ with $g \in \mathcal{G}$ of the automorphism α is a strongly unitary representation of $\mathcal{G}$ on $\mathcal{K}$.
- If $\{V_t \,|\, t \in \mathbb{R}\}$ is a unitary one-parameter group in $\mathcal{M}$, then the canonical implementation of the inner automorphism $M \mapsto V_t M V_t^*$ with $M \in \mathcal{M}$ is given by

$$\alpha_t(M) = U_t M U_t^* \quad \text{with } U_t = V_t J V_t J. \tag{A.36}$$

- If H is the Stone generator of the unitary group $\{V_t \,|\, t \in \mathbb{R}\}$, then the Stone generator of the unitary group $\{U_t \,|\, t \in \mathbb{R}\}$ is given by the (possibly unbounded) Liouville operator L,

$$L = H - JHJ, \quad H \in \mathcal{M}, \; JHJ \in \mathcal{M}'. \tag{A.37}$$

Example: Finite KMS-Systems[267]
Let $\mathcal{M} = \mathcal{B}(\mathcal{H})$ be a matrix algebra acting on a finite-dimensional Hilbert space $\mathcal{H}$ with a bi-cyclic normalized vector $\Theta \in \mathcal{H}$ and $\{\mathcal{M}\Theta\} = \mathcal{H}$, so that

---

[266] Many elementary textbooks still maintain that in the irreducible Hilbert-space representation symmetry groups can be represented by unitary operators. That this claim is wrong has been known since long; compare the short note by von Neumann (1935) and the excellent analysis by Bargmann (1954).

[267] This example is inspired by Araki (1978, pp. 19f) and Emch (1985, p. 180).

$$\langle A^*\Theta | B\Theta \rangle = \mathrm{tr}(A^* B), \quad A, B \in \mathcal{M}. \tag{A.38}$$

Every state functional ρ of $\mathcal{M}$ can be expressed in terms of a positive density matrix $D_\rho \in \mathcal{B}(\mathcal{H})$ as

$$\rho(M) = \mathrm{tr}(D_\rho M) \quad \text{for every } M \in \mathcal{M},\ \mathrm{tr}(D_\rho) = 1. \tag{A.39}$$

The centralizer $\mathcal{M}_\rho$ is a W*-subalgebra of $\mathcal{M}$. If ρ is a *faithful* normal state functional, then $\mathcal{M}_\rho$ is the fixed-point algebra of the modular automorphism group of $\mathcal{M}$, corresponding to ρ (Takesaki 2003, Sect. 8.2).
If ρ is a *non-degenerate* faithful normal state, then $\mathcal{M}_\rho$ is a maximal commutative W*-algebra, $\mathcal{M}_\rho = \mathcal{M}'_\rho$,[268] so that the centralizer $\mathcal{M}_\rho$ can be generated by a self-adjoint operator $H_\rho \in \mathcal{B}(\mathcal{H})$ with discrete spectrum $\{\epsilon_k > 0\}$, $\epsilon_k \neq \epsilon_\ell$ for $k \neq \ell$. Since we assumed that the state ρ is faithful, the corresponding density operator D_ρ is *strictly positive*, $D_\rho > 0$, so that we can define the positive operator H_ρ by

$$H_\rho := F_\rho - \beta^{-1} \ln(D_\rho) \quad \text{where } F_\rho := -\beta^{-1} \ln \mathrm{tr}\big(e^{-\beta H_\rho}\big). \tag{A.40}$$

In order to conform with the traditional way to represent the Gibbs state with inverse temperature $\beta = (kT)^{-1} > 0$, we introduced the normalization $F_\rho > 0$ and the scaling factor $\beta > 0$,

$$D_\rho = e^{-\beta(H_\rho - F_\rho)} = \frac{e^{-\beta H_\rho}}{\mathrm{tr}(e^{-\beta H_\rho})}, \tag{A.41a}$$

$$\mathrm{tr}\big(e^{-\beta H_\rho}\big) = \sum_k e^{-\beta \epsilon_k} < \infty. \tag{A.41b}$$

The modular operator J_ρ and the modular operator Δ_ρ are given by

$$J_\rho M\Theta = M^*\Theta, \quad M \in \mathcal{M}, \tag{A.42a}$$

$$\Delta_\rho = D_\rho J_\rho D_\rho^{-1} J_\rho = e^{-\beta(H_\rho - H'_\rho)}, \tag{A.42b}$$

$$H_\rho \in \mathcal{M}, \quad H'_\rho := J_\rho H_\rho J_\rho \in \mathcal{M}'. \tag{A.42c}$$

Since H'_ρ commutes with every $M \in \mathcal{M}$, the modular automorphism group $\{\sigma_{\rho,\tau} | \tau \in \mathbb{R}\}$ is given by

$$\sigma_{\rho,\tau}(M) := \Delta_\rho^{i\tau} M \Delta_\rho^{-i\tau} = e^{-i\tau\beta H_\rho} M e^{i\tau\beta H_\rho}. \tag{A.43}$$

Since the self-adjoint operator H_ρ generates the modular automorphism group, it can be interpreted as the Hamiltonian with respect to which ρ is a canonical equilibrium state at inverse temperature β. In terms of the one-parameter time evolution group $\{e^{itH_\rho/\hbar} | t \in \mathbb{R}\}$, the modular automorphism group $\{\sigma_{\rho,\tau} | \tau \in \mathbb{R}\}$ can be related to the dynamical automorphism group $\{\alpha_{\rho,t} | t \in \mathbb{R}\}$,

$$\alpha_{\rho,t}(M) := e^{itH_\rho/\hbar} M e^{-itH_\rho/\hbar}, \quad M \in \mathcal{M}, \tag{A.44}$$

by the substitution $t = -\beta\hbar\tau$,

$$\alpha_{\rho,-\beta\hbar\tau}(M) = \sigma_{\rho,\tau}(M), \quad \tau \in \mathbb{R},\ M \in \mathcal{M}. \tag{A.45}$$

[268] If the faithful state ρ is maximally mixed, then $\mathcal{M}_\rho = \mathcal{M}$.

Bibliography

Aboitiz F, Scheibel AB, Fisher RS, Zaidel E (1992) Fiber composition of the human corpus callosum. Brain Res 598:143–153. Quoted in Sect 4.2

Accardi L (1984) Some trends and problems in quantum probability. In: Accardi L, Frigerio A, Gorini V (eds) Quantum probability and applications to the quantum theory of irreversible processes. Springer, Berlin, pp 1–19. Quoted in Sect 3.5

Acín A, Andrianov A, Costa L, Jané E, Latorre JI, Tarrach R (2000) Generalized Schmidt decomposition and classification of three-quantum-bit states. Phys Rev Lett 85:1560–1563. Quoted in Sect 8.5

Acín A, Andrianov A, Jané E, Tarrach R (2001) Three-qubit pure-state canonical forms. J Phys A 34:6725–6739. Quoted in Sect 8.5

Aczel AD (2001) Entanglement. The greatest mystery in physics. Four Walls Eight Windows, New York. Quoted in Sect 6.3

Aitchison J, Brown JAC (1957) The lognormal distribution with special reference to its uses in economics. Cambridge University Press, Cambridge. Quoted in Sect 11.7

Akhiezer NI, Glazman IM (1993) Theory of linear operators in Hilbert space. Dover, New York. Quoted in Sect 11.6

Alexander HG (ed) (1956) The Leibniz–Clark correspondence, vol 1. Manchester University Press, New York. Quoted in Sects 1.4, 8.3, 12.1

Ali ST, Emch GG (1974) Fuzzy observables in quantum mechanics. J Math Phys 15:176–182. Quoted in Sects 3.5, 6.2

Ali ST, Cameli C, Heinosaari T, Toigo A (2009) Commutative POVMs and fuzzy observables. Found Phys 39:593–612. Quoted in Sect 3.5

Amann A (1986) Observables in W*-algebraic quantum mechanics. Fortschr Phys 34:167–215. Quoted in Sect 9.4

Amann A (1987) Integrable ergodic W*-system. J Phys A 20:L1039–L1045. Quoted in Sect 9.4

Amann A (1988) Chirality as a classical observable in algebraic quantum mechanics. In: Amann A, Cederbaum L, Gans W (eds) Fractals, quasicrystals, chaos, knots and algebraic quantum mechanics. Kluwer Academic, Dordrecht, pp 305–325. Quoted in Sect 6.5

Amann A (1991a) Chirality: a superselection rule generated by the molecular environment? J Math Chem 6:1–15. Quoted in Sect 6.5

Amann A (1991b) Molecules coupled to their environment. In: Gans W, Blumen A, Amann A (eds) Large-scale molecular systems. Plenum, New York, pp 3–22. Quoted in Sect 6.5

Amann A (1991c) Theories of molecular chirality: a short review. In: Gans W, Blumen A, Amann A (eds) Large-scale molecular systems. Plenum, New York, pp 23–32. Quoted in Sect 6.5

Ambrose W (1945) Structure theorems for a special class of Banach algebras. Trans Am Math Soc 57:364–386. Quoted in Sect 3.4

H. Primas, *Knowledge and Time*,
DOI 10.1007/978-3-319-47370-3

Amo A, Montero J, Biging G, Cutello V (2004) Fuzzy classification systems. Eur J Oper Res 156:495–507. Quoted in Sect 3.3
Amrein WO (1969) Localizability for particles of mass zero. Helv Phys Acta 42:149–190. Quoted in Sect 6.2
Andersen H, Grush R (2009) A brief history of time consciousness: historical precursors to James and Husserl. J Hist Philos 47:277–307. Quoted in Sect 13.3
Anderson PW (1972) More is different. Science 177:393–396. Quoted in Sect 6.1
Anderson PW (1995) Physics: the opening to complexity. Proc Natl Acad Sci USA 92:6653–6654. Quoted in Sect 6.5
Andrews GE, Berndt BC (2005) Ramanujan's lost notebook part I. Springer, New York. Quoted in Sect 5.10
Andrews GE, Berndt BC (2008) Ramanujan's lost notebook part II. Springer, New York. Quoted in Sect 5.10
Andronov R, Pontriagin L (1937) Systems grossiers. Dokl Akad Nauk SSSR 14:247–251. Quoted in Sect 6.5
Antoniou I, Gialampoukidis I, Ioannidis E (2016) Age and time operator of evolutionary processes. In: Atmanspacher H et al (eds) Quantum interaction QI-15. Springer, Berlin, pp 61–75. Quoted in Sect 12.1
Arai A (2005) Generalized weak Weyl relations and decay of quantum dynamics. Rev Math Phys 17:1071–1109. Quoted in Sect 10.5
Arai A (2007) Spectrum of time operators. Lett Math Phys 80:211–221. Quoted in Sect 10.5
Arai A, Matsuzawa Y (2007) Construction of a Weyl representation from a weak Weyl representation of the canonical commutation relation. Lett Math Phys 83:201–211. Quoted in Sect 10.5
Araki H (1960) Hamiltonian formalism and the canonical commutation relations in quantum field theory. J Math Phys 1:492–504. Quoted in Sect 10.4
Araki H (1968) Multiple time analyticity of a statistical state satisfying the KMS boundary condition. Publ Res Inst Math Sci Kyoto Univ 4:361–371. Quoted in Sect 7.3
Araki H (1974) Some properties of modular conjugation operator of von Neumann algebras and a non-commutative Radon–Nikodym theorem with a chain rule. Pac J Math 50:309–354. Quoted in Appendices A.5, A.6
Araki H (1978) Relative Hamiltonian for states of von Neumann algebras. Proc Steklov Inst Math 1978(1):17–25. Quoted in Appendix A.6
Araki H (1987) On superselection rules. In: Namiki M (ed) Proceedings of the 2nd international symposium foundations of quantum mechanics in the light of new technology. Physical Society of Japan, Tokyo, pp 348–354. Quoted in Sect 6.2
Araki H (1999) Mathematical theory of quantum fields. Oxford University Press, Oxford. Quoted in Sect 7.3
Aravind PK (1997) Borromean entanglement of the GHZ state. In: Cohen RS, Horne MA, Stachel J (eds) Potentiality, entanglement and passion-at-a-distance. Kluwer Academic, Dordrecht, pp 53–59. Quoted in Sect 8.5
Aristotle (1983) Physics, books III–IV. Clarendon Press, Oxford, transl with notes by Hussey E. Quoted in Sect 1.4
Aristotle (2004) On memory. University of Chicago Press, Chicago, transl with notes by Sorabji R. Quoted in Sect 4.1
Arthurs E, Kelly JL (1965) On the simultaneous measurement of a pair of conjugate observables. Bell Syst Tech J 44:725–729. Quoted in Sect 6.2
Arveson WB (1967) Subalgebras of C*-algebras. Acta Math 181:141–224. Quoted in Appendix A.2
Arkeryd LO, Cutland NJ, Henson CW (1997) Nonstandard analysis. Theory and applications. Springer, Dordrecht. Quoted in Sect 5.3
Ash RB, Gardner MF (1975) Topics in stochastic processes. Academic Press, New York. Quoted in Sect 12.4
Aspect A, Dalibard J, Roger G (1982) Experimental test of Bell's inequalities using time-varying analyzers. Phys Rev Lett 49:1804–1807. Quoted in Sect 13.5

Atmanspacher H (2002) Determinism is ontic, determinability is epistemic. In: Atmanspacher H, Bishop RC (eds) Between chance and choice. Imprint Academic, Exeter, pp 49–74. Quoted in Sect 12.3

Atmanspacher H (2009) Contextual emergence. Scholarpedia 4(3):7997. Available at www.scholarpedia.org/article/Contextual_emergence. Quoted in Sect 6.5

Atmanspacher H (2012) Dual-aspect monism à la Pauli and Jung. J Conscious Stud 19(9/10):96–120. Quoted in the Preface

Atmanspacher H (2015) Quantum approaches to consciousness. In: Zalta EN (ed) Stanford encyclopedia of philosophy. Accessible at plato.stanford.edu/entries/qt-consciousness/. Quoted in Sect 6.2

Atmanspacher H, Filk T (2010) A proposed test of temporal nonlocality in bistable perception. J Math Psychol 54:314–321. Quoted in Sect 13.5

Atmanspacher H, Filk T (2012) Determinism, causation, prediction, and the affine time group. J Conscious Stud 19(5/6):75–94. Quoted in Sects 7.1, 12.3

Atmanspacher H, beim Graben P (2009) Extending the philosophical significance of the idea of complementarity. In: Atmanspacher H, Primas H (eds) Recasting reality. Springer, Berlin, pp 99–113. Quoted in Sect 4.1

Augustine (1955) Confessions and Enchiridion. Westminster Press, Philadelphia, ed and transl by Outler AC. Quoted in Sects 1.4, 8.3

Augustine (2002) On the trinity. Cambridge University Press, Cambridge, Matthews GB (ed), transl by McKenna S. Quoted in Sects 1.4, 8.3

Bacciagaluppi G (2015) Leggett–Garg inequalities, pilot waves and contextuality. Int J Quantum Found 1:1–17. Quoted in Sect 13.5

Baker BM (1980) A central decomposition. J Funct Anal 35:1–25. Quoted in Appendix A.4

Balaguer M (1998) Platonism and anti-platonism in mathematics. Oxford University Press, New York. Quoted in Sect 5.8

Banaschewski B (1977) On G. Spencer Brown's laws of form. Notre Dame J Form Log 18:507–509. Quoted in Sect 2.1

Bargmann V (1954) On unitary ray representations of continuous groups. Ann Math 59:1–46. Quoted in Appendix A.6

Bargmann V (1964) Note on Wigner's theorem on symmetry operations. J Math Phys 5:862–868. Quoted in Sect 9.2

Bass J (1974) Stationary functions and their applications to the theory of turbulence. I. Stationary functions. J Math Anal Appl 47:354–399. Quoted in Sect 12.4

Bass J (1984) Fonctions de corrélation, fonctions pseudo-aléatoires et applications. Masson, Paris. Quoted in Sect 12.4

Bass J (1996) Moyennes et algèbres de fonctions. Ann Fond Louis Broglie 21:287–299. Quoted in Sect 12.4

Bastick T (1982) Intuition. How we think and act. Wiley, Chichester. Quoted in Sect 4.1

Baumgärtel H (1995) Operator algebraic methods in quantum field theory, vol 7. Akademie, Berlin. Quoted in Sects 7.3, 8.5

Baumgärtel H, Wollenberg M (1992) Causal nets of operator algebras. Mathematical aspects of algebraic quantum field theory. Akademie, Berlin. Quoted in Appendix A.5

Beaumont JG (2008) Introduction to neuropsychology. Guilford Press, New York. Quoted in Sect 4.2

Bell JS (1966) On the problem of hidden variables in quantum mechanics. Rev Mod Phys 28:447–452. Quoted in Sect 13.5

Bell JS (1975) On wave packet reduction in the Coleman–Hepp model. Helv Phys Acta 48:93–98. Quoted in Sect 6.5

Benioff PA (1972a) Decision procedures in quantum mechanics. J Math Phys 13:908–915. Quoted in Sect 6.2

Benioff PA (1972b) Operator valued measures in quantum mechanics: finite and infinite processes. J Math Phys 13:231–242. Quoted in Sect 6.2

Benioff PA (1972c) Procedures in quantum mechanics without von Neumann's projection axiom. J Math Phys 13:1347–1353. Quoted in Sect 6.2
Benioff P (1973) On definitions of validity applied to quantum theories. Found Phys 3:359–379. Quoted in Sect 6.2
Bennett CH, Grudka A, Horodecki M, Horodecki P, Horodecki R (2011) Postulates for measures of genuine multipartite correlations. Phys Rev A 83:012312. Quoted in Sect 8.5
Berberian SK (1966) Notes on spectral theory. Van Nostrand, Princeton. Quoted in Sect 3.5
Bergson H (1992) The creative mind. Citadel Press, New York, originally published 1934 in French. Quoted in Sect 4.3
Bergson H (1998) Creative evolution. Dover, New York, originally published 1911 in French. Quoted in Sect 4.3
Bergson H (2001) Time and free will: an essay on the immediate data of consciousness. Dover, New York, originally published 1910 in French. Quoted in Sect 13.3
Berkeley G (1710) A treatise concerning the principles of human knowledge. Arron Rhames, Dublin. Quoted in Sect 8.3
Berkeley G (1721) De motu; sive de motus principio & natura, et de causa communicationis motuum. Jacobi Tonson, London. Quoted in Sect 8.3
Berlin I (1953) The hedgehog and the fox: an essay on Tolstoy's view of history. Weidenfeld & Nicolson, New York. Quoted in Sect 5.9
Berlin I (1998) The proper study of mankind: an anthology of essays. Farrar, Straus and Giroux, New York. Quoted in Sect 4.1
Berlin I (1999) The roots of romanticism. Princeton University Press, Princeton. Quoted in Sect 4.1
Bernays P (1946) Review of "Russell's mathematical logic" by Kurt Gödel. J Symb Log 11:75–79. Quoted in Sect 5.8
Bernays P (1971) Zum Symposium über die Grundlagen der Mathematik. Dialectica 25:171–195. Quoted in Sects 5.5, 5.6
Berndt BC (1985) Ramanujan's notebooks. Part I. Springer, New York. Quoted in Sect 5.10
Berndt BC (1989) Ramanujan's notebooks. Part II. Springer, New York. Quoted in Sect 5.10
Berndt BC (1991) Ramanujan's notebooks. Part III. Springer, New York. Quoted in Sect 5.10
Berndt BC (1994) Ramanujan's notebooks. Part IV. Springer, New York. Quoted in Sect 5.10
Berndt BC (1998) Ramanujan's notebooks. Part V. Springer, New York. Quoted in Sect 5.10
Berndt BC, Rankin RA (2000) The books studied by Ramanujan in India. Am Math Mon 107:595–601. Quoted in Sect 5.10
Berne E (1949) The nature of intuition. Psychiatr Q 23:203–226. Quoted in Sect 4.1
Besicovitch AS (1932) Almost periodic functions. Cambridge University Press, Cambridge. Quoted in Sect 12.4
Betsch T (2008) The nature of intuition and its neglect in research on judgment and decision making. In: Plessner H, Bretsch C, Betsch T (eds) Intuition in judgment and decision making. Erlbaum, New York, pp 3–22. Quoted in Sect 5.10
Bhattacharyya A (1943) On a measure of divergence between two statistical populations defined by their probability distributions. Bull Calcutta Math Soc 35:99–109. Quoted in Sect 6.5
Birkhoff GD (1931) Proof of the ergodic theorem. Proc Natl Acad Sci USA 17:656–660. Quoted in Sect 12.4
Birkhoff GD, Koopman BO (1932) Recent contributions to the ergodic theory. Proc Natl Acad Sci USA 18:279–282. Quoted in Sect 12.4
Birkhoff G, von Neumann J (1936) The logic of quantum mechanics. Ann Math 37:823–843. Quoted in Sects 3.2, 4.6
Bishop E (1967) Foundations of constructive analysis. McGraw–Hill, New York. Quoted in Sect 5.8
Bishop RC (2006) The hidden premise in the causal argument for physicalism. Analysis 66:44–52. Quoted in Sect 7.1
Bishop RC, Atmansapcher H (2006) Contextual emergence in the description of properties. Found Phys 36:1753–1777. Quoted in Sect 6.5

Bisognano JJ, Wichmann EH (1975) On the duality condition for a Hermitian scalar field. J Math Phys 16:985–1007. Quoted in Sect 7.3

Bisognano JJ, Wichmann EH (1976) On the duality condition for quantum fields. J Math Phys 17:303–321. Quoted in Sect 7.3

Black F, Scholes M (1973) The pricing of options and corporate liabilities. J Political Econ 81:637–654. Quoted in Sect 11.7

Blumenthal O (1935) Lebensgeschichte. In: David Hilbert. Gesammelte Abhandlungen. Dritter Band. Springer, Berlin, pp 388–429. Quoted in Sect 5.4

Bode HW (1945) Network analysis and feedback amplifier design. Van Nostrand, New York. Quoted in Sect 11.7

Bös W (1976) Direct integrals of selfdual cones and standard forms of von Neumann algebras. Invent Math 37:241–251. Quoted in Appendix A.6

Bogen JE (1969a) The other side of the brain I: dysgraphia and dyscopia following cerebral commissurotomy. Bull Los Angel Neurol Soc 34:73–105. Quoted in Sect 4.2

Bogen JE (1969b) The other side of the brain II: an appositional mind. Bull Los Angel Neurol Soc 34:135–162. Quoted in Sects 4.2, 4.5

Bogen JE (1975) The other side of the brain VII: some educational aspects of hemispheric specialization. UCLA Educ 17:24–32. Quoted in Sect 4.2

Bogen JE, Bogen GM (1969) The other side of the brain III: the corpus callosum and creativity. Bull Los Angel Neurol Soc 34:191–220. Quoted in Sect 4.2

Bogoliubov NN (1970) Lectures on quantum statistics, vol 2. Gordon and Breach, New York. Quoted in Sect 6.4

Bohr H (1932) Fastperiodische Funktionen. Springer, Berlin. Quoted in Sect 12.4

Bohr N (1913) On the constitution of atoms and molecules. Part I. Philos Mag 26:1–25. Quoted in Sect 6.2

Bohr N (1928) The quantum postulate and the recent development of atomic theory. Nat (Suppl) 121:580–590. Quoted in Sect 4.6

Bohr N (1934) Atomic theory and the description of nature. Cambridge University Press, Cambridge. Quoted in Sects 2.4, 4.6, 8.4

Bohr N (1948) On the notions of causality and complementarity. Dialectica 2:312–319. Quoted in Sect 2.4

Bohr N (1949) Discussion with Einstein on epistemological problems in atomic physics. In: Schilpp PA (ed) Albert Einstein: philosopher–scientist. Open Court, Evanston, pp 199–241. Quoted in Sects 2.4, 4.6

Bohr N (1961) The unity of human knowledge. In: Revue de la fondation européenne de la culture, July 1961, pp 63–66. Quoted in Sect 4.6

du Bois-Reymond E (1872) Die Grenzen des Naturerkennens. Veit & Comp, Leipzig. Quoted in Sect 7.1

Boole G (1847) The mathematical analysis of logic, being an essay towards a calculus of deductive reasoning. Macmillan, Cambridge. Quoted in Sect 2.2

Boole G (1848) The calculus of logic. Camb Dublin Math J 3:183–198. Quoted in Sect 2.2

Boole G (1849) Elementary treatise on logic not mathematical. In: Grattan-Guinness I, Bornet G (eds) George Boole. Selected manuscripts on logic and its philosophy. Birkhäuser, Basel, pp 13–41, 1997. Quoted in Sect 2.2

Boole G (1854) An investigation of the laws of thought, on which are founded the mathematical theories of logic and probabilities. Macmillan, London. Quoted in Sect 2.2

Borchers HJ, Yngvason J (2001) On the PCT-theorem in the theory of local observables. In: Longo R (ed) Mathematical physics in mathematics and physics: quantum and operator algebraic aspects. American Mathematical Society, Providence, pp 39–64. Quoted in Sect 7.3

Borel E (1914) Introduction géométrique à quelques théories physiques. Gauthier–Villars, Paris. Quoted in Sect 12.4

Borgnat P, Flandrin P, Amblard PO (2002) Stochastic discrete scale invariance. IEEE Signal Process Lett 9:181–184. Quoted in Sect 10.2

Born M (1921) Kritische Betrachtungen zur traditionellen Darstellung der Thermodynamik. Phys Z 22:218–224, 249–254, 282–286. Quoted in Sect 7.1

Born M (1955a) Continuity, determinism and reality. Dan Vidensk Selsk Math Fys Medd 30(2):1–26. Quoted in Sect 12.3

Born M (1955b) Ist die klassische Mechanik wirklich deterministisch? Phys Bl 11:49–54. Quoted in Sect 12.3

Bottazzini U (2002) "Algebraic truths" vs "geometric fantasies": Weierstrass' response to Riemann. Proc Int Congr Math 3:923–934. Quoted in Sect 1.2

Bourbaki N (1949) Foundations of mathematics for the working mathematician. J Symb Log 14:1–7. Quoted in Sect 5.10

Bourbaki N (1950) The architecture of mathematics. Am Math Mon 57:221–232. Quoted in Sect 5.10

Boyer TH (1975a) General connection between random electrodynamics and quantum electrodynamics for free electromagnetic fields and for dipole oscillator systems. Phys Rev D 11:809–830. Quoted in Sect 6.3

Boyer TH (1975b) Random electrodynamics: the theory of classical electrodynamics with classical electromagnetic zero-point radiation. Phys Rev D 11:790–808. Quoted in Sect 6.3

Bratteli O, Robinson DW (1979) Operator algebras and quantum statistical mechanics. I. C*- and W*-algebras, symmetry groups, decomposition of states. Springer, New York. Quoted in Sect 6.5

Bratteli O, Robinson DW (1981) Operator algebras and quantum statistical mechanics. II. Equilibrium states models in quantum statistical mechanics. Springer, New York. Quoted in Sects 7.3, 7.4, 8.5, 9.4, Appendix A.2

Brillouin L (1933) Le champ self-consistent, pour des électrons liés; la supraconductibilité. J Phys Radium 4:333–361. Quoted in Sect 12.2

Brillouin L (1960) Poincaré and the shortcomings of the Hamilton–Jacobi method for classical or quantized mechanics. Arch Ration Mech Anal 5:76–94. Quoted in Sect 12.3

Brillouin L (1962) Science and information theory. Academic Press, New York. Quoted in Sect 12.3

Broad CD (1925) The mind and its place in nature. Harcourt, Brace and Company, New York. Quoted in Sect 6.5

Brouwer LEJ (1907) On the foundation of mathematics. In: Heyting A (ed) Collected works vol I. Philosophy and foundations of mathematics. North-Holland, Amsterdam, pp 13–101, 1975. Quoted in Sects 3.2, 5.2, 5.6

Brown R (1828) A brief account of microscopical observations made in the month of June, July, and August 1827, on the particles contained in the pollen of plants; and on the general existence of active molecules in organic and inorganic bodies. Philos Mag 4:161–173. Quoted in Sect 10.2

Brown R (1829) Additional remarks on active molecules. Edinb J Sci 1:314–320. Quoted in Sect 10.2

de Bruijn NG (1967) Uncertainty principles in Fourier analysis. In: Shisha O (ed) Inequalities. Academic Press, New York, pp 57–71. Quoted in Sect 9.3

Bruner J (1986) Actual minds, possible worlds. Harvard University Press, Cambridge. Quoted in Sect 4.2

Brunetti R, Guido D, Longo R (2002) Modular localization and Wigner particles. Rev Math Phys 14:759–785. Quoted in Sect 7.3

Brush SG (1968) A history of random processes. I. Brownian movement from Brown to Perrin. Arch Hist Exact Sci 5:1–36. Quoted in Sect 10.2

Brychkov YuA, Prudnikov AP (1989) Integral transforms of generalized functions. Gordon and Breach, New York. Quoted in Sect 10.2

Buchholz D, Wichmann EH (1986) Causal independence and the energy-level density of states in local quantum field theory. Commun Math Phys 106:321–344. Quoted in Sect 7.3

Buchholz D, D'Antoni C, Fredenhagen K (1987) The universal structure of local algebras. Commun Math Phys 111:123–135. Quoted in Sect 7.3

Buchholz D, Solveen C (2013) Unruh effect and the concept of temperature. Class Quantum Gravity 30:085011. Quoted in Sect 7.3
Bunge M (1962) Intuition and science. Prentice Hall, Englewood Cliffs. Quoted in Sect 4.1
Bunge M (1973) Philosophy of physics. Reidel, Dordrecht. Quoted in Sects 6.1, 7.1
Bunge M (1985) Treatise on basic philosophy, volume 7. Epistemology & methodology, III: philosophy of science and technology. Reidel, Dordrecht. Quoted in Sect 7.1
Bureau International des Poids et Mesures (2006) The international system of units (SI), 8th edn. Organisation Intergouvernementale de la Convention du Mètre, Paris. Quoted in Sect 8.3
Burniston Brown C (1953) Is the definition of the word FACT the first problem of philosophy? Philosophy 28:154–159. Quoted in Sect 8.4
Busch P (1986) Unsharp reality and joint measurements for spin observables. Phys Rev D 33:2253–2261. Quoted in Sect 3.5
Busch P (2003) Quantum states and generalized observables: a simple proof of Gleason's theorem. Phys Rev Lett 91:120403. Quoted in Sect 6.5
Busch P (2008) The time energy uncertainty relation. In: Muga JG, Mayato RS, Egusquiza IL (eds) Time in quantum mechanics, vol 1. Springer, Berlin, pp 73–105. Quoted in Sect 1.4
Busch P, Grabowski M, Lahti PJ (1995) Operational quantum physics. Springer, Berlin. Quoted in Sect 3.5
Busch P, Heinosaari T (2008) Approximate joint measurements of qubit observables. Quantum Inf Comput 8:797–818. Quoted in Sect 3.5
Busch P, Lahti PJ (1986) To what extent do position and momentum commute? Phys Lett A 115:259–264. Quoted in Sects 3.5, 11.2
Busch P, Lahti P, Mittelstaedt P (1991) The quantum theory of measurement. Springer, Berlin. Quoted in Sect 3.5
Butterworth S (1930) On the theory of filter amplifiers. Wirel Eng 7:536–541. Quoted in Sect 3.3
Butterworth W (1999) What counts: how every brain is hardwired for math. Free Press, New York. Quoted in Sect 5.2
Calaprice A (2011) The ultimate quotable Einstein. Princeton University Press, Princeton. Quoted in Sect 4.1
Cameron RH, Martin WT (1947) The behavior of measure and measurability under change of scale in Wiener space. Bull Am Math Soc 53:130–137. Quoted in Sect 6.5
Carathéodory C (1909) Untersuchungen über die Grundlagen der Thermodynamik. Math Ann 67:355–386. Quoted in Sects 7.1, 7.4
Carnap R (1928) Der logische Aufbau der Welt. Weltkreis-Verlag, Berlin. Quoted in Sect 4.7
Carnap R (1932) Die physikalische Spache als Universalsprache der Wissenschaft. Erkenntnis 2:432–465. Quoted in Sect 7.1
Carnap R (1934) Die Antinomien und die Unvollständigkeit der Mathematik. Monatshefte Math Phys 41:263–284. Quoted in Sect 5.8
Carnap R (1936) Testability and meaning. Philos Sci 3:419–471. Quoted in Sect 7.1
Carnap R (1937) The logical syntax of language. Kegan Paul, London. Quoted in Sect 5.8
Carnap R (2003) The logical structure of the world and pseudoproblems in philosophy. Open Court, Chicago. Quoted in Sect 4.7
Carstoiu J (1967) Note on electric and magnetic polarization in moving media. Proc Natl Acad Sci USA 57:1536–1541. Quoted in Sect 6.3
Cassinelli G, Varadarajan VS (2002) On Accardi's notion of complementary observables. Infin Dimens Anal Quantum Probab Relat Top 5:135–144. Quoted in Sect 3.5
Cassirer E (1910) Substanzbegriff und Funktionsbegriff: Untersuchungen über die Grundfragen der Erkenntniskritik. Verlag von Bruno Cassirer, Berlin. Quoted in Sect 4.1
Cassirer E (1929) Philosophie der symbolischen Formen. Dritter Teil: Phänomenologie der Erkenntnis. Verlag von Bruno Cassirer, Berlin. Quoted in Sect 4.1
Cassirer E (1957) The philosophy of symbolic forms, volume three: the phenomenology of knowledge. Yale University Press, New Haven, transl by Manheim R. Quoted in Sect 8.4
Chandrasekhar S (1990) Truth and beauty. Aesthetics and motivations in science. University of Chicago Press, Chicago. Quoted in Sect 6.4

Church A (1936a) An unsolvable problem of elementary number theory. Am J Math 58:345–363. Quoted in Sect 5.5

Church A (1936b) A note on the Entscheidungsproblem. J Symb Log 1:40–41. Quoted in Sect 5.5

Clavadetscher-Seeberger E (1983) Eine partielle Prädikatenlogik. Doctoral thesis, ETH Zürich, nr 7347. Quoted in Sect 4.6

Clifton R, Halvorson H (2001) Entanglement and open systems in algebraic quantum field theory. Stud Hist Philos Mod Phys 32:1–31. Quoted in Sect 8.5

Cobern WW (1996) Worldview theory and conceptual change in science education. Sci Educ 80:579–610. Quoted in Sect 2.4

Coffman V, Kundu J, Wootters WK (2000) Distributed entanglement. Phys Rev A 61:052306. Quoted in Sect 8.5

Cohen G (1973) Hemispheric differences in serial versus parallel processing. J Exp Psychol 97:349–356. Quoted in Sect 1.4

Cohen L (1993) The scale representation. IEEE Trans Signal Process 41:3275–3292. Quoted in Sect 10.2

Cohen L (1995) Time-frequency analysis. Prentice Hall, Upper Saddle River. Quoted in Sects 9.3, 10.2

Cohen-Tannoudji C (1968) Optical pumping and interaction of atoms with the electromagnetic field. In: Lévy M (ed) Cargèse lectures in physics, vol 2. Gordon and Breach, New York, pp 347–393. Quoted in Sect 6.3

Coleman AJ (1963) Structure of fermion density matrices. Rev Mod Phys 35:668–687. Quoted in Sect 6.3

Coleman AJ (1968) Induced and subduced representations. In: Loebl EM (ed) Group theory and its applications, vol 1. Academic Press, New York, pp 57–118. Quoted in Sect 8.5

Collingwood RG (1998) An essay on metaphysics. Oxford University Press, Oxford. Quoted in Sect 12.2

Connes A (1973) Une classification des facteurs de type III. Ann Sci Éc Norm Supér 6:133–252. Quoted in Sect 8.5, Appendix A.5

Connes A (1974) Charactérisation des espaces vectoriels ordonnés sous-jacents aux algèbres de von Neumann. Ann Inst Fourier (Grenoble) 24:121–155. Quoted in Appendix A.6

Connes A (1994) Noncommutative geometry. Academic Press, San Diego. Quoted in Sect 8.5, Appendix A.5

Connes A (2001) Noncommutative geometry—year 2000. In: Alon N et al (eds) Visions in mathematics. Birkhäuser, Basel, pp 481–559. Quoted in Sects 5.5, 7.3

Connes A (2005) A view of mathematics. Accessible at www.irem.uhp-nancy.fr/Lomb/maths.pdf. Quoted in Sect 7.4

Connes A (2008) An interview with Alain Connes, part II (conducted by C. Goldstein and G. Skandalis). Eur Math Soc Newsl 67(March):29–33. Quoted in Sect 5.7

Connes A, Rovelli C (1994) Von Neumann algebra automorphisms and time-thermodynamics relation in generally covariant quantum theories. Class Quantum Gravity 11:2899–2917. Quoted in Sect 7.4

Connes A, Størmer E (1978) Homogeneity of the state space of factors of type III_1. J Funct Anal 28:187–196. Quoted in Sect 7.3

Copei F (1950) Der fruchtbare Moment im Bildungsprozess. Quelle & Meyer, Heidelberg. Quoted in Sect 1.4

Cornfeld IP, Fomin SV, Sinai YaG (1982) Ergodic theory, Springer, New York, Quoted in Sects 8.5, 12.1

Corngold S (1973) The commentator's despair: the interpretation of Kafka's metamorphosis. Kennikat Press, Port Washington. Quoted in Sect 4.4

Costa de Beauregard O (1983) Running backwards the Mermin device: causality in EPR correlations. Am J Phys 51:513–516. Quoted in Sect 6.2

Cox DR (1991) Long-range dependence, non-linearity and time irreversibility. J Time Ser Anal 12:329–335. Quoted in Sect 10.2

Cramér H (1961) On the structure of purely non-deterministic stochastic processes. Ark Mat 4:249–266. Quoted in Sect 12.4

Cramér H, Leadbetter MR (1967) Stationary and related stochastic processes. Wiley, New York. Quoted in Sect 12.4

Crick F (1994) The astonishing hypothesis. The scientific search for the soul. Simon and Schuster, London. Quote in Sect 7.1

Crispino LCB, Higuchi A, Matsas GEA (2008) The Unruh effect and its applications. Rev Mod Phys 80:787–838. Quoted in Sect 7.3

Cunningham GW (1916) A study in the philosophy of Bergson. Longmans, New York. Quoted in Sect 4.3

Cunningham GW (1924) Bergson's doctrine of intuition. Philos Rev 33:604–606. Quoted in Sect 4.3

Curie P (1894a) Sur la symétrie dans les phénoménes physiques, symétrie d'un champ électrique et d'un champ magnétique. J Phys Theor Appl 3:393–415. Quoted in Sects 6.4, 9.3, 11.7

Curie P (1894b) Sur la possibilité d'existence de la conductibilité magnétique et du magnétisme libre. Séances Soc Fr Phys (Paris) 76–77. Quoted in Sect 6.3

Cusanus N (1954) Of learned ignorance. Yale University Press, New Haven, transl by Heron G. Quoted in Sect 5.1

Cushen CD, Hudson RL (1971) A quantum-mechanical central limit theorem. J Appl Probab 8:454–469. Quoted in Sect 10.4

Cushing JT (1991) Quantum theory and explanatory discourse: endgame for understanding? Philos Sci 58:337–358. Quoted in Sect 6.2

Czelakowski J (1978) On extending of partial Boolean algebras to partial *-algebras. Colloq Math 40:13–21. Quoted in Sect 3.4

Dales HG, Olivery G (1998) Truth in mathematics, Clarendon, Oxford. Quoted in Sect 5.1

Dalla Chiara ML, Giuntini R (2002) Quantum logics. In: Gabbay D, Guenther F (eds) Handbook of philosophical logic, vol 6. Kluwer Academic, Dordrecht, pp 129–228. Quoted in Sect 3.2

Dalla Chiara ML, Giuntini R, Greechie R (2004) Reasoning in quantum theory: sharp and unsharp quantum logics. Kluwer Academic, Dordrecht. Quoted in Sect 3.3

Davenport G (1995) 7 Greeks. New Directions Publishing Corporation, New York. Quoted in Sect 5.9

Davies EB (1970) On the repeated measurement of continuous observables in quantum mechanics. J Funct Anal 6:318–346. Quoted in Sects 3.5, 6.2

Davies EB (1976) Quantum theory of open systems. Academic Press, London. Quoted in Sects 3.5, 6.2, 12.1

Davies EB (2005a) Pluralism in mathematics. Philos Trans R Soc A 363:2449–2460. Quoted in Sect 5.8

Davies EB (2005b) A defence of mathematical pluralism. Philos Math, Ser III 13:252–276. Quoted in Sect 5.8

Davies EB, Lewis ET (1970) An operational approach to quantum probability. Commun Math Phys 17:239–260. Quoted in Sects 3.5, 6.2, 12.1

Davis PJ, Hersh R, Marchisotto EA (1981) The mathematical experience. Birkhäuser, Boston. Quoted in Sects 4.1, 5.3

Debnath L (2003) Wavelets and signal processing. Birkhäuser, Boston. Quoted in Sect 9.3

Debye P, Hückel E (1923) Zur Theorie der Elektrolyte. Phys Z 24:185–206. Quoted in Sect 6.3

DeCamp WH (1989) The FDA perspective on the development of stereoisomers. Chirality 1:2–6. Quoted in Sect 6.5

Decker T, Janzing D (2005) Implementation of group-covariant positive operator valued measures by orthogonal measurements. J Math Phys 46:012104. Quoted in Sect 3.5

Dedekind R (1888) Was sind und was sollen die Zahlen? Vieweg, Braunschweig. Quoted in Sect 6.4

Dehaene S (1997) The number sense. How the mind creates mathematics. Oxford University Press, New York. Quoted in Sect 5.2

Demoen B, Vanheuverzwijn P (1980) Implementable positive maps on standard forms. J Funct Anal 38:354–365. Quoted in Appendix A.6
Denbigh KG (1981) Three concepts of time. Springer, Berlin. Quoted in Sect 12.2
Deutsch D, Jozsa R (1992) Rapid solution of problems by quantum computation. Proc R Soc Lond A 439:553–558. Quoted in Sect 1.4
Dieudonné JA (1970) The work of Nicholas Bourbaki. Am Math Mon 77:134–145. Quoted in Sect 5.8
Dirac PAM (1926a) On the theory of quantum mechanics. Proc R Soc Lond A 112:661–677. Quoted in Sect 6.2
Dirac PAM (1926b) Relativity quantum mechanics with an application to Compton scattering. Proc R Soc Lond A 111:405–423. Quoted in Sect 1.4
Dirac PAM (1928) The quantum theory of the electron. Proc R Soc Lond A 117:610–624. Quoted in Sect 7.3
Dirac PAM (1930a) The principles of quantum mechanics, Clarendon, Oxford. Quoted in Sect 6.2
Dirac PAM (1930b) Note on exchange phenomena in the Thomas atom. Proc Camb Philos Soc 26:376–385. Quoted in Sect 10.4
Dirac PAM (1931) Quantised singularities in the electromagnetic field. Proc R Soc Lond A 133:60–72. Quoted in Sect 6.3
Dirac PAM (1933) Homogeneous variables in classical dynamics. Proc Camb Philos Soc 29:389–400. Quoted in Sect 8.3
Dixmier J (1948) Position relative de deux variétés linéaires fermées dans un espace de Hilbert. Rev Sci 86:387–399. Quoted in Sect 11.1
Dixmier J (1969) Les algèbres d'opérateurs dans l'espace Hilbertien. (Algèbres de von Neumann). Gauthier–Villars, Paris. Quoted in Appendix A.6
Dörfler M (2001) Time-frequency analysis for music signals. A mathematical approach. J New Music Res 22(5):2–10. Quoted in Sect 9.3
Doetsch G (1950) Handbuch der Laplace-Transformation Band 1. Theorie der Laplace-Transformation. Birkhäuser, Basel. Quoted in Sect 10.2
de Dominicis C (1962a) Renormalization in equilibrium statistical mechanics. In: Caianiello ER (ed) Lectures on the many-body problem, vol 2. Academic Press, New York, pp 203–216. Quoted in Sect 6.3
de Dominicis C (1962b) Variational formulation of equilibrium statistical mechanics. J Math Phys 3:983–1002. Quoted in Sect 6.3
de Dominicis C, Englert F (1967) Potential-correlation duality in statistical mechanics. J Math Phys 8:2143–2146. Quoted in Sect 6.3
de Dominicis C, Martin PC (1964a) Stationary entropy principle and renormalization in normal and superfluid systems. I. Algebraic formulation. J Math Phys 5:14–30. Quoted in Sect 6.3
de Dominicis C, Martin PC (1964b) Stationary entropy principle and renormalization in normal and superfluid systems. II. Diagrammatic formulation. J Math Phys 5:31–59. Quoted in Sect 6.3
Domotor Z (1974) The probability structure of quantum-mechanical systems. Synthese 29:155–185. Quoted in Sect 4.6
Doob JL (1944) The elementary Gaussian processes. Ann Math Stat 15:229–282. Quoted in Sect 12.4
Doob JL (1953) Stochastic processes. Wiley, New York. Quoted in Sect 12.4
Doplicher S, Longo R (1984) Standard and split inclusions of von Neumann algebras. Invent Math 75:493–536. Quoted in Sect 7.3
Doran RS, Belfi VA (1986) Characterizations of C*-algebras: the Gelfand–Naimark theorems. CRC Press, Boca Raton. Quoted in Sect 6.5
Driessler W (1977) On the type of local algebras in quantum field theory. Commun Math Phys 53:295–297. Quoted in Sect 7.3
Duflo M, Moore CC (1976) On the regular representation of a nonunimodular locally compact group. J Funct Anal 21:209–243. Quoted in Sect 10.1
Dür W, Vidal G, Cirac JI (2000) Three qubits can be entangled in two inequivalent ways. Phys Rev A 62:062314. Quoted in Sect 8.5

Dugac P (1976) Richard Dedekind et les fondements des mathématiques (avec de nombreux textes inédits). Vrin, Paris. Quoted in Sect 5.4

Dumitras A, Moschytz G (2007) Understanding fuzzy logic: an interview with Lotfi Zadeh. IEEE Signal Process Mag 24:102–105. Quoted in Sect 3.3

Dutta P, Horn PM (1981) Low-frequency fluctuations in solids; 1/f noise. Rev Mod Phys 53:497–516. Quoted in Sect 10.2

Dyson FJ (1956) Obituary prof. Hermann Weyl. Nature 177:457–488. Quoted in Sect 6.4

Dyson F (2009) Birds and frogs. Not Am Math Soc 56:212–223. Quoted in Sect 5.9

Dzhafarov EN, Zhang R, Kujala JV (2015) Is there contextuality in behavioral and social systems? Philos Trans R Soc A 374:20150099. Quoted in Sect 13.5

Eddington A (1918) Report on the relativity theory of gravitation. Fleetway, London. Quoted in Sect 7.2

Eddington A (1920) Space, time and gravitation. Cambridge University Press, Cambridge. Quoted in Sect 2.3

Eddington AS (1946) Fundamental theory. Cambridge University Press, Cambridge. Quoted in Sect 6.5.

Edwards B (1979) Drawing on the right side of the brain. A course in enhancing creativity and artistic confidence. Tarcher, Los Angeles. Quoted in Sects 1.2, 1.4

Effros EE (1963) Order ideals in a C*-algebra and its dual. Duke J Math 30:391–411. Quoted in Sect 6.5

Einstein A (1905) Zur Elektrodynamik bewegter Körper. Ann Phys 17:891–921. Quoted in Sects 6.3, 8.3, 8.5

Einstein A (1906) Zur Theorie der Brownschen Bewegung. Ann Phys 19:371–381. Quoted in Sect 10.2

Einstein A (1919) Time, space, and gravitation. The Times (London) 28 November 1919, pp 13–14. Quoted in Sect 7.2

Einstein A (1921) Geometrie und Erfahrung. Sitzber Preuss Akad Wiss 1921:123–130. Quoted in Sect 3.1

Einstein A (1931) Cosmic religion, Covici-Friede, New York, Quoted in Sect 4.1

Einstein A (1934) On the method of theoretical physics. Philos Sci 1:163–169. Quoted in Sect 8.5

Einstein A (1936) Physics and reality. J Franklin Inst 221:349–382. Quoted in Sect 4.1

Einstein A (1944) Remarks on Bertrand Russell's theory of knowledge. In: Schilpp PA (ed) The philosophy of Bertrand Russell. Open Court, La Salle, pp 278–291. Quoted in Sect 5.3

Einstein A (1949) Autobiographisches. In: Albert Einstein: philosopher–scientist. Open Court, Evanston, pp 2–95. Quoted in Sect 6.4

Einstein A (1950) On the generalized theory of gravitation. Sci Am 182(4):13–17. Quoted in Sect 4.1

Einstein A (1954) Ideas and opinions. Three Rivers Press, New York. Quoted in Sect 4.1

Einstein A (1955) Erinnerungen—Souvenirs. Schweiz Hochsch 28:145–153 (Sonderheft 100 Jahre ETH). Quoted in Sect 4.7

Einstein A (1997) The collected papers of Albert Einstein, volume 6. The Berlin years: writings 1914–1917. Princeton University Press, Princeton. English translation of selected texts. Quoted in Sect 4.7

Einstein A, Infeld L (1939) The evolution of physics. Cambridge University Press, Cambridge. Quoted in Sect 6.4

Einstein A, Podolsky B, Rosen N (1935) Can quantum-mechanical description of physical reality be considered complete? Phys Rev 47:777–780. Quoted in Sects 1.5, 6.3

Ekstein H (1969) Presymmetry II. Phys Rev 184:1315–1337. Erratum: Phys Rev D 1:1851 (1970). Quoted in Sect 6.3

Ellis GFR (2002) Cosmology and local physics. Int J Mod Phys A 17:2667–2671. Quoted in Sect 7.1

Ellis GFR (2008) On the nature of causation in complex systems. Trans R Soc S Afr 63:69–84. Quoted in Sect 7.1

Ellis GFR (2009) Top-down causation and the human brain. In: Murphy N, Ellis GFR, O'Connor T (eds) Downward causation and the neurobiology of free will. Springer, Berlin, pp 63–81. Quoted in Sect 7.1

Emary C, Lambert N, Nori F (2014) Leggett–Garg inequalities. Rep Prog Phys 77:016001. Quoted in Sect 13.5

Emch GG (1972) Algebraic methods in statistical mechanics and quantum field theory. Wiley, New York. Quoted in Sect 9.2

Emch GG (1976) Generalized K-flows. Commun Math Phys 49:191–215. Quoted in Sect 6.5

Emch GG (1985) KMS structures in geometric quantization. Contemp Math 62:175–186. Quoted in Appendix A.6

Emch G, Jauch JM (1965) Structures logiques et mathématiques en physique quantique. Dialectica 19:259–279. Quoted in Sect 11.1

Eringen AC, Maugin GA (1990a) Electrodynamics of continua I. Foundations and solid media. Springer, New York. Quoted in Sect 6.3

Eringen AC, Maugin GA (1990b) Electrodynamics of continua II. Fluids and complex media. Springer, New York. Quoted in Sect 6.3

d'Espagnat B (1965) Conceptions de la physique contemporaire. Hermann, Paris. Quoted in Sect 6.2

d'Espagnat B (1966) An elementary note about "mixtures". In: De-Shalit A, Feshbach H, van Hove L (eds) Preludes in theoretical physics. In honor of V.F. Weisskopf. North-Holland, Amsterdam, pp 185–191. Quoted in Sect 6.2

d'Espagnat B (1976) Conceptual foundations of quantum mechanics. Benjamin, London. Quoted in Sect 6.2

d'Espagnat B (1979) The quantum theory and reality. Sci Am 241(11):128–140. Quoted in Sect 13.5

Evans DE, Lewis JT (1977) Dilations of irreversible evolutions in algebraic quantum theory. Commun Dublin Inst Adv Stud A 24:1–104. Quoted in Appendix A.2

de Facio B, Taylor DC (1973) Commutativity and causal independence. Phys Rev D 8:2729–2731. Quoted in Sect 7.3

Favrholdt D (1999) Niels Bohr collected works, volume 10: complementarity beyond physics (1928–1962). Elsevier, Amsterdam. Quoted in Sect 4.6

Fechner GT (1860) Elemente der Psychophysik, zweiter Theil. Breitkopf und Härtel, Leipzig. Quoted in Sect 11.7

Feigl H (1953) Notes on causality. In: Feigl H, Brodbeck M (eds) Readings in the philosophy of science. Appleton-Century-Crofts, New York. Quoted in Sect 12.3

Feigl H (1963) Physicalism, unity of science and the foundation of psychology. In: Schilpp PA (ed) The philosophy of Rudolf Carnap. Open Court, La Salle, pp 227–267. Quoted in Sect 7.1

Ferguson ES (1977) The mind's eye: nonverbal thought in technology. Science 197:827–836. Quoted in Sect 4.1

Fetzer JH, Almeder RF (1993) Glossary of epistemology/philosophy of science. Paragon House, New York. Quoted in Sect 12.3

Feyerabend PK (1962) Explanation, reduction and empiricism. In: Feigl H, Maxwell G (eds) Minnesota studies in the philosophy of science III. University of Minnesota Press, Minneapolis, pp 28–97. Quoted in Sect 7.1

Feyerabend P (1994) Quantum theory and our view of the world. In: Hilgevoord J (ed) Physics and our view of the world. Cambridge University Press, Cambridge, pp 149–168. Quoted in Sect 2.4

Feynman R (1965) The character of physical law. British Broadcasting Corporation, London. Quoted in Sect 6.2

Feynman R (1985) QED: the strange theory of light and matter. Princeton University Press, Princeton. Quoted in Sect 6.2

Feynman RP (1998) The meaning of it all. Perseus Books, Reading. Quoted in Sect 6.2

Fierz M (1970) Pauli, Wolfgang. In: Gillispie CC (ed) Dictionary of scientific biography, vol 10. Charles Scribner's Sons, New York, pp 422–425. Quoted in Sect 4.1

Filk T, von Müller A (2009) Quantum physics and consciousness: the quest for a common conceptual framework. Mind Matter 7:59–79. Quoted in Sect 13.5

Finch PD (1969) On the structure of quantum logic. J Symb Log 34:275–282. Quoted in Sects 3.4, 4.6

Flandrin P, Borgnat P, Amblard PO (2003) From stationary to self-similarity, and back: variations on the Lamperti transformation. In: Rangarajan G, Ding M (eds) Processes with long-range correlations. Springer, Berlin, pp 329–335. Quoted in Sect 10.2

Fleck L (1935a) Entstehung und Entwicklung einer wissenschaftlichen Tatsache. Schwabe, Basel. Quoted in Sect 4.1

Fleck L (1935b) Zur Frage der Grundlagen der medizinischen Erkenntnis. Klin Wochenschr 14:1255–1259. Quoted in Sect 4.1

Fleck L (1979) Genesis and development of a scientific fact. University of Chicago Press, Chicago. Quoted in Sect 4.1

Fleck L (1986) Problems of the science of science. In: Cohen RS, Schnelle T (eds) Cognition and fact: materials on Ludwik Fleck. Reidel, Dordrecht, pp 113–127. Quoted in Sect 4.1

Foiaş C, Gehér L, Sz-Nagy B (1960) On the permutability condition of quantum mechanics. Acta Sci Math (Szeged) 21:78–89. Quoted in Sect 10.5

Foiaş C, Gehér L (1963) Über die Weylsche Vertauschungsrelation. Acta Sci Math (Szeged) 24:97–102. Quoted in Sect 10.5

Folland GB (1989) Harmonic analysis in phase space. Princeton University Press, Princeton. Quoted in Sect 9.3

Forsee A (1963) Albert Einstein: theoretical physicist. Macmillan, New York. Quoted in Sect 8.3

Fowler R, Guggenheim EA (1939) Statistical thermodynamics. Cambridge University Press, Cambridge. Quoted in Sects 7.1, 7.4

van Fraassen BC (1970) An introduction to the philosophy of time and space. Random House, New York. Quoted in Sect 12.2

Fraser D (2008) The fate of "particles" in quantum field theories with interactions. Stud Hist Philos Mod Phys 39:841–858. Quoted in Sect 7.3

Fredenhagen K (1993) Global observables in local quantum physics. In: Araki H, Ito LR, Kishimoto A, Ojima I (eds) Quantum and non-commutative analysis. Kluwer Academic, Dordrecht, pp 41–51. Quoted in Sect 7.3

Fredenhagen K (2015) An introduction to quantum field theory. In: Brunetti R et al (eds) Advances in algebraic quantum field theory. Springer, Berlin, pp 1–30. Quoted in Sect 7.3

Freeman WJ, Vitiello G (2016) Matter and mind are entangled in two streams of images guiding behavior and informing the subject through awareness. Mind and Matter 14:7–24. Quoted in Appendix A.6

Frege G (1918/1919) Der Gedanke. Eine logische Untersuchung. Beitr Philos Dtsch Ideal 2:58–77. Quoted in Sect 2.2

Frege G (1952) Grundgesetze der Arithmetik. In: Geach P, Black M (eds) Translations from the philosophical writings of Gottlob Frege. Blackwell, Oxford, pp 137–244. Quoted in Sect 3.1

Frege G (1979) Posthumous writings. Blackwell, Oxford. Quoted in Sect 4.1

Frink O (1938) New algebras of logic. Am Math Mon 45:210–219. Quoted in Sect 3.2

Frisch M (1977) Sketchbook 1946–1949. Harcourt, Brace, Jovanovich, San Diego, transl by Skelton G. Quoted in Sect 4.3

Furstenberg H (1960) Stationary processes and prediction theory. Princeton University Press, Princeton. Quoted in Sect 12.4

Fushchich W, Nikitin AG (1987) Symmetries of Maxwell's equations. Reidel, Dordrecht. Quoted in Sect 6.3

Gabbay D, Guenthner F (eds) (1986) Handbook of philosophical logic, volume III: alternatives to classical logic. Reidel, Dordrecht. Quoted in Sect 3.2

Gabbay DM, Woods J (eds) (2007) Handbook of the history of logic, volume 8: the many-valued and nonmonotonic turn in logic. Elsevier, Amsterdam. Quoted in Sect 3.2

Gabor D (1946) Theory of communication. J Inst Electr Eng (Lond) 93(Pt III):429–457. Quoted in Sects 9.3, 9.4

Gale D (1989) The truth and nothing but the truth. Math Intell 11(3):62–67. Quoted in Sect 5.1
Galton F (1879) The geometric mean, in vital and social statistics. Proc R Soc Lond 29:365–367. Quoted in Sect 11.7
Gardner WA (1988) Statistical spectral analysis: a nonprobabilistic theory. Prentice Hall, Englewood Cliffs. Quoted in Sect 12.4
Gasking D (1955) Causation and recipes. Mind 64:479–487. Quoted in Sect 12.2
Gel'fand IM, Neumark MA (1943) On the imbedding of normed rings into the ring of operators on a Hilbert space. Mat Sb 12:197–217. Quoted in Appendix A.3
Gel'fand I, Neumark MA (1947) Unitary representations of the group of linear transformations of the straight line. C R (Dokl) Acad Sci URSS 55:567–570. Quoted in Sects 9.7, 10.5
Gel'fand IM, Shilov GE (1964) Generalized functions, volume 1: properties and operations. Academic Press, New York. Quoted in Sect 10.2
Gel'fand IM, Vilenkin NYa (1964) Generalized functions, volume 4: applications of harmonic analysis. Academic Press, New York. Quoted in Sects 6.5, 10.2
George C, Prigogine I (1979) Coherence and randomness in quantum theory. Physica A 99:369–382. Quoted in Appendix A.6
George A, Velleman DJ (1998) Two conceptions of natural number. In: Dales HG, Olivery G (eds) Truth in mathematics, Clarendon Oxford, pp 311–327. Quoted in Sect 5.1
Giaquinto M (1983) Hilbert's philosophy of mathematics. Br J Philos Sci 34:119–132. Quoted in Sect 5.5
Gilder L (2008) The age of entanglement. Vintage, New York. Quoted in Sect 13.5
Giles R (1970) Foundations for quantum mechanics. J Math Phys 11:2139–2160. Quoted in Sect 3.5
Gillies DA (1973) An objective theory of probability, Methuen, London. Quoted in Sect 12.3
Gilmour JSL (1989) Two early papers on classification. Plant Syst Evol 167:97–107. Quoted in Sect 3.4
Girard J-Y (1987) Linear logic. Theor Comput Sci 50:1–102. Quoted in Sect 3.2
Girard J-Y (1995) Linear logic: its syntax and semantics. In: Girard J-Y, Lafont Y, Regnier L (eds) Advances in linear logic. Cambridge University Press, Cambridge, pp 1–42. Quoted in Sect 3.2
Gleason AM (1957) Measures on the closed subspaces of a Hilbert space. J Math Mech 6:885–893. Quoted in Sect 6.5
Gödel K (1931) Über formal unentscheidbare Sätze der Principia Mathematica und verwandter Systeme I. Monatshefte Math Phys 38:173–198. Quoted in Sect 5.7
Gödel K (1933) The present situation in the foundations of mathematics. In: Feferman S (ed) Kurt Gödel, collected works, volume III: unpublished essays and lectures. Oxford University Press, Oxford, pp 45–53. Quoted in Sect 5.7
Gödel K (1935) Review of Carnap, Rudolf: Die Antinomien und die Unvollständigkeit der Mathematik. Zent Math Grenzgeb 11:1. Quoted in Sect 5.8
Gödel K (1951) Some basic theorems on the foundations of mathematics and their implications. In: Feferman S (ed) Kurt Gödel, collected works, volume III: unpublished essays and lectures. Oxford University Press, Oxford, pp 304–323. Quoted in Sect 5.3
Gödel K (1953) Is mathematics syntax of language? In: Feferman S (ed) Kurt Gödel, collected works, volume III: unpublished essays and lectures. Oxford University Press, Oxford, pp 334–356. Quoted in Sect 5.8
Gödel K (1961) The modern development of the foundations of mathematics in the light of philosophy. In: Feferman S (ed) Kurt Gödel, collected works, volume III: unpublished essays and lectures. Oxford University Press, Oxford, pp 375–387. Quoted in Sect 5.3
Gödel K (1964) What is Cantor's continuum problem? (Second expanded version). In: Benacerraf P, Putnam H (eds) Philosophy of mathematics: selected readings. Prentice Hall, Englewood Cliffs, pp 258–273. Quoted in Sect 5.10
Gödel K (1995) Kurt Gödel, collected works, volume III: unpublished essays and lectures. Oxford University Press, Oxford. Quoted in Sects 5.7, 5.10, 8.3
Gödel K (2003) Kurt Gödel, collected works, volume V: correspondence H–Z. Clarendon, Oxford. Quoted in Sect 8.3

von Goethe JW (2003) Faust, parts I and II, transl by Kline AS, available at www.poetryintranslation.com/PITBR/German/FaustIScenesIVtoVI.htm. Quoted in Sect 2.1
Goldschmidt HL (1990) Nochmals Dialogik. ETH Zürich, Stiftung Dialogik. Quoted in Sect 6.4
Gordon JP, Louisell WH (1966) Simultaneous measurement of noncommuting observables. In: Kelley BL, Lax B, Tannenwald PE (eds) Physics of quantum electronics. McGraw–Hill, New York, pp 833–840. Quoted in Sect 6.2
Gottwald S (2010) An early approach toward graded identity and graded membership in set theory. Fuzzy Sets Syst 161:2369–2379. Quoted in Sect 3.2
Gove PB (1964) Webster's third new international dictionary. Merriam, Springfield. Quoted in Sect 8.4
Grawert G, Lüders G, Rollnik H (1959) The TCP theorem and its applications. Fortschr Phys 7:291–328. Quoted in Sect 12.2
Greechie RJ (1968) On the structure of orthomodular lattices satisfying the chain condition. J Comb Theory 4:210–218. Quoted in Sects 3.4, 4.6
Greenberger DM, Horne MA, Zeilinger A (1989) Going beyond Bell's theorem. In: Kafatos M (ed) Bell's theorem, quantum theory and conceptions of the universe. Kluwer Academic, Dordrecht, pp 69–72. Quoted in Sect 8.5
Gröchenig K (2001) Foundations of time-frequency analysis. Birkhäuser, Boston. Quoted in Sect 9.3
Grossmann A (1976) Parity operator and quantization of δ-functions. Commun Math Phys 48:191–194. Quoted in Sect 10.4
Grossmann A, Morlet J, Paul T (1986) Transforms associated to square integrable group representations. II. Examples. Ann Instit Henri Poincaré A 45:293–309. Quoted in Sects 9.5, 10.1
Grünbaum A (1973) Philosophical problems of space and time. Reidel, Dordrecht. Quoted in Sect 12.2
Guggenheim EA (1949) Thermodynamics. An advanced treatment for chemists and physicists. North-Holland, Amsterdam. Quoted in Sect 7.1
Gustafson K, Misra B (1976) Canonical commutation relations of quantum mechanics and stochastic regularity. Lett Math Phys 1:275–280. Quoted in Sect 12.4
Haag R (1955) On quantum field theories. Dan Vidensk Selsk Mat Fys Medd 29(12):1–37. Quoted in Sect 6.5
Haag R (1970) Observables and fields. In: Deser S, Grisaru M, Pendleton H (eds) Lectures on elementary particles and quantum field theory, vol 2. MIT Press, Cambridge, pp 1–89. Quoted in Sect 7.3
Haag R (1992) Local quantum physics. Springer, Berlin. Quoted in Sect 7.3, Appendix A.5
Haag R (1999) Objects, events, and localization. In: Blanchard P, Jadczyk AZ (eds) Quantum future. Springer, Berlin, pp 58–79. Quoted in Sect 7.3
Haag R (2010) Some people and some problems met in half a century of commitment to mathematical physics. Eur Phys J H 35:263–307. Quoted in Sect 7.3
Haag R, Hugenholtz NM, Winnink M (1967) On the equilibrium states in quantum statistical mechanics. Commun Math Phys 5:215–236. Quoted in Sect 7.3
Haag R, Kadison RV, Kastler D (1970) Nets of C*-algebras and classification of states. Commun Math Phys 16:81–104. Quoted in Sect 6.5
Haag R, Kastler D (1964) An algebraic approach to quantum field theory. J Math Phys 5:848–861. Quoted in Sects 6.5, 7.3, 12.1
Haag R, Kastler D, Trych-Pohlmeyer EB (1974) Stability and equilibrium states. Commun Math Phys 38:173–193. Quoted in Sect 7.4
Haagerup U (1975) The standard form of von Neumann algebras. Math Scand 37:271–283. Quoted in Appendix A.6
Haagerup U (1987) Connes' bicentralizer problem and uniqueness of the injective factor of type III_1. Acta Math 158:95–148. Quoted in Sect 8.5
Haak S (1996) Deviant logic, fuzzy logic: beyond the formalism. University of Chicago Press, Chicago. Quoted in Sect 3.2

Haar A (1933) Der Massbegriff in der Theorie der kontinuierlichen Gruppen. Ann Math 34:147–169. Quoted in Sect 9.4, Appendix A.5

Hadamard J (1945) An essay on the psychology of invention in the mathematical field. Princeton University Press, Princeton, p 5.10. Quoted in Sects 4.1, 5.3, 5.10

Hahn H (1912) Über die Integrale des Herrn Hellinger und die Orthogonalinvarianten der quadratischen Formen von unendlich vielen Veränderlichen. Monatshefte Math Phys 23:161–224. Quoted in Sect 6.5

Hahn H (1933) Logik, Mathematik und Naturerkennen. Carl Gerold's Sohn, Wien. Quoted in Sect 4.7

Hahn H (1980) The crisis in intuition. In: McGuinness B (ed) Hans Hahn. Empiricism, logic and mathematics. Philosophical papers. Reidel, Dordrecht, pp 73–102. Quoted in Sect 4.7

Haken H (1983) Synergetics. Springer, Berlin. Quoted in Sect 7.2

Halmos PR (1944) The foundations of probability. Am Math Mon 51:493–510. Quoted in Sect 3.4

Halmos PR (1969) Two subspaces. Trans Am Math Soc 144:381–389. Quoted in Sect 11.1

Halvorson H (2001) Locality, localization, and the particle concept. Topics in the foundation of quantum field theory. Dissertation, University of Pittsburgh. Quoted in Sects 7.3, 8.5

Hammer PC (1969) Filters in general. In: Hammer PC (ed) Advances in mathematical system theory. Pennsylvania State University Press, London, pp 107–120. Quoted in Sect 2.5

Hanner O (1950) Deterministic and non-deterministic stationary random processes. Ark Mat 1:161–177. Quoted in Sect 12.4

Hanson NR (1958) Patterns of discovery. Cambridge University Press, Cambridge. Quoted in Sects 8.4, 12.4

Hardegree GM, Frazer PJ (1981) Charting the labyrinth of quantum logics. In: Beltrametti EG, van Fraassen BC (eds) Current issues in quantum logic. Plenum, New York, pp 53–76. Quoted in Sects 3.4, 4.6

Hardy GH (1921) Srinivasa Ramanujan (1887–1920). Proc R Soc Lond A 99:xi–xxix. Quoted in Sect 5.10

Hardy GH (1940) A mathematician's apology. Cambridge University Press, Cambridge. Quoted in Sect 5.7

Hartley RVL (1928) Transmission of information. Bell Syst Tech J 7:535–563. Quoted in Sect 10.2

Hauser K (2006) Gödel's program revisited, part I: the turn to phenomenology. Bull Symb Log 12:529–590. Quoted in Sect 5.5

Havas P (1965) Relativity and causality. In: Bar-Hillel Y (ed) Logic, methodology and philosophy of science. North-Holland, Amsterdam, pp 347–362. Quoted in Sect 12.2

Havas P (1968) Causality requirements and the theory of relativity. Synthese 18:75–102. Quoted in Sect 12.2

Havin V, Jöricke B (1994) The uncertainty principle in harmonic analysis. Springer, Berlin. Quoted in Sect 3.3

Heaviside O (1893) Electromagnetic theory, vol 1. The Electrician Publishing, London. Quoted in Sect 6.3

Hegel GWF (1969) Wissenschaft der Logik II. Die subjektive Logik. Suhrkamp, Frankfurt (Werkausgabe in 20 Bänden, Band 6). Quoted in Sect 6.1

Hegerfeldt GC (1974) Remark on causality and particle localization. Phys Rev D 10:3320–3321. Quoted in Sect 7.3

van Heijenoort J (1967) From Frege to Gödel. A source book in mathematical logic, 1879–1931. Harvard University Press, Cambridge. Quoted in Sects 2.2, 5.4, 5.6

Heinosaari T, Reitzner D, Stano P (2008) Notes on joint measurability of quantum observables. Found Phys 38:1133–1147. Quoted in Sect 3.5

Heisenberg W (1927) Über den anschaulichen Inhalt der quantentheoretischen Kinematik und Mechanik. Z Phys 43:172–198. Quoted in Sect 1.4

Heisenberg W (1959) Wolfgang Paulis philosophische Auffassungen. Naturwissenschaften 46:661–663. Quoted in Sect 11.1

Heisenberg W, Pauli W (1929) Zur Quantendynamik der Wellenfelder. Z Phys 56:1–61. Quoted in Sect 7.3

Heitler W (1954) The quantum theory of radiation. Oxford University Press, Oxford. Quoted in Sect 6.3
Helgason S (1978) Differential geometry, Lie groups, and symmetric spaces. Academic Press, New York. Quoted in Sect 9.4
Hellige JB (1993) Hemispheric asymmetry: what's right and what's left. Harvard University Press, Cambridge. Quoted in Sect 4.2
Hellinger E (1909) Neue Begründung der Theorie quadratischer Formen von unendlich vielen Veränderlichen. J Reine Angew Math 36:210–271. Quoted in Sect 6.5
Hellwig KE, Kraus K (1969) Pure operations and measurements. Commun Math Phys 11:214–220. Quoted in Sect 12.1
Helstrom CW (1976) Quantum detection and estimation theory. Academic Press, New York. Quoted in Sect 6.2
Hepp K (1972) Quantum theory of measurement and macroscopic observables. Helv Phys Acta 45:237–248. Quoted in Sects 6.5, 12.1
Hermann A, von Meyenn K, Weisskopf VF (eds) (1979) Wolfgang Pauli. Wissenschaftlicher Briefwechsel, Band I: 1919–1929. Springer, Berlin. Quoted in Sects 1.4, 6.2, 6.3, 8.3
Hersh R (1994) Fresh breezes on the philosophy of mathematics. In: Ernest P (ed) Mathematics, education and philosophy: an international perspective. Falmer Press, London, pp 11–20. Quoted in Sect 5.1
Hertz H (1890) Über die Beziehung zwischen Licht und Elektrizität. Verlag von Emil Strauß, Bonn. Quoted in Sect 1.1
Hertz H (1892) Untersuchungen über die Ausbreitung der elektrischen Kraft. Johann Ambrosius Barth, Leipzig. Quoted in Sect 7.2
Hertz H (1894) Die Prinzipien der Mechanik, in neuem Zusammenhange dargestellt. Meiner, Leipzig. Quoted in Sect 4.1
Hertz H (1896) Miscellaneous papers. Macmillan, London, transl by Jones DE and Schott GA. Quoted in Sect 7.2
Hertz H (1999) Die Constitution der Materie. Eine Vorlesung über die Grundlagen der Physik aus dem Jahre 1884. Springer, Berlin. Quoted in Sect 4.1
Hewitt E, Ross KA (1970) Abstract harmonic analysis, volume 2: structure and analysis for compact groups, analysis on locally compact Abelian groups. Springer, Berlin. Quoted in Sect 9.4
Heyting A (1931) Die intuitionistische Grundlegung der Mathematik. Erkenntnis 2:106–115. Quoted in Sect 3.2
Heyting A (1934) Mathematische Grundlagenforschung—Intuitionismus—Beweistheorie. Springer, Berlin. Quoted in Sect 3.2
Heyting A (1971) Intuitionism. An introduction. North-Holland, Amsterdam. Quoted in Sects 5.2, 5.6
Hida T (1970) Stationary stochastic processes. Princeton University Press, Princeton. Quoted in Sect 12.4
Hida T (1980) Brownian motion. Springer, New York. Quoted in Sect 6.5
Hilbert D (1899) Grundlagen der Geometrie. Teubner, Leipzig. Quoted in Sect 5.4
Hilbert D (1900) Mathematische Probleme. In: Nachrichten von der Königlichen Gesellschaft der Wissenschaften zu Göttingen. pp 253–297. Quoted in Sect 5.4
Hilbert D (1930) Probleme der Grundlegung der Mathematik. Math Ann 102:1–9. Quoted in Sect 5.4
Hilbert D (1931) Beweis des tertium non datur. In: Nachrichten von der Gesellschaft der Wissenschaften zu Göttingen, Mathematisch-Physikalische Klasse, pp 120–125. Quoted in Sect 2.2
Hilbert D, Ackermann W (1928) Grundzüge der theoretischen Logik. Springer, Berlin. Quoted in Sect 5.4
Hilbert D, Bernays P (1968) Grundlagen der Mathematik I. Springer, Berlin. Quoted in Sect 5.5
Hilbert D, Bernays P (1970) Grundlagen der Mathematik II. Springer, Berlin. Quoted in Sect 5.5
Hilgevoord J (2005) Time in quantum mechanics: a story of confusion. Stud Hist Philos Mod Phys 36:29–60. Quoted in Sect 1.4

Hjalmars S (1962) Some remarks on time and energy as conjugate variables. Nuovo Cimento 25:355–364. Quoted in Sect 8.3

Hogan JA, Lakey JD (2005) Time-frequency and time-scale methods. Birkhäuser, Boston. Quoted in Sect 9.3

Holevo AS (1972a) On the mathematical theory of quantum communication channels. Probl Inf Transm 8:47–54. Quoted in Sect 3.5

Holevo AS (1972b) An analog of the theory of statistical decisions in noncommutative probability theory. Trans Mosc Math Soc 26:133–149. Quoted in Sect 3.5

Holevo AS (1973a) Information-theoretical aspects of quantum measurements. Originally in Russian, English translation in Probl Inf Transm 9:110–118 (1975). Quoted in Sect 3.5

Holevo AS (1973b) Statistical decision theory for quantum systems. J Multivar Anal 3:337–394. Quoted in Sect 3.5

Holevo AS (1978) Investigations in the general theory of statistical decisions. Proc Steklov Inst Math 3:1–140. Quoted in Sect 6.2

Holevo AS (1982) Probabilistic and statistical aspects of quantum theory. North-Holland, Amsterdam. Quoted in Sect 6.2

Holton G (1970) The roots of complementarity. Daedalus 99:1015–1055. Quoted in Sect 2.4

Hooker CA (1975) The logico-algebraic approach to quantum mechanics, volume I: historical evolution. Reidel, Dordrecht. Quoted in Sect 2.4

Horuzhy SS (1990) Introduction to algebraic quantum field theory. Kluwer Academic, Dordrecht. Quoted in Sect 7.3

van Hove L (1952) Les difficultés de divergences pour un modèle particulier de champ quantifié. Physica 18:145–159. Quoted in Sect 6.5

Houtappel RMF, Van Dam H, Wigner EP (1965) The conceptual basis and use of the geometric invariance principles. Rev Mod Phys 37:595–632. Quoted in Sect 7.1

Hughes RIG (1985) Logics based on partial Boolean algebras. J Symb Log 50:558–566. Quoted in Sect 3.4

Hume D (1739) A treatise of human nature, book 1, London. Quoted in Sect 12.2

Hume D (1748) An enquiry concerning human understanding, London. Quoted in Sect 12.2

Hunt BJ (1991) The Maxwellians. Cornell University Press, Ithaca. Quoted in Sect 6.3

Hurst HE (1951) Long-term storage capacity of reservoirs. Trans Am Soc Civ Eng 116:770–799. Quoted in Sect 10.2

Husserl E (1991) Phenomenology of the consciousness of internal time. Kluwer Academic, Dordrecht, transl by Brough J. Quoted in Sect 13.3

Huxley TH (1879) Hume. Macmillan, London. Quoted in Sect 4.1

van Inwagen P (1990) Material beings. Cornell University Press, Ithaca. Quoted in Sect 3.1

van Inwagen P (2009) Indeterminacy and vagueness: logic and metaphysics. Eur J Philos Relig 1:1–19. Quoted in Sect 3.1

Jackson JD (1998) Classical electrodynamics. Wiley, New York. Quoted in Sect 6.3

Jagannathan R, Simon R, Sudarshan ECG, Vasudevan R (1987) Dynamical maps and nonnegative phase-space distribution functions in quantum mechanics. Phys Lett A 120:161–164. Quoted in Sect 10.4

Jain JK (2000) The composite fermion: a quantum particle and its quantum fluids. Phys Today 53(4):39–45. Quoted in Sect 6.3

James W (1890) Principles of psychology, volume 1. Dover, New York. Quoted in Sects 4.1, 4.6, 13.3

Jammer M (1961) Concepts of mass in classical and modern physics. Harvard University Press, Cambridge. Quoted in Sect 8.2

Jammer M (1974) The philosophy of quantum mechanics. Wiley, New York. Quoted in Sect 3.2

Jammer M (2004) The strange story of the concept which inaugurated modern theoretical physics. Found Phys 34:1617–1641. Quoted in Sect 8.3

Jammer M (2006) Concepts of simultaneity. From antiquity to Einstein and beyond. Harvard University Press, Cambridge. Quoted in Sect 8.3

Jauch JM (1960) Systems of observables in quantum mechanics. Helv Phys Acta 33:711–726. Quoted in Sect 6.2

Jauch JM (1968) Foundations of quantum mechanics. Addison–Wesley, Reading. Quoted in Sect 6.3

Jauch JM, Misra B (1961) Supersymmetries and essential observables. Helv Phys Acta 34:699–709. Quoted in Sect 6.2

Jauch JM, Piron C (1967) Generalized localizability. Helv Phys Acta 40:559–570. Quoted in Sect 6.2

Jeffreys H, Jeffreys BS (1956) Methods of mathematical physics. Cambridge University Press, Cambridge. Quoted in Sect 8.5

Johnson NL, Kotz S (1970) Continuous univariate distributions 1. Houghton Mifflin, New York. Quoted in Sect 8.5

Jordan P (1927) Kausalität und Statistik in der modernen Physik. Naturwissenschaften 15:105–110. Quoted in Sect 7.3

Jordan P (1933) Über die Multiplikation quantenmechanischer Größen. Z Phys 80:285–291. Quoted in Sect 9.1

Jordan P (1934) Über die Multiplikation quantenmechanischer Größen II. Z Phys 87:505–512. Quoted in Sect 9.1

Jordan P (1947) Verdrängung und Komplementarität. Stromverlag, Hamburg. Quoted in Sect 5.10

Jordan P, von Neumann J, Wigner E (1934) On an algebraic generalization of the quantum mechanical formalism. Ann Math 35:29–64. Quoted in Sect 6.5

Jordan P, Wigner E (1928) Über das Paulische Äquivalenzverbot. Z Phys 47:631–651. Quoted in Sect 7.3

Jørgensen PT, Muhly PS (1980) Selfadjoint extensions satisfying the Weyl operator commutation relations. J Anal Math 37:46–99. Quoted in Sect 10.5

Jung CG (1971) The collected works of C.G. Jung, volume 6: psychological types. Princeton University Press, Princeton. Quoted in Sect 6.4

Jung CG (1976) C.G. Jung letters, volume 2. Princeton University Press, Princeton, Adler G (ed), transl by Hulen J. Quoted in Sect 4.2

Kadanoff LP, Baym G (1962) Quantum statistical mechanics. Benjamin, New York. Quoted in Sect 6.3

Kadison RV (1951) Isometries of operator algebras. Ann Math 54:325–338. Quoted in Sect 9.2

Kadison RV, Ringrose JR (1983) Fundamentals of the theory of operator algebras. I. Elementary theory. Academic Press, New York. Quoted in Sect 7.3

Kadison RV, Ringrose JR (1986) Fundamentals of the theory of operator algebras. II. Advanced theory. Academic Press, Orlando. Quoted in Sects 7.3, 8.5, 9.7, Appendices A.2, A.5

Kadison RV, Ringrose JR (1992) Fundamentals of the theory of operator algebras. IV. Special topics. Birkhäuser, Boston. Quoted in Appendix A.2

Källen G (1962) Renormalization theory. In: Lectures in theoretical physics, vol 1. Benjamin, New York, pp 169–256. Quoted in Sect 6.3

Kafka F (1972) The metamorphosis. Bantam, New York. transl by Corngold S, originally published 1915 in German. Quoted in Sect 4.4

Kailath T (1974) A view of three decades of linear filtering theory. IEEE Trans Inf Theory 20:146–181. Quoted in Sect 12.4

Kakutani S (1948) On equivalence of infinite product measures. Ann Math 49:214–224. Quoted in Sect 6.5

Kakutani S (1950) Review of "Extrapolation, interpolation and smoothing of stationary time series" by Norbert Wiener. Bull Am Math Soc 56:378–381. Quoted in Sect 12.4

Kalckar J (ed) (1985) Niels Bohr. Collected works, volume 6: foundations of quantum physics I (1926–1932). North-Holland, Amsterdam. Quoted in Sect 4.6

Kalckar J (ed) (1996) Niels Bohr. Collected works, volume 7: foundations of quantum physics II (1933–1958). Elsevier, Amsterdam. Quoted in Sects 4.6, 8.4

Kalisa C, Torrésani B (1991) N-dimensional affine Weyl–Heisenberg wavelets. Ann Instit Henri Poincaré A 59:201–236. Quoted in Sects 9.5, 10.5

Kallianpur G (1961) Some ramifications of Wiener's ideas on nonlinear prediction. In: Masani P (ed) Norbert Wiener. Collected works with commentaries, volume III. MIT Press, Cambridge, pp 402–424. Quoted in Sect 12.4
Kallianpur G, Mandrekar V (1965) Multiplicity and representation theory of purely non-deterministic stochastic processes. Theory Probab Appl 10:553–581. Quoted in Sect 12.4
Kamber F (1964) Die Struktur des Aussagenkalküls in einer physikalischen Theorie. Nachr Akad Wiss Gött Math-Phys Kl 10:103–124. Quoted in Sect 2.4
Kanigel R (1991) The man who knew infinity: a life of the genius Ramanujan. Charles Scribner's Sons, New York. Quoted in Sect 5.10
Kant I (1783) Prolegomena zu einer jeden künftigen Metaphysik, die als Wissenschaft wird auftreten können. Johann Friedrich Hartknoch, Riga. Quoted in Sect 12.2
Kanwal RP (1983) Generalized functions. Aademic Press, New York. Quoted in Sect 9.7
Kappos DA (1969) Probability algebras and stochastic spaces. Academic Press, New York. Quoted in Sect 3.4
Kato T (1966) Perturbation theory for linear operators. Springer, Berlin. Quoted in Sect 11.5
Kauffman LH, Lomonaco SJ (2002) Quantum entanglement and topological entanglement. New J Phys 4:1–73. Quoted in Sect 8.5
Kemeny JG, Oppenheim P (1956) On reduction. Philos Stud 7:6–19. Quoted in Sect 7.1
Keshner MS (1982) 1/f noise. Proc IEEE 70:212–218. Quoted in Sect 10.2
Kim J (1993) Supervenience and mind: selected philosophical essays. Cambridge University Press, Cambridge. Quoted in Sect 12.2
Kirillov AA (1976) Elements of the theory of representations. Springer, Berlin. Quoted in Sect 9.4
Klaua D (1965) Über einen Ansatz zur mehrwertigen Mengenlehre. Mon Dtsch Akad Wiss Berl 7:859–867. Quoted in Sect 3.2
Klaua D (1967) Ein Ansatz zur mehrwertigen Mengenlehre. Math Nachr 33:273–296. Quoted in Sect 3.2
Klein F (1926) Vorlesungen über die Entwicklung der Mathematik im 19. Jahrhundert. Springer, Berlin. Quoted in Sect 1.2
Kline M (1980) Mathematics. The loss of certainty. Oxford University Press, New York. Quoted in Sect 5.10
Koch AL (1966) The logarithm in biology I. Mechanisms generating the log-normal distribution exactly. J Theor Biol 12:276–290. Quoted in Sect 11.7
Koch AL (1969) The logarithm in biology II. Distributions simulating the log-normal. J Theor Biol 23:251–268. Quoted in Sect 11.7
Kochen S, Specker EP (1965a) Logical structures arising in quantum theory. In: Addison J, Henkin L, Tarski A (eds) The theory of models. North-Holland, Amsterdam, pp 177–189. Quoted in Sects 2.4, 3.4
Kochen S, Specker EP (1965b) The calculus of partial propositional functions. In: Bar-Hillel Y (ed) Logic, methodology and philosophy of science. North-Holland, Amsterdam, pp 45–57. Quoted in Sects 2.4, 3.4
Kolmogorov A (1932) Zur Deutung der intuitionistischen Logik. Math Z 35:58–65. Quoted in Sect 3.2
Kolmogorov A (1939) Sur l'interpolation et extrapolation des suites stationaires. C R Hebd Séances Acad Sci Paris 208:2043–2045. Quoted in Sect 12.4
Kolmogorov AN (1940a) Curves in a Hilbert space, invariant under a one-parameter group of motions. Dokl Akad Nauk SSSR 26:6–9. Quoted in Sect 10.2
Kolmogorov AN (1940b) Wiener spirals and some other interesting curves in a Hilbert space. Dokl Akad Nauk SSSR 26:115–118. Quoted in Sect 10.2
Kolmogorov AN (1941a) Interpolation und Extrapolation von stationären zufälligen Folgen. Isv Akad Nauk SSSR, Ser Mat 5:3–14. Quoted in Sect 12.4
Kolmogorov AN (1941b) Stationary sequences in Hilbert space. Bull Mosk Gos Univ Mat 2(6):1–40. Quoted in Sect 12.4
Kolmogorov AN (1941c) The local structure of turbulence in incompressible viscous fluid for very large Reynolds numbers. Proc Acad Sci USSR 30:301–305. Quoted in Sect 10.2

Kolmogorov A (1948) Algèbres de Boole métriques complètes. VI. Zjazd Matematyków Polskich. Ann Soc Pol Math 20:21–30. Quoted in Sect 3.4

Koopman BO (1931) Hamiltonian systems and transformations in Hilbert space. Proc Natl Acad Sci USA 17:315–318. Quoted in Appendix A.6

Kraft C (1955) Some conditions for consistency and uniform consistency of statistical procedures. In: Neyman J, LeCam LM, Scheffé H (eds) University of California publications in statistics, vol 2, no 6. University of California Press, Berkeley, pp 125–141. Quoted in Sect 6.5

Kraus K (1983) States, effects, and operations. Springer, Berlin. Quoted in Sect 12.1

Kraus K (1977) Position observables of the photon. In: Price WC, Chissick SS (eds) The uncertainty principle and foundations of quantum mechanics. Wiley, London, pp 293–320. Quoted in Sect 6.2

Kraut R (2009) Critical monism and pluralism. In: Davies S et al (eds) A companion to aesthetics. Blackwell, Oxford, pp 211–214. Quoted in Sect 4.4

Kreĭn MG (1945a) On a generalization of some investigations of G. Szegö, W.M. Smirnov, and A.N. Kolmogorov. Dokl Akad Nauk SSSR 46:91–94. Quoted in Sect 12.4

Kreĭn MG (1945b) On a problem of extrapolation of A.N. Kolmogorov. Dokl Akad Nauk SSSR 46:306–309. Quoted in Sect 12.4

Krengel U (1971) K-flows are forward deterministic, backward completely nondeterministic stationary point processes. J Math Anal Appl 35:611–620. Quoted in Sect 12.4

Krengel U (1973) Recent results on generators in ergodic theory. In: Transactions of the sixth Prague conference on information theory, statistical decision functions, random processes. Academia, Prague, pp 465–482. Quoted in Sect 12.4

Kripal J (2011) Coincidentia Oppositorum (an interview with Jeffrey Kripal by Harald Atmanspacher). Mind and Matter 9:205–210. Quoted in Sect 13.2

Kruszynski P, Napiórkowski K (1973) On the independence of local algebras II. Rep Math Phys 4:303–306. Quoted in Sect 6.3

Kubo R (1957) Statistical-mechanical theory of irreversible processes I. General theory and simple applications to magnetic and conduction problems. J Phys Soc Jpn 12:570–586. Quoted in Sect 6.5

Küpfmüller K (1924) Über Einschwingvorgänge in Wellenfiltern. Elektr Nachr-Tech 1:141–152. Quoted in Sect 3.3

Kuhn TS (1962) The structure of scientific revolutions. University of Chicago Press, Chicago. Quoted in Sect 4.1

Kujala JV, Dzhafarov EN (2016) Probabilistic contextuality in EPR/Bohm-type systems with signaling allowed. In: Dzhafarov EN, Jordan JS, Zhang R, Cervantes V (eds) Contextuality from quantum physics to psychology. World Scientific, Singapore, pp 287–308. Quoted in Sect 13.5

Kuroda S (1951) Intuitionistische Untersuchungen der formalistischen Logik. Nagoya Math J 2:35–47. Quoted in Sect 5.10

Lahti P (2003) Coexistence and joint measurability in quantum mechanics. Int J Theor Phys 42:893–906. Quoted in Sect 3.5

Lakatos I (1976) Proofs and refutations. The logic of mathematical discovery. Cambridge University Press, Cambridge. Quoted in Sect 5.4

Lakoff G (1973) Hedges: a study in meaning criteria and the logic of fuzzy concepts. J Philos Log 2:458–508. Quoted in Sect 3.2

Lamperti J (1962) Semi-stable stochastic processes. Trans Am Math Soc 104:62–78. Quoted in Sect 10.2

Lanczos C (1949) The variational principles of mechanics. University of Toronto Press, Toronto. Quoted in Sect 8.3

Landau HJ, Pollak HO (1961) Prolate spherical wave functions, Fourier analysis and uncertainty II. Bell Syst Tech J 40:65–84. Quoted in Sect 10.2

Landau HJ, Pollak HO (1962) Prolate spherical wave functions, Fourier analysis and uncertainty III: the dimension of the space of essentially time- and bandlimited signals. Bell Syst Tech J 41:1295–1336. Quoted in Sect 10.2

Landau LD (1927) Das Dämpfungsproblem in der Quantenmechanik. Z Phys 45:430–441. Quoted in Sect 6.2
Landau LD, Lifschitz EM (1980) Statistical physics. Pergamon, Oxford. Quoted in Sect 6.2
Landau LJ (1987a) On the non-classical structure of the vacuum. Phys Lett A 123:115–118. Quoted in Sect 7.3
Landau LJ (1987b) On the violation of Bell's inequality in quantum theory. Phys Lett A 120:54–56. Quoted in Sect 7.3
Landes MW (1924) A suggested interpretation of Bergson's doctrine of intuition. Philos Rev 33:450–462. Quoted in Sect 4.3
Landsman NP (1995) Observation and superselection in quantum mechanics. Stud Hist Philos Mod Phys 26:45–73. Quoted in Sect 6.5
Landsman NP (1996) Essay review: local quantum physics. Stud Hist Philos Mod Phys 27:511–524. Erratum: Stud Hist Philos Mod Phys 28:i–ii (1997). Quoted in Sect 7.3
Langer S (1978) Philosophy in a new key. A study in the symbolism of reason, rite, and art. Harvard University Press, Cambridge. Quoted in Sect 2.1
Langer LL (1991) Holocaust testimonies. The ruins of memory. Yale University Press, New Haven. Quoted in Sect 13.4
Langer LL (1995) Admitting the Holocaust. Oxford University Press, New York. Quoted in Sect 13.4
Langer LL (1998) Preempting the Holocaust. Yale University Press, New Haven. Quoted in Sect 13.4
Lanzmann C (1994) Holocauste, la représentation impossible. Le Monde (Supplément Arts-Spectacles), 3 March. Quoted in Sect 13.4
de Laplace PS (1814) Essai philosophique sur les probabilités, Paris. Quoted in Sect 12.3
von Laue M (1955) Ist die klassische Physik wirklich deterministisch? Phys Bl 11:269–270. Quoted in Sect 12.3
Lax PD, Phillips RS (1967) Scattering theory. Academic Press, New York. Quoted in Sect 12.1
Lederman LM (1993) The god particle: if the universe is the answer, what is the question? Houghton Mifflin, Boston. Quoted in Sect 6.2
Lee TD (1970) Symmetry principles in physics. Report BNL 50261 (T-591); Physics-TID-4500. Brookhaven National Laboratory, New York. Quoted in Sect 6.4
Lee TD (1981) Particle physics and introduction to field theory. Science Press, Beijing. Quoted in Sect 6.4
Lee TD, Yang CN (1959) Many-body problem in quantum statistical mechanics. I. General formulation. Phys Rev 113:1165–1177. Quoted in Sect 6.3
Leggett AJ, Garg A (1985) Quantum mechanics versus macroscopic realism: is the flux there when nobody looks? Phys Rev Lett 54:857–860. Quoted in Sect 13.5
Leibniz GW (1989) The metaphysical foundations of mathematics. In: Loemker LL (ed) Philosophical papers and letters. Kluwer Academic, Dordrecht, pp 666–674. Quoted in Sect 12.2
Lenard A (1972) The numerical range of a pair of projections. J Funct Anal 10:410–423. Quoted in Sect 11.1
Lenker TD (1979) Carathéodory's concept of temperature. Synthese 42:167–171. Quoted in Sect 7.4
Levy-Agresti J, Sperry RW (1968) Differential perceptual capacities in major and minor hemispheres. Proc Natl Acad Sci USA 61:1151. Quoted in Sect 4.5
Lévi-Strauss C (1969) The raw and the cooked. University of Chicago Press, Chicago. Quoted in Sect 4.4
Lewis CS (1944) The abolition of man. Harper, San Francisco. Quoted in Sect 4.1
Liang Y-C, Spekkens RW, Wiseman HM (2011) Specker's parable of the overprotective seer: a road to contextuality, nonlocality and complementarity. Phys Rep 506:1–39. Quoted in Sect 2.4
Liebermann M (2005) Max Liebermann. Zeichnen heisst Weglassen—Arbeiten auf Papier. Hatje Cantz, Ostfildern. Quoted in Sect 3.1
Ligeti G (2011) Of foreign lands and strange sounds. Boydell Press, Woodbridge. Quoted in Sect 10.2

Littlewood JE (1929) Collected papers of Srinivasa Ramanujan. Math Gaz 14(200):425–428. Quoted in Sect 5.10

Lloyd DR (2010) Symmetry and beauty in Plato. Symmetry 2:455–465. Quoted in Sect 6.4

Lock PF, Hardegree GM (1984) Connections among quantum logics. Part 1: quantum propositional logics. Int J Theor Phys 24:43–53. Quoted in Sect 3.4

Lockhart CM, Misra B (1986) Irreversibility and measurement in quantum mechanics. Physica A 136:47–76. Quoted in Sects 6.5, 12.1

Loemker LL (ed) (1969) Gottfried Wilhelm Leibniz: philosophical papers and letters. Reidel, Dordrecht. Quoted in Sect 8.3

London F, Bauer E (1939) La théorie de l'observation en mécanique quantique. Hermann, Paris. Quoted in Sect 6.2

Lonergan B (1992) Insight. A study of human understanding. University of Toronto Press, Toronto. Quoted in Sect 4.3

Loomis LH (1952) Note on a theorem by Mackey. Duke Math J 19:641–645. Quoted in Sects 6.5, 9.4

Loomis LH (1953) An introduction to abstract harmonic analysis. Van Nostrand, Princeton. Quoted in Sect 3.4

Lorentz HA (1909) The theory of electrons and its applications to the phenomena of light and radiant heat. Teubner, Leipzig. Quoted in Sect 6.3

Loś J (1955) On the axiomatic treatment of probability. Colloq Math (Wroc) 3:125–137. Quoted in Sect 3.4

Lowe V (1949) Influence of Bergson, James and Alexander on Whitehead. J Hist Ideas 10:267–296. Quoted in Sect 13.3

Ludwig G (1964) Versuch einer axiomatischen Grundlegung der Quantenmechanik und allgemeinerer physikalischer Theorien. Z Phys 181:233–260. Quoted in Sects 3.5, 6.2

Ludwig G (1970) Deutung des Begriffs "physikalische Theorie" und axiomatische Grundlegung der Hilbertraumstruktur der Quantenmechanik durch Hauptsätze des Messens. Springer, Berlin. Quoted in Sect 6.2

Ludwig G (1976) Einführung in die Grundlagen der Theoretischen Physik Band 3: Quantentheorie. Vieweg, Braunschweig. Quoted in Sect 3.5

Ludwig G (1979) Einführung in die Grundlagen der Theoretischen Physik Band 4: Makrosysteme, Physik und Mensch. Vieweg, Braunschweig. Quoted in Sect 3.5

Ludwig G (1983) Foundations of quantum mechanics I. Springer, New York. Quoted in Sect 6.2

Ludwig G (1985a) An axiomatic basis for quantum mechanics, volume 1: derivation of Hilbert space structure. Springer, Berlin. Quoted in Sect 6.2

Ludwig G (1985b) Foundations of quantum mechanics II. Springer, New York. Quoted in Sect 6.2

Ludwig G (1987) An axiomatic basis for quantum mechanics, volume 2: quantum mechanics and macrosystems. Springer, Berlin. Quoted in Sect 6.2

Lüders G (1954) On the equivalence of invariance under time reversal and under particle-antiparticle conjugation for relativistic field theories. K Dan Vidensk Selsk 28(5):1–17. Quoted in Sect 7.3

Lüders G, Zumino B (1958) Connection between spin and statistics. Phys Rev 110:1450–1453. Quoted in Sect 7.3

Lukasiewicz J (1920) On three-valued logic. Ruch Filoz 5:169–172. Quoted in Sect 3.2

Lukasiewicz J, Tarski A (1930) Untersuchungen über den Aussagenkalkül. C R Séances Soc Sci Lett Vars, cl III 23:30–50. Quoted in Sect 3.2

Maak W (1950) Fastperiodische Funktionen. Springer, Berlin. Quoted in Sect 12.4

Mach E (1866) Bemerkungen über die Entwicklung der Raumvorstellungen. Fichtes Z Philos Philos Krit 49:227–232. Quoted in Sect 8.3

Mach E (1902) The science of mechanics. A critical and historical account of its development, 2nd edn. Open Court, Chicago. Quoted in Sect 12.2

Mach E (1919) The science of mechanics. A critical and historical account of its development, 4th edn. Open Court, Chicago. Quoted in Sect 8.3

Mackey GW (1949) A theorem of Stone and von Neumann. Duke Math J 16:313–326. Quoted in Sects 6.5, 9.4
Mackey GW (1974) Ergodic theory and its significance for statistical mechanics and probability theory. Adv Math 12:178–268. Quoted in Sect 12.4
Madden C (2007) Fractals in music. Introductory mathematics for musical analysis. High Art Press, Salt Lake City. Quoted in Sect 10.2
Maddy P (1990) Realism in mathematics. Clarendon, Oxford. Quoted in Sect 5.2
Maeda S (1990) Probability measures on projections in von Neumann algebras. Rev Math Phys 1:235–290. Quoted in Sect 6.5
Maejima M, Sato K (1999) Semi-selfsimilar processes. J Theor Probab 12:347–383. Quoted in Sect 10.2
Mahler G (2016) Temporal non-locality: fact or fiction? In: Atmanspacher H et al (eds) Quantum interaction QI 16. Springer, Berlin, pp 243–254
Mancosu P (1998) From Brouwer to Hilbert. Oxford University Press, New York. Quoted in Sect 5.6
Mandelbrot B (1963) The variation of certain speculative prices. J Bus 36:394–419. Quoted in Sect 11.6
Mandelbrot B (1967) Some noises with 1/f spectrum, a bridge between direct current and white noise. IEEE Trans Inf Theory 13:289–298. Quoted in Sect 10.2
Mandelbrot BB (1982) The fractal geometry of nature. Freeman, San Francisco. Quoted in Sect 10.2
Mandelbrot BB (1999) Multifractals and 1/f noise. Springer, New York. Quoted in Sect 10.2
Mandelbrot BB, van Ness JW (1968) Fractional Brownian motions, fractional noises and applications. SIAM Rev 10:422–437. Quoted in Sect 10.2
Manin YI (2007) Mathematics as metaphor. Selected essays. American Mathematical Society, Providence. Quoted in Sect 5.5
Manuceau J (1968) C*-algèbre de relations de commutation. Ann Instit Henri Poincaré A 8:139–161. Quoted in Sect 9.4
Manuceau J, Sirugue M, Testard D, Verbeure A (1973) The smallest C*- algebra for canonical commutation relations. Commun Math Phys 32:231–243. Quoted in Sect 9.4
Margenau H (1936) Quantum-mechanical description. Phys Rev 49:240–242. Quoted in Sect 6.2
Margenau H (1950) The nature of physical reality. McGraw–Hill, New York. Quoted in Sect 6.2
Margenau H (1958) Philosophical problems concerning the meaning of measurement in physics. Philos Sci 25:23–33. Quoted in Sect 6.2
Margenau H (1963a) Measurements and quantum states, part I. Philos Sci 30:1–16. Quoted in Sect 6.2
Margenau H (1963b) Measurements in quantum mechanics. Ann Phys 23:469–485. Quoted in Sect 6.2
Margolus N (1990) Parallel quantum computation. In: Zurek WH (ed) Complexity, entropy and the physics of information. Addison–Wesley, Reading, pp 273–286. Quoted in Sect 1.4
Martin PC, Schwinger J (1959) Theory of many-particle systems I. Phys Rev 115:1342–1373. Quoted in Sects 6.3, 6.5
Masani P (1963) Review of "stationary processes and prediction theory". Bull Am Math Soc 69:195–207. Quoted in Sect 12.4
Masani P (1979) Commentary on the memoire on generalized harmonic analysis. In: Masani P (ed) Norbert Wiener. Collected works with commentaries, volume II. MIT Press, Cambridge, pp 333–379. Quoted in Sect 12.4
Masani P (ed) (1981) Norbert Wiener. Collected works with commentaries, volume III. MIT Press, Cambridge. Quoted in Sect 12.4
Masani RR (1990) Norbert Wiener, 1894–1964. Birkhäuser, Basel. Quoted in Sect 12.4
Masani P, Wiener N (1959a) Nonlinear prediction. In: Grenander U (ed) Probability and statistics. Almqvist & Wiksell, Stockholm, pp 190–212. Quoted in Sect 12.4
Masani P, Wiener N (1959b) On bivariate stationary processes and the factorization of matrix-value functions. Theory Probab Appl 4:322–331. Quoted in Sect 12.4

Maslow AH (1966) The psychology of science: a reconnaissance. Maurice Bassett Publishing, Chapel Hill. Quoted in Sect 5.1
Matusita K (1951) On the theory of statistical decision functions. Ann Inst Stat Math 3:17–35. Quoted in Sect 6.5
Maxwell JC (1873) A treatise on electricity and magnetism. Clarendon, Oxford. Quoted in Sect 6.3
Mazzola G (1985) Gruppen und Kategorien in der Musik. Heldermann, Berlin. Quoted in Sect 9.3
Mazzola G (2002) The topos of music. Birkhäuser, Basel. Quoted in Sect 10.2
McAllister D (1879) The law of the geometric mean. Proc R Soc Lond 29:367–376. Quoted in Sect 11.7
McCulloch WS, Pitts W (1943) A logical calculus of the ideas immanent in nervous activity. Bull Math Biophys 5:115–133. Quoted in Sect 2.3
McKenna J, Blatt JM (1962) The expectation value of a many-body Hamiltonian in the quasi-chemical equilibrium theory. Prog Theor Phys (Kyoto) 27:511–528. Quoted in Sect 6.3
Meixner J (1968) Entropie im Nichtgleichgewicht. Rheol Acta 7:8–13. Quoted in Sect 12.1
Menger K (1942) Statistical metric. Proc Natl Acad Sci USA 28:535–537. Quoted in Sect 3.2
Menger K (1951) Ensembles flous et fonctions aléatoires. C R Hebd Séances Acad Sci Paris 232:2001–2003. Quoted in Sect 3.2
Menzies P, Price H (1993) Causation as a secondary quality. Br J Philos Sci 44:187–203. Quoted in Sect 12.2
von Meyenn K (ed) (1985) Wolfgang Pauli. Wissenschaftlicher Briefwechsel, Band II: 1930–1939. Springer, Berlin. Quoted in Sect 7.2
von Meyenn K (ed) (2001) Wolfgang Pauli. Wissenschaftlicher Briefwechsel, Band IV/Teil III: 1955–1956. Springer, Berlin. Quoted in Sect 6.5
von Meyenn K (2009) Wolfgang Pauli's philosophical ideas viewed from the perspective of his correspondence. In: Atmanspacher H, Primas H (eds) Recasting reality. Springer, Berlin, pp 11–32. Quoted in Sect 6.4
Middleton D (1960) Statistical communication theory. McGraw–Hill, New York. Quoted in Sect 6.5
Mie G (1912) Grundlagen einer Theorie der Materie. Erste Mitteilung. Ann Phys 37:511–534. Quoted in Sect 6.3
Mie G (1941) Lehrbuch der Elektrizität und des Magnetismus. Enke, Stuttgart. Quoted in Sect 6.3
Mill JS (1882) A system of logic, ratiocinative and inductive. Harper, New York. Quoted in Sect 12.2
Minkowski H (1908) Die Grundgleichungen für die electromagnetischen Vorgänge in bewegten Körpern. In: Nachrichten von der Königlichen Gesellschaft der Wissenschaften und der Georg-August-Universität zu Göttingen, Mathematisch-Physikalische Klasse, pp 53–111. Quoted in Sect 6.3
Misra B, Sudarshan ECG (1977) The Zeno's paradox in quantum theory. J Math Phys 18:756–763. Quoted in Sect 13.5
Mitcham C (1978) Types of technology. Rese Philos Technol 1:229–294. Quoted in Sect 4.1
Mitzenmacher M (2003) A brief history of generative models for power law and lognormal distributions. Internet Math 1:226–251. Quoted in Sect 11.7
Montero B (2003) Varieties of causal closure. In: Walter S, Heckmann H-D (eds) Physicalism and mental causation. Imprint Academic, Exeter, pp 173–187. Quoted in Sect 7.1
Moyal JE (1949) Quantum mechanics as a statistical theory. Proc Camb Philos Soc 45:99–124. Quoted in Sect 10.4, Appendix A.6
von Müller A (2010) Thought and reality. In: Glatzeder BM, Goel V, von A (eds) Toward a theory of thinking. Springer, Berlin, pp 59–70. Quoted in Sect 13.1
von Müller A, Filk T (eds) (2015) Re-thinking time at the interface of physics and philosophy. Springer, Berlin. Quoted in the Preface
Müller-Herold U (1980) Disjointness of β-KMS states with different chemical potential. Lett Math Phys 4:45–88. Quoted in Sect 6.5
Mund J, Schroer B, Yngvason J (2006) String-localized quantum fields and modular localization. Commun Math Phys 268:621–672. Quoted in Sect 7.3

Münster A (1959) Prinzipien der statistischen Mechanik. In: Flügge S (ed) Handbuch der Physik Band III/2: Prinzipien der Thermodynamik und Statistik. Springer, Berlin, pp 176–412. Quoted in Sect 6.2
Murray FJ, von Neumann J (1936) On rings of operators. Ann Math 37:116–229. Quoted in Sects 6.3, 6.5, 8.5, Appendix A.1
Murray FJ, von Neumann J (1937) On rings of operators II. Trans Am Math Soc 41:208–248. Quoted in Sect 8.5
Murray FJ, von Neumann J (1940) On rings of operators III. Ann Math 41:94–161. Quoted in Sect 8.5
Murray FJ, von Neumann J (1943) On rings of operators IV. Ann Math 44:716–808. Quoted in Sect 8.5
Musil R (1995) The man without qualities. Vintage, New York, transl by Wilkins S. Quoted in Sect 4.1
Nagel E (1961) The structure of science. Harcourt, Brace & World, New York. Quoted in Sect 7.1
Nagel E, Newman J (2001) Gödel's proof. New York University Press, New York. Quoted in Sect 5.5
Nagel T (2012) Mind and cosmos: why the materialist neo-Darwinian conception of nature is almost certainly false. Oxford University Press, Oxford. Quoted in Sect 8.5
Nahin PJ (1988) Oliver Heaviside. Johns Hopkins University Press, Baltimore. Quoted in Sect 6.3
Napiórkowski K (1972) On the independence of local algebras. Rep Math Phys 3:33–35. Quoted in Sect 6.3
Narnhofer H, Thirring W (1990) Algebraic K-systems. Lett Math Phys 20:231–250. Erratum: Lett Math Phys 22:81 (1991). Quoted in Sect 12.1
Nasr SH (1987) Islamic art and spirituality. SUNY Press, Albany. Quoted in Sect 4.4
Negoitã CV, Ralescu DA (1975) Applications of fuzzy sets to system analysis. Birkhäuser, Basel. Quoted in Sect 3.3
Neumann H (1972) Transformation properties of observables. Helv Phys Acta 45:811–819. Quoted in Sect 6.2
von Neumann J (1927) Wahrscheinlichkeitstheoretischer Aufbau der Quantenmechanik. In: Nachrichten von der Gesellschaft der Wissenschaften Göttingen, Mathematisch-Physikalische Klasse, pp 245–272. Quoted in Sect 6.2
von Neumann J (1931) Die Eindeutigkeit der Schrödingerschen Operatoren. Math Ann 104:570–578. Quoted in Sects 3.5, 9.7
von Neumann J (1932a) Mathematische Grundlagen der Quantenmechanik. Springer, Berlin, Quoted in Sects 6.2, 6.3, 6.5, 8.4
von Neumann J (1932b) Zur Operatorenmethode in der klassischen Mechanik. Ann Math 33:587–642. Quoted in Sects 8.5, 12.4
von Neumann J (1932c) Proof of the quasi-ergodic hypothesis. Proc Natl Acad Sci USA 18:70–82. Quoted in Sect 12.4
von Neumann J (1935) Representations and ray-representations in quantum mechanics. Bull Am Math Soc 41:305. Quoted in Appendix A.6
von Neumann J (1936) On an algebraic generalization of the quantum mechanical formalism. Mat Sb 1:415–484. Quoted in Sect 6.5
von Neumann J (1938) On infinite direct products. Compos Math 6:1–77. Quoted in Sect 6.5
von Neumann J (1949) On rings of operators. Reduction theory. Ann Math 50:401–485. Quoted in Sect 6.5, Appendix A.1
von Neumann J (1951) The general and logical theory of automata. In: Jeffress LA (ed) Cerebral mechanisms in behavior. Wiley, New York, pp 1–41. Quoted in Sect 2.3
von Neumann J (1956) Probabilistic logics and the synthesis of reliable organisms from unreliable components. In: Shannon CE, McCarthy J (eds) Automata studies. Princeton University Press, Princeton, pp 43–98. Quoted in Sect 4.3
von Neumann J (1958) The computer and the brain. Yale University Press, New Haven. Quoted in Sect 1.4

von Neumann J (2005) John von Neumann. Selected letters. American Mathematical Society, Providence. Quoted in Sect 6.5
Neumark MA (1940) Self-adjoint extensions of the second kind of a symmetric operator. Izv Akad Nauk SSSR, Ser Mat 4:35–104. Quoted in Sect 3.5
Neumark MA (1943a) On a representation of additive operator set functions. C R Acad Sci URSS 41:359–361. Quoted in Sect 3.5
Neumark MA (1943b) On spectral functions of a symmetric operator. Izv Akad Nauk SSSR, Ser Mat 7:285–296. Quoted in Sect 3.5
Neurath O (1932) Soziologie im Physikalismus. Erkenntnis 2:393–431. Quoted in Sect 7.1
Neurath O (1983) Philosophical papers 1913–1946. Reidel, Dordrecht. Quoted in Sect 7.1
Newton I (1729) The mathematical principles of natural philosophy. University of California Press, Berkeley, Cajori F (ed), transl by Motte A. Quoted in Sects 7.1, 8.3
Newton I (1730) Opticks. London. Reprinted by Dover, New York (1952). Quoted in Sect 6.1
Nielsen MA, Chuang IL (2000) Quantum computation and quantum information. Cambridge University Press, Cambridge. Quoted in Sect 6.3
Nielsen JA, Zielinski BA, Ferguson MA, Lainhart JE, Anderson JS (2013) An evaluation of the left-brain vs. right-brain hypothesis with resting state functional connectivity magnetic resonance imaging. PLoS ONE 8(8):e71275. Quoted in Sect 4.2
Noether E (1918) Invariante Variationsprobleme. In: Nachrichten von der Gesellschaft der Wissenschaften zu Göttingen, Mathematisch-Physikalische Klasse, pp 235–257. Quoted in Sects 6.3, 8.2
Nonaka I, Takeuchi H (1995) The knowledge-creating company. Oxford University Press, New York. Quoted in Sect 4.1
Oden GC (1977a) Integration of fuzzy logical information. J Exp Psychol 3:565–575. Quoted in Sect 3.3
Oden GC (1977b) Fuzziness in semantic memory: choosing exemplars of subjective categories. Mem Cogn 5:198–204. Quoted in Sect 3.3
Olbers W (1900) Briefwechsel zischen Olbers und Gauss. In: Schilling C (ed) Wilhelm Olbers. Sein Leben und seine Werke, zweiter Band. Springer, Berlin. Quoted in Sect 5.10
Olver FWJ, Lozier DW, Boisvert RF, Clark CW (2010) NIST handbook of mathematical functions. Cambridge University Press, New York. Quoted in Sect 11.7
Ornstein R (1997) The right mind. Making sense of the hemispheres. Harcourt Brace, New York. Quoted in Sect 4.2
Ozaktas HM, Zalevsky Z, Kutay MA (2001) The fractional Fourier transform. Wiley, Chichester. Quoted in Sect 9.4
Ozawa M (1984) Quantum measuring processes of continuous observables. J Math Phys 25:79–87. Quoted in Sect 6.2
Padoa A (1901) Essai d'une théorie algébrique des nombre entiers, précédé d'une introduction logique à une théorie déductive quelconque. In: Bibliothèque du Congrès international de philosophie Paris 1900. Colin, Paris, pp 309–365. Quoted in Sect 2.2
Paley REAC, Wiener N (1933) Notes on the theory and application of Fourier transforms I–II. Trans Am Math Soc 35:348–355. Quoted in Sect 3.3
Paley REAC, Wiener N (1934) Fourier transforms in the complex domain. American Mathematical Society, Providence. Quoted in Sects 3.3, 12.4
Panofsky E (1939) Studies in iconology. Oxford University Press, Oxford. Quoted in Sect 1.4
Papineau D (2001) The rise of physicalism. In: Gillett C, Loewer BM (eds) Physicalism and its discontents. Cambridge University Press, Cambridge, pp 3–36. Quoted in Sect 7.1
Pareyon G (2011) On musical self-similarity. International Semiotics Institute at Imatra, Helsinki. Quoted in Sect 10.2
Parthasarathy KR (2004) On estimating the state of a finite level quantum system. Infin Dimens Anal Quantum Probab Relat Top 7:607–617. Quoted in Sect 3.5
Pauli W (1921) Relativitätstheorie. In: Encyklopädie der mathematischen Wissenschaften, vol V. Teubner, Leipzig, pp 539–775. Quoted in Sect 6.3
Pauli W (1927) Über Gasentartung und Paramagnetismus. Z Phys 41:81–102. Quoted in Sect 6.2

Pauli W (1933) Die allgemeinen Prinzipien der Wellenmechanik. In: Geiger H, Scheel K (eds) Handbuch der Physik, Band 24, 1. Teil. Springer, Berlin, pp 83–272. Quoted in Sect 6.2
Pauli W (1948) Editorial. Dialectica 2:307–311. Quoted in Sect 4.6
Pauli W (1952) Der Einfluss archetypischer Vorstellungen auf die Bildung naturwissenschaftlicher Theorien bei Kepler. In: Jung CG, Pauli W (eds) Naturerklärung und Psyche. Rascher, Zürich, pp 109–194. Quoted in Sects 4.1, 6.4
Pauli W (1954) Matter. In: Muschel H (ed) Man's right to knowledge. Columbia University Press, New York, pp 10–18. Quoted in Sects 8.2, 11.2
Pauli W (1955) Exclusion principle, Lorentz group and reflection of space-time and charge. In: Pauli W (ed) Niels Bohr and the development of physics. Pergamon, London, pp 30–51. Quoted in Sect 7.3
Pauli W (1956) Schlusswort durch den Präsidenten der Konferenz "Fünfzig Jahre Relativitätstheorie". Helv Phys Acta, Suppl IV:261–267. Quoted in Sect 6.3
Pauli W (1957) Phänomen und physikalische Realität. Dialectica 11:36–48. Quoted in Sects 4.1, 6.4
Pauli W (1980) The general principles of quantum mechanics. Springer, Berlin, transl by Achuthan P and Venkatesan K. Quoted in Sect 6.2
Pauli W (1994) Writings on physics and philosophy. Springer, Berlin. Quoted in Sects 4.1, 6.4
Peck SN (1979) Canonical Schwartz spaces and generalized operators. J Math Phys 20:336–343. Quoted in Sect 10.4
Pedersen GK (1979) C*-algebras and their automorphism groups. Academic Press, London. Quoted in Appendix A.4
Peirce CS (1933) Collected papers of Charles Sanders Peirce, vols III and IV: exact logic and the simplest mathematics. Harvard University Press, Cambridge. Quoted in Sect 3.2
Perez JF, Wilde IF (1977) Localization and causality in relativistic quantum mechanics. Phys Rev D 16:315–317. Quoted in Sect 7.3
Perrin J (1909) Mouvement brownien et réalité moléculaire. Ann Chim Phys 18:1–114. Quoted in Sect 10.2
Perrin J (1913) Les atomes. Alcan, Paris. Quoted in Sect 10.2
Petz D (1990) An invitation to the algebra of canonical commutation relations. Leuven University Press, Leuven. Quoted in Sect 9.4
Petz D (2007) Complementarity in quantum mechanics. Rep Math Phys 59:209–224. Quoted in Sect 3.5
Pfaffelhuber E (1971) Generalized impulse response and causality. IEEE Trans Circuit Theory 18:218–223. Quoted in Sect 3.3
Pfeifer P (1980) Chiral molecules—a superselection rule induced by the radiation field. Thesis ETH Zürich, no 6551, ok Gotthard S+D AG, Zürich. Quoted in Sect 6.5
Pfeifer P (1983a) Dynamical aspects of broken inversion symmetry in chiral molecules. In: Maruani J, Serre J (eds) Symmetries and properties of non-rigid molecules: a comprehensive survey. Elsevier, Amsterdam, pp 379–388. Quoted in Sect 6.5
Pfeifer P (1983b) Molecular structure derived from first principles quantum mechanics: two examples. In: Hinze P (ed) Energy storage and redistribution in molecules. Plenum, New York, pp 315–326. Quoted in Sect 6.5
Picht G (1966) Grundlinien einer Philosophie der Musik. Merkur 20:710–728. Quoted in Sect 9.3
Piron C (1964) Axiomatique quantique. Helv Phys Acta 37:439–468. Quoted in Sect 11.1
Planck M (1909) Die Einheit des physikalischen Weltbildes. Phys Z 10:62–75. Quoted in Sect 6.4
Plato (1891) Apology of Socrates, transl by Jowett B, accessible at http://classics.mit.edu/Plato/apology/html. Quoted in Sect 6.4
von Plato J (1982) The significance of the ergodic decomposition of stationary measures for the interpretation of probability. Synthese 53:419–432. Quoted in Sect 12.4
Platt JR (1956) Amplification aspects of biological response and mental activity. Am Sci 44:180–197. Quoted in Sect 2.3
Pohl RW (1949) Einführung in die Elektrizitätslehre. Springer, Berlin. Quoted in Sect 6.3
Poincaré H (1898) La measure du temps. Rev Métaphys Morale 6:1–13. Quoted in Sects 1.2, 8.5

Poincaré H (1900) La théorie de Lorentz et le principe de réaction. Arch Neerl Sci Exactes Nat 5:252–278. Quoted in Sect 6.3
Poincaré H (1904a) La valeur de la science. Flammarion, Paris. Quoted in Sects 5.2, 5.9, 5.10
Poincaré H (1904b) Les définitions générales en mathématiques. Enseign Math 6:267–283. Quoted in Sect 5.10
Poincaré H (1905) Science and hypothesis. Walter Scott, London. Quoted in Sect 13.2
Poincaré H (1914) Science and method. Nelson and Sons, London. Quoted in Sect 5.3
Polanyi M (1946) Science, faith and society. University of Chicago Press, Chicago. Quoted in Sect 4.1
Polanyi M (1958) Personal knowledge. Towards a post-critical philosophy. Routledge and Kegan Paul, London. Quoted in Sects 2.4, 4.1
Polanyi M (1962) Tacit knowing: its bearing on some problems of philosophy. Rev Mod Phys 34:601–616. Quoted in Sect 4.1
Polanyi M (1966) The tacit dimension. Doubleday, Garden City. Quoted in Sect 4.1
Polya G (1945) How to solve it. Princeton University Press, Princeton. Quoted in Sect 1.3
Polya G (1954) Patterns of plausible inference. Princeton University Press, Princeton. Quoted in Sect 1.3
Pool JCT (1966) Mathematical aspects of the Weyl correspondence. J Math Phys 7:66–76. Quoted in Sect 10.4
Post EL (1921) Introduction to a general theory of elementary propositions. Am J Math 43:165–185. Quoted in Sect 3.2
Pratt V (1992) Linear logic for generalized quantum mechanics. In: Matzke DJ (ed) Proceedings of the IEEE workshop on physics and computation. IEEE Computer Society Press, Los Alamitos, pp 166–180. Quoted in Sect 3.2
Price H (1992) Agency and causal asymmetry. Mind 101:501–520. Quoted in Sect 12.2
Priest G (2002) Paraconsistent logic. In: Gabbay D, Guenther F (eds) Handbook of philosophical logic, vol 6 Kluwer Academic, Dordrecht, pp 287–293. Quoted in Sect 3.2
Prigogine I (1980) From being to becoming. Freeman, San Francisco. Quoted in Appendix A.6
Primas H (1975) Pattern recognition in molecular quantum mechanics. I. Background dependence of molecular states. Theor Chim Acta 39:127–148. Quoted in Sect 6.3
Primas H (1977) Theory reduction and non-Boolean theories. J Math Biol 4:281–301. Quoted in Sects 6.2, 6.5
Primas H (1978) Kinematical symmetries in molecular quantum mechanics. In: Kramer P, Rieckers A (eds) Group theoretical methods in physics. Springer, Berlin, pp 72–91. Quoted in Sect 6.5
Primas H (1980) Foundations of theoretical chemistry. In: Woolley RG (ed) Quantum dynamics of molecules. The new experimental challenge to theorists. Plenum, New York, pp 39–113. Quoted in Sects 6.4, 8.5
Primas H (1981) Chemistry, quantum mechanics, and reductionism. Springer, Berlin. Quoted in Sects 6.2, 6.4, 6.5, 7.1, 8.5, and in the Preface
Primas H (1984) Verschränkte Systeme und Komplementarität. In: Kanitscheider B (ed) Moderne Naturphilosophie. Königshausen und Neumann, Würzburg, pp 243–260. Quoted in Sect 8.5
Primas H (1985) Kann Chemie auf Physik reduziert werden? Erster Teil: Das Molekulare Programm, Zweiter Teil: Die Chemie der Makrowelt. Chem Unserer Zeit 19:109–119, 160–166. Quoted in Sect 7.1
Primas H (1990a) Mathematical and philosophical questions in the theory of open and macroscopic quantum systems. In: Miller AI (ed) Sixty-two years of uncertainty. Plenum, New York, pp 233–257. Quoted in the Preface
Primas H (1990b) The measurement process in the individual interpretation of quantum mechanics. In: Cini M, Lévy-Leblond JM (eds) Quantum theory without reduction. Adam Hilger, Bristol, pp 49–68. Quoted in Sect 6.2
Primas H (1993) The Cartesian cut, the Heisenberg cut, and disentangled observers. In: Laurikainen KV, Montonen C (eds) Symposia on the foundations of modern physics 1992. The Copenhagen interpretation and Wolfgang Pauli. World Scientific, Singapore, pp 245–269. Quoted in Sect 6.3

Primas H (1997) The representation of facts in physical theories. In: Atmanspacher H, Ruhnau E (eds) Time, temporality, now. Springer, Berlin, pp 241–263. Quoted in Sects 6.5, 8.4, 12.1, 12.2

Primas H (1998) Emergence in exact natural science. Acta Polytech Scand 91:83–98. Quoted in Sects 6.2, 6.5

Primas H (2000) Asymptotically disjoint quantum states. In: Blanchard P et al (eds) Decoherence: theoretical, experimental, and conceptual problems. Springer, Berlin, pp 161–178. Quoted in Sects 6.2, 6.5

Primas H (2003) Time-entanglement between mind and matter. Mind and Matter 1:81–119. Quoted in Sect 8.5 and in the Preface

Primas H (2007) Non-Boolean descriptions for mind-matter problems. Mind and Matter 5:7–44. Quoted in Sects 1.1, 8.5

Primas H (2009) Complementarity of mind and matter. In: Atmanspacher H, Primas H (eds) Recasting reality. Springer, Berlin, pp 171–209. Quoted in Sect 8.5 and in the Preface

Primas H, Müller-Herold U (1978) Quantum mechanical system theory: a unifying framework for observations and stochastic processes in quantum mechanics. Adv Chem Phys 38:1–107. Quoted in Sect 12.1

Prior A (1957) Time and modality. Clarendon, Oxford. Quoted in Sect 3.2

Prosser RT (1963) On the ideal structure of operator algebras. Mem Am Math Soc 45:1–28. Quoted in Sect 6.5

Prugovečki E (1967) On a theory of measurement of incompatible observables in quantum mechanics. Can J Phys 45:2173–2219. Quoted in Sect 3.5

Prugovečki E (1973) A postulational framework for theories of simultaneous measurement of several observables. Found Phys 3:3–18. Quoted in Sect 3.5

Prugovečki E (1984) Stochastic quantum mechanics and quantum spacetime. Reidel, Dordrecht. Quoted in Sect 3.5

Pushkarov DI (1991) Quasiparticle theory of defects in solids. World Scientific, Singapore. Quoted in Sect 6.3

Putnam CR (1967) Commutation properties of Hilbert space operators and related topics. Springer, Berlin. Quoted in Sects 3.5, 10.5.

Putnam H (1981) Reason, truth and history. Cambridge University Press, Cambridge. Quoted in Sects 6.2, 8.4

Putnam H (1995) Words and life. Harvard University Press, Cambridge. Quoted in Sect 8.4

Rachev ST (2003) Handbook of heavy tailed distributions in finance. Elsevier, Amsterdam. Quoted in Sect 11.6

Raggio GA, Rieckers A (1983) Coherence and incompatibility in W*-algebraic quantum theory. Int J Theor Phys 22:267–291. Quoted in Sects 8.5, 11.1

Rajagopalan M, Narahari Achar BN, Bellur S (1993) The man who knew infinity. Not Am Math Soc 40(2):99–100. Quoted in Sect 5.10

Ramo H (1999) An Aristotelian human time-space manifold: from chronochora to kairotopos. Time Soc 8:309–328. Quoted in Sect 1.4

Rasiowa H (1974) An algebraic approach to non-classical logics. North-Holland, Amsterdam. Quoted in Sect 3.2

Redhead M (1995) More ado about nothing. Found Phys 25:123–137. Quoted in Sect 7.3

Redlich O (1968) Fundamental thermodynamics since Carathéodory. Rev Mod Phys 40:556–563. Quoted in Sect 7.4

Reed M, Simon B (1975) Methods of modern mathematical physics II: Fourier analysis, self-adjointness. Academic Press, New York. Quoted in Appendix A.6

Reeh H, Schlieder S (1961) Bemerkungen zur Unitäräquivalenz von Lorentzinvarianten Feldern. Nuovo Cimento 22:1051–1068. Quoted in Sect 7.3

Reichenbach H (1938) Experience and prediction. An analysis of the foundations and the structure of knowledge. University of Chicago Press, Chicago. Quoted in Sect 4.7

Reichenbach H (1956) The direction of time. University of California Press, Berkeley. Quoted in Sect 12.2

Rényi A (1966) On the amount of missing information and the Neyman–Pearson lemma. In: Research papers in statistics. Festschrift for J. Neyman. Wiley, London, pp 281–288. Quoted in Sect 6.5

Rényi A (1967) Statistics and information theory. Studia Sci Math Hung 2:249–256. Quoted in Sect 6.5

Roberts JE (2004) More lectures on algebraic quantum field theory. In: Doplicher S, Longo R (eds) Noncommutative geometry. Springer, Berlin, pp 321–338. Quoted in Sect 7.3

Robinson A (1966) Non-standard analysis. North-Holland, Amsterdam. Quoted in Sect 5.3

Robinson K (ed) (2009) Deleuze, Whitehead, Bergson: rhizomatic connections. Palgrave Macmillan, New York. Quoted in Sect 13.3

Rogalevicz V (1988) Any orthomodular poset is a pasting of Boolean algebras. Comment Math Univ Carol 29:557–558. Quoted in Sect 6.4

Root WL (1963) Singular Gaussian measures in detection theory. In: Rosenblatt M (ed) Proceedings of the symposium on time series analysis. Wiley, New York, pp 292–315. Quoted in Sect 6.5

Root WL (1964) Stability in signal detection problems. In: Bellman R (ed) Stochastic processes in mathematical physics and engineering. American Mathematical Society, Providence, pp 247–263. Quoted in Sect 6.5

Root WL (1968) The detection of signals in Gaussian noise. In: Balakrishnan AV et al (eds) Communication theory. McGraw–Hill, New York, pp 160–191. Quoted in Sect 6.5

Rosch E (1973a) On the internal structure of perceptual and semantic categories. In: Moore T (ed) Cognitive development and acquisition of language. Academic Press, New York, pp 111–144. Quoted in Sect 3.3

Rosch E (1973b) Natural categories. Cogn Psychol 4:328–350. Quoted in Sect 3.3

Rosch E (1975) Cognitive representations of semantic categories. J Exp Psychol 104:192–233. Quoted in Sect 3.3

Rosen R (1960) A quantum-theoretic approach to genetic problems. Bull Math Biophys 22:227–255. Quoted in Sect 6.3

Rosen R (1985) Anticipatory systems. Pergamon, Oxford. Quoted in Sect 12.2

Rosen R (1991) Life itself: A comprehensive inquiry into the nature, origin, and fabrication of life. Columbia University Press, New York. Quoted in Sect 4.1

Rosenblatt M (1971) Markov processes: structure and asymptotic behavior. Springer, Berlin. Quoted in Sect 12.4

Rosenblueth A, Wiener N, Bigelow J (1943) Behavior, purpose and teleology. Philos Sci 10:18–24. Quoted in Sect 12.2

Roubens M (1978) Pattern classification problems and fuzzy sets. Fuzzy Sets Syst 1:239–253. Quoted in Sect 3.3

Rousseau R, Daele A, van Vanheeswijck L (1977) A necessary and sufficient condition for a von Neumann algebra to be in standard form. J Lond Math Soc 15:147–154. Quoted in Appendix A.6

Rovelli C (1993a) Statistical mechanics of gravity and the thermodynamical origin of time. Class Quantum Gravity 10:1549–1566. Quoted in Sect 7.4

Rovelli C (1993b) The statistical state of the universe. Class Quantum Gravity 10:1567–1578. Quoted in Sect 7.4

Rovelli C (1996) Relational quantum mechanics. Int J Theor Phys 35:1637–1678. Quoted in Sect 8.3

Rovelli C (2001) Quantum spacetime: what do we know? In: Callender C, Huggett N (eds) Physics meets philosophy at the Planck scale: contemporary theories in quantum gravity. Cambridge University Press, Cambridge, pp 101–122. Quoted in Sect 7.4

Rovelli C (2011) Forget time. Found Phys 41:1475–1490. Quoted in Sect 7.4

Rozanov YuA (1967) Innovation processes. Wiley, New York. Quoted in Sect 12.4

Ruitenburg W (2001) Review of "truth in mathematics" by H.G. Dales and G. Oliveri. Rev Mod Log 8:153–163. Quoted in Sect 5.1

Russell B (1913) On the notion of cause. Proc Aristot Soc 13:1–26. Quoted in Sects 3.1, 12.2

Russell B (1919) Introduction to mathematical philosophy. Macmillan, London. Quoted in Sect 5.4
Russell B (1994) Philosophical essays. Routledge, London. Quoted in Sect 6.1
Russell JS (1844) Report on waves. In: Proceedings of the 14th meeting of the British association for the advancement of science. John Murray, London, pp 311–390. Quoted in Sect 6.2
Sakai S (1971) C*-algebras and W*-algebras. Springer, Berlin. Quoted in Sects 6.3, 8.5, Appendix A.1
Scagliola F (2012) Toward a rational practice of arithmetic. A model for the musical (and multimedia) composition. In: Emmer M (ed) Imagine math. Between culture and mathematics. Springer, Berlin, pp 283–288. Quoted in Sect 9.3
Scarpellini B (1978a) Un problème de théorie de la prédiction. C R Hebd Séances 'Acad Sci Paris A 286:1251–1254. Quoted in Sect 12.4
Scarpellini B (1978b) Fourier analysis on dynamical systems. J Differ Equ 28:309–326. Quoted in Sect 12.4
Scarpellini B (1978c) Entropy and nonlinear prediction. Z Wahrscheinlichkeitstheor Verw Geb 50:165–178. Quoted in Sect 12.4
Scarpellini B (1978d) Predicting the future of functions on flows. Math Syst Theory 12:281–296. Quoted in Sect 12.4
Scheibe E (1988) Paul Feyerabend und die rationalen Rekonstruktionen. In: Hoyningen-Huene P, Hirsch G (eds) Wozu Wissenschaftsphilosophie? de Gruyter, Berlin, pp 149–171. Quoted in Sect 7.1
Schickore J, Steinle F (eds) (2006) Revisiting discovery and justification: historical and philosophical perspectives on the context distinction. Springer, Berlin. Quoted in Sect 4.7
Schlick M (1961) Causality in contemporary physics (I). Br J Philos Sci 12:177–193. Quoted in Sect 12.2
Schlick M (1985) General theory of knowledge. Open Court, Chicago. transl by Blumberg AE. Quoted in Sect 7.1
Schlögl F (1960) Die Mannigfaltigkeit informationsfremder Zustände in der Quantentheorie. Z Phys 159:411–427. Quoted in Sect 3.4
Schmidt E (1907a) Zur Theorie der linearen und nichtlinearen Integralgleichungen I. Math Ann 63:433–476. Quoted in Sect 6.3
Schmidt E (1907b) Zur Theorie der linearen und nichtlinearen Integralgleichungen II. Math Ann 64:161–174. Quoted in Sect 6.3
Schmidt E (1908) Über die Auflösung linearer Gleichungen mit unendlich vielen Unbekannten. Rend Circ Mat Palermo 25:53–77. Quoted in Sect 6.3
Schmitt LM (1990) Direct integrals of standard forms of W*-algebras. Rocky Mt J Math 20:561–574. Quoted in Appendix A.6
Schmüdgen K (1983) On the Heisenberg commutation relation I. J Funct Anal 50:8–49. Quoted in Sect 10.5
Schreiber M (1958) A functional calculus for general operators in Hilbert space. Trans Am Math Soc 87:108–118. Quoted in Sect 3.5
de Schreye D (1983) Integrable, ergodic actions of Abelian groups on von Neumann algebras. Math Scand 53:265–280. Quoted in Sect 9.4
Schroeder M (1991) Fractals, chaos, power laws: minutes from an infinite paradise. Freeman, New York. Quoted in Sect 10.2
Schröder W (1984) A hierarchy of mixing properties for non-commutative K-systems. In: Accardi L, Frigerio A, Gorini V (eds) Quantum probability and applications to the quantum theory of irreversible processes. Springer, Berlin, pp 340–351. Quoted in Sect 12.1
Schrödinger E (1931) Spezielle Relativitätstheorie und Quantenmechanik. In: Sitzungsberichte der Preussischen Akademie der Wissenschaften, Physikalisch-Mathematische Klasse, pp 238–247. Quoted in Sects 8.3, 8.5
Schrödinger E (1932) Sur la théorie relativiste de l'électron et l'interprétation de la mécanique quantique. Ann Inst Henri Poincaré 2:269–310. Quoted in Sect 8.5
Schrödinger E (1935a) Die gegenwärtige Situation in der Quantenmechanik. Naturwissenschaften 23:807–812, 823–828, 844–849. Quoted in Sect 6.3

Schrödinger E (1935b) Discussion of probability relations between separated systems. Proc Camb Philos Soc 31:555–563. Quoted in Sect 6.3
Schrödinger E (1936) Probability relations between separated systems. Proc Camb Philos Soc 32:446–452. Quoted in Sect 6.3
Schroer B (2000) Basic quantum theory and measurement from the viewpoint of local quantum physics. In: Doebner H-D et al (eds) Trends in quantum mechanics. World Scientific, Singapore, pp 274–286. Quoted in Sect 7.3
Schroer B (2006) String theory and the crisis in particle physics. Int J Mod Phys D 17:2373–2431. Quoted in Sect 7.3
Schroer B (2007) New constructions in local quantum physics. In: de Monvel AB et al (ed) Rigorous quantum field theory. Birkhäuser, Basel, pp 283–300. Quoted in Sect 7.3
Schroer B (2010a) Localization and the interface between quantum mechanics, quantum field theory and quantum gravity I. The two antagonistic localizations and their asymptotic compatibility. Stud Hist Philos Mod Phys 41:104–127. Quoted in Sect 7.3
Schroer B (2010b) Localization and the interface between quantum mechanics, quantum field theory and quantum gravity II. The search of the interface between QFT and QG. Stud Hist Philos Mod Phys 41:293–308. Quoted in Sect 7.3
Schroer B (2010c) Pascual Jordan's legacy and the ongoing research in quantum field theory. Eur Phys J H 35:377–434. Quoted in Sect 7.3
Schwinger J (1959) The algebra of microscopic measurement. Proc Natl Acad Sci USA 45:1542–1553. Quoted in Sect 3.4
Schwinger J (1960) Unitary operator bases. Proc Natl Acad Sci USA 46:570–579. Quoted in Sect 3.5
Schwinger J, DeRaad LL, Kimball AM, Tsai W (1998) Classical electrodynamics. Perseus, Reading. Quoted in Sect 6.3
Segal I (1947) Irreducible representations of operator algebras. Bull Am Math Soc 53:73–88. Quoted in Sect 6.5 and in Appendix A.3
Segal IE (1961) Foundations of the theory of dynamical systems of infinitely many degrees of freedom II. Can J Math 13:1–18. Quoted in Sect 10.4
Segal IE (1963) Transforms for operators and symplectic automorphisms over locally compact Abelian groups. Math Scand 13:31–43. Quoted in Sect 10.4
Segal IE (1964) Quantum fields and analysis in the solution manifolds of differential equations. In: Martin WT, Segal IE (eds) Proceedings of a conference on the theory and applications of analysis in function space. MIT Press, Cambridge, pp 129–153. Quoted in Sect 7.3
Segalowitz SJ (1983) Two sides of the brain. Prentice Hall, Englewood Cliffs. Quoted in Sect 4.2
Seising R (2007) Pioneers of vagueness, haziness, and fuzziness in the 20th century. In: Nikravesh M, Kacprzyk J, Zadeh LA (eds) Forging new frontiers: fuzzy pioneers I. Springer, Berlin, pp 55–81. Quoted in Sect 3.3
Sellerio AL, Daniele Mari D, Gremaud G (2012) Fractional Brownian motion and anomalous diffusion in vibrated granular materials. J Stat Mech 2012:1–18. Quoted in Sect 10.2
Sewell GL (1973) States and dynamics of infinitely extended physical systems. Commun Math Phys 33:43–51. Quoted in Sect 6.5
She CY, Heffner H (1966) Simultaneous measurement of noncommuting observables. Phys Rev 152:1103–1110. Quoted in Sect 6.2
Silberstein M, McGeever J (1999) The search for ontological emergence. Philos Q 49:182–200. Quoted in Sect 6.5
Silverman LK (2002) Upside-down brilliance. The visual-spatial learner. DeLeon, Denver. Quoted in Sect 4.5
Simon HA (1962) The architecture of complexity. Proc Am Philos Soc 106:467–482. Quoted in Sect 6.1
Sinowjew AA (1970a) Über mehrwertige Logik. Vieweg, Braunschweig. Quoted in Sect 3.2
Sinowjew AA (1970b) Komplexe Logik. Vieweg, Braunschweig. Quoted in Sect 3.2
Sipiora P (2002) The ancient concept of kairos. In: Sipiora P, Baumlin JS (eds) Rhetoric and kairos. SUNY Press, Albany, pp 1–22. Quoted in Sect 1.4

Slepian D (1976) On bandwidth. Proc IEEE 64:292–300. Quoted in Sect 10.2
Slepian D (1983) Some comments on Fourier analysis, uncertainty and modeling. SIAM Rev 25:379–393. Quoted in Sects 3.3, 10.2
Slepian D, Pollak HO (1961) Prolate spheroidal wave functions, Fourier analysis and uncertainty I. Bell Syst Tech J 40:43–64. Quoted in Sect 10.2
Smirnow WL (1967) Lehrgang der höheren Mathematik, Teil V. VEB Deutscher Verlag der Wissenschaften, Berlin. Quoted in Sect 11.5
Smith NJJ (2008) Vagueness and degrees of truth. Oxford University Press, Oxford. Quoted in Sect 3.2
Smoluchowsky M (1906) Zur kinetischen Theorie der Brownschen Molekularbewegung und der Suspensionen. Ann Phys 21:756–780. Quoted in Sect 10.2
Soanes C, Stevenson A (eds) (2008) The concise Oxford dictionary. Oxford University Press, Oxford. Quoted in Sect 4.1
Sommerfeld A (1914) Über die Fortpflanzung des Lichtes in dispergierenden Medien. Ann Phys 44:177–202. Quoted in Sect 3.3
Sommerfeld A (1949) Elektrodynamik. Geest und Porting, Leipzig. Quoted in Sect 6.3
Specker E (1960) Die Logik nicht gleichzeitig entscheidbarer Aussagen. Dialectica 14:239–246. Quoted in Sect 2.4
Spekkens RW (2005) Contextuality for preparations, transformations, and unsharp measurements. Phys Rev A 71:052108. Quoted in Sect 3.5
Speiser A (1952) Die mathematische Denkweise. Birkhäuser, Basel. Quoted in Sect 9.3
Spencer Brown G (1969) Laws of form. Allen and Unwin, London. Quoted in Sects 2.1, 2.5
Sperry RW (1961) Cerebral organization and behavior: the split brain behaves in many respects like two separate brains, providing new research possibilities. Science 133:1749–1757. Quoted in Sect 4.2
Sperry RW (1973) Lateral specialization of cerebral function in the surgically separated hemispheres. In: McGuigan FJ, Schoonover RA (eds) The psychophysiology of thinking. Academic Press, New York, pp 209–229. Quoted in Sect 4.2
Sperry RW (1984) Consciousness, personal identity and the divided brain. Neuropsychologia 22:661–673. Quoted in Sect 4.2
Spurrett D (1999) The completeness of physics. Thesis at the University of Natal, full text available at http://cogprints.org/3379/1/DSTHESIS.pdf. Quoted in Sect 7.1
Stano P, Reizner D, Heinosaari T (2008) Coexistence of qubit effects. Phys Rev A 78:012315. Quoted in Sect 3.5
Stein G (1938) Picasso. Batsford, London. Quoted in Sect 6.2
Steiner G (1970) Language and silence. Essays on language, literature, and the inhuman. Atheneum, London. Quoted in Sect 13.4
Stepanov VV (1925) Über einige Verallgemeinerungen der fastperiodischen Funktionen. Math Ann 45:473–498. Quoted in Sect 12.4
Stinespring WF (1955) Positive functions of C*-algebras. Proc Am Math Soc 6:211–216. Quoted in Appendix A.2
Stone MH (1930) Linear transformations in Hilbert space III. Operational methods and group theory. Proc Natl Acad Sci USA 16:172–175. Quoted in Sects 3.5, 9.3, 9.7
Stone MH (1932) On one-parameter unitary groups in Hilbert space. Ann Math 33:643–648. Quoted in Sect 9.3
Størmer E (1974) Spectra of ergodic transformations. J Funct Anal 15:202–215. Quoted in Sect 9.4
Strătilă S (1981) Modular theory in operator algebras. Abacus Press, Kent. Quoted in Sect 7.3
Strătilă S, Zsidó L (1979) Lectures on von Neumann algebras. Abacus Press, Kent. Quoted in Sect 7.3 and Appendix A.6
Stratton JA (1941) Electromagnetic theory. McGraw–Hill, New York. Quoted in Sect 6.3
Strauss M (1936a) Komplementarität und Kausalität im Lichte der logischen Syntax. Erkenntnis 6:335–339. Quoted in Sects 2.4, 4.6
Strauss M (1936b) Zur Begründung der statistischen Transformationstheorie der Quantenphysik. Sitz Preuss Akad Wiss, Phys-Math Kl 27:382–398. Quoted in Sects 2.4, 4.6

Strauss M (1938) Mathematics as logical syntax—a method to formalize the language of a physical theory. Erkenntnis 7:147–153. Quoted in Sects 2.4, 4.6

Strauss M (1967) Grundlagen der modernen Physik. In: Ley H, Löther R (eds) Mikrokosmos–makrokosmos, vol 2. Akademie Verlag, Berlin, pp 55–92. Quoted in Sects 2.4, 3.4, 4.6

Strauss M (1970) Intertheory relations. In: Weingartner P, Zecha G (eds) Induction, physics and ethics. Reidel, Dordrecht, pp 220–284. Quoted in Sect 2.4, 4.6

Strauss M (1972) Modern physics and its philosophy. selected papers in the logic, history, and philosophy of science. Reidel, Dordrecht. Quoted in Sect 4.6

Strauss M (1973) Logics for quantum mechanics. Found Phys 3:265–276. Quoted in Sects 2.4, 4.6

Streater RF, Wightman AS (1964) PCT, spin and statistics, and all that. Benjamin, New York, Quoted in Sects 6.2, 7.3

Strocchi F (1985) Elements of quantum mechanics of infinite systems. World Scientific, Singapore. Quoted in Sect 6.2

Strocchi F (2008) Symmetry breaking. Springer, Berlin. Quoted in Sect 6.4

Sudarshan ECG (1983) Perception of quantum systems. In: van der Merwe A (ed) Old and new questions in physics, cosmology, philosophy, and theoretical biology. Plenum, New York, pp 457–467. Quoted in Sect 13.5

Sugita A (2006) Borromean entanglement revisited. In: Kawauchi A (ed) Knot theory for scientific objects. Osaka Municipal Universities Press, Osaka. Available at arXiv:0704.1712. Quoted in Sect 8.5

Summers SJ (1990) On the independence of local algebras in quantum field theory. Rev Math Phys 2:201–247. Quoted in Sect 7.3

Summers SJ (2009) Subsystems and independence in relativistic microscopic physics. Stud Hist Philos Mod Phys 40:133–141. Quoted in Sects 7.3, 8.5

Summers SJ, Werner RF (1985) The vacuum violates Bell's inequalities. Phys Lett A 110:257–259. Quoted in Sect 7.3

Summers SJ, Werner RF (1987) Maximal violation of Bell's inequalities is generic in quantum field theory. Commun Math Phys 110:247–259. Quoted in Sect 7.3

Summers SJ, Werner RF (1988) Maximal violation of Bell's inequalities for algebras of observables in tangent spacetime regions. Ann Inst Henri Poincaré 49:215–243. Quoted in Sect 7.3

Summers SJ, Werner RF (1995) On Bell's inequalities and algebraic invariants. Lett Math Phys 33:321–334. Quoted in Sect 7.3

Sunder VS (1987) An invitation to von Neumann algebras. Springer, New York. Quoted in Appendix A.5

Suppes P, Zanotti M (1981) When are probabilistic explanations possible? Synthese 48:191–199. Quoted in Sect 13.5

Swift J (1995) Gulliver's travels: complete, authoritative text with biographical and historical contexts. Palgrave Macmillan, New York, originally published in 1726. Quoted in Sect 7.3

Synge JL (1960) Classical dynamics. In: Flügge S (ed) Principles of classical mechanics and field theory. Encyclopedia of physics, volume III/1. Springer, Berlin, pp 1–225. Quoted in Sect 8.3

Szegö G (1915) Ein Grenzwertsatz über die Toeplitzschen Determinanten einer reellen positiven Funktion. Math Ann 76:490–503. Quoted in Sect 12.4

Szegö G (1920) Beiträge zur Theorie der Toeplitzschen Formen. Math Z 6:167–202. Quoted in Sect 12.4

Szent-Györgi A (1962) The persistence of the caveman. Saturday Rev 45:11. Quoted in Sect 2.3

Tait WW (2001) Gödel's unpublished papers on foundations of mathematics. Philos Math 9:87–126. Quoted in Sect 5.7

Takesaki M (1958) On the direct product of W*-factors. Tohoku Math J 10:116–119. Quoted in Sect 6.3

Takesaki M (1970) Tomita's theory of modular Hilbert algebras and its applications. Springer, Berlin. Quoted in Sects 6.5, 7.3, Appendices A.5, A.6

Takesaki M (1979) Theory of operator algebras I. Springer, New York. Quoted in Sects 6.2, 6.5, 7.3, 8.5, Appendix A.2

Takesaki M (2003) Theory of operator algebras II. Springer, Berlin. Quoted in Sects 7.3, 8.5, Appendices A.5, A.6
Temple G (1932) The physical principles of quantum theory. Proc R Soc Lond A 138:479–494. Quoted in Sect 3.4
Temple G (1934) The general principles of quantum theory. Methuen, London. Quoted in Sect 3.4
Thangavelu S (1998) Harmonic analysis on the Heisenberg group. Birkhäuser, Boston. Quoted in Sect 9.3
Thomson W (1891) Electrical units of measurement. In: Popular lectures and addresses, vol I: constitution of matter. Macmillan, London, pp 80–143. Quoted in Sect 13.2
Titchmarsh EC (1937) Introduction to the theory of Fourier integrals. Oxford University Press, London. Quoted in Sect 10.2
Tjøstheim D (1976) A commutation relation for wide sense stationary processes. SIAM J Appl Math 30:115–122. Quoted in Sect 12.4
Tomita M (1967) On canonical forms of von Neumann algebras (in Japanese). In: Fifth functional analysis symposium, Sendai, Mathematical Institute, Tohoku University. Compare Math Rev 0284822 (44 #2046) by Sakai S. Quoted in Sect 7.3 and Appendix A.5
Tomiyama J (1957) On the projection of norm one in W*-algebras. Proc Jpn Acad 33:608–612. Quoted in Appendix A.2
Torrésani B (1991) Wavelets associated with representations of the affine Weyl–Heisenberg group. J Math Phys 32:1273–1279. Quoted in Sects 9.5, 10.5
Toulmin S (1970) Physical reality: philosophical essays on twentieth-century physics. Harper & Row, New York. Quoted in Sect 6.4
Trenn TJ (1981) On the question of the foundations of medical knowledge. J Med Philos 6:237–255. Quoted in Sect 4.1
Trimble MR (2007) The soul in the brain. The cerebral basis of language, art, and belief. Johns Hopkins University Press, Baltimore. Quoted in Sect 4.2
Truesdell C, Toupin RA (1960) The classical field theories. In: Flügge S (ed) Principles of classical mechanics and field theory. Encyclopedia of physics, vol III/1. Springer, Berlin, pp 226–793. Quoted in Sect 6.3
Tsoukas H (2003) Do we really understand tacit knowledge? In: Handbook of organizational learning and knowledge management. Blackwell, Oxford, pp 410–427. Quoted in Sect 4.1
Turing AM (1936) On computable numbers, with an application to the Entscheidungsproblem. Proc Lond Math Soc 42:230–256. Corrections: Proc Lond Math Soc 43:544–546 (1943). Quoted in Sect 5.5
Turing AM (1939) Systems of logic based on ordinals. Proc Lond Math Soc 45:161–228. Quoted in Sect 5.8
Uhlenbeck GE, Ornstein LS (1930) On the theory of the Brownian motion. Phys Rev 36:823–841. Quoted in Sects 10.2, 12.4
Ulam S (1958) John von Neumann, 1903–1957. Bull Am Math Soc 64(3/2):1–49. Quoted in Sect 5.8
Ulam SM (1976) Adventures of a mathematician. Charles Scribner's Sons, New York. Quoted in Sect 4.1
Umezawa H, Matsumoto H, Tachiki M (1981) Thermo field dynamics and condensed states. North-Holland, Amsterdam. Quoted in Appendix A.6
Unruh WG (1976) Notes on black-hole evaporation. Phys Rev D 14:870–892. Quoted in Sect 7.3
Varadarajan VS (1968) Geometry of quantum theory. Van Nostrand, Princeton. Quoted in Sect 6.2
Valley GE, Wallman H (1948) Vacuum tube amplifiers. McGraw–Hill, New York. Quoted in Sect 3.3
Vervaat W (1985) Sample path properties of self-similar processes with stationary increments. Ann Probab 13:1–27. Quoted in Sect 10.2
Vico G (1982) On the ancient wisdom of the Italians taken from the origins of the Latin language. In: Pompa L (ed) Vico: selected writings. Cambridge University Press, Cambridge, pp 49–80. Quoted in Sect 8.4

Vilenkin NJ (1968) Special functions and the theory of group representations. American Mathematical Society, Providence. Quoted in Sect 9.7
Vitiello G (1995) Dissipation and memory capacity in the quantum brain model. Int J Mod Phys B 9:973–989. Quoted in Appendix A.6
Vitiello G (2001) My double unveiled. The dissipative quantum model of brain. Benjamins, Amsterdam. Quoted in Appendix A.6
Voss RF, Clarke J (1978a) "1/f noise" in music and speech. Nature 258:317–318. Quoted in Sect 10.2
Voss RF, Clarke J (1978b) "1/f noise" in music: music from 1/f noise. J Acoust Soc Am 63:258–263. Quoted in Sect 10.2
van der Waals JD (1927) Lehrbuch der Thermostatik, erster Teil: Allgemeine Thermostatik. Ambrosius Barth, Leipzig. Quoted in Sect 7.1
Wang H (1987) Reflections on Kurt Gödel. MIT Press, Cambridge. Quoted in Sects 5.8, 5.10
Wang H (1996) A logical journey. From Gödel to philosophy. MIT Press, Cambridge. Quoted in Sects 4.1, 5.5
Watanabe S (1969) Knowing and guessing. A quantitative study of inference and information. Wiley, New York. Quoted in Sect 12.2
Watanabe S (1986) Separation of an acausal subsystem from a causal total system. In: Namiki M (ed) Proceedings of the 2nd international symposium: foundations of quantum mechanics in the light of new technology. Physical Society of Japan, Japan, pp 276–281. Quoted in Sect 12.2
Weber EH (1834) Pulsu, resorptione, auditu et tactu. Koehler, Lipsiae. Quoted in Sect 11.7
Weideman JAC (1995) Computing the Hilbert transform on the real line. Math Comput 64:745–762. Quoted in Sect 11.7
Weinberg S (1972) Gravitation and cosmology: principles and applications of the general theory of relativity. Wiley, New York. Quoted in Sect 7.3
von Weizsäcker CF (1958) Der zweite Hauptsatz und der Unterschied von Vergangenheit und Zukunft. Ann Phys 428:275–283. Quoted in Sect 12.1
von Weizsäcker CF (1992) Zeit und Wissen. Hanser, München. Quoted in the Preface
von Weizsäcker CF (2006) The structure of physics. Springer, Dordrecht, edited, revised and enlarged by Görnitz T and Lyre H. Quoted in Sect 12.1
Werner RF (1987) Local preparability of states and the split property in quantum field theory. Lett Math Phys 13:325–329. Quoted in Sect 7.3
Wertheimer M (1946) Productive thinking. Harper & Brothers, London. Quoted in Sect 4.1
Wertheimer M (1985) A Gestalt perspective on computer simulations of cognitive processes. Comput Hum Behav 1:19–33. Quoted in Sect 1.2
Weyl H (1925) Die heutige Erkenntnislage in der Mathematik. Symposion 1:1–32. Quoted in Sect 5.6
Weyl H (1927) Quantenmechanik und Gruppentheorie. Z Phys 46:1–46. Quoted in Sects 3.5, 6.2, 9.3, 9.4
Weyl H (1928a) Gruppentheorie und Quantenmechanik. Hirzel, Leipzig. Quoted in Sects 6.2, 9.3, 9.4
Weyl H (1928b) Diskussionsbemerkungen zu dem zweiten Hilbertschen Vortrag über die Grundlagen der Mathematik. Abh Math Semin Univ Hamb 6:86–88, 97. Quoted in Sects 5.6, 6.4
Weyl H (1931) The theory of groups and quantum mechanics, Methuen, London. Quoted in Sects 3.5, 9.3
Weyl H (1934) Mind and matter. University of Pennsylvania Press, Philadelphia. Quoted in Sect 6.4
Weyl H (1940) The ghost of modality. In: Farber M (ed) Philosophical essays in memory of Edmund Husserl. Harvard University Press, Cambridge, pp 278–303. Quoted in Sect 3.2
Weyl H (1946) Encomium. Science 103:216–218. Quoted in Sect 6.4
Weyl H (1949a) Philosophy of mathematics and natural science. Princeton University Press, Princeton. Quoted in Sect 5.6
Weyl H (1949b) Wissenschaft als symbolische Konstruktion des Menschen. In: Fröbe-Kapteyn O (ed) Eranos-Jahrbuch 1948, Band XVI. Rhein-Verlag, Zürich, pp 375–431. Quoted in Sect 6.4

Weyl H (1952) Symmetry. Princeton University Press, Princeton. Quoted in Sects 6.4, 9.3
Weyl H (1955) Address on the unity of knowledge. In: Leary L (ed) The unity of knowledge. Doubleday, New York, pp 1–8. Quoted in Sects 4.1, 4.7
Weyl H (1985) Axiomatic versus constructive procedures in mathematics. Math Intell 7(4):10–17. Quoted in Sect 6.4
Weyl H (2009) Mind and matter. Selected writing on philosophy, mathematics, and physics. Princeton University Press, Princeton. Quoted in Sects 4.1, 6.4
Whaples G (1952) Carathéodory's temperature equations. J Ration Mech Anal 1:301–307. Quoted in Sect 7.4
White MJ (1969) Laterality differences in perception. Psychol Bull 72:387–405. Quoted in Sect 1.4
Whitehead AN (1929) Process and reality. Free Press, New York. Quoted in Sect 13.3
Whittaker ET (1904) A treatise on the analytical dynamics of particles and rigid bodies. Cambridge University Press, Cambridge. Quoted in Sect 8.3
Wick GC, Wightman AS, Wigner EP (1952) The intrinsic parity of elementary particles. Phys Rev 88:101–105. Quoted in Sect 6.2
Wiener N (1930) Generalized harmonic analysis. Acta Math 55:117–258. Quoted in Sect 12.4
Wiener N (1942) Extrapolation, interpolation, and smoothing of stationary times series, with engineering applications. Classified report to section D2, National Defense Research Committee. Quoted in Sect 12.4
Wiener N (1948) Cybernetics, or control and communication in the animal and the machine. MIT Press, New York. Quoted in Sects 9.1, 12.2
Wiener N (1949) Extrapolation, interpolation, and smoothing of stationary times series, with engineering applications. Wiley, New York. Quoted in Sects 9.3, 12.4
Wiener N (1956) I am a mathematician. Victor Gollancz, London. Quoted in Sects 3.3, 9.3, 12.4
Wiener N, Akutowicz EJ (1957) The definition and ergodic properties of the stochastic adjoint of a unitary transformation. Rend Circ Mat Palermo 6:205–217. Quoted in Sect 12.4
Wiener N, Winter A (1958) Random time. Nature 181:561–562. Quoted in Sect 8.5
Wiener PP (1951) Leibniz selections. Charles Scribner's Sons, New York. Quoted in Sect 12.2
Wiesbrock H-W (1992) A comment on a recent work of Borchers. Lett Math Phys 25:157–159. Quoted in Appendix A.5
Wiesel E (1978) Trivializing the holocaust: semi-fact and semi-fiction. New York Times. April 26. Quoted in Sect 13.4
Wightman AS (1959) Relativistic invariance and quantum mechanics. Suppl Nuovo Cim 14:81–94. Quoted in Sect 6.2
Wightman AS (1995) Superselection rules, old and new. Nuovo Cimento B 110:751–769. Quoted in Sect 6.2
Wightman AS, Glance N (1989) Superselection rules in molecules. Nucl Phys B, Proc Suppl 6:202–206. Quoted in Sect 6.2
Wigner E (1932a) On the quantum correction for thermodynamic equilibrium. Phys Rev 40:749–759. Quoted in Appendix A.6
Wigner EP (1932b) Über die Operation der Zeitumkehr in der Quantenmechanik. Nachr Ges Wiss Gött, Math-Phys Kl 31:546–559. Quoted in Sect 10.2
Wigner EP (1939) On unitary representations of the inhomogeneous Lorentz group. Ann Math 40:149–204. Quoted in Sects 6.2, 7.3
Wigner EP (1952) Die Messung quantenmechanischer Operatoren. Z Phys 133:101–108. Quoted in Sect 6.2
Wigner EP (1959) Group theory. Academic Press, New York. Quoted in Sect 10.2
Wigner EP (1962) Remarks on the mind-body question. In: Good IJ (ed) The scientist speculates. Heinemann, London, pp 284–301. Quoted in Sect 6.2
Wigner EP (1963) The problem of measurement. Am J Phys 31:6–15. Quoted in Sect 8.4
Wigner EP (1964a) Two kinds of reality. Monist 48:248–264. Quoted in Sect 6.2
Wigner EP (1964b) Events, laws of nature, and invariance principles. Science 145:995–999. Quoted in Sect 7.1

Wigner EP (1969) Epistemology of quantum mechanics. In: Contemporary physics vol II. Atomic Energy Agency, Vienna, pp 431–437. Quoted in Sect 8.4
Wilczek F (2002) Setting standards. Nature 415:265. Quoted in Sect 6.3
Wittenberg A (1953) Über adäquate Problemstellung in der mathematischen Grundlagenforschung. Dialectica 7:232–254. Quoted in Sect 5.7
Wittenberg A (1956) Warum kein Platonismus? Eine Antwort an Herrn Prof. Finsler. Dialectica 10:256–261. Quoted in Sect 5.7
Wittgenstein L (1969) Tractatus logico-philosophicus. In: Ludwig Wittgenstein. Schriften 1. Suhrkamp, Frankfurt. Quoted in Sects 1.4, 8.3, 8.4
Wold H (1938) A study in the analysis of stationary times series. Almquist and Wiksell, Stockholm. Quoted in Sect 12.4
Woodward J (2003) Making things happen: a theory of causal explanation. Oxford University Press, New York. Quoted in Sect 12.2
Wootters WK (1984) "Time" replaced by quantum correlations. Int J Theor Phys 23:701–711. Quoted in Sect 8.3
Yaglom AM (1961) Second-order homogeneous random fields. In: Proceedings of the fourth Berkeley symposium on mathematical statistics and probability, vol 2. University of California Press, Berkeley, pp 593–622. Quoted in Sect 10.2
Yaglom AM (1962) An introduction to the theory of stationary random functions Prentice Hall, Englewood Cliffs. Quoted in Sects 10.2, 12.4
Yandell B (2001) The honors class: Hilbert's problems and their solvers. Peters, Natick. Quoted in Sect 5.10
Yazici B (1997) Stochastic deconvolution over groups. IEEE Trans Inf Theory 50:494–510. Quoted in Sect 10.2
Yazici B, Kashhyap RL (1997) A class of second-order stationary self-similar processes for 1/f phenomena. IEEE Trans Signal Process 45:396–410. Quoted in Sect 10.2
Yngvason J (2005) The role of type III factors in quantum field theory. Rep Math Phys 55:135–147. Quoted in Sect 8.5
Yu S, Liu N, Li L, Oh CH (2010) Joint measurement of two unsharp observables of a qubit. Phys Rev A 81:062116. Quoted in Sect 3.5
Zach R (2007) Hilbert's program then and now. In: Jacquette D (ed) Philosophy of logic. Elsevier, Amsterdam, pp 411–447. Quoted in Sect 5.5
Zadeh LA (1965) Fuzzy sets. Inf Control 8:338–353. Quoted in Sects 3.2, 3.3
Zamolodchikov AB (1985) Infinite extra symmetries in two-dimensional conformal quantum field theory. Akad Nauk SSSR Teor Mat Fiz 65:347–359. Quoted in Sect 6.5
Zumbahlen H (2008) Linear circuit design handbook. Elsevier, Oxford. Quoted in Sect 3.3
Zurek WH (1991) Quantum measurements and the environment-induced transition from quantum to classical. In: Ashtekar A, Stachel J (eds) Conceptual problems of quantum gravity. Birkhäuser, Boston, pp 43–62. Quoted in Sect 6.5
Zwicky F (1933) On a new type of reasoning and some of its possible consequences. Phys Rev 43:1031–1033. Quoted in Sect 3.2

Index

H. Primas, *Knowledge and Time*,
DOI 10.1007/978-3-319-47370-3

MIX
Papier aus verantwortungsvollen Quellen
Paper from responsible sources
FSC® C105338

If you have any concerns about our products,
you can contact us on
ProductSafety@springernature.com

In case Publisher is established outside the EU,
the EU authorized representative is:
Springer Nature Customer Service Center GmbH
Europaplatz 3, 69115 Heidelberg, Germany

Printed by Libri Plureos GmbH
in Hamburg, Germany